Geobotany 10

Series Editor

M.J.A. WERGER

Disturbance in Grasslands

Causes, effects and processes

edited by

J. VAN ANDEL, J.P. BAKKER AND R.W. SNAYDON

1987 **DR W. JUNK PUBLISHERS**
a member of the KLUWER ACADEMIC PUBLISHERS GROUP
DORDRECHT / BOSTON / LANCASTER

Distributors

for the United States and Canada: Kluwer Academic Publishers, P.O. Box 358, Accord Station, Hingham, MA 02018-0358, USA
for the UK and Ireland: Kluwer Academic Publishers, MTP Press Limited, Falcon House, Queen Square, Lancaster LA1 1RN, UK
for all other countries: Kluwer Academic Publishers Group, Distribution Center, P.O. Box 322, 3300 AH Dordrecht, The Netherlands

ISBN 90-6193-640-3 (this volume)
ISBN 90-6193-895-3 (series)

PRINTED IN THE NETHERLANDS

PREFACE

This volume is the result of a Symposium, held in October, 1985, to celebrate the Silver Jubilee of the Department of Plant Ecology of the University of Groningen.

The Department of Plant Ecology was founded by Professor Dingeman Bakker and has, since its inception, been involved with studies of grasslands, and especially with the effects of human interference on grassland communities. It has been especially involved in an almost unique enterprise to convert areas of grassland from intensive agricultural production to nature reserves. With over-production of most food commodities in the EEC, and increasing interest in various types of 'set aside' schemes, this initiative now seems to be remarkably prescient.

The Netherlands have a long history of human manipulation of the environment. Indeed, the Dutch environment is probably more man-made than that of any other country. In view of this, and the Laboratory's interest in managed grasslands, it is not surprising that 'Disturbance in Grasslands' was selected as the topic for the Laboratory's Jubilee Symposium. Although both the Symposium and the resulting volume, have a strong Dutch component, the organisers invited a number of contributors from Europe and elsewhere to provide key chapters. The result is a volume which covers many aspects of disturbance in a variety of grasslands. Disturbance is considered at both the community and the population level, and the principles and processes underlying those responses are explored. A variety of disturbances are considered, though the main emphasis is on those factors of particular interest to the Department of Plant Ecology, especially defoliation, soil fertility, drainage and salinity.

The organisers of the Symposium would like to thank the Royal Netherlands Academy of Arts and Sciences, the Faculty of Mathematics and Natural Sciences and the Subfaculty of Biology of the University of Groningen for their financial support of this Symposium, C.W. Biewenga for assisting with the organisation, and Dr. R.W. Snaydon for his contribution towards the editing of this volume.

PROLOGUE

The importance of disturbance

Disturbance, both man-induced and natural, has received increasing attention from ecologists in the last decade. This interest has culminated in several recent books (e.g. Cairns et al., 1977; Mooney & Godron, 1983; Pickett & White, 1985) and reviews (e.g. White, 1979; Sousa, 1984; Rapport et al., 1985). The increasing interest in disturbance has probably been stimulated by a recognition, sometimes perhaps not conscious, that more can be learned about the functioning of communities and populations by studying their response to disturbance than by studying these systems in equilibrium. Such studies may be of systems that have been naturally disturbed, unintentionally disturbed by man, or intentionally disturbed by man. Ideally, studies should be made of systems that have been experimentally disturbed, where the treatments have been imposed after careful consideration, and a valid experimental design has been used with replication. These ideal conditions are rarely achieved, and we may be forced to observe the effects of some previously imposed, unreplicated disturbance; sometimes there may not be even a valid control treatment with which to compare the disturbed community or population. In spite of the fact that we may often be forced to study far from ideal situations, disturbance is still a valuable tool in ecological and genecological study, allowing us to study the underlying processes at work within communities and populations.

Defining disturbance

There are several underlying difficulties in considering disturbance in plant communities and populations. The first of these is the difficulty of nomenclature. The early stages of most developing fields of science are marked by a proliferation of jargon; studies of disturbance in communities and populations are no exception. The problem is heightened by the fact that most of the jargon (e.g. disturbance, perturbation, and stress) has been borrowed from everyday speech, often with subtle changes in meaning. In the first two chapters of this volume, the authors tackle this problem and, largely as a result of discussions at the symposium, opt to simplify the nomenclature by dispensing with synonyms, or terms which were used inconsistently. Simplifying the nomenclature, however, does not overcome the difficulties of defining disturbance per se. It is extremely difficult to distinguish disturbance from normally occurring variation in environmental conditions. At one extreme, a disturbance may occur that has never previously existed, e.g. the application of a novel man-made herbicide; in this case adaptation will not have already occurred but, as the example of herbicides indicates, adaptation may still occur as a result of intense selection for mutants. At the other extreme, environmental variation may be such an important attribute of the habitat, that communities and populations are fully adapted to the variation, and their very survival may depend upon it. Between these extremes there is a complete spectrum.

Another difficulty is that of measuring disturbance or even, in some cases, recognising its occurrence. In theory, like most biological processes, disturbance can be recognised, and perhaps measured, by its causes or by its effects. Indeed, disturbance is normally defined in terms of either its cause or its effects, and is commonly recognised by its effects; changes in composition or structure of communities and/or populations are probably the most effective way of finding whether disturbance has occurred. While the recognition of disturbance is difficult, the processes of response to disturbance are even more difficult to discover. Although most of the papers in this volume deal with the causes and effects of disturbance, many also attempt to define the processes of response to disturbance.

Disturbance in grasslands

The are several advantages in studying disturbance in grasslands. Firstly, the responses to disturbance in grassland are usually rapid, so that experiments can be carried out in reasonably short time spans. Whereas responses to disturbance may take decades, or even centuries, to develop in woodlands, they can often be detected within weeks in grasslands, and a new equilibrium may be reached in a few years. The reasons for this rapid response are twofold: (i) individuals of most species in grassland are phenotypically very plastic, so that changes in plant sizes or form are usually apparent in a few weeks; (ii) individuals of most grassland species are usually short-lived, so that changes in the structure of both populations and communities can occur in a few years. The second advantage of studying disturbance in grassland is that a great diversity of grassland types is available for study, ranging from natural to man-made, from botanically simple to botanically complex, and in a diversity of habitats varying, for example, from semi-arid to water-logged. In addition, grasslands are abundant, easily accessible and, because of their agricultural importance, have been widely studied.

Various types of disturbance have been studied in the context of grasslands. The most common, and those widely considered in this volume, can be broadly categorised into particular changes in soil conditions, and defoliation regime. Disturbance in soil conditions might include: (i) increase or decrease in nutrients supply (e.g. increase or decrease in application of fertilizer); (ii) increase or decrease in soil water content (e.g. drainage or flooding, or change in tidal pattern); (iii) change in soil salinity (e.g. change in tidal pattern). Disturbance of the defoliation regime may include: (i) change in frequency, intensity or seasonality of defoliation; (ii) change in the type of defoliation (e.g. cutting versus grazing). Most of these forms of potential disturbance are considered at least once in this volume.

The response to disturbance can be measured at each level of biological organisation from the cellular level to the ecosystem level. Here we are concerned with levels from the individuals upwards. The volume is broadly divided into responses at the community level (parts II and III), with some consideration of the ecosystem level, and responses at the level of the individual and population (part IV). Chapters 1 and 2 consider the general principles relating to those two groupings.

References

Cairns, J., Dickson, K.L. & Hendricks, E.E. (1977) Recovery and Restoration of Damage to Ecosystems. Charlottsville, University Press of Virginia.

Mooney, H.A. & Godron, M. (1983) Disturbance and Ecosystems. Berlin, Verlag-Springer.

Pickett, S.T.A. & White, P.S. (1985) The Ecology of Natural Disturbance and Patch Dynamics. New York, Academic Press.

Rapport, D.J., Regier, H.A. & Hutchinson, T.C. (1985) Ecosystem behaviour under stress. American Naturalist, 125, 617–640.

Sousa, W.P. (1984) The role of disturbance in natural communities. Annual Review of Ecology and Systematics, 15, 353–391.

White, P.S. (1979) Pattern, process and natural disturbance in vegetation. Botanical Review, 45, 229–299.

Table of contents

List of contributors

G.J.R. Allersma, P. Campersingel 65, 9713 AD Groningen, The Netherlands

J.P. Bakker, Department of Plant Ecology, University of Groningen, Biological Centre, P.O. Box 14, 9750 AA Haren (Gn), The Netherlands

W.G. Beeftink, Delta Institute for Hydrobiological Research, Vierstraat 28, 4401 EA Yerseke, The Netherlands

A. Biere, Department of Plant Ecology, University of Groningen, Biological Centre, P.O. Box 14, 9750 AA Haren (Gn), The Netherlands

P.C. de Hullu, State Forestry Service, Department Flora and Fauna, P.O. Box 200.20, 3502 LA Utrecht, The Netherlands

R.H. Drent, Zoological Laboratory, University of Groningen, Biological Centre, P.O. Box 14, 9750 AA Haren (Gn), The Netherlands

P. Dijkstra, Department of Plant Physiology, University of Groningen, Biological Centre, P.O. Box 14, 9750 AA Haren (Gn), The Netherlands

W.H.O. Ernst, Department of Ecology and Ecotoxicology, Free University, De Boelelaan 1087, 1081 HV Amsterdam, The Netherlands

A. Fischer, Institut für Pflanzenökologie der Justus-Liebig-Universität, Heinrich Buff-Ring 38, D-6300 Giessen, BRD

L.F.M. Fresco, Department of Plant Ecology, University of Groningen, Biological Centre, P.O. Box 14, 9750 AA Haren (Gn), The Netherlands

A.P. Grootjans, Department of Plant Ecology, University of Groningen, Biological Centre, P.O. Box 14, 9750 AA Haren (Gn), The Netherlands

C. Kik, Department of Plant Ecology, University of Groningen, Biological Centre, P.O. Box 14, 9750 AA Haren (Gn), The Netherlands

F. Klötzli, Geobotanisches Institut ETH, Stiftung Rübel, Zürichbergstrasse 38, CH-8044 Zürich, Switzerland

M.M. Kwak, Department of Plant Ecology, University of Groningen, Biological Centre, P.O. Box 14, 9750 AA Haren (Gn), The Netherlands

H. Lambers, Department of Plant Ecology, University of Utrecht, Lange Nieuwstraat 106, 3512 PN Utrecht, The Netherlands

M.J.J.E. Loonen, Department of Plant Ecology, University of Groningen, Biological Centre, P.O. Box 14, 9750 AA Haren (Gn), The Netherlands

L.A.P. Lotz, Institute for Ecological Research, Duinzoom 20a, 3233 EG Oostvoorne, The Netherlands

J. Miles, Institute of Terrestrial Ecology, Banchory Research Station, Banchory, Kincardineshire, AB3 4BY, UK

T. Moesker, Department of Plant Ecology, University of Groningen, Biological Centre, P.O. Box 14, 9750 AA Haren (Gn), The Netherlands

P. Oosterveld, Research Institute for Nature Management, P.O. Box 46, 3956 ZR Leersum, The Netherlands

D.M. Pegtel, Department of Plant Ecology, University of Groningen, P.O. Box 14, 9750 AA Haren (Gn), The Netherlands

H. Poorter, Department of Plant Ecology, University of Utrecht, Lange Nieuwstraat 106, 3512 PN Utrecht, The Netherlands

H.H.T. Prins, Zoological Laboratory, University of Groningen, Biological Centre, P.O. Box 14, 9750 AA Haren (Gn), The Netherlands

R.W. Snaydon, Department of Agricultural Botany, Plant Science Laboratories, University of Reading, Whiteknights, Reading RG6 2AS, UK

D.C.P. Thalen, Research Institute for Nature Management, P.O. Box 46, 3956 ZR Leersum, The Netherlands

K.M. Urbanska, Geobotanisches Institut ETH, Stiftung Rübel, Zürichbergstrasse 38, CH-8044 Zürich, Switzerland

J. van Andel, Department of Plant Ecology, University of Groningen, Biological Centre, P.O. Box 14, 9750 AA Haren (Gn), The Netherlands

J.P. van den Bergh, Centre for Agrobiological Research, Bornsesteeg 65, 6708 PD Wageningen, The Netherlands

H.P.M. van Laarhoven, Department of Plant Ecology, University of Groningen, Biological Centre, P.O. Box 14, 9750 AA Haren (Gn), The Netherlands

J.C. Zadoks, Laboratory of Phytopathology, Agricultural University, Binnenhaven 9, 6709 PD Wageningen, The Netherlands

PART 1

Disturbance in perspective

CHAPTER 1

Disturbance of grasslands
Outline of the theme

J. van Andel & J.P. van den Bergh

Detailed contents

In this introduction to the theme 'Disturbance of Grasslands' we will review some of the relevant terminology, before going on to consider some of the more important concepts concerning responses to disturbance in plant communities. The terminology of 'disturbance' is very confused, firstly, because synonyms, such as 'perturbation', have gained popularity and, secondly, because causes (i.e. disturbance or perturbation) sometimes tend to be confused with effects (e.g. stress). Various authors have provided definitions, often with little agreement amongst them. We will try to clarify basic definitions concerning 'disturbance', pointing out where possible areas of agreement between authors of the present book. Disturbance is considered here largely in the context of grassland. In this chapter we shall consider responses of species in the community and interpret the mechanisms of response largely in terms of competition for resources. Finally, we will discuss the problem of reversibility of change, and the implication that a biological system can return to its original state.

1. The concept of disturbance

1.1 Disturbance

In an ecological context, the term 'disturbance' indicates change in the conditions of an organism, population or community caused by an external agency, often man. Such changes usually imply a shift towards sub-optimal conditions, since we can usually assume that the organisms were previously adapted to the existing environmental conditions. Many definitions of 'disturbance' have been given in the literature, almost all of them confuse cause (change in the environment) and effect (responses of organisms, populations or communities), mainly because disturbance is identified by its effects. For example, Bazzaz (1983) defined disturbance as 'a sudden change in the resource base of a unit of landscape that is expressed as a readily detectable change in population response'. Grime's (1979) definition 'the mechanisms which limit the plant biomass by

J. van Andel et al., (eds.), Disturbance in Grasslands. ISBN 90-6193-640-3.

causing its partial or total destruction', places emphasis on mechanisms rather than causes. Grubb (1985a) accepted this definition, with the caveat that senescence of individuals is excluded, because disturbance must be initiated from outside the plant community. The definition given by Sousa (1984) places the emphasis strongly upon the effects: 'Disturbance is a discrete, punctuated killing, displacement, or damaging of one or more individuals (or colonies) that directly or indirectly creates an opportunity for new individuals (or colonies) to become established'. With regard to the causes of disturbance he stated that 'disturbance lies near one extreme of the continuum of natural perturbations that affect organisms'. Rykiel (1985) criticized this type of reasoning (i.e. perturbations cause perturbation) and stated that 'the challenge is to define terms that distinguish between cause and effect in so far as possible'. He considered disturbance to be a cause, i.e. a physical force or agent.

1.2 Stress

Rapport et al. (1985) unfortunately, defined the term 'stress' in a similar way to disturbance: 'Stress denotes an external force or actor, or stimulus that causes changes in the ecosystem, or causes the ecosystem to respond, or entrains ecosystem dysfunctions that may exhibit symptoms'. This definition, derived from considerations of stress physiology in animals, mainly from Selye (1973), does obviously categorize stress as a cause; it therefore confuses stress with disturbance. Rapport et al. (1985) have created more confusion by using the terms response, perturbation, dysfunction, and distress syndrome as approximate equivalents, thus further intermingling causes and effects. Grime (1979) also defined stress as a cause, i.e. 'external constraints which limit the rate of dry matter production of all or part of the vegetation'. Whittaker (1975) used the term stress (e.g. overgrazing causing retrogression) in a similar way to Rapport et al. (1985) and Grime (1979), i.e. as a causative agent. On the contrary, Rykiel (1985) defined stress as an effect: 'the physiological response of an individual, or the functional response of a system caused by disturbance or other ecological process'. Unfortunately, he considered stress a particular type of perturbation.

1.3 Perturbation

According to Vogl (1980) 'perturbation is a term used to describe biological or environmental events that produce successional upset, disaster, stress or disturbance to organisms or systems. These events are generally sudden, brief and violent, and are usually considered to be unpredictable and unexpected, even if they are sometimes repetitive. Perturbations can be either unnatural or a natural part of a system.' On the other hand, McIntosh (1980) criticized the use of the term perturbation in ecology; he saw it as part of the increasing use of 'newspeak' in ecology, viz. 'many recent papers on succession replace the perfectly descriptive long-used and apt word 'disturbance' by 'perturbation' (and the bastard non-word perturbate), with similarly little added insight into succesion. Perturbation came into ecology in the late 1960s (Brookhaven Symposium, 1969), carrying with it an aura of precision from mathematics and physics, where it has a highly specific meaning. In physics it is synonymous with small vibration. It implies a system which differs in minor respects only from a system whose behaviour is well understood and completely described – an unusual case in ecology' (McIntosh, 1980).

1.4 Ecological and evolutionary aspects

So far we have considered the terms disturbance, stress, and perturbation in an ecological context. Notwithstanding the evident confusion in this context, some ecologists have not hesitated to apply these terms in an evolutionary context; this adds further to the confusion. Grime (1979) suggested that stress and disturbance may have been the basis for the evolution of three primary 'strategies' of plants. In an evolutionary context, terms like 'stress-tolerators' (Grime, 1979) and 'perturbation-dependent systems' (Vogl, 1980) may be useful to explain the adaptive origin of particular characteristics but, in an ecological sense, they are contradictory. A stress-tolerator does not experience stress any longer, because it is adapted to the stress (Ernst, 1983). Sousa (1984) stated that 'the levels of environmental fluctuation to which a present-day species responds with effective homeostatic mechanisms probably represented a far greater hazard early in the species' evolutionary history. An environmental fluctuation that once caused disturbance does so no longer. The dynamic evolutionary relationship between an organism and the environmental stresses it encounters is ongoing and subject to numerous constraints. The evolutionary moderation of stress is exhibited in its most extreme form by species that depend on environmental disruption for the completion of their life cycles and the persistence of their populations. Allen & Starr (1982) argue that at this point the disruptive event ceases to be a disturbance at all.'

1.5 Cause and effect

Obviously one of the main problems of using the terms distubance, stress, and perturbation is the confusion of cause and effect (cf. Rykiel, 1985). This largely springs from the fact that the phenomena cannot be measured independent of organisms or communities. Although ecologists can measure temperature, light intensity, or soil water content by specifically designed instruments, disturbance or perturbation can only be measured by its effect on living systems (plants, populations, communities). This implies the risk of tautological reasoning (cf. Van der Steen & Scholten, 1985). In a similar context, Jain (1983) came to the conclusion that 'an attempt to standardize various terms and empirical measures in spatial and temporal environments is essential. How else can we utilize theories of evolution and adaptation under heterogeneous environments with the ecologists' model of patchiness, disturbance, stress, and cycles?' A few years ago, White & Pickett (1985) started to tackle this problem.

1.6 Agreement on definitions

The authors of the present book have agreed to discard the term perturbation in an ecological context, because (i) the term is already used in the physical sciences, to indicate a small vibration in an existing equilibrium-type situation, and this strongly conflicts with its use in ecology, where it is normally used to indicate extreme events; (ii) perturbation is largely synonymous with disturbance, and perhaps also with stress, so giving rise to confusion; (iii) in ecology the term perturbation has been used in the sense of both cause and effect, once again leading to confusion. A general consensus has been reached, between the authors of this volume, to use the term disturbance in the following sense: '*Disturbance is a change in conditions which interferes with the normal functioning of a given biological system*'. The term stress will be used sparingly, and then in an ecophysiological sense only, to indicate a specific negative response of individuals to a disturbance (or stress-factor). Although we agree with Rykiel's (1985) analysis of conceptual problems, we reject his definition of disturbance, since he confuses it with stress and perturbation.

The authors will differentiate between the kind, frequency, intensity, and scale (or extent) of disturbance, in agreement with Connell & Slatyer (1977), Reiners (1983), Sousa (1984), and Grubb (1985a). The *kind* of disturbance depends on the environmental factor concerned, whether biotic or abiotic. The *degree* of disturbance is determined by the difference between the new conditions and the previous, 'normal' conditions. According to Rykiel (1985), both a system and a reference state must be defined; neither need be static. *Frequency* of disturbance is important, because isolated, recurrent, and continuous disturbances have different ecological effects (cf. Grubb 1985a), while the pattern of recurrence (regular, irregular), and the duration of each disturbance, may influence the subsequent response. The *scale* of the disturbance can be distinguished in terms of spatial and temporal patterns, but also in terms of the level of organization of the biological system under investigation (cf. Rykiel, 1985), i.e. the ecosystem, the community, the population, or the individual. In this way one is able to clearly describe *responses* of biological systems to disturbances, to adequately analyze the *mechanisms* of these responses, and to present coherent interpretations about ecological and/or evolutionary *functions* (consequences).

One might argue that dissection of disturbance, and of the resulting responses, undermines the search for unifying concepts. However, as Rosenberg (1985) comments 'it is the aswers to the narrower questions that should determine our answer to the broader one. Any other procedure substitutes preconception for analysis by letting a conviction about the broad question govern answers to the narrower ones.' Our objective is to provide a unifying concept concerning the role of disturbance.

1.7 Operational analysis

Our first task is to provide an operational analysis of the role of disturbance, both in an ecological and an evolutionary context, as opposed to the previous theoretical approaches. This chapter is particularly concerned with communities, and especially with vegetation succession in grasslands, while Chapter 2 is concerned with populations. There are two main ways in which the responses of communities or populations to disturbance can be determined. First, the system may be disturbed, either experimentally or naturally, usually by alteration of the environment. Secondly, organisms (communities, populations, ecotypes or clones) may be transplanted to another type of habitat. In all cases, comparison with the 'normal' (control, reference) condition is essential. Analysis of response, without sufficient knowledge (or clearly defined preassumptions) about the 'normal' conditions or processes, may lead to misinterpretations; studies should therefore always include controls. This is not to suggest that the normal state represents some kind of static equilibrium. Sousa (1984) correctly observed that there is no clear demarcation between assemblages in a state of equilibrium and those that are not; one must accept that most 'normal' biological systems are in a state of dynamic equilibrium, responding to naturally occurring changes in the environment. These natural changes may, or may not, be regarded as 'disturbance', depending on the context.

2. Experimental disturbance of permanent grassland

Disturbance is a change in conditions which interferes with the normal functioning of a biological system (see 1.6). One example of such a disturbance is an experiment, started in 1958 near Wageningen, where the management regime of an old pasture on heavy river clay soil was changed. The pasture had previously been grazed (at a low stocking rate) and sometimes cut for hay. The predominant species were *Holcus lanatus, Anthoxanthum odoratum, Festuca rubra* and *Agrostis spp.* After 1958, part of the field remained under grazing (pasture → pasture) and in another part cut annually for hay without grazing (pasture → hayfield). In addition, several fertilizers were applied annually (Elberse et al., 1983). These changes in management regime may be considered either as a single disturbance, from one state to another, or as a continuous disturbance, at least until the system adapted to the new regime. The treatment 'pasture → pasture' without fertilizer application acts as the control.

2.1 Responses

Soil conditions did not change when swards remained grazed and no fertilizer was applied (pasture → pasture), but the P and K content of soil decreased when hay was cut (pasture → hayfield), because of the annual removal of the hay crop. This process was aggrevated by liming. Application of lime increased pH (4.3 to 6.7) and stimulated soil mineralization (% organic matter decreased from 22.4 to 17.0). This resulted in increased herbage yields, averaging 1.4 t. dry matter per year (Fig. 1), and a greater removal of N, P and K (Table 1). Application of N, P and K to the hayfield greatly increased the P and K con-

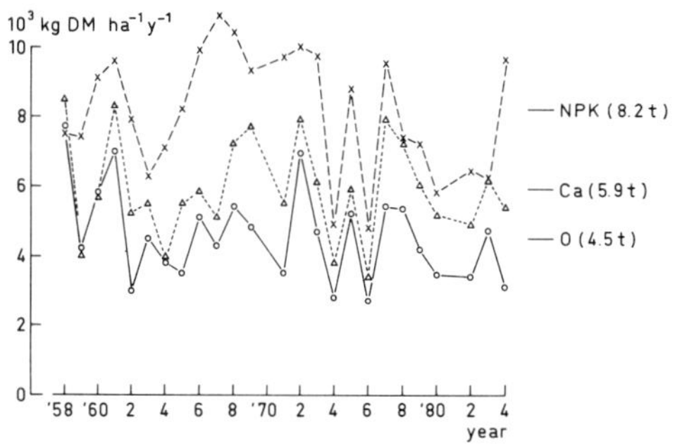

Fig. 1. The effect of lime and NPK fertilizer on the dry matter yield of a permanent pasture during a period of 26 years. The average yield of each treatment over the period is shown in brackets.

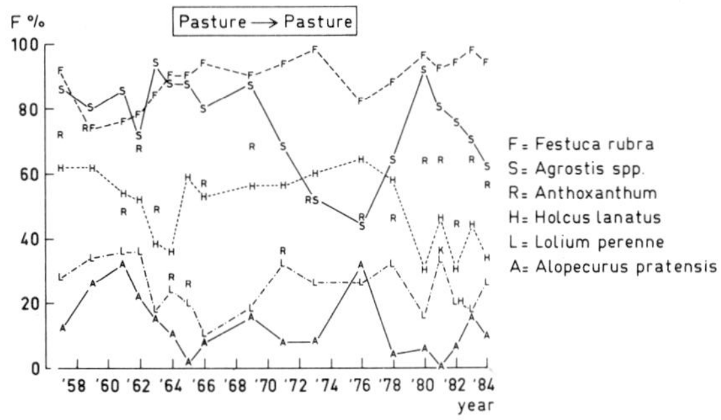

Fig. 2. Changes in frequency (%) of plant species on the control plots (pasture → pasture).

tents of the soil, the herbage yields (Fig. 1), and the amount of N, P and K harvested (Table 1).

Although the botanical composition of the vegetation varied considerably from year to year, no sytematic changes were observed on the control plots (pasture → pasture, Fig. 2). The change from pasture to hay production caused a rapid change in botanical composition. *Lolium perenne* and *Poa trivialis*, which are characteristic pasture species, disappeared within 10 years, and *Trifolium repens* disappeared within 20 years (Fig. 3). *Holcus lanatus* increased slightly at first but then declined to a low level, while *Alopecurus pratensis* showed large fluctuations. The dominant species *Festuca rubra* and *Agrostis spp.* did not estab-

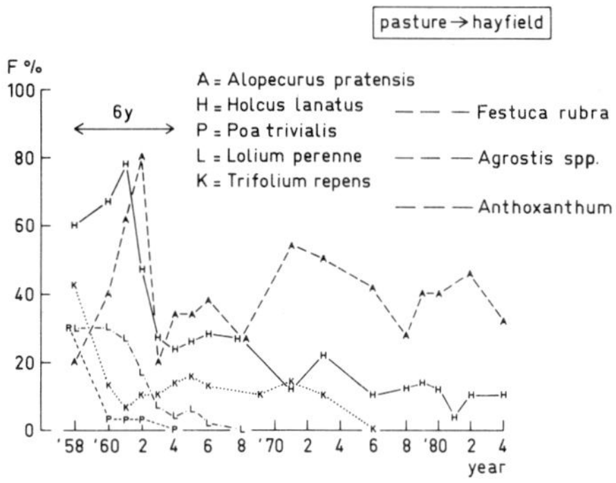

Fig. 3. Changes in frequency (%) of plant species on unfertilized plots following a change from pasture to hay management in 1958. The average level of the rather constant frequencies of *Festuca rubra, Agrostis spp.* and *Anthoxanthum odoratum* are indicated by — — —.

soil data / treatment	pH KCl '59	pH KCl '85	org. m. (%) '59	org. m. (%) '85	P (P_2O_5 ‰₀) '59	P (P_2O_5 ‰₀) '85	K (K_2O ‰₀) '59	K (K_2O ‰₀) '85	harvested (kg ha^{-1} $year^{-1}$) (last 5 years average) N	P	K
pasture ↓ pasture	4.8	4.9	23.9	23.4	7	7	31	28	?	?	?
pasture ↓ hayfield	4.3	3.8	21.2	20.9	7	5	28	18	67	5.3	37
pasture ↓ hayfield + Ca	4.3	<u>6.7</u>	22.4	<u>17.0</u>	8	3	29	12	92	9.9	49
pasture ↓ hayfield + NPK	4.3	3.8	21.8	20.7	6	<u>50</u>	31	<u>72</u>	131 (160)	20 (50)	195 (330) given

Table 1. The effect of cutting hay and of applying lime and NPK fertilizer on soil conditions (0–5 cm), and on the harvested amounts of minerals from hayfields in a long term experiment at Wageningen. The experiment started in 1958.

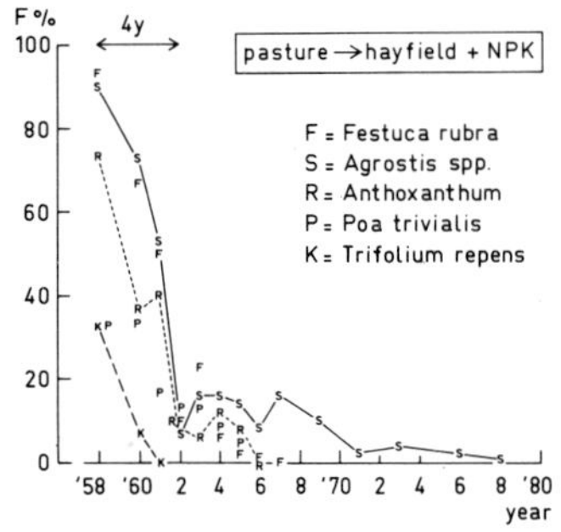

Fig. 4. Changes in frequency (%) of some decreasing plant species on fertilized plots (NPK) following a change from pasture to hay management in 1958.

lish a mono-culture on unfertilized plots, either grazed or hayed (Figs. 2 and 3), it seems obvious that there was enough 'space' left for other species. The ecological interpretation of this space is still obscure. Competition for light is probably slight in both treatments.

Changing from grazing to hay production with NPK application caused a rapid reduction in the

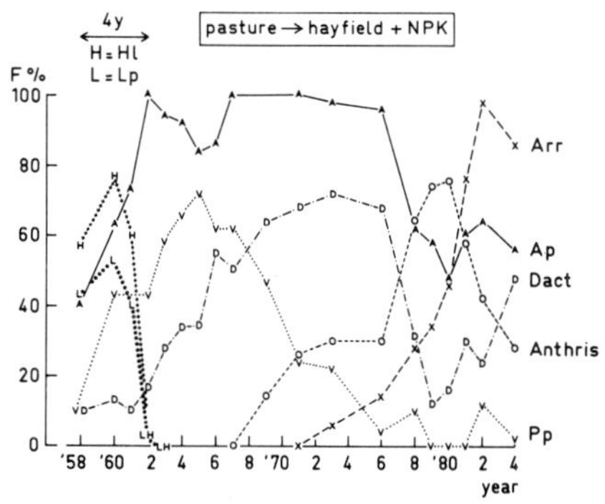

Fig. 5. Changes in frequency (%) of some fluctuating plant species on fertilized plots (NPK) following a change from pasture to hay management in 1958.
Arr = *Arrhenatherum elatius*, Ap = *Alopecurus pratensis*, Dact = *Dactylis glomerata*, Anthris = *Anthriscus sylvestris*, Pp = *Poa pratensis*, Hl = *Holcus lanatus*, Lp = *Lolium perenne*.

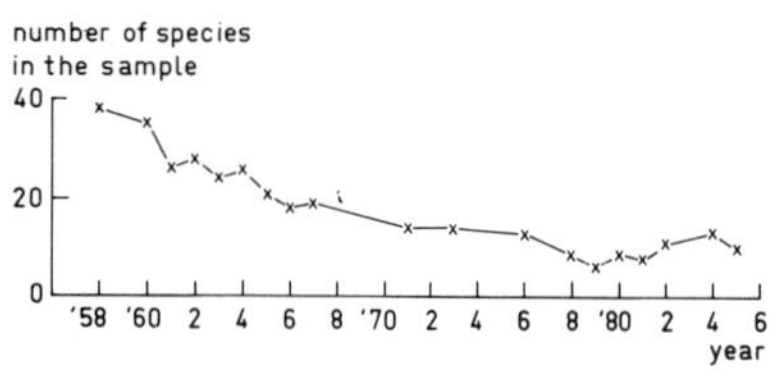

Fig. 6. Change in the number of species on NPK-fertilized plots following a change from pasture to hay management in 1958.

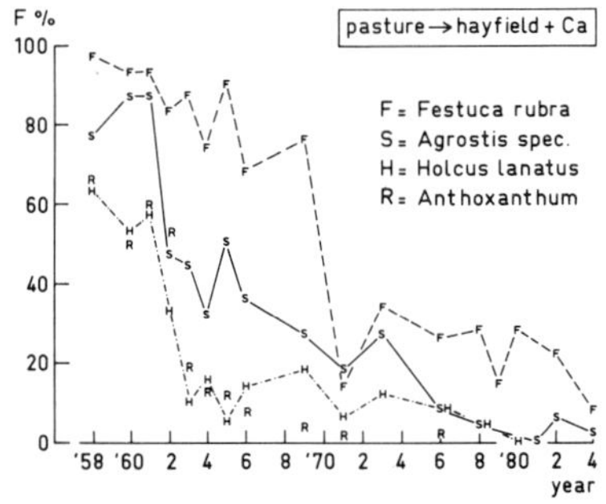

Fig. 7. Changes in frequency (%) of four steadily decreasing plant species on lime-fertilized plots following a change from pasture to hay management in 1958.
Anthoxanthum odoratum behaves similar to *Holcus lanatus*.

frequency of *Festuca rubra, Poa trivialis* and *Anthoxanthum odoratum* (Fig. 4), but a rapid increase in *Alopecurus pratensis, Poa pratensis* and *Dactylis glomerata,* though all species later declined in frequency as *Anthriscus sylvestris* and *Arrhenatherum elatius* increased (Fig. 5). The total number of species decreased from 38 to less than 10 within 20 years (Fig. 6). These changes may well have been induced partly by competition for light.

With the change from grazing to hay production with the application of lime, changes in botanical composition occurred much more gradually. The frequency of *Festuca rubra, Agrostis spp., Holcus lanatus* and *Anthoxanthum odoratum* declined slowly over about 15 years (Fig. 7), while *Arrhenatherum elatius, Alopecurus pratensis* and *Poa pratensis* increased slowly (Fig. 8).

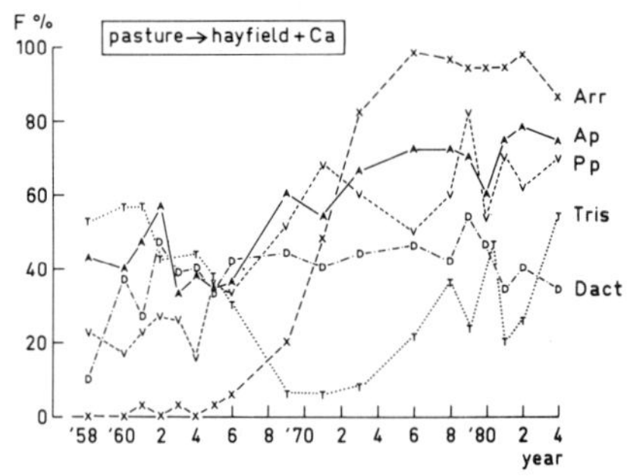

Fig. 8. Changes in frequency (%) of four steadily increasing plant species on lime-fertilized plots following a change from pasture to hay management in 1958. *Trisetum flavescens* (Tris) decreases at first and increases later on. (For other symbols see Fig. 5)

2.2 Mechanisms of response

Competition for resources plays a major part in the 'resource-ratio hypothesis of plant succession' as presented by Tilman (1982, 1985). According to this hypothesis, succession results from a gradient, through time, in the relative availabilities of limiting resources. Tilman suggested that habitat gradients from poor soil but high availability of light at the soil surface to habitats with rich soils but low light availability, are an important feature of plant succession. These gradients, and the evolution of plant life histories in response to them, may be the cause of otherwise difficult-to-explain similarity between primary and secondary succession. Plants that are superior competitors for soil resources normally have a higher ratio of root biomass to shoot biomass, are shorter, grow more rapidly, and reproduce at an earlier age than plants that are superior competitors for light. According to Tilman's idea, succession is a directional and repeatable process only in so far as the resource-supply trajectory is repeatable and directional. Retrogressive succession is associated with a decrease in the availability of soil resources and an increasing light penetration. In its most general form the resource-ratio hypothesis states that plant species are specialized on different proportions of limiting resources, and that the composition of a plant community should change whenever the relative supply of the limiting resources changes.

In the context of Tilman's hypothesis, disturbance may be defined as a sudden change in the existing ratio of available below-ground and above-ground resources. We can interpret the mechanisms of response to disturbance in the aforementioned experiments using Tilman's hypothesis.

The increase of *Alopecurus pratensis* and *Holcus lanatus* on the treatment pasture → hayfield (Fig. 3) was probably caused by the more open aspect of hayfields after cutting, linked with the fact that both species had abundant seed production. The subsequent decline in the frequency of these species could have been caused by a decline in soil nutrients (Table 1), though *Alopecurus pratensis* was previously abundant (F = 40%, Fig. 2) in spite of low soil fertility. The main effect of lime was to slowly activate mineralization and so release nutrients; as a result, tall species (e.g. *Arrhenatherum elatius*) with a high growth rate slowly increased, probably by competition for light.

Competition between *Arrhenatherum elatius* and *Festuca rubra* has been studied at low and high fertility (Elberse et al., 1987). Initially *Arrhenatherum elatius* suppressed *Festuca rubra* at low and high fertility, because of its more rapid growth rate; thereafter, however, *Festuca rubra* suppressed *Arrhenatherum elatius* on the unfertilized soil. The ultimate superiority of *Festuca rubra* on the infertile soil may reflect the fact that this species had more roots than *Arrhenatherum elatius* and so could better exploit the limited soil resources. Conversely *Arrhenatherum elatius* with its large shoot mass, relative to root, would be at a disadvantage, especially since competition for light would probably be negligible under such conditions. Moreover, under unfertile conditions *Festuca rubra* had the advantage over *Arrhenatherum elatius* by forming a dense sod under the 5 cm cutting level, whereas *Arrhenatherum elatius* with its erect growth habit lost nearly all its above-ground parts after cutting.

Annual application of NPK fertilizer increased available soil nutrients but also reduced the soil pH to 3.8 (Table 1). These changes, together with infrequent cutting, led to the initial dominance of the tall species *Alopecurus pratensis* and *Dactylis glomerata* (Fig. 5), but the ultimate dominance of the taller *Arrhenatherum elatius* and, briefly, *Anthriscus sylvestris*.

The general patterns of change in species composition of the field plots (Figs. 1-8) probably reflect, to a considerable extent, the relative importance of shoot and root competition in plots differing in grazing/cutting management and in soil nutrient availability.

3. Reversibility

The experiment just considered may be regarded

as an example of 'continious disturbance'. After the sudden change in management (disturbance as an event) the different treatments were continued year after year and, as a result, the grassland communities diverged. However, after a certain stage, the treatment can no longer be considered 'disturbance' since the community adapts to the new regimes. Adaptation to disturbance may take place to such an extent that the response appears irreversible.

3.1 Successional series

Grime (1979) presented the corridor model of species richness in grasslands, with a maximum density of species at moderate standing crop + litter mass. According to this model, artificial fertilization should turn a species-rich grassland into species-poor grassland (dominance of a few species) as shown in Fig. 6. Cessation of fertilizer application might then cause a return to a species-rich community. Two important questions then arise: (i) does the species composition of the re-enriched vegetation resemble the original state, and (ii) is the genetic structure of the re-established populations similar to the initial state? In other words, is the reversal true recovery? These questions have particular importance in relation to vegetation restoration and wildlife management, where, depending on the nature of change, management could be termed restoration or evolutionary manipulation. Grubb (1985b) has stated that 'the study of maintenance of species richness should never be divorced from a study of its origin'.

The successional pathway of a community also determines the chance of reversibility. The 'successional rain' (cf. Klötzli, 1981) is particularly important. Dispersal of diaspores – either in space or in time (seed bank) – of species that disappeared is a prerequisite for recovery. In the grassland study just considered, the hayfield species do not produce a persistent seed bank (Bakker, 1983; Bakker et al., 1979). When such a field is disturbed by manuring, the number of species is reduced (Fig. 6) and the recovery of hayfield species, after cessation of manuring, depends on the spatial immigration of propagules. Similarly, Gartner et al. (1983) showed that natural revegetation of a disturbed Alaskan tussock grassland, mainly from buried seeds, resulted in a cover of only two of the original species. Complete recovery may still be possible in the long run by slow dispersal of the other species. Disturbance sometimes results in some kind of cyclic succession. Van der Valk (1981) showed that disturbance of a prairie glacial marsh (Eagle Lake, Iowa) by the sequential events of water drawdown, reflooding, standing water, and muskrat 'eat out', could explain the observed cyclic succession, though it is remarkable that the necessary seed bank of at least 50 different species persisted, especially since this included submersed aquatics, emergents, and mud-flat annuals. In systems ecology (e.g. Odum, 1971; Van der Maarel & Werger, 1978; Lepś et al., 1982; Baumová, 1985) it has been postulated that later stages of succession are more 'resistant' than earlier stages, but less 'resilient' in their response to disturbance. This means that a community would become less disturbance-prone in the course of succession, but, if disturbed, there would be less chance of recovery at later stages. Retrogressive succession, caused by disturbance of a progressive succession, may result in an alternative route of succession, rather than in re-initiation of the previous sere. Similarly, Conrad (1983) stated that the highest level of community organization is the most sensitive with regard to disturbance. His view of 'the minimal tendency of adaptability' implies only a reinitiation of *processes* after the collaps of a highly organized community, resulting again in increasing complexity and specialization. But this reinitiation of processes does not necessarily imply the re-establishment of particular species. Even if the existing populations of species are able to become re-established, they may be selected for specific tolerance mechanisms (through ecotypic differentiation) or avoidance mechanisms (short life cycle, seed dormancy, persistent seed bank) to cope with the disturbance.

These considerations have influenced many theories of vegetation succession. In general, response to disturbance is not usually reversible and changes to another type of community or population is more common. Disturbance, therefore, can be considered a common factor in the development of plant communities and/or plant species (e.g. Connell & Slatyer, 1977; Grime, 1979; Watt, 1979). Grubb (1985a) criticized the adoption of the all-pervading role of disturbance as one of the primary factors causing differentiation among plant populations in particular types of vegetation. His criticism stems from the fact that disturbance can be of very different types, differing in frequency, extent and intensity. 'A basic division is into continual and periodic disturbance. Continual disturbance produces distinctive, long-persistent types of vegetation, whereas periodic disturbance produces transient gaps in which regeneration occurs.' In addition, the scale of disturbance has to be taken into account. Small-scale disturbance within a plant community may result in local extinction of individual plants, but this does not disturb the plant community. On the contrary, the species diversity of the community may increase. These comments do not suggest that disturbance should not be considered a common factor in the development of communities and populations. We would like to emphasize only that a more differentiated approach allows an operational analysis of the concept of disturbance.

3.2 Gradients

Disturbance of spatial gradients in the environment is of particular interest. Schat (1982) showed that populations of plant species in gradients of dune slacks, exposed to natural fluctuations in both water table and salt concentration, make opportunistic use of suitable parts of the gradient. If the normal fluctuations in the gradient are disturbed (e.g. by lowering the groundwater table), extinction of some populations may occur, since the particular conditions they require no longer exist along the gradient. Similar phenomena were observed by Van der Valk & Davis (1976) around a prairie pothole, where the water table fluctuated relatively rapidly in response to short-term variations of the climate. Excessive variations in water level, caused by drought or abnormally high rainfall, shifted the position of many species along the coenocline and caused the virtual elimination of some species. The overall result was that species richness increased in the submersed and emergent zones, but decreased in the meadow zone.

Although many species may be able to persist in fluctuating environmental gradients, by a shift of their position, other species depend on a fixed position in a gradient. Van Leeuwen (1966) and Van Wirdum (1980) suggested that the occurrence of rare fen-meadow species is indirectly correlated with the presence of stable gradients in groundwater types. Grootjans (1985) showed that the distribution of many marsh plant species was significantly correlated with the distribution of groundwater types, differing in trophic status. This does not mean that any change in groundwater type should be regarded as a disturbance. Rare species in particular (e.g. *Dactylorhiza incarnata, Carex appropinquata, Valeriana dioica*) appeared to be related to those small zones of the gradient that were alternatively influenced by base-rich groundwater, discharging from deep aquifers, and base-poor, rainfed groundwater. As far as these species are concerned, stability means maintenance of natural fluctuations. Drainage and/or pollution leads eventually to an irreversible situation all along the gradient.

3.3 Dependence of disturbance

Grazing has a detrimental effect on communities with little history of grazing, but is necessary to maintain communities with a long history of grazing (e.g. Naveh & Whittaker, 1980). This may be interpreted in two ways: (i) grazing is a disturbance for the former system, but not for the latter; (ii) grazing is, by definition, a disturbance and the former system is disturbance-prone while the latter is disturbance-dependent. We hold, in view of the definition of disturbance given in section 1.6, to the first interpretation, and therefore have to reconsider the term 'disturbance-dependent system' (Vogl, 1980). Fire-dependent garrigues and grass savannas, inundation-dependent salt marshes, and many other similar communities have been named 'disturbance-dependent systems' (e.g. Vogl, 1980). Comparisons between biological systems may, indeed, eventually lead to an ordination of systems on an axis from disturbance-prone to disturbance-dependent. However, as already stated, this would require the definition and measurement of disturbance independent of responses. According to our definition (section 1.6) disturbance is a cause (a change in condition) measured by its effects (change in normal functioning of a system). At first sight; this seems to open up a trap of tautological reasoning, but the choice of a reference (control, normal) system eliminates this problem. At the same time, however, the question arises whether a particular system should be considered as a reference or as a disturbed system. It depends on the criteria to define a status quo. The question whether a community, which has become adapted to disturbance, is 'disturbance-dependent' (relative to the original community) or a part of a new status quo, may seem semantic, but answering this question is crucial to avoid confusion, because of the use of the term disturbance in both an ecological and evolutionary context. An example may elucidate our approach: the garrigue may be regarded as a disturbed community, in which the original community dominated by *Quercus ilex* has been replaced by or dominated by *Quercus coccifera*. Because the garrigue has become adapted to disturbance, fire is no longer a disturbance factor. Further disturbance, e.g. fire prevention, causes changes to another type of community.

Mechanisms of adaptation to disturbance are of particular interest. Variability is the sine qua non of adaptability (cf. Conrad, 1983), whether of the genome (gene pool diversity), the organism (plasticity), the population (fluctuation in numbers, niche narrowing, genetic flexibility), or the community ('routability' of matter and energy flow). Adaptability at one organizational level is achieved by flexibility and/or variation at one or more lower levels. For example, plant communities, such as grasslands, cope with changed envirionmental conditions by changes in the proportions of the constituent species. Something similar is true, though over a more narrow range of environmental changes, for the relative stability of the population and changes in individual genotypes. The next chapter considers population responses to disturbance.

References

Allen, T.F.H. & Starr, T.B. (1982). Hierarchy. University Chicago Press, Chicago.

Bakker, J.P. (1983). Seed bank species diversity in moist grassland communities. Acta Bot. Neerl. 32: 246.

Bakker, J.P., Dekker, M. & De Vries, Y. (1979). The effects of different management practices on a grassland community and the resulting fate of seedlings. Acta Bot. Neerl. 29: 469–482.

Baumová, H. (1985). Influence of disturbance by mowing on the vegetation of old-fields in the Bohemian Karst. Folia Geobot. Phytotax., Praha 20: 245–265.

Bazzaz, F.A. (1983). Characteristics of populations in relation to disturbance in natural and man-modified ecosystems. In: Mooney, H.A. & Godron, M. (Eds.). Disturbance and Ecosystems. Ecological Studies 44, pp 259–275. Springer-Verlag, Berlin-Heidelberg.

Brookhaven Symposia in Biology (1969). Diversity and Stability in Ecological Systems, No. 22. Brookhaven National Laboratory, Upton, New York.

Connell, J.H. & Slatyer, R.O. (1977). Mechanisms of succession in natural communities and their role in community stability and organization. Amer. Nat. 111: 1119–1144.

Conrad, M. (1983). Adaptability – The Significance of Variability from Molecule to Ecosystem. Plenum Press, New York, London.

Elberse, W.Th., Van den Bergh, J.P. & Dirven, J.G.P. (1983). Effects of use and mineral supply on the botanical composition and yield of old grassland on heavy-clay soil. Neth. J. Agric. Sci. 31: 63–88.

Elberse, W. Th., Van den Bergh, J.P. & Stamhuis, M. (1987). Growth and development of *Arrhenatherum elatius* and *Festuca rubra* in monocultures and mixtures at low and high nutrient levels. Neth. J. Agric. Sci. 33 (in prep.).

Ernst, W.H.O. (1983). Ökologische Anpassungsstrategien an Bodenfaktoren. Ber. Deutsch. Bot. Ges. 96: 49–71.

Gartner, B.L., Stuart Chapin III, F. & Shaver, G.R. (1983). Demographic patterns of seedling establishment and growth of native graminoids in an Alaskan tundra disturbance. J. Appl. Ecol. 20: 965–980.

Grime, J.P. (1979). Plant Strategies and Vegetation Processes. Whiley, Chichester.

Grootjans, A.P. (1985). Changes of Groundwater Regime in Wet Meadows. Thesis, University of Groningen, Netherlands.

Grubb, J.P. (1985a). Plant populations in relation to habitat, disturbance and competition: problems of generalization. In: White, J. (Ed.). The Population Structure of Vegetation. Handbook Vegetation Science. No. 3, pp 595–621. Junk, Dordrecht.

Grubb, J.P. (1985b). Problems posed by sparse and patchily distributed species in species-rich plant communities. In: Diamond, J. & Case, T.J. (Eds.). Community Ecology, pp 207–225. Harper & Row, New York.

Jain, S. (1983). Genetic characteristics of populations. In: Mooney, H.A. & Godron. M. (Eds). Disturbance and Ecosystems. Ecological Studies 44, pp 240– 258. Springer-Verlag, Berlin-Heidelberg.

Klötzli, F.A. (1981). Zur Reaktion verpflantzer Ökosysteme der Feuchgebiete. Dat. Dok. Umweltsch. 31: 107–117. Univ. Hohenheim, Stuttgart.

Lepś, J., Osbornová-Kosinová, J. & Rejmánek, M. (1982). Community stability, complexity and species life history strategies. Vegetatio 50: 53–63.

McIntosh, R.P. (1980). The relationship between succession and the recovery process in ecosystems. In: Cairns jr., J. (Ed.). The Recovery Process in Damaged Ecosystems, pp 11–62. Ann Arbor Science, Michigan.

Naveh, Z. & Whittaker, R.H. (1980). Structural and floristic diversity of shrublands and woodlands in northern Israel and other Mediterranean areas. Vegetatio 41: 171–190.

Odum, E.P. (1971). Fundamentals of Ecology. Saunders, Philadelphia.

Rapport, D.J., Regier, H.A. & Hutchinson, T.C. (1985). Ecosystem behavior under stress. Amer. Nat. 125: 617–640.

Reiners, W.A. (1983). Disturbance and basic properties of ecosystem energetics. In: Mooney, H.A. & Godron, M. (Eds). Disturbance and Ecosystems. Ecological Studies 44, pp 83–98. Springer-Verlag, Berlin-Heidelberg.

Rosenberg, A. (1985). The Structure of Biological Science. Cambridge University Press, New York.

Rykiel Jr., E.J. (1985). Towards a definition of ecological disturbance. Austr. J. Ecol. 10: 361–365.

Schat, H. (1982). On the ecology of some dune slack plants. Thesis, Free University, Amsterdam.

Selye, H. (1973). The evolution of the stress concept. Amer. Sci. 61: 692–699.

Sousa, W.P. (1984). The role of disturbance in natural communities. Ann. Rev. Ecol. Syst. 15: 353–391.

Tilman, D. (1982) Resource competition and community structure. Monographs in Population Biology. No. 17. Princeton Univ. Press, Princeton-New Yersey.

Tilman, D. (1985). The resource-ratio hypothesis of plant succession. Amer. Nat. 125: 827-852.

Van der Maarel, E. & Werger, M.J.A. (1978). On the treatment of succession data. In: Faliński, J.B. (Ed.). Vegetation Dynamics, Phytocenosis 7: 257–278.

Van der Steen, W.J. & Scholten, M. (1985). Stress and stress tolerance: and exercise in definitions. Acta Biotheoretica 34: 81–90.

Van der Valk, A.G. (1981). Succession in wetlands: A Gleasonian approach. Ecology 62: 688–696.

Van der Valk, A.G. & Davis, C.G. (1976). Changes in the composition, structure, and production of plant communities along a perturbed wetland coenocline. Vegetatio 32: 87–96.

Van Leeuwen, Chr. G. (1966). Schets van de belangrijkste natuurlijke milieus in Nederland. In: Tweede nota van de ruimtelijke ordening in Nederland. Den Haag. pp 108–109.

Van Wirdum, G. (1980). Eenvoudige beschrijving van de waterkwaliteitsverandering gedurende de hydrologische kringloop t.b.v. de natuurbescherming. In: Hooghart, J.C. (Ed.). Waterkwaliteit in grondwaterstromingsstelsels. CHO-TNO rapport 5: 118–143.

Vogl, R.J. (1980). The ecological factors that produce perturbation-dependent ecosystems. In: Cairns jr., J. (ed.). The Recovery Process in Damaged Ecosystems, pp 63–94. Ann Arbor Science, Michigan.

Watt, P.S. (1979). Pattern, process, and natural disturbance in vegetation. Bot. Rev. 45: 229–299.

White, P.S. & Pickett, S.T.A. (1985). Natural disturbance and patch dynamics: an introduction. In: Pickett, S.T.A. & White, P.S. (Eds.). The Ecology of Natural Disturbance and Patch Dynamics, pp 3–13. Acad. Press, New York.

Whittaker, R.H. (1975). Communities and Ecosystems. 385 pp. MacMillan Publishing Co., Inc., New York.

CHAPTER 2

Population responses to environmental disturbance

R.W. Snaydon

Detailed contents

1. Introduction

The term ‘population’ is used in several biological contexts and, if confusion is to be prevented, we need first to define the context. Leaving aside the misuse of the term as a synonym for ‘density’, there are two main uses of the term ‘population’. Firstly, it is used in a genetical context, sometimes termed ‘Mendelian population’, to indicate a group of individuals, normally of the same species, which interact reproductively, i.e. exchange genes. Secondly, it is used in an ecological context to indicate a group of individuals of the same species that interact ecologically, mainly by competition. In each case, unless the group of individuals is geographically isolated, it is usually not possible to clearly define the boundary of the population. As a result, the group is ill-defined, and the term population is therefore normally used very loosely.

In this review, I shall mainly use the term population in its genetical context, i.e. Mendelian populations, giving special attention to the genetic structure of populations, but considering Mendelian populations from an ecological viewpoint (Snaydon, 1985).

Plant populations are dynamic entities, their genetic structure responding to various types of disturbance. The rate and form of the response will, of course, depend both on the nature of the disturbance, and on biological attributes of the species, e.g. life history characteristics and ecological tolerance. The response may also appear to differ depending on which attributes are studied, e.g. morphological attributes as opposed to physiological or biochemical attributes.

The aim of this introductory chapter is to review the various responses of populations to environmental disturbance and variation, so setting the background to chapters 14–20 of this volume. Several reviews have already been published on responses of populations to environmental change (Lewontin, 1957; Levins, 1968; Slobodkin & Rapoport, 1974; Jain, 1979; Lloyd, 1984);

J. van Andel et al., (eds.), Disturbance in Grasslands. ISBN 90-6193-640-3.

some reviews (e.g. Bradshaw, 1965, Hedrick et al., 1976; Loveless & Hamrick, 1984) have concentrated on particular types of response. Although most of these reviews have concentrated on adaptive (or supposed adaptive) responses, and their evolutionary significance, they have usually been written by geneticists, and from a genetical viewpoint. Even when the review title suggests a strong ecological context (e.g. Loveless & Hamrick, 1984), many of the factors considered (e.g. breeding system, floral morphology, mode of reproduction, pollination mechanism and seed dispersal) are hardly ecological and, surprisingly, environmental variation is rarely considered. I intend to take a more ecological approach, concentrating on the diverse nature of environmental variation and consequent differences in populations response, emphasising the underlying principles.

Van Andel and van den Bergh (Chapter 1, section 1.6) have already considered the definition of disturbance and related terms. They have defined disturbance as 'A change in conditions which interferes with the existing normal funtioning of a given biological system', this definition is accepted here. They also point out (Chapter 1, section 3.3) the difficulty of defining what constitutes 'a change of conditions', since various kinds of change may already be an inherent part of the system, to which the system is adapted, and which may actually be required for normal function. In some cases, therefore, disturbance may actually entail a cessation in the inherent pattern of change. As a result, it may be very difficult to distinguish between change which contitutes disturbance, and change which is an inherent part of the system; as van Andel and van den Bergh point out, the status of the change can only be determined by studying the response of the system and comparing it with an undisturbed control.

It follows from these considerations that any study of disturbance must start from a sound knowledge of the status quo, which is almost certain to include inherent variation in time or space. Furthermore, knowledge of equivalent systems in other states is likely to increase our understanding of probable responses to disturbance, since systems are likely to shift from one state to another. This chapter, therefore, initially considers population responses to environmental change in both time and space, regardless of whether the change is inherent or constitutes disturbance, and I shall normally use the neutral term 'environmental variation', rather than disturbance. This consideration of 'environmental variation' then sets a background against which the effects of disturbance on plant populations can be considered.

2. Forms of environmental variation

The form taken by environmental variation (or disturbance) is important in determining the nature of population response. In particular, we need to distinguish between variation in space and in time, but also such aspects as the pattern of change, in space or time, and the rapidity with which change occurs.

Although this volume is mainly concerned with changes induced by fertilizer use and defoliation, I shall not restrict consideration to those factors, since the principles are sometimes better demonstrated by other factors, such as climate.

2.1 Variation in time

Three main attributes of environmental variation in time influence population response: (i) pattern of variation (e.g. random v cyclic), (ii) frequency or periodicity of variation, (iii) amplitude or magnitude of variation. Note that frequency and am-

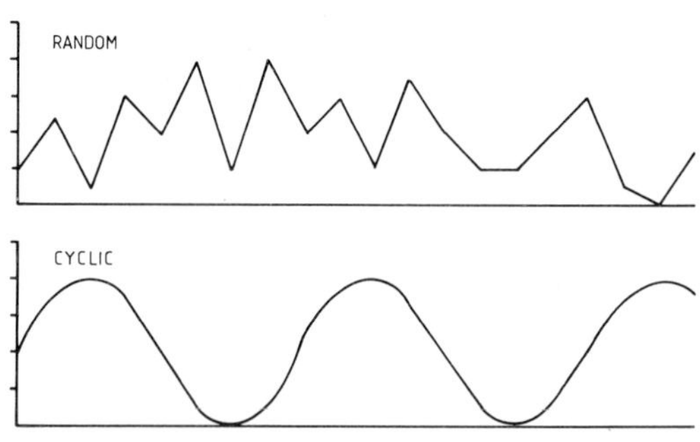

Fig. 1. Examples of (a) random variation and (b) cyclic variation of an environmental factor with time.

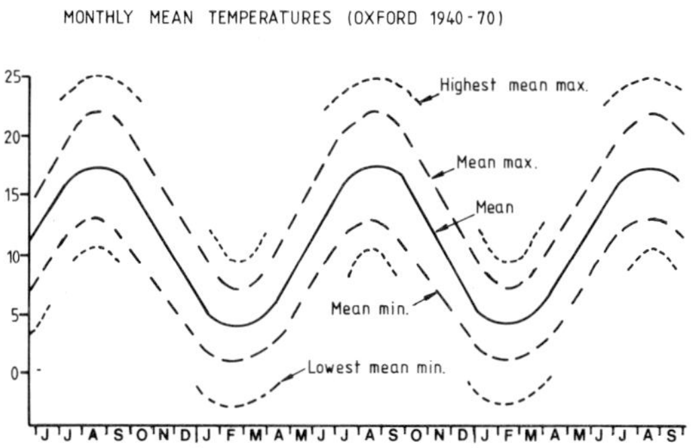

Fig. 2. Seasonal variation in temperature at Oxford. The monthly mean temperature (----), and the monthly mean maximum and minimum temperatures (– –) are shown for the period 1940-70. The highest monthly maximum and minimum temperatures (– – –) are also shown for that period.

plitude together determines what might otherwise be regarded as a fourth attribute, i.e. rate of change. The contrasting patterns of environmental variation, such as cyclic v random (Fig. 1), are especially apparent for climatic factors. Some factors, such as photoperiod, are totally cyclic; other factors, such as temperature, have underlying cyclic patterns, e.g. diurnal and seasonal cycles, but have a strong random component superimposed on this.

Variations in temperature at Oxford (Fig. 2) provide an example of cyclic and random variation. The seasonal cycle of mean temperature is clearly apparent, while the average diurnal cycle is also visible as the difference between mean monthly maxima and minima. In addition, random variation is apparent in the highest maxima and minima for the winter and summer months. You will notice how the highest mid-winter maxima equal the lowest mid-summer minima. If plants are to persist, they must be able to tolerate these cycles and random fluctuations, and adapted plants may actually require these variation and fluctuations; the mechanisms of tolerance and adaptation will be considered later (Section 3.4).

In addition to diurnal and annual cycles, cyclic variation occurs over longer time spans, though usually partly hidden by random variation. Cycles of about (100) years have been reported from meteorological records (Mason, 1976). Longer cycles (500–2000 years) termed ‘climatic periods’, have been inferred from the palynological record (Pennington, 1969), while the ‘ice ages’ have cycles of about 50,000 years. The consequence of cyclic variation with these various frequences will be considered later (Section 3.2).

The amplitude (or magnitude) of cyclic variation is not necessarily related to frequency (or periodicity). Diurnal, annual and longer term varia-

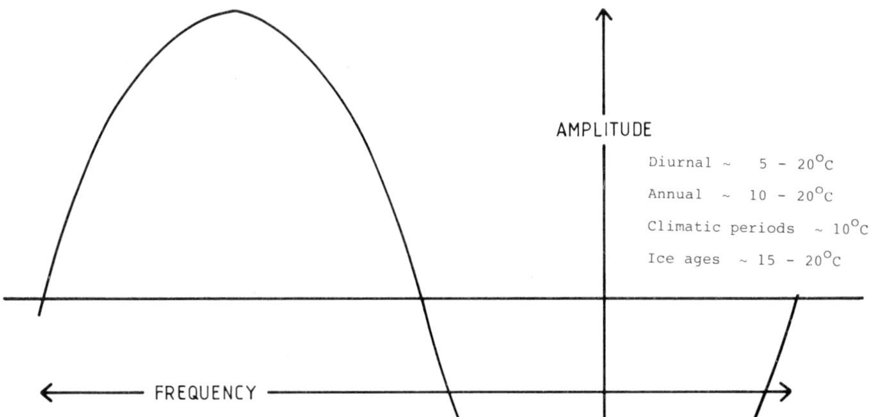

Fig. 3. The amplitude and frequency of temperature variation in climatic cycles of various frequencies.

tion in mean temperature can be of equal magnitude (Fig. 3), depending on the site selected. This has interesting implications for plant populations.

In general, the aerial environment shows greater variation in time, both cyclic and random, than does the soil environment; nevertheless the soil environment shows considerable variation in time. For example, diurnal and seasonal cycles of soil temperature occur; these reflect, but lag behind, the equivalent aerial cycles. The amplitude of variation in the soil depends on depth; the amplitude at the soil surface exceeds that of the aerial environment, but rapidly decreases with depth. Soil water content also follows a seasonal cycle, but with a strong random component, reflecting random variation in rainfall. There are also strong seasonal cycles in the availability of various mineral nutrients (Simpson, 1962; Usher, 1970; Davey & Taylor, 1974; Gupta & Rorison, 1975; Veresoglou & Fitter, 1984). In agricultural grasslands this pattern may be heavily modified by applications of fertilizer, which produce large variations in nutrient availability (e.g. MacDuff & White, 1984). In grazed grasslands, there are also rapid, but highly localised, variations in soil nutrient content, caused by animal excreta (During & McNaught, 1961; Barrow, 1967; MacDiarmid & Watkin, 1972; During & Weeda, 1973).

2.2 *Variation in space*

Three main attributes of environmental variation in space influence population response: (i) dimension of variation (vertical or horizontal), (ii) amplitude or magnitude of variation, and (iii) scale (i.e. distance over which variation occurs). The gradient (i.e. rapidity of change in space-abrupt or clinal) will obviously depend on (ii) and (iii).

Environmental factors change vertically within vegetation; both above and below ground; two examples are given in Fig. 4. In each of these cases the vertical variation is largely plant-induced; in the case of light, it is caused by light interception (Stern, 1962; Monteith, 1965), while in the case of phosphate it is caused by the longer-term extraction of P by roots, and the subsequent

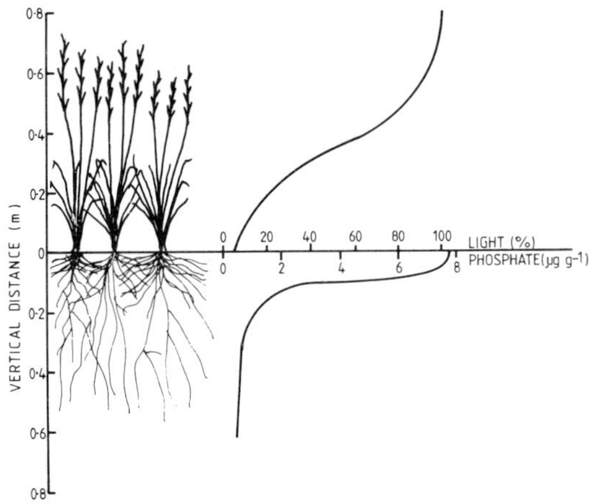

Fig. 4. Vertical variation of light intensity within a vegetation canopy, and of available phosphate concentration in the soil profile.

deposition of this P on the soil surface in dead shoot material and animal faeces. Various other factors vary vertically above-ground (e.g. temperature, relative humidity, and wind speed), and below-ground (e.g. temperature, water content, atmospheric composition, and other nutrients). The magnitude and scale varies, depending on the factor considered and the type of vegetation.

The importance of lateral variability in the environment has almost certainly been underestimated in both ecological and genetic studies of populations and communities. Although variation in climatic (aerial) factors occurs, especially in relation to aspect and shelter, the most important variations probably occur in soil factors. Surprisingly large variations occur within distances of less than 1 m (Table 1); these differences are not substantially less than those found over much larger distances (>1 ha). Some of this sharp variation is the result of animal excreta (see above), but other factors of soil deposition and devel-

Table 1. Variation in the content of extractable potassium and phosphate in soils, expressed as coefficients of variation (%) over various distances (Beckett & Webster, 1971).

Nutrient	1 m^2	0.01 ha	<1 ha
Extractable K	10–40	10–110	20–140
Extractable P	7–45	12–120	10–130

opment also contribute to the variation since soil, unlike air, is largely immiscible and some nutrients (e.g. phosphate) are very immobile.

3. Population responses to environmental variation

Responses to environmental change are commonly considered in the context of adaptation. Biologists perhaps too readily assume that responses are adaptive. Although it is likely that natural selection over millenia has tended to favour adaptive response, we should not assume that all responses are adaptive. In fact, responses range from highly adaptive through adaptively neutral to misadaptive. The latter kind of response might be expected where the plant has not commonly encountered the environmental factor. For example, many dicotyledenous species convert non-toxic MCPB and 2,4–DB into toxic MCPA and 2,4–D, respectively (Roberts, 1982). There are various reasons why adaptation is so readily assumed. Firstly, there is a large element of wishful thinking. Secondly, there are the difficulties of measuring adaptation, e.g. what to measure and what units to use; the most common measure, based on reproductive fitness, has the disadvantages that it is relative, rather than absolute, and is diffucult to use at organisational levels below (and above) that of the individual. Thirdly, adaptation may be specific to a given organisational level, so that apparent misadaptation at one level (e.g. individual plant mortality) may be adaptive at a higher level (i.e. the population).

3.1 Forms of response

Responses to environmental variation can be broadly divided into (i) genetic responses and (ii) environmentally induced modification. Each of these categories can be further subdivided. Genetic responses can be subdivided into: form (anatomy and morphology) and function (biochemistry and physiology). However, a more useful subdivision is on the basis of level of biological organisation (Table 2). Genetic variation at any particular level of biological organisation usually occurs in response to environmental variation of a particular form (Section 2); this relationship will be considered further in Sections 3.2 and 3.3.

Table 2. Genetic variation at various levels of biological organisation.

Within plant	Within population	Between populations	Between species
Heterozygosity Somatic variation	polymorphism	population differentiation	species differentiation

Environmentally induced modification can be subdivided on the basis of form (i.e. phenotypic plasticity) and function (i.e. acclimation) (Table 3). Changes in form (i.e. plastic response) can occur at a range of levels of biological organisation from the sub-cellular to the whole plant (Table 3a). Similarly, functional response can be observed at various levels of organisation from the molecular (e.g. abundance and activity of individual enzymes) to the whole plant (e.g. plant growth rate or water loss) (Table 3b).

3.2 Response to temporal variation

The nature of the response to temporal changes

Table 3. Types of environmental modification of plants, classified on the basis of form (a), function (b), and of biological level of organisation.

(a) *Form (plasticity)*

Cell	Tissue	Plant
Sub-cellular structure	anatomy	morphology

(b) *Function (acclimation)*

Molecular (cell)	Physiol. process (organ)	Whole plant (individual)
Enzyme activity	photosynthesis	growth rate
	P uptake	water loss

in environmental conditions, whether genetic or environmentally induced, will depend on: (i) pattern of variation, (ii) frequency of variation, and (iii) amplitude of variation (see Section 2.1). The nature of the response, in relation to (i) and (ii), is shown in Table 4. To a large extent, the rate of response is dependent upon the biological processes involved. In general, environmentally induced modification, in response to environmental change, occurs more rapidly than genetic responses. For example, some types of physiological acclimation can occur within a few days, and anatomical and morphological changes take several weeks, while population differentiation normally takes several generations, and speciation (not shown in Table 4) takes even longer.

Responses usually occur more rapidly, and are more effective, if environmental change is cyclic, i.e. follows a predictable pattern. In these cases there can be inbuilt rhythms or cycles in plant function, which may be cued by environmental stimuli or triggers. Diurnal cycles (e.g. stomatal aperture) and annual cycles of response (e.g. flowering and leaf shedding) represent good examples of cyclic patterns in plant response to the climatic cycles considered in Section 2.1.

Rapid (‹1 day) random fluctuations in environmental conditions certainly give rise to changes in plant function, such as photosynthetic rates and water uptake, but most of these changes cannot be regarded as adaptive. In general, plants must tolerate these short-term changes.

If changes occur over several days, adaptive changes in plant function can occur. For example, acclimation in response to changes in temperature have been demonstrated in several species (e.g. Mooney & Shropshire, 1967; Scott, 1970; Billings et al., 1971; Bird et al., 1977; Mooney, 1977; Slatyer, 1977). In each case, growing a plant at a given temperature, sometimes for as little as 48 hours, increased its ability to photosynthesise at that temperature. The ability to acclimate differs between species (e.g. Scott, 1970; Bird et al., 1977) and between populations within species (e.g. Mooney & Shropshire, 1967; Billings et al., 1971; Mooney, 1977; Slatyer, 1977).

The phenomena of cold and drought hardening (Levitt, 1980) constitute other examples of acclimation in response to climatic factors. There is also some evidence of acclimation to mineral nutrients (e.g. Clarkson et al., 1978). The mechnism of some of these acclimations are at least partially understood; for example, cold and drought, hardening is partially achieved by the accumulation of osmotica, such as sugars and proline (Levitt, 1980; Paleg & Aspinall, 1981). Acclimation to temperature is often accompanied by changes in the amount of particular enzymes; there are several possible mechanisms for this (Hochachka & Mero, 1973). The possible link with isozyme variation and heterozygosity has yet to be clearly demonstrated for plant populations, though some progress has been made for animal populations (Hendrick et al., 1976).

Acclimation involving anatomical or morphological changes, often termed phenotypic plasticity (Brandshaw, 1965), occur over weeks or months, rather than days. The adaptive signifi-

Table 4. Adaptive responses in relation to the frequency and form of temporal variation.

Form	Very rapid (≤1 day)	Rapid (1–10 days)	Seasonal (1–12 months)	≃ Lifespan (1–100 yr)	>> Lifespan (10–1000 yr)
Random	tolerance	acclimation	acclimation & phenotypic plasticity	polymorphism	population differentiation
Cyclic	diurnal physiol./biochem. cycles	(not applicable)	cued developmental cycles	polymorphism	population differentiation

Note that both genetic variation (Table 2) and environmental modification (Table 3) are considered.

cance of such environmentally-induced morphological modification, has received considerable attention. Not all morphological responses to environmental changes are adaptive; many may actually indicate a lack of adaptation (Bradshaw, 1965). Other problems in studying phenotypic plasticity include difficulties of measuring plasticity, and the fact that the degree of plasticity for various characters may not be correlated (Scheiner & Goodhart, 1984; Schlichting & Levin, 1984).

Apparently adaptive changes in both form and function can occur concurrently. This is particularly evident in the case of response to light intensity (Boardman, 1977). Species (Bazzaz & Carlson, 1982) and populations (Bjorkman & Holmgren, 1963) differ in their ability to respond adaptively to light intensity, and these differences appear to be correlated with the light environment of their native habitats.

Adaptive changes in both form and function also occur in response to soil factors. Root morphology, for example, changes in response to nutrient availability and water supply. The root: shoot ratio (Gadgil & Gadgil, 1979) is particularly responsive to nutrient availability (e.g. Bradshaw et al., 1964; Chapin & Bieleski, 1982; Mouat, 1983), but root branching is also affected (Section 3.3). Sometimes these morphological changes are accompanied by functional changes (Clarkson & Hanson, 1980; Mouat, 1983).

Some morphological changes follow a developmental sequence; this is particularly true of heterophylly. Juvenile leaf forms occur, for example, in most *Eucalyptus* species. The deciduous habit might also be considered an extreme form of heterophylly, occurring in response to cyclic variation in climatic conditions, and cued by environmental triggers, such as photoperiod.

Heterozygosity, another mechanism giving rise to wider physiological tolerance of individuals (e.g. Griffing & Langridge, 1962), may provide the switch mechanism, which is tripped by environmental factors into alternative pathways; this mechanism has particular relevance in the case of allozymes.

Environmental variation with a frequency similar to the lifespan of individual plants acts as 'disruptive selection' (Thoday, 1972), and normally gives rise to polymorphism (Mather, 1955; Thoday, 1959). Surprisingly little is known about the effects of temporal variation on polymorphism within plant populations, compared with animal populations (Hedrick et al., 1976). Evidence on animal populations comes from two sources: (a) experimental studies, and (b) field surveys. Experimental studies with insects (e.g. Thoday, 1972) have shown that disruptive selection increases genetic variation within populations, while field studies have shown seasonal variations in enzyme polymorphism (e.g. Berger, 1971; Hebart, 1974), in chromosomal polymorphism (e.g. Dobzhansky & Ayala, 1973), and in colour polymorphism (e.g. Timofeeff-Ressovsky, 1940). Correlations between the magnitude of genetic variation within populations and temporal variability of the native environment have been found for animal populations (Bryant, 1974) and plant populations (Snaydon, 1973), though the plastic nature of plant size and morphology, compared with that in animals (Bradshaw, 1972), makes phenotypic plasticity a more effective mechanism then polymorphism for plants to respond to short-term temporal variation. Environmental variation with a frequency longer than the lifespan of individuals will give rise to population differentiation. Obviously there is a gradient from polymorphism to population differentiation; population differentiation involves a larger number of genes or characters, greater linkage disequilibrium, and a greater degree of fixation of morphs. In general, population differentiation is likely to occur when environmental condition change more slowly in relation to lifespan. Significant population differentiation, compared with intra-population variation (i.e. polymorphism), can certainly occur within several generations (e.g. Snaydon, 1978; Snaydon & Davies, 1982).

Environmental change over much longer periods (100–1000 generations?) may result in speciation though, because of the long time span involved, little is known about this process, and it lies outside the remit of this volume.

3.3 Response to spatial variation

Population response to spatial variation (Table 5) depends on the dimension, amplitude, scale and gradient of the variation (Section 2.2). Environmental variation in time (Section 2.1) may well interact with variations in space, but such interactions are not considered in detail here, largely because of the lack of information.

We have already seen that small scale variation (‹1 m) in soil conditions are widespread, and of a surprisingly large magnitude (Table 1). Where this small-scale variation occurs within the area inhabited by a single plant, the individual must be able to adjust to that variation. There are several possible mechnisms of adjustment (Table 5). Plants are able to tolerate such small-scale spatial variations in some environmental factors by the simple expedient of translocating resources, such as metabolites, mineral nutrients and water, from plant parts receiving adequate supplies to other parts receiving inadequate supplies. However, one of the most common adaptive responses to small-scale spatial variation in soil conditions is phenotypic plasticity. For example, high concentrations of N, P and, to a much lesser extent, K increase root branching and root extension rate in that zone, so increasing rooting density. This phenomenon has been demonstrated both in culture solution (e.g. Wiersum, 1958; Drew, 1975) and in soils (e.g. Duncan & Ohlrogge, 1958; Passioura & Wetselaar, 1972; Stryker et al., 1974). Localised variations in soil temeprature can have a similar effect (Crossett et al., 1975).

Small-scale variations in the aerial environment largely occur in a vertical dimension, especially within the canopy; light intensity is especially variable (Section 2.2). This vertical variation has special significance for the developing plant, since seeds germinate at ground level, then grow through the vertical gradient of environmental conditions. Species differ in their ability to germinate and establish within this microclimatic gradient (Grime & Jeffrey, 1965; Fenner, 1978). Similarly, tillers and leaves are often produced at the base of the plant, and grow up through the microclimatic gradient. In some species, the photosynthetic capacity of the mature leaf is determined by the light intensity it experiences during early development, so that leaves developing at the base of dense stands have a low photosynthetic capacity (e.g. Beuerlein & Pendleton, 1971; Woledge, 1977). However, in other species, the photosynthetic capacity of the mature leaf is determined at a later stage of development, so that leaves developing under a dense canopy have high photosynthetic capacities, as long as they emerge from the canopy (Dennis & Woledge, 1983).

Spatial variation of about the size of an individual or rather larger will, like temporal variation with approximately the frequency of the life cycle (Section 3.2), act as disruptive selection and lead to polymorphism. Various studies have demonstrated this polymorphism, though in practice it has proved difficult, especially in plant populations, to distinguish between polymorphism and micro-scale population differentiation (see be-

Table 5. Adaptive responses to horizontal and vertical spatial variation over various distances.

Dimension	Very small scale (< individual)	Small scale (1–10 × individual)	Large scale (> 10 × individual)
Horizontal	tolerance acclimation phenotypic plasticity	polymorphism (clinal differentiation)	population differentiation
Vertical	tolerance acclimation developmental sequence phenotypic plasticity	(not applicable)	(not applicable)

low). Studies of polymorphism in animal population have been reviewed by Hedrick et al., (1976), and polymorphism in plant populations by Loveless & Hamrick (1984).

The extent of polymorphism, or micro-scale population differentiation, will be determined by the relative magnitudes of: (a) disruptive selection and (b) gene flow (Spieth, 1979). Disruptive selection is determined by the pattern and magnitude of environmental heterogeneity, while gene flow within and between populations is determined by such factors as breeding system, longevity, pollination and seed dispersal mechanisms, and population density (Loveless & Hamrick, 1984). In the case of plant populations, it would seem that the magnitude and importance of disruptive selection is greater than had previously been supposed, while the magnitude of gene flow is probably less (Jain & Bradshaw, 1966; Ehrlich & Raven, 1969; Endler, 1973). Most gene flow occurs over quite short distances, in both wind-pollinated species (e.g. Bateman, 1947a; Griffiths, 1950) and insect-pollinated species (e.g. Bateman, 1947b; Handel, 1983a). As a result of this restricted gene flow, the 'neighbourhood size', which is the effective size of the Mendelian population, is small (e.g. Kerster & Levin, 1968; Beattie & Culver, 1979; Bos et al., 1986). This has important implications for both population polymorphism and micro-scale population differentiation.

The factors affecting gene flow have been considered by Handel (1983b). One of the most important factors is the breeding system; inbreeding will obviously restrict gene flow and so might be expected to reduce genetic variation within populations, but increase variation between populations (Loveless & Hamrick, 1984). Although this seems to be generally true for isozyme variation (Loveless & Hamrick, 1984), it may not be true for morphological variation (e.g. Kanneberg & Allard, 1967; Helgadottir & Snaydon, 1986); indeed, variation within populations may sometimes be greater in inbreeding species (Hillel et al., 1973) and facultative apomicts (Loveless & Hamrick, 1984). The apparently conflicting results of studies based on isozyme and morphological variability may perhaps reflect the fact that isozyme variation may be more selectively neutral than morphological variation.

The difficulty of distinguishing between polymorphism and micro-scale population differentiations has already been mentioned. Evidence of the non-random distribution of individual genotypes of perennials, some within areas of 100 m^2 (Harbard, 1961 & 62; Silander, 1979), indicates the difficulty of distinguishing between polymorphism and micro-scale population differentiation. However, the fact that considerable genetic variation exists within micro-populations in areas ‹1 m^2 (Cahn & Harper, 1976; Snaydon & Davies, 1976; McNeilly & Roose, 1984), which presumably cannot be further sub-divided ecologically, indicates that true polymorphism does exist.

The cause of this genetic variation, or polymorphism, are less easy to identify. In some cases, the amount of genetic variation within well-defined populations can be correlated with the magnitude of environmental heterogeneity within the areas from which the populations were collected (e.g. Snaydon, 1973). The age of the population may also be important; for example, there is evidence that genetic diversity initially decreases rapidly in sown swards, because of the high initial mortality rate (Charles, 1961). Thereafter, mortality continues at a slower rate (Jewiss, 1972) and, since there may be little replacement by seedling establishment (Harper, 1977), biotype depletion continues (Snaydon, 1980; Aarssen & Turkington, 1985). As a result, old swards may become dominated by relatively few genotypes. For example, Harberd (1961) found that 50% of ramets of *Festuca rubra*, collected from a 100 m^2 quadrat, belonged to a single genotype, and 10 genotypes constituted 87% of the sward, though he recognised a further 170 genotypes in the quadrat. Similarly, McNeilly & Roose (1984) found that 38% if the ramets of *Lolium perenne* collected from a 1 m^2 quadrat, belonged to a single genotype, and that the three most common genotypes constituted 84% of the sward, though a total of 7 genotypes occurred in the 1 m^2 quadrat. However, other factors, such as somatic mutation (Whitham & Slobodchikoff, 1981), may increase polymorphism in vegetatively reproducing species. There is obviously scope for further study of the nature of genetic variation within populations, and the way in which this is related to spatial variation in the habitat.

Various studies over the past two decades have shown that population differentiation occurs over surprisingly short distances: sometimes over distances of less than 1 m (e.g. Snaydon & Davies, 1976). Such genetic variation would previously have been considered polymorphism. This micro-scale population differentiation might be expected to be most pronounced in perennial species, especially those with well-developed vegetative reproduction. The long life-cycle of such species would tend to reduce genetic recombination and gene flow, though this may be partly offset by the fact that they are usually outbreeding. In addition, vegetative spread would allow adapted genotypes to exploit suitable areas of the habitat. Certainly, such micro-scale differentiation occurs in various perennial plant species (e.g. Aston & Bradshaw, 1966; Watson, 1969; Schaal, 1975; Warwick & Briggs, 1978; 1979; Silander & Antonovics, 1979; Turkington & Harper, 1979). However, it also occurs in short-lived perennials with little capacity for vegetation spread (e.g. Snaydon & Davies, 1976), and even in annual species (e.g. Hamrick & Allard, 1972 & 1975; Linhart, 1974; Hamrick & Holden, 1979; Schemske, 1984), which have greater opportunities for recombination and gene flow, and no capability for vegetative spread.

Innumerable examples have been published of population differentiation in response to environmental variation on spatially larger scales (e.g. Heslop-Harrison, 1964). Loveless & Hamrick (1984) have reviewed some of the factors associated with greater isozyme differentiation among plant populations, though less study has been made of morphological or physiological differentiation, which is ecologically more important. Isozyme variation between populations was greater in inbreeders than outbreeders, greater in annuals than in perennials, greater in species of early successional stages than late successional stages, and greater in monocarpic species than in polycarpic species; these four attributes, of course, tend to be correlated, and it is difficult to isolate their separate effects. Perhaps surprisingly, between-population variation was not closely correlated with geographic range.

4. Population responses to soil variation

Having considered the general principles of population response to environmental variation, I shall now consider response to the groups of factors which are the main focus of this volume, i.e. soil conditions and defoliation. We have already seen (section 2) that soil conditions tend to be relatively stable in time, but sharply variable in space, while defoliation is extremely variable in time, but can also be variable in space.

Since this volume is concerned with grasslands, I shall mainly consider soil variation in the context of the two factors which cause the greatest variation in soil condition in grasslands – animal excreta and fertilizer application. The introduction of grazing animals will greatly increase temporal and spatial variation in soil fertility, because of the effects of excreta (section 2); both the introduction of animals (i.e. increase in variation) and the removal of animals (i.e. reduction in variation) constitutes disturbance.

4.1 Animal excreta

The ecological effects of urine, on the yield and botanical composition of pastures, are localised and generally short-lived (During &McNaught, 1961). The effects on populations have only recently received attention; the main effects are apparently phenotypic, with large plastic responses in morphology (Rebuffo, 1986). Genetic changes over short distances, caused by excreta, are unlikely because of the short-lived nature of the effect, but differences over larger distances (e.g. between fields) might be expected, due to average differences in nutrient turnover.

The ecological effects of dung (MacDiarmid & Watkin, 1972; During & Weedsa, 1973) are more prolonged (1—2 years), and are usually exaggerated by the fact that animals reject the herbage around dung (Marsh & Campling, 1976). Although the species composition around dung pats changes (Norman & Green, 1958), and large plastic responses in morphology occur (Rebuffo, 1986), it is again unlikely that the genetic structure of species population changes over such short distances. It seems likely that the environmental variation, in space and time, caused by animal excreta, would increase genetic polymorphism within pasture species; the comparison by Mc Neilly & Roose (1984) of cut and grazed pasture seems to bear this out.

4.2 Application of fertilizers

In general, the effects of fertilizers are less localised and longer lived than those of excreta. As a result, sequential applications of fertilizers induce population differentiation; both morphological differences (e.g. Charles, 1961; Snaydon, 1970; Snaydon & Davies, 1972; Helgadottir & Snaydon, 1986) and physiological differences (e.g. Charles, 1971; Davies & Snaydon, 1973) have been demonstrated. This population differentiation can occur within 10 years (Snaydon & Davies, 1982) and, provided the boundary is sufficiently sharp, over distances of less than 1 m (Snaydon & Davies, 1976). However, in some other cases, no population differentiation has been detected; for example, Harris & Lazenby (1977) found no differences in response to N fertilizer after 3½ years.

Applications of mineral toxins, such as heavy metals (Antonovics et al., 1971), can also induce population differentiation. Again, the differences can evolve quite rapidly (e.g. Wu et al., 1975), and over quite short distances (e.g. McNeilly, 1968).

In most of the examples quoted, significant differences have been found between population means, when compared with the genetic variation existing within populations. However considerable genetic variation within populations has also been found. There is some evidence (e.g. Snaydon, 1970; Snaydon & Davies, 1972; Helgadottir & Snaydon, 1986) that fertilizer applications affect the amount of genetic variation, i.e. polymorphism, within populations of grass species. These differences in genetic variation may, in part, be due to the fact that fertilizer applications change the sensitivity to climatic variation, since Snaydon (1973) found that the year-to-year variation in herbage yield differed between fertilizer treatments, and the amount of genetic variation within populations was correlated with yield variation.

There is also considerable evidence of morphological plasticity in response to fertilizer applications. This response occurs both in the root systems (Section 2.2) and in the shoot system. The latter may sometimes be an indirect response, following changes in the surrounding vegetation. For example, N fertilizer applications increase the rate of growth and the height of grass in swards; and this may then increase the leaf size and petiole length of *Trifolium repens* in the sward (e.g. Snaydon, 1985).

5. Population responses to defoliation

The phsyiological and ecological effects of defoliation on pastures has been reviewed by Jameson (1963), Humphreys (1966), Youngner (1973) and Harris (1978), and mechanisms of adaptation to grazing have been reviewed by Hodgkinson & Williams (1983).

Defoliation (i.e. the removal of parts of a plant) is totally different in form from other environmental factors. It is usually frequent, relative to the life-span, it occurs instantaneously, and effects only part of the plant, the fraction being variable in size and location.

5.1. Morphological plasticity

In view of the frequency and localised nature of defoliation, it is not surprising that some of the most commonly observed responses to defoliation involve phenotypic plasticity. Perhaps the best examples of phenotypic plasticity in responses to defoliation are tillering in grasses (Langer, 1963; Jewiss, 1972), and changes in leaf size and petiole length in clovers (Snaydon, 1985). Frequent and severe defoliation increases tillering in grasses (Langer, 1963; Jewiss, 1972), and the response appears to be adaptive. Similarly, frequent defoliation reduces the leaf size and petiole length of *Trifolium repens* (e.g. Wilman & Asiegbu, 1982); again the response appears to be adaptive. King et al. (1978) have studied the effects of a single defoliation, either complete or partial, on the leaf size, petiole length, and stolon growth of *T. repens*. They found that both complete and partial defoliation affected leaf size and petiole length, and interpreted the results largely in terms of the acropetal movement of assimilates along stolons. In general, morphological plasticity in response to defoliation has no received the attention that it deserved, in view of its probable importance as an adaptive mechanism.

5.2 Polymorphism

The effects of defoliation on polymorphism has received even less attention than phenotypic plasticity. There is no doubt that considerable genetic variation exists within population of various pasture species (e.g. Harbard, 1961 & 1962; Chan & Harper, 1976; Snaydon & Davies, 1976; McNeilly & Roose, 1984), but little indication of whether this differs substantially from that of other vegetation types, or what are the main factors determining the variation. McNeilly & Roose (1984) found more genotypes per unit area in grazed swards than in equivalent cut swards, but equal numbers of genotypes in swards grazed by sheep and by cattle. The difference between grazed and cut swards might be due either to grazing *per se* or to the effects of excreta (Section 4.1).

5.3 Population differentiation

There is ample evidence that population differentiation occurs in response to defoliation (e.g. Stapledon, 1928; van Dyke, 1955; Bradshaw, 1959; Charles, 1961; Watson, 1969; Harris & Brougham, 1970; Charles, 1970; Law et al., 1977; Warwick & Briggs, 1978; Pike et al., 1979). In general, intense defoliation leads to a genetic shift towards a more prostrate habit, greater tillering, reduced and later flowering, and greater pereniality. There is considerable circumstantial evidence that these differences are adaptive, though little direct evidence.

Significant differences between populations, in response to defoliation, have been detected over distances of a few metres (e.g. Watson, 1969; Warwick & Briggs, 1978), and over periods of about 10 years (e.g. Charles, 1961 & 1970; Harris & Brougham, 1970; Pike et al., 1979).

6. Response to disturbance

Throughout most of the chapter, I have not distinguished between inherent environmental variation and disturbance. It is now necessary to consider possible differences in the response of populations to these two forms of environmental change. The essential question is 'do populations respond differently to environmental change, depending on whether they have experienced change before?' Just as in the case of plant communities (chapter 1), populations are likely to become adapted to inherent environmental variation; indeed, they may often become dependent upon it. I have already considered the nature of such adaptation to such environmental variation (Section 3). Adaptive responses, as we have seen, are dependent upon the type of environmental variation, and can occur at various levels of biological organisation, from the molecular level to the whole population. In general, our knowledge of population response to disturbance is currently inadequate, but the nature of responses can probably be deduced from a knowledge of responses to inherent environmental variation (Section 3).

Disturbance can involve: (i) change to a new level of the environmental factor and/or, (ii) imposition of a new pattern of environmental variation. A new pattern (ii) may involve: (a) imposition of new variation, (e.g. introduction of grazing animals), (b) change in the previous pattern (e.g. more frequent cutting or grazing), or (c) removal of variation (e.g. removal of grazing animals). In addition, disturbance may vary in magnitude or speed of change. In view of the very large number of types of change, combinations of changes, and the variety of environmental factors involved, I shall not attempt to review the diversity of possible population responses; to some extent it has been covered in Section 3. However, I shall try to draw some general conclusions about such responses.

Adaptive responses to disturbance will depend upon the presence of the necessary genetic variation; this is true regardless of the level of organisation at which the response occurs. For example, adaptive phenotypic response (acclimation and phenotypic plasticity) depends on a genetically determined ability to respond (Bradshaw, 1965), some of this may be attributable to heterozygosity or to isozymes (Section 3.2). Similarly, changes in the genetic structure of populations depends on the presence of the necessary genetic variability within or between plants of the population. As a result, adaptive response to disturbance is more likely to occur if the population has previously experienced the new conditions (or similar conditions) at some time, and consequently carries a residuum of appropriate individuals or genes. Such previous experience, however, might be regarded as placing the new conditions within the context of inherent variation, and therefore no longer regarded as disturbance. Obviously, the distinction between inherent variation and disturbance is likely to be very blurred.

Adaptive response is, however, not totally dependent upon prior experience of particular conditions. In some cases we can identify responses to conditions, especially man-made conditions, which could not have previously been experienced. A good example of this is tolerance or resistance to man-made herbicides, such as the atrazine group; resistance may already exist within whole species (pre-adaptation), or may occur within individual mutant genotypes and then spread through whole populations (Le Baron & Gressel, 1982).

Adaption to previously variable conditions may be one of the most important attributes allowing adaption response to new disturbance. The various mechanisms of adaptation to environmental variation, such as phenotypic plasticity or polymorphism (Section 3), are likely to allow populations to respond adaptively to disturbance, especially if the disturbance is merely a change in mean level of a factor, and involves the same environmental factor(s) that previously varied. Does adaptation to one variable environmental factor also confer ability to respond adaptively to disturbance in another factor? Although our knowledge of the factors causing genetic variations in populations, and of population responses to disturbance, is currently inadequate to answer this question, it seems likely that some such influence is possible.

7. Conclusions

The response of plant populations to environmental variation depends on the pattern and magnitude of environmental variation and on biological attributes of the plant species.

Environmental variation in time and space need to be considered separately, Variation in time can be categorised on the basis of form (e.g. random v cyclical), frequency, and amplitude. Variations in space can be categorised on the basis of dimension (horizontal v vertical), scale, and amplitude.

Plant responses to environmental variation can be either genetic or environmentally induced; they occur over a wide spectrum of levels of organisation, from the molecular level to the population (and species) level. The type of plant response is broadly related to the type and pattern of environmental variation. The most common responses to rapid environmental changes in time (i.e. substantially less than lifespan), and to rapid changes in space (i.e. less than plant size), are phenotypic plasticity and physiological acclimation. Temporal changes that more or less equal the lifespan, and spatial changes that more or less equal plant size, are usually associated with polymorphism within populations. Temporal changes with a frequency substantially longer than the lifespan, and spatial changes over distance substantially greater than the size of an individual plant, give rise to population differentiation. Many responses, involving phenotypic plasticity, physiological acclimation, polymorphism and population differentiation, appear to be adaptive. However, not all responses are adaptive, and the adaptive nature of responses has rarely been demonstrated directly; some reponses may actually be misadaptive.

Examples are considered of responses, ranging from phenotypic plasticity and acclimation to population differentiation, induced by animal excreta, fertilizer applications and defoliation.

In general, responses at the higher levels of organisation (i.e. population differentiation and speciation) have received the greatest attention and have been studied for a longer time. More recently, polymorphism, phenotypic plasticity and physiological acclimation have received some attention, but information on the latter responses is still very inadequate.

Response to disturbance is likely to take the same form as response to inherent environmental variation, i.e. it is likely to occur at various levels of biological organisation and depend on such factors as the magnitude and speed of change in the environment. Adaptive response is more likely to occur if the population has already experienced similar variation previously, but adaptive responses to novel environmental variation are known.

References

Aarssen, L.W. & Turkington, R. (1985). Within species diversity in natural populations of *Holcus lanatus, Lolium perenne* and *Trifolium repens* from four different-aged pastures. J. Ecol. 73: 869–886.

Antonovics, J., Bradshaw, A.D. & Turner, R.G. (1971). Heavy metal tolerance in plants. Ad. Ecol. Res. 7: 1–85.

Aston, J.L. & Bradshaw, A.D. (1966). Evolution in closely adjacent plant populations. II *Agrostis stolonifera* in maritime habitats. Heredity 21: 649–664.

Bakker, J.P., de Leeuw, J. & van Wieren, S.E. (1984). Micro-patterns in grassland vegetation created and sustained by sheep-grazing. Vegetation 55: 153–161.

Barrow, M.J. (1967). Some aspects of the effects of grazing on the nutrition of pastures. J. Aust. Inst. agric. Sci. 33: 254–266.

Bazzaz, F.A. & Carlson, R.W. (1982). Photosynthetic acclimation to variability in the light environment of early and late successional plants. Oecologia (Berl) 54: 313–316.

Bateman, A.J. (1947a). Contamination of seed crops. II Wind pollination. Heredity 1: 235–246.

Bateman, A.J. (1947b). Contamination of seed crops. I Insect pollination. J. Genet. 48: 257–275.

Beattie, A.J. & Culver, D.C. (1979). Neighbourhood size in *Viola*. Evolution 33: 1226–1229.

Beckett, P.H.T. & Webster, R. (1971). Soil variability: a review. Soil & Fert. 34: 1–15.

Berger, E.M. (1971). Temporal survey of allelic variation in natural and laboratory populations of *Drosophila melanogaster*. Genetics 67: 121–136.

Beuerlein, J.E. & Pendleton. J.W. (1971). Photosynthesis rates and light saturation curves of individual soybean leaves under field conditions. Crop Sci. 11: 217–219.

Billings, W.D., Godfrey, P.J., Chabot, B.F. & Bourque, D.P. (1971). Metabolic acclimation to temperature in arctic and alpine ecotypes of *Oxyria digyna*. Arct. Alp. Res. 3: 277–289.

Bird, I.F., Cornelius, M.J. & Keys, A.J. (1977). Effects of temperature on photosynthesis of maize and wheat. J. exp. Bot. 28: 519–524.

Bjorkman, D. & Holmgren, P. (1963). Adaptability of the photosynthetic apparatus to light intensity in ecotypes from exposed and shaded habitats. Physiologia Pl. 16: 889–914.

Boardman, N.K. (1977). Comparative photosynthesis of sun and shade plants. A. Rev. Pl. Physiol. 28: 355–377.

Bos, M., Harmens, H. & Vrieling (1966). Gene flow in *Plantago*. I Gene flow and neighbourhood size in *P. lanceolata*. Heredity 56: 43–54.

Bradshaw, A.D. (1959). Population differentiation in *Agrostis tenuis*. 1 Morphological differentiation. New Phytol. 58: 310–315.

Bradshaw, A.D. (1965). Evolutionary significance of phenotypic plasticity in plants. Adv. Genet. 13: 115–155.

Bradshaw, A.D. (1972). Some of the evolutionary consequences of being a plant. Evol. Biol. 5: 25–47.

Bradshaw, A.D., Chadwick, M.J., Jowett, D. & Snaydon, R.W. (1964). Experimental investigations into the mineral nutrition of several grass species. IV Nitrogen. J. Ecol. 52: 665–676.

Bryant, E.H. (1974). On the adaptive significance of enzyme polymorphism in relation to environmental variability. Am. Nat. 108: 1–19.

Cahn, M.A. & Harper, J.L. (1976). The biology of the leaf mark polymorphism in *Trifolium repens*. I Distribution of phenotypes at a local scale. Heredity 37: 309–325.

Chapin, F.S. & Bieleski, R.L. (1982). Mild phosphorus stress in barley and a related low-phosphorus-adapted barley grass. Physiol. Plant. 54: 309–317.

Charles, A.H. (1961). Differential survival of cultivars of *Lolium, Dactylis* and *Phleum*. J. Br. Grassld. Soc. 16: 69–75.

Charles, A.H. (1970). Ryegrass populations from intensively managed leys. 1 Seedling and spaced plant characters. J. agric. Sci. 75: 103–107.

Charles, A.H. (1971). Ryegrass populations from intensively managed leys. II Reaction to nitrogen and *Poa trivialis*. J. agric. Sci. 76: 233–241.

Clarkson, D.T. & Hanson, J.B. (1980). The mineral nutrition of higher plants. Ann. Rev. Plant. Physiol. 31: 239–298.

Clarkson, D.T., Sanderson, J. & Scattergood, C.B. (1978). Influence of phosphate-stress on phosphate absorption and translocation by various parts of the root system of *Hordeum vulgare*. Planta 139: 47–53.

Clements, R.J., Haywood, M.D. & Blyth, D.E. (1983). Genetic adaptation in pasture plants. In: McIver, J.G. & Bray, R.A. (Eds.) Genetic Resources of Forage Plants, pp 101–115. CSIRO, Melbourne.

Crossett, R.N., Campbell, D.J. & Stewart, H.E. (1975) Compensatory growth in cereal root systems. Pl. Soil 42: 673–683.

Davies, M.S. & Snaydon, R.W. (1973). Physiological differences among populations of *Anthoxanthus odoratum* collected from the Park Grass Experiment. I Response to calcium & II Response to aluminium. J. appl. Ecol. 10: 33–55.

Davey, A.J. & Taylor, K. (1974). Seasonal patterns of nitrogen availability in contrasting soils in the Chiltern Hills. J. Ecol. 62: 793–807.

Dennis, W. & Woledge, J. (1983). The effect of shade during leaf expansion on photosynthesis of white clover leaves. Ann. Bot. 51: 111–118.

Dobzhansky, Th. & Ayala, F.J. (1973). Temporal frequency changes of enzyme and chromosomal polymorphism in natural populations of *Drosophila*. Proc. Natl. Acad. Sci. 70: 680–683.

Drew, M.C. (1975). Comparison of the effects of a localised supply of phosphate, nitrate, ammonium and potassium on the growth of the seminal root system, and the shoot, in barley. New Phytol. 75: 479–490.

Duncan, W.G. & Ohlrogge, A.J. (1958). Principles of nutrient uptake from fertilizer bands. II Root development in the band. *Agron. J.* 50: 605–608.

During, C. & McNaught, K.J. (1961). Effects of cow urine on growth of pasture and uptake of nutrients NZ J. agric. Res. 4: 591–605.

During, C. & Weeda, W.C. (1973). Some effects of cattle dung on soil properties, pasture production, and nutrient uptake I & II. NZ J. agric. Res. 16: 423–438.

Ehrlich, P.R. & Raven, P.H. (1969). Differentiation of populations. Science 165: 1228–1232.

Endler, J.A. (1973). Gene flow and population differentiation. Science 179: 243–250.

Fenner, M. (1978). A comparison of the abilities of colonizers and close-turf species to establish from seed in artificial swards. J. Ecol. 66: 953–963.

Gadgil, M. & Gadgil, S. (1979). Adaptive significance of the relation between root and shoot growth. Jour. Ind. Inst. Sci. 61C: 25–40.

Griffing, B. Langridge, J. (1963). Phenotypic stability of growth in the self-fertilizing species *Arabidopsis thaliana*. In: Statistical Genetics and Plant Breeding, pp 368–394. Nat. Acad. Sci, New York.

Grime, J.P. & Jeffery, D.W. (1965). Seedling establishment in vertical gradients of sunlight. J. Ecol. 53: 621–642.

Gupta, P.L. & Rorison, I.H. (1975). Seasonal differences in the availability of nutrients down a podzolic profile. J. Ecol. 63: 521–534.

Hamrick, J.L. (1979). Genetic variation and longevity. In: Silbrig, O.T., Jain, S., Johnson, G.B. & Raven, P.H. (Eds.) Topics in Plant Population Biology, pp 84–113. Columbia University Press, New York.

Hamrick, J.L. & Allard, R.W. (1972). Microgeographical variation in allozyme frequencies in *Avena barbata*. Proc. Nat. Acad. Sci. 69: 2100–2104.

Hamrick, J.L. & Allard, R.W. (1975). Correlations between quantitative characteristics and allozyme genotypes in *Avena barbata*. Evolution 29: 438–442.

Hamrick, J.L. & Holden, L.R. (1979). Influences of microhabitat heterogeneity on gene frequency distribution and genetic phase disquilibrium in *Avena barbata*. Evolution 33: 521–533.

Handel, S.N. (1983a). Contrasting gene flow paterns and genetic subdivision in adjacent populations of *Cucumis sativa*. Evolution 37: 760–771.

Handel, S.N. (1983b). Pollination ecology, plant population structure, and gene flow. In: Real, L. (Ed.) Population Biology. pp 163–211. Academic Press, New York.

Harberd, D.J. (1961). Observations on population structure and longevity of *Festuca rubra*. New Phytol. 60: 184–206.

Harberd, D.J. (1962). Some observations on natural clones in *Festuca ovina*. New Phytol. 61: 81–100.

Harris, W. (1976). Defoliation as a determinant of growth, persistence and composition of pasture. In: Wilson, J.K. (Ed.) Plant Relations in Pastures, pp 67–85. CSIRO, Melbourne.

Harris, W. & Brougham, R.W. (1970). The effect of grazing on the persistence of genotypes in a ryegrass population. NZ Jl. agric. Res. B: 263–278.

Harris, W. & Lazenby, A. (1977). Nitrogen responses of tall fescue and ryegrass survivor populations from a rate-of-nitrogen experiment. NZ Jl. agric. Res. 20: 327–330.

Harris, W., Pandey, K.K., Gray, Y.S. & Couchman, P.K. (1979). Observations on the spread of perennial ryegrass by stolons in a lawn. NZ. Jl. agric. Res. 22: 61–69.

Hebert, P.D.N. (1974). Ecological differences between genotypes in natural populations of *Daphnia magna*. Heredity 33: 227–237.

Hedrick, P.W., Ginevan. M.E. & Ewing, E.P. (1976). Genetic polymorphism in heterogeneous environments. Ann. Rev. Ecol. Syst. 7: 1–32.

Helgadottir, A. & Snaydon, R.W. (1986). Patterns of genetic variation among populations of *Poa pratensis* and *Agrostis capillaris* from Britain and Iceland. J. apl. Ecol. 23: 703–719.

Heslop-Harrison, J. (1964). Forty years of genecology, Adv. Ecol. Res. 2: 159–247.

Hillel, J. Feldman, M.W. & Sinchen, G. (1973). Mating systems and population structure in two closely related species of the wheat group. I Variation between and within populations. Heredity 30: 141–167.

Hochachka, P.N. & Somero, G.N. (1973). Strategies of Biochemical Adaptation. Saunders, Philadelphia.

Hodgkinson, K.C. & Williams, O.B. (1983). Adaption to grazing in forage plants. In. McIver, J.G. & Bray, R.A. (Eds.) Genetic Resources of Forage Plants, pp 85–100. CSIRO, Melbourne.

Humphrey, L.R. (1986). Pasture defoliation: a review. J. Aust. Inst. agric. Sci. 32: 93–105.

Jain, S. (1979). Adaptive strategies: polymorphism, plasticity and homeostasis. In: Solbrig, O.T., Jain, S., Johnson, G.B. & Raven, P.H. (Eds.). Topics in Plant Population Biology, pp 160–187, Columbia Univ. Press, New York.

Jain, S.K. & Bradshaw, A.D. (1966). Evolutionary divergence among adjacent plant populations. I The evidence and its theoretical analysis. Heredity 21: 407–441.

Jameson, D.A. (1963). Responses of individual plants to harvesting. Bot. Rev. 29: 582–594.

Jewiss, O.R. (1972). Tillering in grasses – its significance and control. J. Br. Grassld. Soc. 27: 65–82.

Kannenberg, L.W. & Allard, R.W. (1967). Population studies in predominantly self-pollinated species. XI Genetic divergence among members of the *Festuca microstachys* complex. *Evolution* 22: 517–528.

Kerster, H.W. & Levin, D.A. (1968). Neighbourhood size in *Lethospermum carolinensa*. Genetics 60: 577–587.

King, J., Lamb, W.K. & McGregor, M.T. (1978). Effects of partial and complete defoliation on regrowth of white clover plants. J. Brit. Grassld. Soc. 33: 49–55.

Langer, R.H.M. (1963). Tillering in herbage grasses. Herb. Abstr. 33: 141–148.

Law, R., Bradshaw, A.D. & Putwain, P.D. (1977). Life history variation in *Poa annua*. Evolution 31: 233–246.

Le Baron, H. & Gressel, J. (1982). Herbicide Resistance in Plants, Wiley, Chichester.

Levin, D.A. & Kerster, H.W. (1968). Local gene dispersal in *Phlox*. Evolution 22: 130–139.

Levins, R. (1968). Evolution in changing environments. Princeton Univ. Press, Princeton, New Jersey.

Levitt, J. (1980). Responses of plants to environmental stresses, Vols I & II. Academic Press, New York.

Lewontin, R.C. (1957). The adaptation of populations to varying environments. Cold Spring Harbor Symp. Quant. Biol. 22: 395–408.

Linhart, Y.B. (1974). Intra-population differentiation in annual plants 1. *Veronica perigrina* raised under non-competitive conditions. Evolution 28: 232–243.

Lloyd, D.G. (1984). Variation strategies of plants in heterogeneous environments. Biol. J. Linn. Soc. 21: 357–385.

Loveless, M.D. & Hamrick, J.L. (1984), Ecological determinants of genetic structure in plant populations. Ann. Rev. Syst. Ecol. 15: 65–95.

MacDiarmid, B.N. & Watkin, B.R. (1972). The cattle dung patch. 2. The effect of a dung patch on the chemical status of the soil. J. Brit. Grassld. Soc. 26: 43–48.

MacDuff, J.H. & White, R.E. (1984). Components of the nitrogen cycle measured for cropped and grassland soil-plant systems. Pl. Soil 76: 35–47.

McNeilly, T. (1968). Evolution in closely adjacent plant populations. III *Agrostis tenuis* on a small copper mine. Heredity 23: 99–108.

McNeilly, T. & Roose, M.L. (1984). The distribution of perennial ryegrass genotypes in swards. New Phytol. 98: 503–513.

Marsh, R. & Campling, R.C. (1970). Fouling of pastures by dung. Herb. Abstr. 40: 123–130.

Mason, B.J. (1976). Towards the understanding and prediction of climatic variation. Quart J.R. Met. Soc. 102: 473–498.

Mather. K. (1955). Polymorphism as an outcome of disruptive selection. Evolution 9: 52–61.

Monteith, J.L. (1965). Light distribution and photosynthesis in field crops. Ann. Bot. 29: 17–37.

Mooney, H.A. & Shropshire, R. (1967). Population variability in temperature related photosynthetic acclimation. Oecol. Plant. 2: 1–13.

Mouat, M.C.H. (1983). Competitive adaptations by plants to nutrient shortage through modification of root growth and surface change. NZ. Jl. agric. Res. 26: 327–332.

Norman. M.J. & Green, J.O. (1958). The local influence of cattle dung and urine upon the yield and botanical composition of permanent pasture, J. Brit. Grassl. Soc. 13: 39–45.

Paleg, L.G. & Aspinåll, D. (1981). The physiology and biochemistry of drought resistance in plants. Academic Press, Sydney.

Passioura, J.B. & Wetselaar (1972). Consequences of banding nitrogen fertilizers in soil. II Effects on the growth of wheat roots. Pl. Soil 36: 461–473.

Pennington, W. (1969). The History of British Vegetation. English Univ. Press, London.

Pike, A., McNeilly, T. & Putwain, P.D. (1979). Survivor populations of S23 perennial ryegrass from zero-grazed and set-stocked swards. Grass For. Sci. 34: 89–94.

Rebuffo, M. (1986). The effects of excreta on white clover-perennial ryegrass pasture under continuous cattle grazing. MPhil thesis. University of Reading. 252 pp.

Roberts, H.A. (1982). Weed Control Handbook: Principles. Blackwell Scientific Publ, Oxford.

Schaal, B.A. (1975). Population structure and local differentiation in *Liatris cylindrica*. Amer. Natur. 109: 511–528.

Scheiner, S.M. & Goodhart, C.J. (1984). The comparison of phenotypic and genetic variation in populations of the grass *Danthonia spicata*. Evolution 38: 845–855.

Schemske, D.W. (1984). Population structure and local selection in *Impatiens pallida*, a selfing annual. Evolution 38: 817–832.

Schlichting, C.D. & Levin, D.A. (1984). Phenotypic plasticity of annual phlox: tests of some hypotheses. Amer. J. Bot. 71: 252–260.

Scott, D. (1970). CO_2 exchange of plants. 3 Temperature acclimation of three species. NZ J. Bot. 8: 369–379.

Simpson, J.R. (1962). Mineral nitrogen fluctuations in soils under improved pasture in Southern New South Wales. Aust. J.agric. Res 13: 1059–1072.

Silander, J.A. (1979). Microevolution and clone structure in *Spartina patens*. Science 203: 658–660.

Silander, J.A. & Antonovics, J. (1979). The genetic basis of the ecological amplitude of *Spartina patens*. I Morphometric and physiological traits. Evolution 33: 1114–1127.

Slatyer, R.O. (1977). Altitudinal variation in the photosynthetic characteristics of snow gum, *Eucalyptus pausciflora*. IV Temperature response of four populations grown in different temperatures. Aust. J. Plant. Physiol. 4: 583–593.

Slobodkin. L.B. & Rapoport, A. (1974). An optimum strategy of evolution. Q. Rev. Biol. 49: 181–200.

Spieth, P.T. (1979). Environmental heterogeneity: a problem of contradictory selection pressures, gene flow, and local polymorphism. Am. Nat. 113; 247–260.

Snaydon, R.W. (1970). Rapid population diferentiation in a mosaic environment. I The response of *Anthoxanthum odoratum* populations to soils. Evolution 24: 257–269.

Snaydon, R.W. (1973). Ecological factors, genetic variation and speciation in plants. In: Heywood, V.H. Taxonomy and Ecology, pp 1–29. Academic Press, London.

Snaydon, R.W. (1978). Genetic changes in pasture populations. In: Wilson, J.R. (Ed.) Plant Relations in Pastures, pp 253–269. CSIRO, Melbourne.

Snaydon, R.W. (1980). Plant demography in agricultural systems. In: Solbrig, O.T. (Ed.) Demography and Evolution in Plant Populations, pp 131–160. Blackwell, Oxford.

Snaydon, R.W. (1985). Aspects of the ecological genetics of pasture species. In: Haeck, J. & Woldendorp, J.W. (Eds.) Structure and Functioning of Plant Populations 2, pp 127–152. North-Holland Publ., Amsterdam.

Snaydon, R.W. & Davies, M.S. (1972). Rapid population differentiation in a mosaic environment. II Morphological variation in *Anthoxanthum odoratum*. Evolution 26: 390–405.

Snaydon, R.W. & Davies, M.S. (1976). Rapid population differentiation in a mosaic environment. IV Population of *Anthoxanthum odoratum* at sharp boundaries. Heredity 37: 9–25.

Snaydon, R.W. & Davies, T.M. (1982). Rapid divergence of plant populations in response to recent changes in soil conditions. Evolution 36: 289–297.

Stapledon, R.G. (1928). Cocksfoot grass (*Dactylis glomerata*): ecotypes in relation to the biotic factor. O. Ecol. 16: 71–104.

Stern, W.R. (1962). Light measurements in pastures. Herb. Abstr. 32: 1–6.

Stryker, R.B., Gilliam, J.W. & Jackson, W.A. (1974). Nonuniform phosphorus distribution in the root zone of corn: growth and phosphorus uptake. Soil Sci. Soc. Am. Proc. 38: 334–340.

Thoday, J.M. (1959). Effects of disruption selection. I Genetic flexibility. Heredity 13: 187–203.

Thoday, J.M. (1972). Disruptive selection. Proc. R. Soc. Lond. B. 182: 109–143.

Timofeeff-Ressovsky, N.W. (1940). Zur analyse des polymorphisms bei *Adalia pibunctata*. Biol. Zeutrahbl. 60: 130–137.

Turkington, R. & Harper, J.L. (1979). The growth, distribution and neighbour relationships of *Trifolium repens* in a permanent pasture. IV Fine-scale biotic differentiation. J. Ecol. 67: 245–254.

Usher, M.E. (1970). Pattern and seasonal variability in the environment of a Scots pine forest soil. J. Ecol. 58: 669–679.

Van Dyke, G.E. (1955). The influences of sward-age and management on the type of timothy and cocksfoot. Euphytica 4: 83–93.

Veresoglou, D.S. & Fitter, A.H. (1984). Spatial and temporal patterns of growth and nutrient uptake of five co-existing grasses. J. Ecol. 72: 259–272.

Warwick, S.I. & Briggs, D. (1978). The genecology of lawn weeds. I. Population differentiation in *Poa annua* in a mosaic environment of bowling green lawns and flower beds. New Phytol. 81: 711–723.

Watson, P.J. (1969). Evolution in closely adjacent plant populations VI An entomophilous species, *Potentilla erecta*, in two contrasting habitats. Heredity 24: 407–422.

Whitham, T.G. & Slobodchikoff, C.N. (1981). Evolution by individual plant-herbivore interactions and mosaics of genetic variability: the adaptive significance of somatic mutation in plants. Oecologia (Berl). 45: 287–292.

Wiersum, L.K. (1958). Density of root branching as affected by substrate and separate ions. Acta Botan. Neerl. 7: 174–90.

Wilman, D. & Asiegbu, J.E. (1982). The effects of variety, cutting interval and nitrogen application on the morphology and development of stolons and leaves of white clover. Grass Forage Sci. 37: 15–27.

Woledge, J. (1977). The effects of shading and cutting treatments on the photosynthetic rate of ryegrass leaves. Ann. Bot. 41: 1279–1286.

Wu, L., Bradshaw, A.D. & Thurman, D.A. (1975). The potential for evolution of heavy metal tolerance in plants. III The rapid evolution of copper tolerance in *Agrostis stolonifera*. Heredity 34: 165–187.

Youngner, V.B. (1973). Physiology of defoliation and regrowth. In: Youngner, V.B. & McKell, C.M. (Eds.) The Biology and Utilisation of Grasses, pp 292–303. Academic Press, New York.

PART 2

Vegetation response to changing soil conditions

Part 2 is concerned with changing soil conditions and the resulting vegetation response. The complex interrelationships between soil and vegetation are considered by Miles (Chapter 3), especially in the context of plant-induced soil changes. Plant-induced soil changes have been most commonly considered in relation to vegetation succession but, as Miles and the other authors in Part 2 show, various types of disturbance, such as changes in fertilizer (Pegtel), hydrology (Grootjans) and grazing intensity can also cause considerable changes in soil conditions. More severe disturbances (e.g. Fischer and Beeftink) might be expected to cause even greater effects, though efforts are often made to minimise those effects (Klötzli).

All plants influence soil properties, both directly and indirectly, and so affect soil fertility. The longer that individuals or stands persist at the same spot, the greater the effects on the soil are likely to be, though ultimately an equilibrium condition is likely to develop. Disturbance usually changes both the vegetation and the soil, because of the interactions between plants and soil, it may take decades before a new equilibrium is reached following human interference.

Miles (Chapter 3) describes the vegetational changes brought about by selective grazing at low stocking density resulting in a patchy vegetation and litter accumulation, or by cessation of grazing resulting in tree invasion and subsequent changes in soil conditions. Pegtel (Chapter 4) discusses vegetational changes after cessation of fertilizer applications on formerly intensively used agricultural grassland. The example used by Grootjans (Chapter 5) involves drainage of a river valley system, with consequent changes in groundwater quality. The disturbance discussed by Klötzli (Chapter 6) involves the removal of complete wet grassland communities transplanting them into similar hydrological conditions. Beeftink (Chapter 7) describes the vegetational responses to change in tidal conditions, caused by building barriers in estuarine environments. The vegetational development from a complete new substrate, i.e. bare soil resulting from the reconstruction of terraces in vineyard areas, is discussed by Fischer (Chapter 8).

The processes by which soil chemical or physical attributes cause vegetational changes are hardly considered, because so little is known, due to the complexity of the problem. This problem is well illustrated by the way in which Miles (Chapter 3) places emphasis on changes in soil pH, partly for simplicity, partly because most information is on this parameter, and partly because pH is such a good indicator of many other soil properties and process rates. However, pH is not normally a causal factor but a secondary attribute. Grootjans (Chapter 5) also describes changes in soil conditions after disturbance where groundwater quality is the complex attribute. Drainage causes a diminished supply of mineral rich deep groundwater and hence an increased influence of mineral poor rain water. Again, it is not possible to define the causal relationships between soil or groundwater parameters and the performance of plant species. Pegtel (Chapter 4) points out that direct evidence for the vegetational changes in semi-natural grasslands, caused by hay-making without fertilizer applications, cannot be clearly related to specific soil factors. He therefore introduces the phytometric method using bio-assays, as an alternative to direct (chemical) estimation of nutrient availability. Klötzli (Chapter 6) uses indicator values, following Ellenberg (1979) to illustrate changes in the mineralization of peaty soil after the turfs are transplanted. Beeftink (Chapter 7) describes changes in soil conditions only indirectly, by fitting the newly developed vegetation types into well-known successional series of salt marsh communities, and relating this to changes in salinity, and then concludes whether progressive or retrogressive succession has occurred. Fischer (Chapter 8) discusses the succession on a denunded soil, indicating a general maturation of the soil during the changes from pioneer to more mature communities.

In studying vegetational changes a number of features should be kept in mind. Firstly no sharp transition from one vegetation type to another occurs, and Klötzli (Chapter 6), proposes the term 'successional rain'. As a result studies of vegetational response should be long-term if they are to distinguish between unidirectional change and fluctuations caused by such factors as weather conditions (cf. Beeftink, Chapter 7). Secondly, in addition to fluctuations in time, patchiness in space may complicate our interpretation of response. For example Miles (Chapter 3) also stresses the mutual influence of vegetation and soil, over distances of only 20 cm. Similarly, Pegtel (Chapter 4) suggests that the canopy structure is a very important phenomenon in explaining vegetational changes, this topic will be discussed in more detail in Part 3. Thirdly, the availability of species in the area is important in determining vegetational change. In particular the seed bank and seed dispersal are important factors considered by Fischer (Chapter 8), while Grootjans (Chapter 5) stresses the importance of the emergence of hybrids in disturbed environments, this will be treated further in Part 4. Finally, the role of soils as a buffer changing human or natural interference and vegetational response needs to be recognized. It will become clear from the chapters in Part 2 that the speed of vegetational responses to changing soil conditions varies. Miles (Chapter 3) mentions that well drained, poorly buffered, siliceous soils are most liable to rapid modification, while soils well-buffered by a high clay content, or by free calcium carbonate, and soils strongly influenced by groundwater, change much more slowly. The different reactions of various soil types on the influence of acid deposition could serve as an example.

Vegetational responses per se are often very well described. A gap in our knowledge, however, is the cause of vegetational responses. The character of the vegetation present appears to determine its reaction upon a change in environmental conditions.

Grootjans et al. (1985) pointed out that a lowering of the groundwater table in a poorly fertilized Calthion palustris hay meadow resulted in high N-mineralization of the fen peat soil. Most of the mineralized NO_3^- was absorbed by plant roots with a subsequent increased yield. Lowering of the groundwater table in unfertilized fen meadows, also resulted in high N-mineralization of the fen peat, but the produced NO_3^- was not absorbed by the plant roots, hence no vegetational response could be recorded (Grootjans et al., 1986).

Obviously the speed with which communities respond to changes in environmental conditions will depend on: (a) the form of the change in environmental conditions, (b) the buffering capacity of the environment (e.g. the soil), and (c) the buffering capacity of the vegetation, which will be determined by the lifespan of plants, the type of seed bank and speed with which invading species arrive. As a result, the indicator values of plant species should be regarded critically, recognizing the possibility of a time lag or even the absence of a vegetational response at least in the short term. This highlights the need for long-lasting studies, especially if a new equilibrium is to be defined. The chapters in this Part 2 also indicate that knowledge of vegetation response to changing soil conditions is still inadequate to construct satisfactory models and allow predictions of the likely response to changing environmental conditions.

References

Ellenberg, H. (1979). Zeigerwerte der Gefässpflanzen Mitteleuropas. Scripta Geobotanica 9. Göltze, Göttingen.

Grootjans, A.P., Schipper, P.C. & van der Windt, H.J. (1985). Influence of drainage on N-mineralization and vegetation response in wet meadows. I Calthion palustris stands. Oecologia Plantarum 6: 403–417.

Grootjans, A.P., Schipper, P.C. & van der Windt, H.J. (1986). Influence of drainage on N-mineralization and vegetation response in wet meadows. II. Cirsio-Molinietum stands. Oecologia Plantarum 7: 3–14.

CHAPTER 3

Soil variation caused by plants: a mechanism of floristic change in grassland?

J. Miles

Detailed contents

1. Introduction

There is an abundant literature correlating the distribution of different plant species and vegetation types with variations in soil properties. Subsequent studies to elucidate the reasons for such correlations have mostly been of the effects of soil on plants. The reciprocal hypothesis that vegetation influences soil properties, and can bring about at least some of the characteristic associations of soil with vegetation, has received less attention. This is surprising, since the importance of vegetation as a factor in soil development has long been recognized (Müller, 1879, 1884; Hilgard, 1906; Glinka, 1914). Indeed, Clements (1916) regarded the effects of vegetation on soil as the main driving force of primary succession, though he was uncertain of its role in secondary succession.

However, there is now enough evidence to suggest that many seral and 'climax' species and vegetation types often markedly influence labile soil properties (Miles, 1985a). Some plant-soil correlations may well be the result of plant-induced soil changes, as Ernst (1978) has suggested for the case of *Deschampsia flexuosa* on acid soils.

In this chapter, I first briefly discuss the susceptibility of soil to rapid change, caused by changes in the overlying vegetation. I then consider the mechanisms of change in soil conditions and their implications for plants, before discussing the reported and likely occurrence of plant-induced changes in grassland soils, and their consequent effects on floristic composition. Since no substantive study of such changes in grasslands has yet been published, many of the suggestions are made by analogy from studies of woodland and heathland.

2. Background to soil change

All plants influence soil properties, both directly and indirectly, and hence affect soil productivity. Plant species vary in such attributes as: (i) nutrient demand and uptake; (ii) nutrient return in lit-

J. van Andel et al., (eds.), Disturbance in Grasslands. ISBN 90-6193-640-3.

ter, dead roots, exudates and leachates; (iii) the chemical composition and acidity of throughfall and stemflow; (iv) the nutrient and secondary chemical content of their parts and litter, which affects their rate of decomposition. Marked changes in nutrient mineralization rates, and in labile soil properties such as pH, cation exchange capacity, and exchangeable base content and saturation can result from these differences. The longer that individuals or stands persist at the same spot, the greater the effects on the soil can be, although marked changes can occur in less than 10 years.

2.1 Susceptibility of soil to change

Temperate zone soils vary greatly in the extent to which chemical properties of the mineral horizons can change when the vegetation changes. Well drained, poorly buffered, siliceous soils are most liable to rapid modification. Soils well buffered by a high clay content or by free calcium carbonate, and soils strongly influenced by ground water, change very slowly. Poorly buffered but poorly drained soils with seasonally perched water tables, and well drained, sandy soils containing significant amounts of base-rich minerals are intermediate in their response. However, the amount, composition and properties of any surface accumulations of organic matter can vary markedly with differences in vegetation, regardless of the nature of the underlying mineral horizons. For example, highly buffered clay soils and limestone can have surface organic matter layers of pH 4.5.

For the rest of Section 2, emphasis is placed on changes in soil pH, partly for simplicity, partly because there is most information on this attribute, and partly because pH is such a good indicator of many other soil properties and process rates.

2.2 Rates of soil change

Change in susceptible soils can be rapid. For example, reductions in the pH of surface soil of 0.5 to 1.5 units in 10 to 40 years, from initial values of between 5 and 6, have been recorded under the ericaceous species *Calluna vulgaris* and *Erica cinerea* (Wilson, 1960; Grubb et al., 1969; Miles, 1985a), and under the leguminous species *Ulex europaeus* and *Trifolium subterraneum* (Grubb et al., 1969; Williams, 1980; Cumming & Williams, 1983). Similar changes in pH, over somewhat longer timescales, have been reported to occur under many tree species (Miles, 1985a). Changes of this magnitude, over this pH range, can strongly influence the species composition and richness of vegetation (Miles, 1985b).

2.3 Mechanisms of plant-induced change in soil pH

Plants can modify soil pH in several ways. Perhaps the commonest way results from changes in the amount of surface organic matter. For example, if decomposition slows, or above-ground litter fall increases, or there is a switch from a mull-producing to a mor-producing decomposition system, the accumulating surface mass of decomposing organic matter increases cation exchange capacity. In the absence of an increased base supply, this decreases base saturation and hence pH. The opposite trend of decreasing surface organic matter decreases cation exchange capacity, and base saturation and pH are both increased. Other mechanisms of soil acidification include immobilization of exchangeable bases in vegetation and litter, addition of organic acids, and accelerated leaching of bases by plants (Grubb et al., 1969; Bruckert & Rouiller, 1979). Nitrogen transformations within the soil also modify pH (Helyar, 1976); for example, nitrification and nitrogen fixation, including that by *Rhizobium* and actinomycete endophytes in root nodules, generate acidity (Troelstra et al., 1985).

2.4 Relation of pH to the soil properties and processes

Soil pH is a critical variable, influencing the distribution and performance of plants, and of soil microbes and fauna. Soil pH affects plant distribution, plant physiology and litter decomposi-

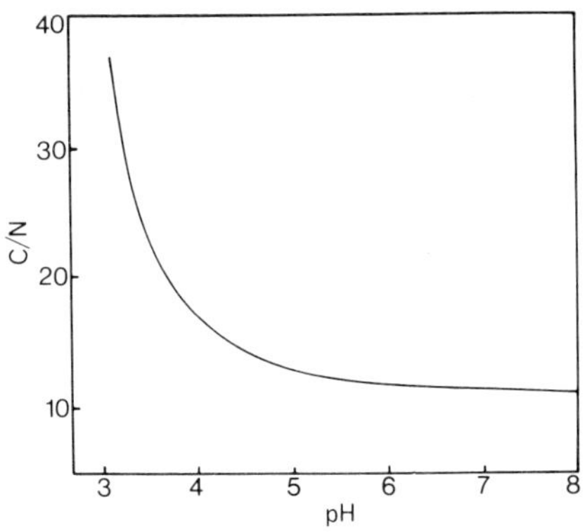

Fig. 1. Mean relationship between pH and carbon/nitrogen ratio in surface soil at 28 grassland, woodland and mire sites in Britain (after Howard, 1969).

tion, and so affects the carbon/nitrogen (C/N) ratio of litter and humus (Fig. 1). The C/N ratio in turn affects the rate at which organic matter decomposes, and hence, for example, the rate at which nitrogen is mineralized (Fig. 2). Integrating Figs. 1 & 2 (Fig. 3) shows that the pH values within the range 3.5 to 5.5 are likely to be particularly critical for early successional species, which tend to be heavily dependent on inorganic nitrogen sources. However, many mycorrhizal fungi, perhaps particularly ectomycorrhizas, can utilize soluble organic forms of nitrogen (Lundeberg, 1970; Bowen, 1973; Stribley & Read, 1974; Bowen & Smith, 1981). This has the effect of raising the critical C/N levels, so that many trees for example, grow well at C/N levels of 40 to 50

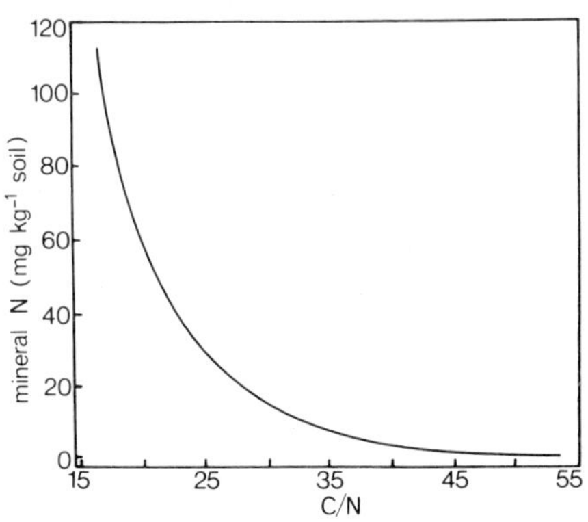

Fig. 2. Mean relationship between the carbon/nitrogen ratio of surface mineral soil at 94 woodland and heathland sites in Saxony and the concentration of mineral nitrogen after incubation at 18° C for 8 weeks (after Kriebitzsch, 1978).

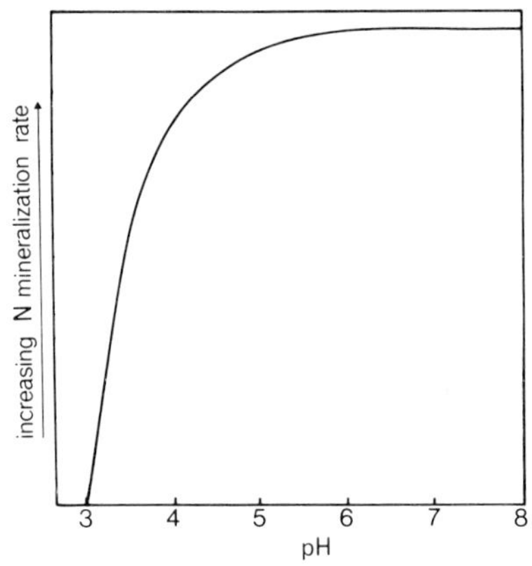

Fig. 3. General relationship between pH and nitrogen mineralization rates in surface soils (after Howard, 1969; Kriebitzsch, 1978; Miles, unpublished).

which few herbaceous species would tolerate, though the negative relationship between C/N levels and nitrogen uptake still holds (Fig. 4).

Another effect of pH is on nitrogen fixation by free-living organisms. Nitrogenase operates best in a narrow pH range around neutrality (Burris, 1966). *Azotobacter* spp. and many blue-green algae, for example, either will not grow or will not fix nitrogen in acid soils (Stewart, 1969; Mulder, 1975), although *Beijerinckia* spp. can fix nitrogen in very acid conditions (Becking, 1961). In general however, nitrogen fixation tends to increase with increasing pH, at least up to pH 5.0 or 5.5 (Fig. 5).

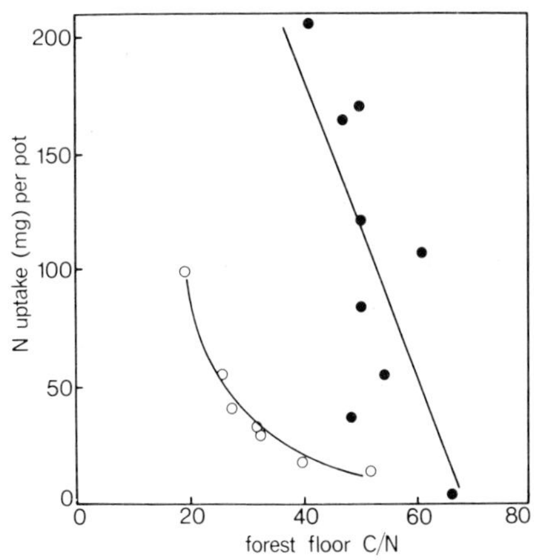

Fig. 4. Nitrogen uptake by seedlings of *Pseudotsuga menziesii* (•) and *Betula papyrifera* (o) grown on forest floor organic matter with different carbon/nitrogen ratios (from Heal *et al.*, 1982, courtesy of the Royal Society).

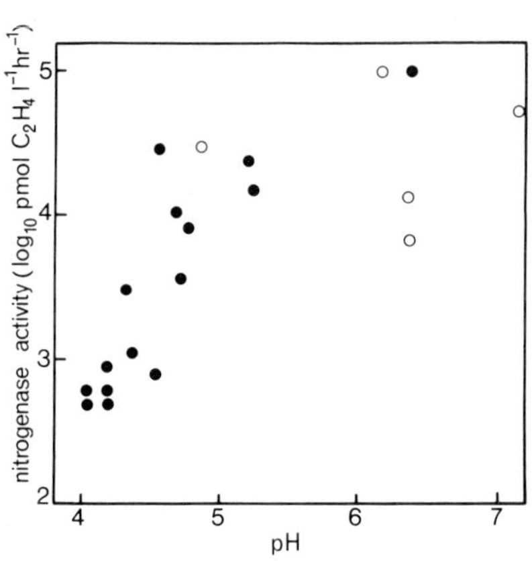

Fig. 5. Relationship between pH and mean summer/autumn nitrogenase activity of the top 3 cm of soil from 15 coniferous (●) and 5 deciduous (o) woodland sites in central Sweden (after Nohrstedt, 1985).

Table 1. Changes in soil properties (0 to 23 cm) during secondary succession on the abandoned Geescroft field, ROthamsted (From Jenkinson, 1971).

	Sampling date and successional phase		
	April 1883 Arable land	April 1904 Scrub	April 1965 *Quercus robur/ Fraxinus excelsior*-dominant woodland
pH	7.1	6.1	4.5
% C	1.04	1.37	1.98
% Organic N	0.115	0.131	0.166
% Organic P	0.025	0.029	0.033
C/N ratio	9.0	10.5	11.9
C/P ratio	42	47	60

3. Soil change under woody vegetation

Considerably more is known about the prevalence and degree of soil changes induced by woody vegetation than those caused by grassland (Miles, 1985a). For example, many conifers can cause reductions in surface-soil pH of up to 1 unit in less than 100 years, and can also cause pronounced podsolization, when grown in place of natural deciduous woodland. Indeed, almost any kind of forest that establishes on former agricultural land, whether broadleaved or coniferous, will probably acidify the topsoil (Krug & Frink, 1983). A good example is the abandoned Geescroft arable field at Rothamsted, which had a lessivated brown soil without free calcium carbonate. Here the pH fell by 2.6 units from near neutrality in 80 years (Table 1). Ericaceous species, and species with nitrogen-fixing root nodules, generally also acidify soils and the former can promote podzolization (Gimingham, 1960; Grubb et al., 1969; Låg, 1959, 1971; Duchaufour, 1977; Miles, 1985a).

The most intensively studied woody species are *Betula pendula* and *B. pubescens*. Study of 11 sites in Scotland and northern England, where either or both species had colonized *Calluna vulgaris*-dominated heathland, showed a general increase in pH, exchangeable calcium, total phosphorus, earthworm numbers, and rates of nitrogen mineralization and cellulose decomposition (Miles & Young, 1980; Miles, 1981); Table 2 gives results from one site. Mull or mull-like moder humus was formed instead of mor, and the appearance and structure of the soil profiles also changed (Miles, 1985b), with a trend from humus-iron podzol to brown podzolic soil characteristics. Similar trends of change occurred under *Betula* growing on primary substrates in two quarries.

Contrasting effects may occur within a stand. In East Falkland, soil under *Empetrum rubrum* bushes was consistently more acid than unvegetated soil, but less acid under *Perezia recurvata* (Table 3). Soil pH under *P. recurvata* bushes depended on their sizes, and hence their ages, mean pH values were 4.9, 5.2 and 5.8 for bushes of <15, 15–20 and >20 cm mean diameter respectively, suggesting the soil became more alkaline as the plants aged.

4. Soil change under herbaceous species and grassland

4.1 Effects of species

Herbaceous species and vegetation seem mostly

Table 2. Soil analyses from *Calluna* heathland and adjacent, first generation, successional stands of *Betula pendula* of different ages near Advie, Morayshire (From Miles & Young, 1980; Miles, 1981).

	Calluna heathland	*Betula pendula* aged: 18 years	26 years	38 years	90 years	LSD at 5% level
* pH	3.8	3.9	4.0	4.7	4.9	0.1
** Exchangeable H (mg dm^{-3})	117	108	109	101	89	25
** Exchangeable Ca (mg dm^{-3})	196	201	207	489	319	83
** Total P (mg dm^{-3})	151	210	196	240	232	82
** C/N ratio	30	26	19	22	15	11
** C/P ratio	500	320	280	270	170	200
** C/K ratio	440	430	410	460	310	210
** N mineralization after 14 days incubation (mg dm^{-3} $week^{-1}$)	−1.3	25	41	45	40	10
Cellulose decomposition estimated by tensile strength loss in kg of cotton strips buried for 10 weeks						
(i) at 0–4 cm	4.5	16	22	23	22	3.6
(ii) at 16–20 cm	3.2	6.0	9.0	8.9	17	4.2
Mean number of earthworms per 1 m^2 (by formalin extraction)	1	5	27	127	78	26

* Soil from 0–5 cm
** Soil from 0–15 cm

to produce and maintain brown soils with mull humus. Jenny (1980) and Ugolini & Edmonds (1983) reviewed a number of studies which showed that north American prairie grasslands were less acid, had higher base saturation and lesser accumulations of surface organic matter than most adjacent forest soils, and lacked eluviated Ea horizons. Stands of *Pteridium aquilinum* are characteristically associated with brown soil profiles, lacking Ea horizons (Hunter, 1962).

Table 3. Variation in soil pH at 0–3 cm in September 1983 under different species within an *Empetrum rubrum* heath, L'Antioja Valley, East Falkland.

Species	* Mean pH	Number of samples	pH range
Perezia recurvata	5.3_a	(53)	4.7–6.3
Bare mineral soil	4.9_b	(18)	4.6–5.4
Baccharis magellanica	4.8_b	(35)	4.3–5.4
Bare *Empetrum*-mor humus	4.6_c	(18)	4.3–4.7
Empetrum rubrum	4.5_c	(56)	4.1–5.0
Cortaderia pilosa	4.1_d	(25)	3.7–4.4

* Figures with a different suffix are significantly different ($P < 0.05$).

Pteridium seems, like *Betula* spp., to be able to colonize podzols and subsequently to reduce soil acidity, produce mull or moder humus in place of mor, and depodzolize profiles, i.e. obliterate the eluviated Ea horizon (Mitchell, 1973; Jarvis & Duncan, 1976). Surface soil under *Fagus sylvatica* stands with a field layer of *Calamagrostis epigejos* and *Brachypodium sylvaticum* had significantly higher pH values, base saturation and nitrogen mineralization rates than soils without a field layer (Faille, 1977; Lemée, 1975). Ten & Fedorova (1975) found similar changes under stands of *Polygonum sachalinense* and *P. weyrichii* in the Kurin Islands and Sakhalin.

In contrast, leguminous species, such as *Trifolium subterraneum, T. repens* and *Medicago sativa*, tend rapidly to acidify surface soils (Rixon, 1966; Nyatsanga & Pierre, 1973; Williams, 1980; Cumming & Williams, 1983), by increasing soil organic matter content, and because nitrogen fixation results in a net production of protons (Troelstra et al., 1985). Some grasses have also been shown to acidify soils relative to many dicotyledonous species. Scott (1975) examined the pH of the top 5 cm of soil under frequently occurring species at three tussock-grassland sites in South Island, New Zealand, and found variations of up to 0.5 unit. In particular, the soil under *Agrostis tenuis* and *Festuca novae-zelandiae* was consistently more acid than that under *Rumex acetosella, Hieracium praealtum, Hypochoeris radicata* and *Crepis capillaris*. Pot experiments also showed that *Rumex acetosella* made the soil significantly more alkaline than either *Trifolium repens* or *Lolium multiflorum x perenne* (Harris, 1971, 1972).

In an *Empetrum rubrum* heath in East Falkland, soil under the grass *Cortaderia pilosa* was markedly more acid than that under any other species (Table 3). In Europe, the grasses *Deschampsia flexuosa, Molinia caerulea* and *Nardus stricta* are commonly associated with acid soils with mor humus. All tend to produce mats of dead material at the soil surface; *Nardus* has been shown to lower pH, and *D. flexuosa* to promote podzolization (King, 1962; Pigott, 1970). The tropical grass *Imperata cylindrica*, which can, like *Pteridium aquilinum*, form dominant and persistent stands after forest clearance, was reported by Vageler (1933) to develop micropodzols under individual tussocks, with strong ironpan development.

4.2 Modifying effects of management

Any factor that modifies nutrient cycling rates, or soil organic matter content (and hence cation exchange capacity) is likely to vary the relative pedogenic effects of different species. Management practices such as grazing, hay making and fire can have a critical effect on soil conditions in grasslands.

Selective grazing causes increasing structural heterogeneity in grassland swards with time except when very heavily grazed (Hunter, 1962; Nicholson et al., 1970; Bakker & Ruyter, 1981). Bakker et al. (1984) have described how a mosaic of small, lightly-grazed patches, 0.5 to 3 m in diameter, developed from an initially uniform *Holcus lanatus*-dominated sward during 9 years of grazing at an average stocking rate of 3 sheep ha^{-1}. Such relatively ungrazed patches tend to develop a mat of decaying surface organic matter (Nicholson et al., 1970). Kirkwood (1964) described the contrasting surface soil properties under areas with and without such mats, which developed during 15 years in 3 adjacent reseeded pastures in Wales (Table 4). Exchangeable calcium, pH and extractable phosphorus decreased markedly, while C/N ratios tended to increase, indicating reduced rates of nitrogen mineralization.

Similar changes can occur when grazing is replaced by hay making (van Andel & van den Bergh, chapter 1), and when grasslands are abandoned (i.e. are no longer managed in any way), or where there are very low natural grazing pressures. Gisi & Oertli (1981) described how, in surface soil of abandoned meadows in the Swiss Jura and Alps, pH decreased from around 6 to 5 and 4.5 respectively, and C/N from around 10 to 15 and 23 respectively, while exchangeable calcium and potassium also decreased.

Table 4. Soil analyses (0–10 cm) from under patches with and without surface mats of decaying litter in 3 adjacent upland pasture sites in Montgomeryshire, Wales, 15 years after ploughing and reseeding (From Kirkwood, 1964).

	Ciderhouse Hill, on a peaty gley podzol		Ciderhouse Hill, on a brown podzolic soil		Llwyn Madoc Hill, on a brown podzolic soil	
	Non-mat	Mat	Non-mat	Mat	Non-mat	Mat
pH	5.4	4.9	5.5	5.1	6.0	4.9
% Exchangeable Ca	0.49	0.22	2.60	0.92	2.46	0.27
% C	19.9	26.5	11.9	12.4	8.6	17.5
% N	0.99	1.15	0.64	0.80	0.68	0.64
C/N ratio	20.1	23.1	18.6	15.5	12.6	27.2
Truog-extractable P (mg 100 g^{-1} dry soil)	4.5	2.0	3.1	1.4	2.7	1.6

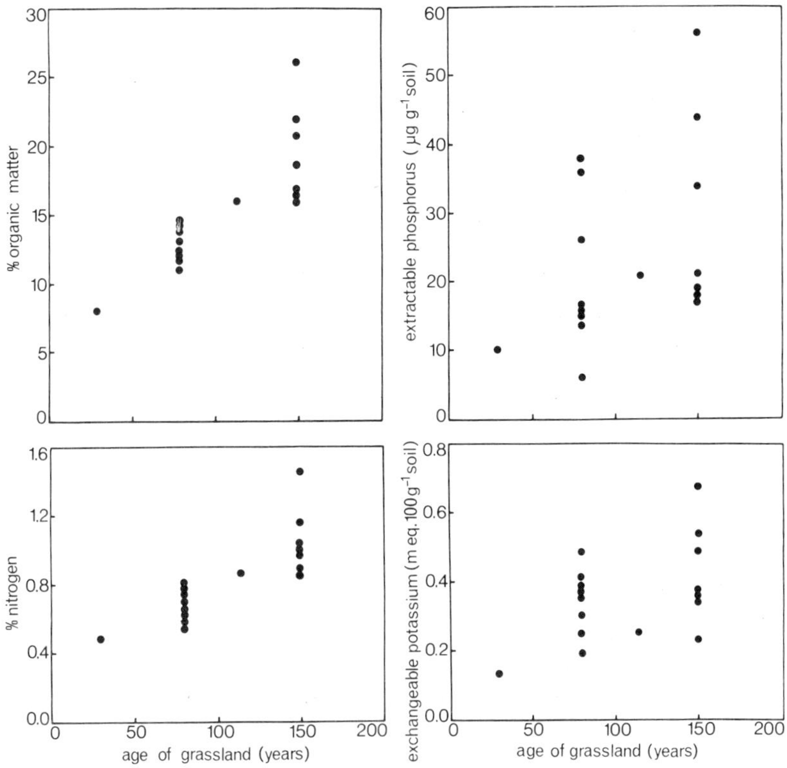

Fig. 6. Relationships between the mean age of 18 semi-natural chalk grasslands on Salisbury Plain, southern England, and the organic matter, nitrogen, 0.5M sodium bicarbonate-extractable phosphorus and exchangeable potassium content of the top 10 cm of soil (after Wells *et al.*, 1976).

Such changes are not inevitable, however, at least within decades. Stählin & Büring (1971) found no change in the soil pH of abandoned grasslands and arable fields at 81 sites with well buffered soils in Hesse, and little or no change in C/N. Wells et al. (1976) examined 17 grasslands of differing ages with rendzina soils in Salisbury Plain, southern England, which showed increasing amounts of organic matter, nitrogen, exchangeable metallic cations, and bicarbonate-extractable phosphorus in the top 10 cm of soil with age (Fig. 6), and also showed increasing rates of mineralization of nitrogen, but no change in C/N ratios. All soils had free calcium carbonate, so no significant decreases in pH or C/N would be expected.

The soil changes that occur with reduced grazing pressures are reversible. Increasing grazing eliminates surface mats or organic matter, reduces C/N ratios, increases base saturation, and may even bring about depodzolization with time (Floate, 1972).

Because various physiognomically different, but successionally related, vegetation types may at different times occupy the same site, grassland soil dynamics should be considered in the context of these potential alternative cover types. Figs. 7–9 show the main semi-natural vegetation types of poorly buffered, freely drained, siliceous soils in Britain, with the known successional transitions between them. The diagrams are arranged with vegetation types tending to promote surface acidification, mor humus or mat formation, and sometimes also podzolization, on the left, and types with the opposite effects on the right. The successions are caused mainly by variations in grazing pressures and burning frequencies, though particular transitions depend *inter alia* on the nearness of sources of seed (e.g. of *Calluna vulgaris, Betula* spp., *Pinus sylvestris, Juniperus*

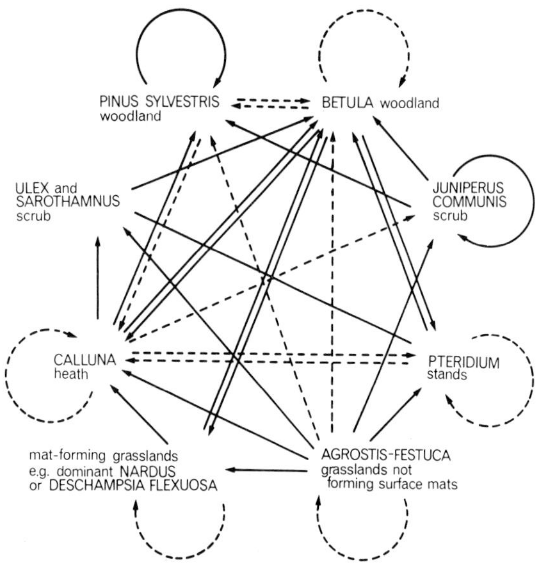

Fig. 7. Successional transitions that occur between common types of semi-natural vegetation on well-drained, acid mineral soils with low grazing pressures (<1 sheep equivalent ha^{-1}) and without burning; solid arrows represent the most frequent transitions, dashed arrows less common transitions. Circular arrows represent self-regeneration, i.e. absence of successional change (From Miles, 1985a, courtesy of the British Society of Soil Science).

communis) or ramets (e.g. of *Pteridium aquilinum*), or the presence of buried, viable seeds (e.g. of *Calluna, Cytisus scoparius, Ulex europaeus*).

At low grazing intensities (Fig. 7), succession from *Agrostis-Festuca* grassland is to mat-forming swards if no physiognomically dominant species invades, though there are equal numbers of potential transitions to acidifying and alkalizing vegetation types, and the aggregate successional trend is to woodland. At high grazing pressures (Fig. 8), the predominant trend is towards *Agrostis-Festuca* grassland, and so towards mull humus formation and the maintenance or development of brown, uneluviated A horizons. However, at intermediate levels of grazing, the situation is more complex. There are roughly equal numbers of potential transitions to acidifying and alkalizing vegetation types, but there is no predominant successional trend. This means that shifting mosaics will tend to develop and be maintained, and that on aggregate the contrasting soil trends will tend to cancel out.

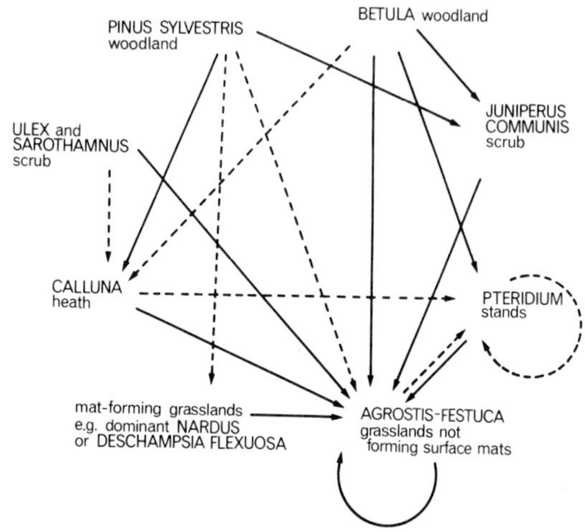

Fig. 8. Successional transitions between common types of semi-natural vegetation on well-drained, acid mineral soils that occur with high grazing pressures (>2–3 sheep equivalents ha^{-1}) and/or frequent burning; conventions as in Fig. 7 (from Miles, 1985a, courtesy of the British Society of Soil Science).

5. Floristic consequences of soil change

The previous sections have discussed various plant-induced soil changes known to occur, especially those associated with herbaceous species and swards. However, the critical question is the extent to which such changes influence plant performance and the floristic composition of vegeta-

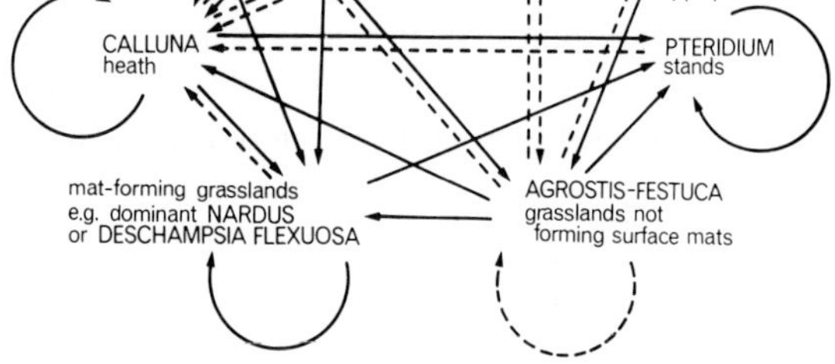

Fig. 9. Successional transitions between common types of semi-natural vegetation on well-drained, acid mineral soils that occur at intermediate levels of grazing (1–2 sheep equivalents ha^{-1}) and burning; conventions as in Fig. 7 (From Miles, 1985, courtesy of the British Society of Soil Science).

tion. The changes in pH that have been recorded in acidic soils, often of 0.5–1.0 unit or more, will, for example, influence nitrogen mineralization rates (see Section 2.4), and so would be expected to affect plant growth rates. A decrease in pH from 5 to 4 would be critical for species intolerant of high levels of exchangeable aluminium, while an increase from 4 to 5 would subject slow growing but aluminium-tolerant species, such as *Deschampsia flexuosa*, to competition from faster growing, aluminium-intolerant species (Rorison, 1969; Grime & Hunt, 1975).

5.1 Known effects of soil change

Excluding allelopathic effects, there seem to have been very few attempts to look for effects of plant-induced soil changes on plant performance and floristic composition (i.e. 'reaction' *sensu* Clements (1916)). Harris (1971, 1972) found that the soils in which he grew *Lolium multiflorum* x *perenne, Rumex acetosella* and *Trifolium repens* for 3 months had differential effects on the subsequent growth of a second sowing, when all possible factorial combinations of the species were studied. It is not clear if these were the result of the kinds of soil fertility changes discussed here or were allelopathic effects of the kind shown among various grassland species (including *T. repens* and *L. perenne*) by Newman & Rovira (1975). However, *T. repens* grew better in soil made more alkaline by *R. acetosella*, which was the expected response (Snaydon, 1962).

The best available evidence is on successional stands of *Betula* colonizing *Calluna* heathland (Section 3). More vascular species appeared in the field layer as soil conditions improved (Table 5a), and the performance of experimentally sown species improved progressively as the soil changed towards mull conditions (Table 5b & c).

Table 5. Performance of vascular species in *Calluna* heathland and adjacent, first generation, successional stands of *Betula pendula* of different ages near Advie, Morayshire: (a) numbers of naturally occurring species present in the field layer; (b) persistence of species 2 years after germinating from experimental sowings of seed (brackets indicate weak and unhealthy plants); (c) mean shoot dry weights (mg) of 8-week-old plants grown in a glasshouse in soil taken from the different stands (From Miles & Young, 1980; Miles, 1981, 1986).

	Calluna heathland	*Betula pendula* aged: 18 years	26 years	38 years	90 years	
(a) Number of species present	12	20	29	24	30	
(b) *Calluna vulgaris*	+	+	+	+		
Deschampsia flexuosa	+	+	+	+		
Luzula sylvatica	+	+	+	+		
Festuca rubra	(+)	+	+	+		
Holcus lanatus	(+)	(+)	+	+		
Galium saxatile		(+)	+	+		
Rumex acetosa		(+)	+	+		
Ranunculus acris			(+)	+		
Rubus idaeus			(+)	+		
Geranium sylvaticum				+		
Primula vulgaris				+		
Prunella vulgaris				+		
Viola riviniana				+		
(c)						LSD at 5% level
Luzula sylvatica	2.5	7.7	22	8.3	9.8	13
Raphanus sativus	8.9	18	43	59	66	18
Rumex acetosa	0.9	5.8	28	33	28	12

Thus there seems little doubt that the changed soil conditions under the *Betula* stands were largely responsible for the increase in species richness of the field layers.

5.2 Possible effects of soil change

Where should evidence be sought for the effects of plant-induced soil change on the floristic composition of grasslands? The studies reviewed in Section 4 suggest two situations in particular: first, sites with changing grazing pressures, especially those associated with developing or decaying mats of organic matter, and second, patches long occupied by the same individual or clone.

One example of soil change and vegetation response, apparently initiated by decreasing grazing pressures, was noted during a reconnaissance of Gruinard Island, northwest Scotland. The island was formerly grazed by sheep, but for forty years the only significant grazing has been from a fluctuating population of rabbits (*Oryctolagus cuniculus*). The vegetation of an area of long-abandoned lazy-bed cultivations, flushed by ground water from a line of springs, was mainly *Juncus effusus-Sphagnum recurvum* community (Birse & Robertson, 1976; Birse, 1980), with dominant *Juncus effusus*, and with *Sphagnum* spp., *Hylocomium splendens* and *Polytrichum commune* occupying most of the intertussock space. Up to 30 cm of spongy, fibrous peat had developed in the area, and *P. commune* had locally formed large tussocks, up to 50 cm high, with *Erica tetralix* plants established on some tussocks that were decaying. Within the flushed area were patches with little or no peat, and with either a grassy cover or small stands of *Ranunculus repens-Juncus effusus* community (Birse & Robertson, 1976). This circumstantial evidence suggests that the cessation of sheep grazing had caused a succession from marshy grassland to *R. repens-J. effusus* community, to *J. effusus-S. recurvum* community, and to local dominance of *P. commune*. Mean pH values in the top 3 cm of soil along this sequence were 6.1, 6.0, 5.4 and 4.8 respectively, suggesting the hypotheses (1) that organic matter accumulation under the *R. repens-J. effusus* community caused surface acidification which allowed the various *Sphagnum* spp. to colonize and spread, eventually creating the *J. effusus-S. recurvum* community that is usually associated with stream sides, and (2) that development of more acidic peat allowed *E. tetralix* to establish.

The above example is an extreme and unusual case. Could plant-induced soil change be partly reponsible for less dramatic floristic changes? Harris Glen on the island of Rhum, Inner Hebrides, was grazed until 1957 by a few feral goats, several hundred red deer (*Cervus elaphus*) and a flock of 1700 sheep. The sheep were then removed and, in subsequent years, large areas of *Agrostis-Festuca* grassland (association Achilleo-Festucetum tenuifolia (Birse & Robertson, 1976)) became progressively ranker and more tussocky. During 1971–72, a herd of cattle was introduced to control these tussocks, and by 1980 most tussocks had disappeared.

Excluding species with broad tolerance to soil acidity (i.e. those rated 'x' by Ellenberg, 1974), the main species decreasing in abundance after cattle grazing began were *Agrostis canina, Anthoxanthum odoratum, Calluna vulgaris, Carex demissa, Carex pilulifera, Festuca ovina, Galium saxatile, Luzula campestris, Nardus stricta, Polygala serpyllifolia* and *Viola riviniana*, while the main species that increased were *Agrostis tenuis, Carex pulicaris, Euphrasia brevipila, Hypochoeris radicata, Koeleria gracilis, Pedicularis sylvatica, Plantago maritima, Prunella vulgaris, Scutellaria galericulata, Sieglingia decumbens* and *Thymus drucei*. Using Ellenberg's (1974) species-indicator values for soil reaction, the species decreasing and increasing in cover with cattle grazing had mean indicator values of 2.7 and 4.6 respectively ($P<0.05$). The increase in annual hemiparasites must reflect the decreased sward height, but could the increases in any other species reflect the probable increase in pH and nutrient cycling rates as tussocks and litter declined? The significant difference in soil reaction indicator values suggests this may have been the case. Unfortunately no soil data were collected in this study, but the hypothesis is worth examining elsewhere because floristic changes of this order are commonplace.

Soil changes and plant reaction should also be sought under long-lived individuals and clones. Individual genets of many herbaceous species can live for decades (Tamm, 1956, 1972a & b; Harper & White, 1974; Callaghan & Emanuelsson, 1985), or even for centuries (Harberd, 1961, 1962, 1967; Oinonen, 1967, 1969, 1971) and possibly millennia (Penzes, 1958); in some cases they may persist for long periods on the same spot. Although it is not known how rapidly soil changes may occur under herbaceous species in the field, Harris (1971, 1972) got pH changes in pot experiments in only 3 months, while many shrubs and trees cause marked change in decades or less (Miles, 1985a). Therefore changes are likely under many herbaceous species.

Finally, is it possible that plant-induced changes in soil 'fertility' can account for some characteristic associations of grassland species, such as those identified by de Vries (1954), Agnew (cited by Greig-Smith, 1983) and Aarssen et al. (1979), and for some correlations between species abundance and soil conditions (e.g. Rorison, 1969)? Surprisingly little attention has been given to these possibilities, though there is evidence that allelopathy causes positive and negative associations between certain species (Newman & Rovira, 1975). Most ecologists have opted only for the alternative explanation that soil conditions determine the distribution of plant species.

6. Conclusions

This paper has reviewed evidence about changes in soil properties resulting from the growth of particular herbaceous species and assemblages, and modified by management. Changes in surface acidity and in nutrient cycling rates are particularly likely when grazing pressures markedly alter. Though little evidence has been sought about species or vegetation response to such soil changes in grasslands, plant-induced soil changes do largely drive the succession in the field layer when *Betula pendula* and *B. pubescens* colonize *Calluna vulgaris*-dominant heathland in upland Britain. Therefore, because substantial soil changes can occur in grasslands, and because many herbaceous species can persist at the same spot for as long as many woody species that have been shown to change soil properties, further evidence should be sought for plant-induced soil change in grasslands, and for consequential floristic responses. Soil variation caused by plants is a neglected topic and, in susceptible soils, may be a not infrequent source of floristic change and succession.

References

Aarssen, L.W., Turkington, R. & Cavers, P.B. (1979). Neighbour relationships in grass-legume communities. III. Temporal stability and community evolution. Can. J. Bot. 57: 2695–2703.

Bakker, J.P., Leeuw, J. de & Wieren, S.E. van (1984). Micro-patterns in grassland vegetation created and sustained by sheep grazing. Vegetatio 55: 153–161.

Bakker, J.P. & Ruyter, J.C. (1981). Effects of five years of grazing on a salt-marsh vegetation. A study with sequential mapping. Vegetatio 44: 81–100.

Becking, J.H. (1961). Studies on nitrogen-fixing bacteria of the genus *Beijerinckia*. Pl. Soil. 14: 49–81.

Birse, E.L. (1980). Plant Communities of Scotland. A Preliminary Phytocoenonia. Soil Surv. Scotland Bull. 4. Macaulay Institute for Soil Research, Aberdeen.

Birse, E.L. & Robertson, J.S. (1976). Plant Communities and Soils of the Lowland and Southern Upland Regions of Scotland. Macaulay Institute for Soil Research, Aberdeen.

Bowen, G.D. (1973). Mineral nutrition of ectomycorrhizae. In: Marks, G.C. & Kozlowski, T.T. (Eds.), Ectomycorrhizae. Their Ecology and Physiology, pp 151–205. Academic Press, New York.

Bowen, G.D. & Smith, S.E. (1981). The effects of mycorrhizas on nitrogen uptake by plants. In: Clark, F.E. & Rosswall, T. (Eds.), Terrestrial Nitrogen Cycles. Ecol. Bull., Stokh. 33: 237–247.

Bruckert, S. & Rouiller, J. (1979). Mécanisms de regulation du pH des Sols. In: Bonneau, M. & Souchier, B. (Eds.), Pédologie. 1. Constituants et Propriétés du Sol, pp 354–365. Masson, Paris.

Burris, R.H. (1966). Biological nitrogen fixation. Ann. Rev. Pl. Physiol. 17: 155–184.

Callaghan, T.V. & Emanuelsson, U. (1985). Population structure and processes of tundra plants and vegetation. In: White, J. (Ed.), The Population Structure of Vegetation, pp 399–439. Dr W. Junk, Dordrecht.

Clements, F.E. (1916). Plant succession. An analysis of the development of vegetation. Carnegie Inst. Wash. Publ. No. 242.

Cumming, R.W. & Williams, C.H. (1983). Soil pH in the Pejar Catchment area, Goulburn. J. Soil Conserv. N.S.W. 39: 106–112.

Duchaufour, P. (1977). Pédologie. I. Pédogenèse et Classification. Paris: Masson.

Ellenberg, H. (1974). Zeigerwerte der Gefässpflanzen Mitteleuropas. Scripta Geobot. 9.

Enwezor, W.O. (1976). The mineralization of nitrogen and phosphorus in organic materials of varying C:N and C:P ratios. Pl. Soil 44: 237–240.

Ernst, W. (1978). Discrepancy between ecological and physiological optima of plant species. A re-interpretation. Oecol. Plant. 13: 175–188.

Faille, A. (1977). Action des peuplements de *Calamagrostis epigeios* (L.) Roth. dans la dynamique des écosystèmes de la Fôret de Fontainebleau. II. Influence sur quelques caractères des humus et leurs activités microbiennes. Rev. Ecol. Biol. Sol 14: 289–306.

Floate, M.J.S. (1972). Plant nutrient cycling in hill land. Proc. North of England Soils Discuss. Group 7: 11–27.

Gimingham, C.H. (1960). Biological Flora of the British Isles: *Calluna vulgaris* (L.) Hull. J. Ecol. 48: 455–483.

Gisi, U. & Oertli, J.J. (1981). Oekologische Entwicklung in Brachland verglichen mit Kulturwiesen. I. Physikalisch-chemische Veränderungen im Boden. Oecol. Plant. 16: 7–21.

Glinka, K.D. (1914). Die Typen der Bodenbildung, ihre Klassifikation und geographische Verbreitung. Borntraeger, Berlin.

Greig-Smith, P. (1983). Quantitative Plant Ecology. 3rd edn. Blackwell Scientific Publications, Oxford.

Grime, J.P. & Hunt, R. (1975). Relative growth rate: its range and adaptive significance in a local flora. J. Ecol. 63: 393–422.

Grubb, P.J., Green, H.E. & Merrifield, R.C.J. (1969). The ecology of chalk heath: its relevance to the calcicole-calcifuge and soil acidification problems. J. Ecol. 57: 175–213.

Harberd, D.J. (1961). Observations on population structure and longevity of *Festuca rubra* L. New Phytol. 60: 184–206.

Harberd, D.J. (1962). Some observations on natural clones in *Festuca ovina*. New Phytol. 61: 85–110.

Harberd, D.J. (1967). Some observations on natural clones in *Holcus mollis*. New Phytol. 66: 401–408.

Harper, J.L. & White, J. (1974). The demography of plants. Ann. Rev. Ecol. Syst. 5: 419–463.

Harris, W. (1971). The effects of fertiliser and lime on the competitive reactions of *Rumex acetosella* L. with *Trifolium repens* L. and *Lolium* sp. N.Z. J. agric. Res. 14: 185–207.

Harris, W. (1972). Shading, defoliation, temperature, growth stage, and residual fertility effects on competition between *Rumex acetosella, Trifolium repens* and *Lolium (multiflorum x perenne)*. N.Z. J. agric. Res. 15: 687–705.

Heal, O.W., Swift, M.J. & Anderson, J.M. (1982). Nitrogen cycling in United Kingdom forests: the relevance of basic ecological research. Phil. Trans. R. Soc. Lond. B296: 427–444.

Helyar, K.R. (1976). Nitrogen cycling and soil acidification. J. Aust. Inst. agric. Sci. 42: 217–221.

Hilgard, E.W. (1906). Soils. Macmillan, New York.

Howard, P.J.A. (1969). The classification of humus types in relation to soil ecosystems. In: Sheals, J.G. (Ed), The Soil Ecosystem, pp 41–54. The Systematics Association (Publ. No. 8), London.

Hunter, R.F. (1962). Hill sheep and their pasture: a study of sheep-grazing in south-east Scotland. J. Ecol. 50: 651–680.

Jarvis, M.C. & Duncan, H.J. (1976). Profile distribution of organic carbon, iron, aluminium and manganese in soils under bracken and heather. Pl. Soil 44: 129–140.

Jenkinson, D.S. (1971). The accumulation of organic matter in soil left undisturbed. Rep. Rothamsted exp. Sta. for 1970: 113–137.

Jenny, H. (1980). The Soil Resource. Origin and Behaviour. Springer-Verlag, New York.

King, J. (1962). The *Festuca-Agrostis* grassland complex in S.E. Scotland. J. Ecol. 50: 321–355.

Kirkwood, R.C. (1964). A study of some factors causing mat formation of reseeded hill pasture swards. J. Br. Grassld Soc. 19: 387–395.

Kriebitzsch, W.-U. (1978). Stickstoffnachlieferung in sauren Waldböden Nordwestdeutschlands. Scripta Geobot. 14.

Krug, E.C. & Frink, C.R. (1983). Effects of acid rain on soil and water. Bull. Conn. agric. Exp. Stn 811.

Lemée, G. (1975). Recherches sur les écosystèmes des reserves biologiques de la Fôret de Fontainebleau. III. Influence des peuplements graminées sur les caractères et l'activité biologique du mull acide. Rev. Écol. Biol. Sol 12: 157–167.

Lundeberg, G. (1970). Utilization of various nitrogen sources, in particular bound soil nitrogen, by mycorrhizal fungi. Stud. for. Suec. 79.

Låg, J. (1959). Influence of forest stand and ground cover vegetation on soil formation. Agrochimica 4: 72–77.

Låg, J. (1971). Some relationships between soil conditions and distribution of different forest vegetation. Acta agr. fenn. 123: 118–125.

Miles, J. (1981). Effects of Birch on Moorlands. Institute of Terrestrial Ecology, Cambridge.

Miles, J. (1985a). The pedogenic effects of different species and vegetation types and the implications of succession. J. Soil Sci. 36: 571–584.

Miles, J. (1985b). Soil in the ecosystem. In: Fitter, A.H., Atkinson, D., Read, D.J. & Usher, M.B. (Eds.), Ecological Interactions in Soil, Br. Ecol. Soc. Special Publ. 4, pp 407–427. Blackwell Scientific Publications, Oxford.

Miles, J. (1986). What are the effects of trees on soils? In: Jenkins, D. (Ed.), Trees and Wildlife in Scotland. pp 55–62. Institute of Terrestrial Ecology, Huntingdon.

Miles, J. & Young, W.F. (1980). The effects on heathland and moorland soils in Scotland and northern England following colonization by birch (*Betula* spp.). Bull. Écol. 11: 233–242.

Mitchell, J. (1973). The bracken problem. In: Tivy, J. (Ed.), The Organic Resources of Scotland: Their Nature and

Evaluation, pp 98–108. Oliver & Boyd, Edinburgh.

Mulder, E.G. (1975). Physiology and ecology of free-living, nitrogen-fixing bacteria. In: Stewart, W.D.P. (Ed.), Nitrogen Fixation by Free-Living Micro-Organisms, pp 3–28. Cambridge University Press, Cambridge.

Müller, P.E. (1879). Studier over Skovjord, som bidrag til skovdyrkningens theori. I. Om bøgemuld og bøgemor paa sand og ler. Tidsskr. Skovbrug 3: 1–124.

Müller, P.E. (1884). Studier over Skovjord, som bidrag til skovdyrkningens theori. II. Om muld og mor i egeskove og paa heder. Tidsskr. Skovkbrug 7: 1–232.

Newman, E.I. & Rovira, A.D. (1975). Allelopathy among some British grassland species. J. Ecol. 63: 727–737.

Nicholson, I.A., Paterson, I.S. & Currie, A. (1970). A study of vegetational dynamics: selection by sheep and cattle in *Nardus* pasture. In: Watson, A. (Ed.), Animal Populations in Relation to their Food Resources, Symp. Br. Ecol. Soc. 10, pp 129–143. Blackwell Scientific Publications, Oxford.

Nohrstedt, H.-Ö. (1985). Nonsymbiotic nitrogen fixation in the topsoil of some forest stands in central Sweden. Can. J. For. Res. 15: 715–722.

Nyatsanga, T. & Pierre, W.H. (1973). Effect of nitrogen fixation by legumes on soil acidity. Agron. J. 65: 936–940.

Oinonen, E. (1967). The correlation between the size of Finnish bracken (*Pteridium aquilinum* (L.) Kuhn) clones and certain periods of site history. Acta For. Fenn. 83 (2).

Oinonen, E. (1969). The time table of vegetative spreading of the lily-of-the-valley (*Convallaria majalis* L.) and the wood small-reed (*Calamagrostis epigeios* (L.) Roth.) in southern Finland. Acta For. Fenn. 97.

Oinonen, E. (1971). The time table of vegetative spreading in oak fern (*Carpogymnia dryopteris* (L.) Löve & Löve) and may-lily (*Maianthemum bifolium* (L.) F.W. Schmidt) in southern Finland. Acta For. Fenn. 118.

Penzes, A. (1958). Survival of stoloniferous plant colonies (polycormans) of a relict character. Biologia 13: 253–264.

Pigott, C.D. (1970). Soil formation and development on the carboniferous limestone of Derbyshire. II. The relation of soil development to vegetation on the plateau near Coombs Dale. J. Ecol. 58: 529–541.

Rixon, A.J. (1966). Soil fertility changes in a red-brown earth under irrigated pastures. 1. Changes in organic carbon, carbon/nitrogen ratio, cation exchange capacity and pH. Aust. J. agric. Res. 17: 303–316.

Rorison, I.H. (1969). Ecological inferences from laboratory experiments on mineral nutrition. In: Rorison, I.H. (Ed.), Aspects of the Mineral Nutrition of Plants, Symp. Br. Ecol. Soc. 9, pp 155–175. Blackwell Scientific Publications, Oxford.

Scott, D. (1975). Variation in soil pH under tussock grassland species. N.Z. J. exp. Agric. 3: 143–145.

Snaydon, R.W. (1962). Micro-distribution of *Trifolium repens* L. and its relation to soil factors. J. Ecol. 50: 133–143.

Stählin, A. & Buring, H. (1971). Sozialbrache auf Äckern und Wiesen in pflanzensoziologischer und Ökologischer Sicht. Z. Acker- und Pflanzenbau 133: 200–214.

Stewart, W.D.P. (1969). Biological and ecological aspects of nitrogen fixation by free-living micro-organisms. Proc. R. Soc. Lond. B172: 367–388.

Stribley, D.P. & Read, D.J. (1974). The biology of mycorrhiza in the Ericaceae. IV. The effect of mycorrhizal infection on uptake of ^{15}N from labelled soil by *Vaccinium macrocarpon* Ait. New Phytol. 73: 1149–1155.

Tamm, C.O. (1956). Further observations on the survival and flowering of some perennial herbs. 1. Oikos 7: 274–292.

Tamm, C.O. (1972a). Survival and flowering of some perennial herbs. II. The behaviour of some orchids on permanent plots. Oikos 23: 23–28.

Tamm, C.O. (1972b). Survival and flowering of perennial herbs. III. The behaviour of *Primula veris* on permanent plots. Oikos 23: 159–166.

Ten, K.M. & Fedorova, L.V. (1975). Role of tall herbaceous plants in soil formation. Soviet Soil Sci. 7: 41–45.

Troelstra, S.R., Dijk, K. van & Blacquière, T. (1985). Effects of N source on proton excretion, ionic balance and growth of *Alnus glutinosa* (L.) Gaertner: comparison of N_2 fixation with single and mixed sources of NO_3 and NH_4. Pl. Soil 84: 361–385.

Ugolini, F.C. & Edmonds, R.L. (1983). Soil biology. In: Wilding, L.P., Smeck, N.E. & Hall, G.F. (Eds), Pedogenesis and Soil Taxonomy. I. Concepts and Interactions, pp 193–231. Elsevier, Amsterdam.

Vageler, P. (1933). An Introduction to Tropical Soils. Macmillan, London.

Vries, D.M. de (1954). Constellation of frequent herbage plants, based on their correlation in occurrence. Vegetatio 5–6: 105–111.

Wells, T.C.E., Sheail, J., Ball, D.F. & Ward, L.K. (1976). Ecological studies on the Porton Ranges: relationships between vegetation, soils and land-use history. J. Ecol. 64: 589–626.

Williams, C.H. (1980). Soil acidification under clover pasture. Aust. J. exp. Agric. Anim. Husb. 20: 561–567.

Wilson, K. (1960). The time factor in the development of dune soils at South Haven Peninsula, Dorset. J. Ecol. 48: 341–359.

CHAPTER 4

Soil fertility and the composition of semi-natural grassland

D.M. Pegtel

Detailed contents

1. Introduction

It is commonly recognized that soil nutritional factors have a profound effect on the botanical composition of grasslands, especially in acidic, calcareous, and saline habitats. As a result of frequent application of fertilizers, semi-natural grasslands have gradually been transformed into intensively used cultivated grasslands; this has led to marked changes in their botanical composition. The numerous slow-growing species typical of low fertility sites, are gradually replaced by a few species with rapid growth and a high nutrient demand, so there is a decrease in species diversity.

Currently, the costs of amelioration are not always justified and some marginal grasslands are abandoned; in the Netherlands, these are being acquired for landscape and nature conservation purposes. These grasslands are being managed to lower standing crop production and increase species diversity. One such management regime, which simulates earlier management practice, involves hay-making once a year without the addition of fertilizers. The off-take of nutrients in hay amounts to only a few percent of the soil nutrient store (Bakker, Chapter 12); consequently, this practice did not clearly affect the nutrient status of the soil. The practice has certainly resulted in a rapid decrease in above-ground production, an increase in species number, and an increase in the ratio of 'poor/rich' species (Bakker, Chapter 12); this is a reversal of the trends occurring when applications of fertilizers were first made. Vegetational change need not be focussed entirely on botanical composition, but may also consider changes in the genetical constitution of plant populations (Snaydon, Chapter 2).

It has been assumed that these changes in botanical composition directly attribute to soil chemical conditions, rather than to other abiotic factors such as light or temperature. However, the nutritional status of the soil is difficult to assess, because the elements occur in various com-

J. van Andel et al., (eds.), Disturbance in Grasslands. ISBN 90–6193–640–3.

binations, the fraction available to plants is normally a small proportion of the total and changes seasonally, and the various plant species have different nutritional requirements which fluctuate with time. This complex situation is further modified by man's activities, such as mowing, grazing, draining, liming, moulding, or adding sand.

Considering the theme 'disturbance', it is necessary to quantify both causes and effects. This chapter deals with the problem how to quantify nutrient availability to plants, in order to be able to interpret 'normal' and 'disturbed' soil conditions, related to plant responses. There are three essentially different approaches to measure the available nutrient content of the soil: (a) soil analysis, (b) plant analysis, (c) pot and field experiments. In this chapter, two of these techniques (soil analysis and plant experiments) are used to explore the relationship between botanical composition and soil nutrient status.

2. Spatial comparison of semi-natural grassland and some soil fertility parameters

The study areas are semi-natural grasslands in the nature reserve 'Stroomdallandschap Drentsche A' (53°05′N, 6°40′E) in the Netherlands. These grasslands were originally reclaimed, centuries ago, from alder forest. They border the brooklet Anloër Diepje, a middle course-tributary of the brook Drentsche A. Water moves into the valley, both laterally and by upward movement of groundwater, from the adjacent elevated plateau and from sandy levees alongside the brooklet (Grootjans, 1980).

The main soil units discerned were black 'Beek' earth soil and 'Made' peat soil (Van Heuveln, 1980). Black 'Beek' earth soil profile (synonymous with humic-sandy gley soil) is characterized by a black humic toplayer of loamy fine sand, up to 30 cm thick overlying grey sand of pleistocene origin, which is often mottled brown (rust patches). 'Made' peat soil profile is characterized by a 20 cm thick toplayer of clay-poor peat overlying pure peat (De Bakker, 1979; Van Heuveln, 1980).

The number of species occurring in the grassland has increased since the use of fertilizers was discontinued on these wet semi-natural grasslands (Table 1). More species and more standing crop + litter occurred on 'Made' peat soil than on black 'Beek' earth soil; contrary to expectation (Grime, 1979), the number of species was not correlated with standing crop + litter. Fig. 1 shows that the contribution of grasses and litter decreased, while mosses increased when fertilization was discontinued. Rushes also increased markedly on black 'Beek' earth soil, but were almost entirely absent from the 'Made' peat soil. Conversely, sedges increased on 'Made' peat soil but were essentially absent on black 'Beek' earth soil. No consistent change in the contribution of forbs was apparent on each soil.

Also supprisingly, standing crop + litter on both soils apparently increased with time after fertilization was discontinued (Table 1). This is not related to the various parameters measured in the soil layer of 0–5 cm (Table 2). These parameters are positively correlated with the organic matter content. Although the soil units discerned are classified as either black 'Beek' earth or 'Made' peat, the variation within each unit was substantial; this spatial heterogeneity may account for the rather erratic results apparent in Table 1 and Fig. 1. The spatial heterogeneity

Table 1. Species number per unit area and standing crop + litter (tons ha^{-1}) of once annually (July/August) cut grassland plots ($0.5 \times 0.5\ m^2$) in April and June 1981 on black 'Beek' earth soil, and on 'Made' peat soil 3 years, 15 years, and 30 years after last being fertilized; n.d. = not determined (Jansen et al., 1981).

Soil type	Years since fertilized	Species number			Standing crop	
		April	June	Total	April	June
Black 'Beek' earth soil	3	19	20	25	1.7	2.6
	15	n.d.	22	–	n.d.	2.6
	30	28	26	36	2.2	5.8
'Made' peat soil	3	15	16	19	2.1	6.3
	15	n.d.	28	–	n.d.	3.8
	30	33	29	40	2.4	7.2

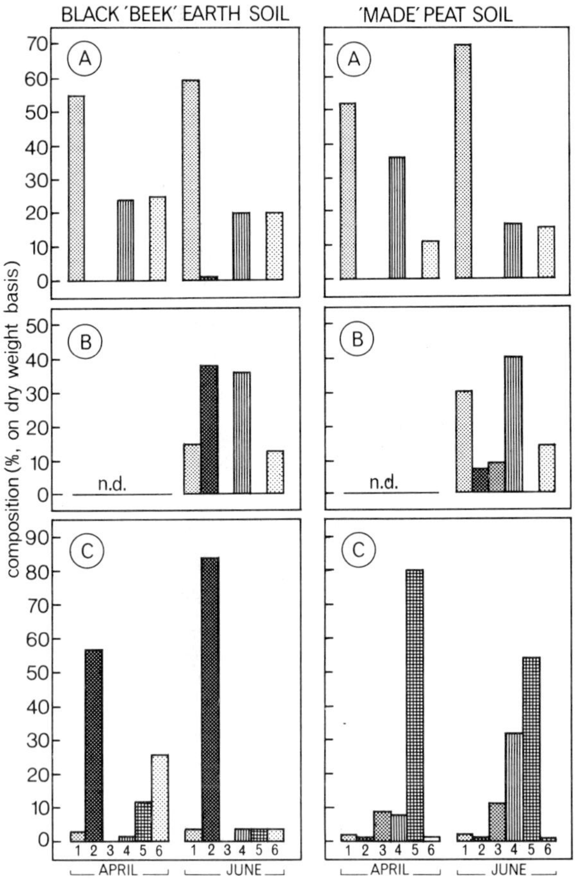

Fig. 1. Botanical composition of grassland in April and June 1981 (a) 3 years, (b) 15 years, and (c) 30 years after fertilization was discontinued on black 'Beek' earth soil and 'Made' peat soil; all plots were cut once annually (July/August). The components are (1) Grasses, (2) Rushes, (3) Sedges, (4) Forbs, (5) Mosses, (6) Litter. All values are based on dry weight yield; n.d. = not determined (Jansen et al., 1981).

may be caused by groundwater regime (Grootjans, 1980; Grootjans & Ten Klooster, 1980), or may be the result of previous agricultural management, or even earlier bog ore-digging (Booy, 1986). Van Heuveln (1980) concluded that soil spectra versus vegetation spectra only give few indications about the present-day relationships between them. This may endorse the general view of Ellenberg (1956) that every plant community only reflects the soil factors on which it is dependent nowadays. Ideally, it would be better to compare the effects of change of management on initially uniform grassland, thus excluding possible effects of pre-existing variation.

3. Effect of prolonged hay-making without fertilization

An attempt to overcome the difficulties of non-uniformity was made by starting with an apparently uniform grassland, located in the nature reserve 'Stroomdallandschap Drentsche A'. The area (0.5 ha), which was acquired by the State in 1972, had a black 'Beek' earth profile. Hay-making and grazing had previously been practised for several decades and application of fertilizers had ceased in 1973 (Bakker & De Vries, 1985). Nine different management regimes are being compared, but only the effect of no cutting versus twice a year hay-making (July and September) will be considered here.

From 1973 onwards, annual changes in the vegetation were recorded in two permanent (2 × 2 m^2) plots at each management site; the results have been published elsewhere (Bakker & De Vries, 1985).

The above-ground standing crop + litter in July declined sharply 5 years after fertilizer use was discontinued (Fig. 2), though there was considerable variation from year to year. The above-ground standing crop +litter in September was less affected. The changes in standing phytomass were not correlated with the concentrations of the major nutrients (N, P, K) in the toplayer (Fig. 3). There were no apparent differences in the content of soil N, P and K between grassland cut twice a year and uncut grassland (Fig. 3). This lack of correlation between phytomass and soil nutrient content is not unique, for example, Saunders & Metson (1971) found no correlation between spring pasture growth and plant available P, while Al-Farraj et al. (1984) found no consistent relationship between concentrations of N, P and K in peat samples, and net in situ above-ground standing crop production of herbaceous vegetation in 16 stands of waterlogged rich-fen. There are several possible reasons for the lack of correlation between phytomass and soil nutrient

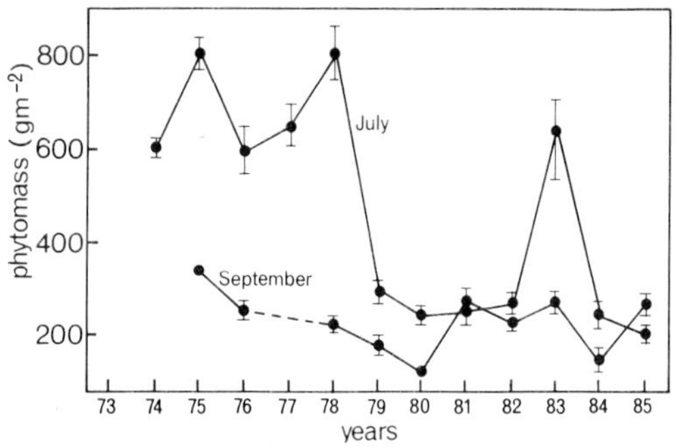

Fig. 2. Changes with time of standing crop + litter (g m^{-2}) of grassland cut in July and September after fertilizer application was discontinued in 1973. Bars denote S.E. (Bakker, partly published).

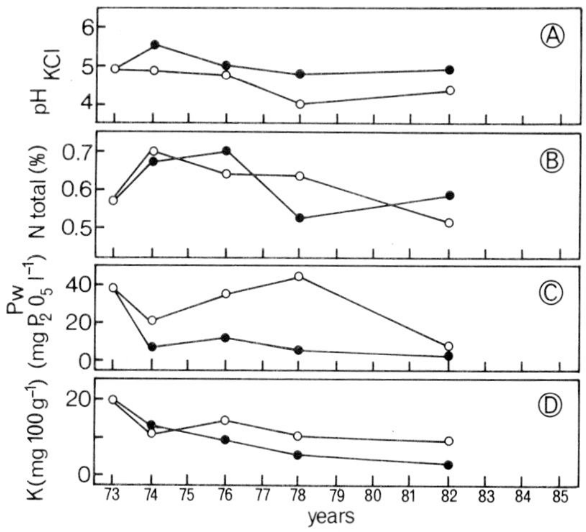

Fig. 3. Changes in (a) pH, extracted in KC1, (b) total N, modified Kjeldahl, (c) P extracted in water, Dutch procedure, (d) K extracted in HC1 and oxalic acid, in the topsoil (0–5 cm) of black 'Beek' earth in the years following the discontinuation of fertilizer application in 1973. The grassland was either cut in July and September (●), or left uncut (○) (Bakker, partly published).

content. Firstly, the soil nutrient content determined by extraction methods may not represent nutrient availability to plants. Nutrient availability in the soil depends upon other soil characteristics, various plant characteristics, and the specific conditions under which a plant grows. Chemical, physical and biological components of the systems and their interactions, are involved in the availability of a nutrient element in the soil and ultimately affect the uptake of an element by plants. In addition, the soil is a dynamic system, and the availability of nutrients varies, especially seasonally, being the result of (a) the release by weathering of parent materials, solubilization of insoluble compounds, decay of organic materials in the soil, leaching from plants, deposition by rain, and (b) the absorption by plant roots. Ultimately, the plant is the only agent that can determine nutrient availability.

Table 2. The composition of the toplayer (0–5 cm) of black 'Beek' earth soil and 'Made' peat soil beneath once annually cut (July/August) grassland 3 years, 15 years, and 30 years after discontinuation of fertilization (Jansen et al., 1981).

Soil type	Years since fertilized	Moisture %	Organic[1] matter %	pH_{KCl}[2]	$N_{mineral}$[3] mg kg^{-1}	N_{total}[4] %	P_w[5] mg P_2O_5 l^{-1}	P_{total}[6] mg P_2O_5 100 g^{-1}	$K_{extr.}$[7] mg 100 g^{-1}
Black 'Beek' earth soil	3	27.4	6.9	4.9	9.4	0.23	5.1	181.8	2.4
	15	36.2	9.2	5.1	11.7	0.37	7.3	106.9	3.2
	30	30.0	5.8	4.5	9.4	0.23	1.7	56.8	2.4
'Made' peat soil	3	38.2	12.1	5.3	10.7	0.39	12.8	194.7	4.1
	15	55.7	23.8	4.8	15.8	0.99	n.d.*	247.3	5.1
	30	81.7	61.2	4.9	44.0	2.41	n.d.*	267.5	8.8

* not detectable

[1] weight loss on ignition (950°C); [2] extracted in KC1 (1:1.5 (w/v), field-moist soil); [3] extracted in KC1 (1:2.4 (w/v), field-moist soil); [4] modified Kjeldahl; [5] extracted in water (1:60 (v/v), Dutch procedure); [6] digested in $HClO_3$, HNO_3, and H_2SO_4; [7] extracted in HC1 and oxalic acid.

4. Soil factors

4.1 Soil heterogeneity

It has often been stressed that most soil sampling is inadequate, mainly because the soil is spatially heterogeneous. Yet this heterogeneity may be very important in determining the distribution and coexistence of species in mixed grassland (Fitter, 1982). Relatively large differences in the nutrient content and pH of soil occur over distances of only a few centimetres (Beckett & Webster, 1971). This spatial variability may prevent detection of seasonal trends in nutrient availability (Ball & Williams, 1968), and may obscure changes over the years.

Soil sampling of grassland is usually restricted to the top 10 cm, where most of the root mass usually occurs, and where the highest concentrations of plant nutrients occur (Gupta & Rorison, 1975; Ernst, 1978). The abundance of roots usually decreases sharply with depth. Maximum root activity may not occur in this zone and may also vary seasonally (Fitter, 1986). Restriction of soil sampling to the toplayer is therefore questionable.

4.2 Seasonal variation in nutrient availability

Seasonal fluctuations in the availabilities of various plant nutrients occur in a variety of soil types (Table 3). The timing and magnitude of these changes seem to be dependent upon the nutrient element considered (Vaugh, Center & Jones, 1986). Dickinson (1984) concluded that nutrient availability in grassland soils is largely dependent upon the rate of mineralization of organic residues.

Grassland species differ in the time of maximum root activity, though most occurs in spring and summer (Fitter, 1986). Autumn or winter soil sampling could therefore be the least satisfactory for assessing soil fertility status.

4.3 Extraction methods

The rate of mass-flow and/or diffusion determines the nutrient supply to the root. Soils clearly vary in their ability to supply nutrients to plants. The total concentration of elements in the soil is normally a very unsatisfactory measure of nutrient availability and most methods of estimating available nutrient content involve a suitable extractant.

The suitability of extractants has largely been determined empirically, often in an agricultural context, by finding the closest correlation between (shoot) response, usually of annually grown crops in monoculture and the extractable nutrient content of the soil (Smith & Cook, 1953; Williams & Knight, 1963; Chapman, 1966; Cornforth & Walmsley, 1971; Walmsley & Cornforth, 1973). These correlations are usually based on nutrient uptake and/or yield response curves. Ultimately, soil sample analysis serves to predict nutrient availability for agricultural production.

Most of the crops used have a high growth rate and a high demand for nutrients, and the results may not be applicable to mesotrophic and oligotrophic species of mixed grasslands. Fresco et al. (1984) tested the correlation between 5 P-extraction methods and the vegetation composition of permanent moist hay fields. They concluded that soil extraction by potassium fluoride was the most suitable.

Table 3. Examples of seasonal variation in availabilities of various nutrients.

Variation in availability	Soil type	Authors
K, Ca, Mg, N, P	Mull/moder humus Mor humus	Frankland et al. (1963)
Organic P	Yellow-grey earth Yellow-brown earth	Saunders & Metson (1971)
Organic P	Dark brown	Dormaar (1972)
N	Chalk Brown earth Acid mull humus	Davy & Taylor (1974)
P, N, pH	Podzolic	Gupta & Rorison (1975)
N	Acidic Calcareous	Taylor et al. (1982)
K, Ca, Mg, N, P, Mn, Fe	Black 'Beek' earth Old arable	De Both et al. (1983)

Wild plant species differ quite markedly in response to N, P and Ca concentrations in nutrient solution. Some (eutrophic) species showed marked response, while others (mesotrophic/oligotrophic) responded weakly, or did respond negatively at high concentrations (Bradshaw et al., 1964; Jefferies & Willis, 1964; Rorison, 1968).

At present, there is insufficient information about the relation between soil tests for plant available nutrients and plant response in soils, deficient in nutrients and with species able to tolerate these deficient conditions. It seems possible that the currently used chemical extraction methods used for fertile soils may have limited value for predicting the nutrient requirements of low-responsive plant species in less fertile soils.

The soil water status, and its seasonal pattern of variation, probably has important consequences for the availability of nutrients in soils. For example, soil water status affects the availability of phosphate and other nutrients to plants (Williams & Simpson, 1965; Islam & Islam, 1973; Ernst, 1978; Gambrell & Patrick, 1978). In particular chemical extraction methods may give a poor indication of the availability of elements in waterlogged soils (Gambrell & Patrick, 1978).

5. Plant factors

The availability of nutrients to plants ultimately depends on the properties of the plant itself, which vary between species, populations and genotypes. Factors such as the size of the root system and its ability to absorb, transport, metabolize and redistribute nutrients, are likely to be important.

5.1 Modification of the rhizosphere

Plants differ in their ability to modify the root environment; in particular they can vary the aeration, pH and micro-organism community of the rhizosphere.

The ability of bog plants to tolerate reducing conditions in the soil is linked with the capacity of the root to oxidize the rhizosphere by means of 0_2 translocated from the shoots (Martin, 1968; Jones & Etherington, 1970; Jones, 1972; Green & Etherington, 1977; Hook & Scholtens, 1978). By this means soluble Fe^{2+}-ions are oxidized to water insoluble Fe^{3+}, both in the soil (Armstrong & Boatman, 1967) and in the intercellular spaces of the root cortex (Armstrong, 1964).

In plants which utilize NO_3^-, uptake of anions is usually greater than the uptake of cations (Riley & Barber, 1971); the resulting release of HCO_3^- increases the rhizosphere pH. This can lead, for example, to a reduction in the availability of B (Barber, 1984) and Al (Foy et al., 1978) to plants. Conversely, utilization of NH_4^+ leads to the loss of H^+-ions by the roots, because cation uptake exceeds anion uptake (Riley & Barber, 1971). This results in the acidification of the rhizosphere, which may enhance P-uptake (Miller et al., 1970; Riley & Barber, 1971; Van Ray & Van Diest, 1979) and Mn-uptake (Godo & Resienauer, 1980).

Other effects on the rhizosphere are non-element-specific. For example, preferential uptake of Ca may increase the mobilization of insoluble, inorganic (rock) phosphate by some dicotyledonous plants (Deist et al., 1971).

Some species and genotypes when Fe-deprived, are able to excrete H^+-ions and Fe-reductants (phenolic compounds) into the rhizosphere and so increase Fe-uptake by reducting Fe^{3+} to Fe^{2+} (Brown, 1978; Olsen et al., 1981). The response is temporary and only persists as long as Fe-stress occurs.

Micro-organisms are abundant in the rhizosphere, feeding on organic exudates from the roots and plant root debris (Rovira & Davey, 1974). In particular, there is an abundance of ammonifying and denitrifying bacteria in the rhizosphere (Rouatt et al., 1960); this may have important consequences for N-supply to plants. Rhizosphere micro-organisms may also hydrolyse various insoluble P-compounds (Barber et al., 1968; Woolhouse, 1969).

Symbiotic associations between micro-organisms and roots have proved to be important in supplying plants with essential nutrients. The symbiotic associations capable of fixing atmospheric N is characteristic of leguminous plants, and of (anaerobiosis-resistant) non-leguminous plants, such as *Alnus* spec. and *Myrica* spec. Various types of mycorrhizal associations may enhance the uptake of various nutrients by extending the effective root length of the host plant grown on nutrient-poor soils. Inaccessible, immobile P in the soil may be more efficiently exploited and extracted by this extended root system (Moser & Haselwandter, 1983).

5.2 Chemical composition of plants

No single method of soil analysis adequately integrates the various factors involved in determining the availability of a particular nutrient element in soils, neither does it accommodate the wide differences between plant species and genotypes in nutrient requirement.

The analysis of plant material offers an alternative diagnostic method in assessing the nutritional status of plants. The asymptotic relationship between tissue nutrient concentration and yield may probably provide a better assessment of nutrient requirements of plants. Because yield increases as tissue concentrations increase, except at very low yields (the 'Steenbjerg effect'), up to a 'critical' tissue concentration at which yield is unaffected by nutrient concentration, tissue analysis may be a useful measure of soil nutrient status in the range of low nutrient availability. In the higher range of availability, soil analysis may be more appropriate since tissue analysis is not sensitive in that range.

Dirven (1984) analysed the nutrient content of various plant species in 76 old grassland plots (Table 4). Herbs generally had higher contents of all major elements (K, Na, Ca, Mg, N, P, S and Cl) than grasses, a conclusion also reached by Wöhlbier & Kirchgessner (1957). These general differences in mineral composition between herbs and grasses may explain why monocotyledonous plants comprise 90% of the dry weight of mixed grassland on P-, N-, and K-deficient soil (Pegtel, 1983).

Because of both the variation in mineral com-

Table 4. The mineral composition of the shoots of various grassland species grown on various sites in the Netherlands. The highest values are shown in bold, and the lowest in italics. n = number of samples analysed (Dirven, 1984).

Species	n	K	Na	Ca	Mg	N	P	S	Cl
		% of dry shoot							
Lolium multiflorum	1	**4.77**	0.33	**0.76**	**0.22**	**4.43**	0.52	0.39	1.83
Molinia caerulea	5	*1.09*	0.09	0.30	0.18	2.56	0.15	0.38	*0.58*
Holcus lanatus	24	2.97	**0.46**	0.54	0.16	2.69	0.37	0.40	1.72
Brachypodium pinnatum	3	1.71	*0.02*	0.70	0.14	1.84	*0.11*	**0.55**	0.59
Phragmites australis	1	2.01	0.06	*0.27*	0.14	2.74	0.24	0.25	1.06
Danthonia decumbens	2	1.15	0.31	0.30	*0.08*	2.29	0.15	0.27	0.72
Holcus mollis	1	1.84	0.03	0.36	0.11	*1.50*	0.25	0.37	0.87
Alopecurus geniculatus	2	4.41	0.17	0.44	0.18	4.06	**0.57**	0.43	**2.75**
Glyceria maxima	3	3.38	0.05	0.41	0.11	3.09	0.38	*0.23*	1.72
Stellaria media	3	**8.06**	0.51	0.87	0.37	3.81	0.67	0.35	1.20
Plantago lanceolata	6	*1.13*	0.98	1.95	0.31	1.97	0.33	0.64	1.52
Hypochaeris radicata	3	1.91	**1.94**	1.65	0.28	2.42	0.21	0.60	2.22
Anthriscus sylvestris	3	4.11	*0.03*	2.38	0.27	3.66	0.59	0.32	0.66
Urtica dioica	3	3.25	0.05	**4.73**	**0.57**	4.62	0.62	0.52	0.63
Rumex obtusifolius	4	5.03	0.12	*0.63*	0.36	4.85	0.57	0.36	1.35
Symphytum officinale	2	4.43	0.61	1.41	*0.19*	3.36	0.31	0.27	1.47
Rorippa palustris	1	5.52	0.08	2.37	0.28	**5.82**	**0.75**	**1.24**	0.80
Sanguisorba minor	1	1.73	0.03	2.24	0.55	*1.66*	*0.12*	0.45	*0.56*
Galium cruciata	1	3.63	0.03	1.70	0.28	2.91	0.39	*0.22*	1.68
Cirsium dissectum	3	1.20	1.54	3.03	0.43	2.42	0.15	0.51	**6.24**

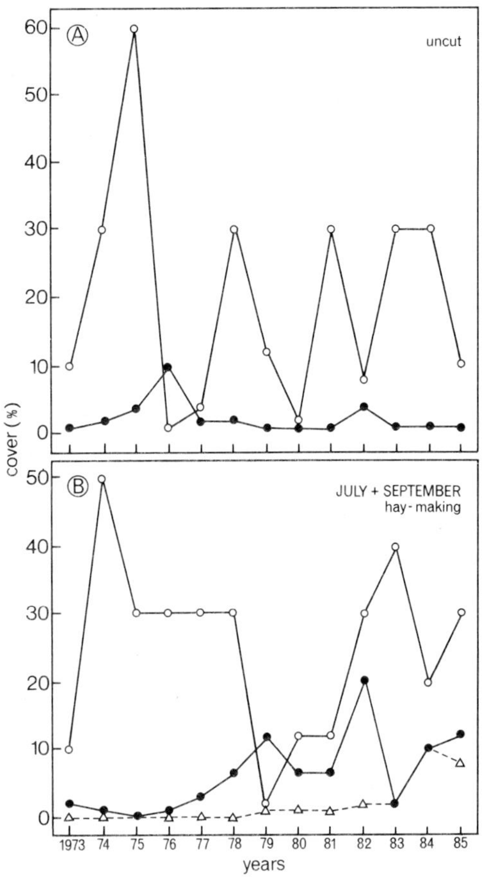

Fig. 4. Changes with time in the cover of (○) *Holcus lanatus,* (●) *Rumex acetosa,* and (△) *Anthoxanthum odoratum* in (a) uncut grassland and (b) grassland cut twice a year (July and September) after discontinuation of fertilizer application in 1973 (Bakker, unpublished).

position of the various component species and the seasonal and annual fluctuations in the botanical composition of mixed grassland, the resulting requirement of the sward may likely vary accordingly. Bakker et al. (1980) and Bakker & De Vries (1985) observed seasonal and annual fluctuations in the composition of grassland. Fluctuations in the abundance of *Holcus lanatus, Rumex acetosa* and *Anthoxanthum odoratum* are shown in Fig. 4; the abundance of *Holcus lanatus* was particularly variable and may reflect its short-lived perennial habit. These three species are employed to assay the nutrient-status of disturbed soils (Section 6.2). The mineral composition of the shoots of field-grown plants is variable (Table 5); it results from various soil- and plant-related factors (Bates, 1971). This type of variability obscures the determination of the species- and genotype-specific nutrients requirements. The analysis of the mineral composition of field-grown plants seems therefore to be only useful for comparative purposes at the level of a plant community on a specific site.

6. Phytometric assessment of fertility

Although soil fertility can be assessed by measuring herbage production, a more suitable method is to measure the productivity of undistrubed turfs taken from the sward or to grow test species on the soil (Pegtel, 1983). In both cases, individual nutrients can be applied to measure plant response.

The phytometric method has been applied to a wide variety of soils using *Anthoxanthum odoratum* (Wells & Saunders, 1960); to woodland mull soils using *Urtica dioica* (Pigott & Taylor, 1964); to a range of soil types in the Sheffield area using various species of different ecological tolerances (Rorison, 1967); to serpentine soils using oats (Proctor, 1971); to dune soils using *Rumex acetosa* (Atkinson, 1973) and *Sonchus arvensis* (Pegtel, 1976); to distinct horizons of podzolic soils using *Rumex acetosa* (Gupta & Rorison, 1975) and *Deschampsia flexuosa, Chamaenerion angustifolium*, *Rumex acetosella,* and *Senecio sylvaticus* (Ernst & Nelissen, 1979); to woodland soils using *Betula verrucosa* and *Acer pseudoplatanus* (Harrison & Helliwell, 1979); to sandy loam soils using *Betula papyrifera* (Safford, 1982); to peat soils using *Agrostis canina, Carex hostiana, Cirsium dissectum* and *Prunella vulgaris* (Pegtel, 1983); and to fen peat soil using *Epilobium hirsutum* (Al-Farraj et al., 1984).

The results, however, must be interpreted with care, since the addition of one nutrient may improve the supply of another or may ameliorate toxicity. Conversely, the added nutrients may become inorganically fixed (P; NH_4; K). Confusion may also arise when multiple deficiencies occur (Pegtel, 1983).

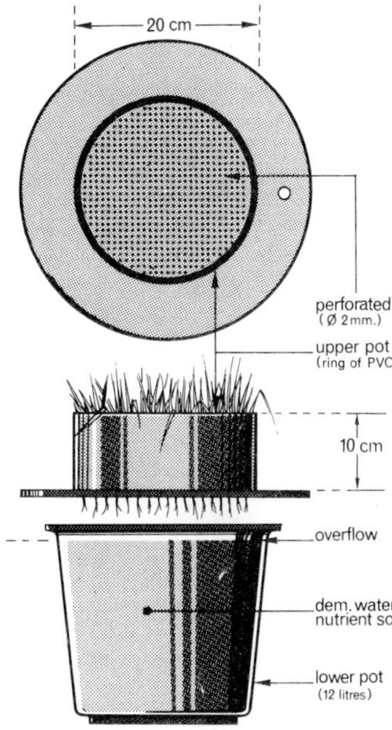

Fig. 5. Bouma-Janssen double pot technique for rapid stepwise screening nutrient deficiencies of (aerobic) soils.

6.1 Pot experiments with undisturbed turfs

I have used the phytometric approach to study the nutrient status of undisturbed turfs collected from black 'Beek' earth soil of which some fertility parameters are shown in Table 6. The Bouma-Janssen method was used (Janssen, 1974; Muller, 1979) to avoid the possibility of P-fixation by the soil because Fe- and Al-contents are high (Table 6). The upper pot (a PVC-ring of 20 cm diameter with a perforated bottom) contains the turf and allows the roots to grow into the lower pot, which was initially filled with demineralized water (Fig. 5). When the roots had grown into the lower pot, water was replaced by nutrient solutions (20 mM total ions), changed at 14 days intervals. The turfs were clipped at the start, and at intervals of 6 weeks. The clippings were handsorted and dried.

Shoot growth was least when root systems grew into water or N-deficient nutrient solution, but K-

Table 5. The mineral composition of the shoots of *Anthoxanthum odoratum, Holcus lanatus, Rumex acetosa,* and *Urtica dioica.*

Species	K	Na	Ca	Mg	N	P	S	Cl	Source
	% of dry shoot								
Anthoxanthum	1.90	0.04	0.33	0.09	1.60	0.09	–	–	1
odoratum	2.10	0.37	0.51	0.15	2.43	0.32	0.42	1.17	3
	1.72	0.06	0.33	0.17	–	0.24	–	–	4
Holcus	2.10	0.07	0.80	0.11	1.70	0.13	–	–	1
lanatus	–	–	0.43	–	1.46	0.27	–	–	2
	2.47	0.16	0.26	0.15	–	0.34	0.22	1.05	2
	2.97	0.46	0.54	0.16	2.69	0.37	0.40	1.72	3
	2.01	0.07	0.32	0.18	–	0.24	–	–	4
	1.82	–	0.39	–	1.31	0.20	–	–	5
	1.57	0.02	0.26	0.16	–	0.27	0.21	0.42	7
Rumex	2.61	0.88	0.72	0.47	4.14	0.50	0.37	0.60	3
acetosa	2.10	0.14	0.55	0.35	–	0.17	–	–	4
	2.47	0.20	0.63	0.32	–	0.38	–	–	4
	1.52	0.26	0.57	0.33	–	0.29	0.10	0.41	6
	2.38	0.02	0.42	0.24	–	0.31	0.16	0.11	7
Urtica	2.70	0.05	7.10	0.53	3.90	0.44	–	–	1
dioica	3.25	0.05	4.73	0.57	4.62	0.62	0.52	0.63	3

Source: 1) Allen, 1974; 2) Beddows, 1961; 3) Dirven, 1984; 4) Denudt & Lambert, 1976; 5) Holst, 1974; 6) Strigel, 1912; 7) Wöhlbier & Kirchgessner, 1957.

Table 6. Some characteristics of black 'Beek' earth soil (0–5 cm) beneath uncut and twice annually cut in July and September of grassland 10 years after discontinuation of fertilizer application (Olde Loohuis & Wemmenhove, 1984).

Parameters	Uncut	July and September hay-making
Moisture (%)	48.6	49.5
Organic matter[1] (%)	16.8	15.8
pH_{KCl}[2]	4.3	5.1
NO_3^-[3] (mg N kg^{-1})	2.2	1.8
NH_4^+[4] (mg N kg^{-1})	25.2	24.7
N_{total}[5] (mg N kg^{-1})	5300	4600
P_w[6] (mg P_2O_5 l^{-1})	15.3	12.2
P-Al[7] (mg P_2O_5 100 g^{-1})	17.5	16.3
$P_{organic}$[8] (mg P_2O_5 100 g^{-1})	114.3	123.0
P_{total}[9] (mg P_2O_5 100 g^{-1})	176.5	210.5
$K_{extr.}$[10] (mg 100 g^{-1})	7.4	8.0
K_{total}[11] (mg 100 g^{-1})	68.1	75.3
$Fe_{extr.}$[12] (mg 100 g^{-1})	51.7	92.9
Fe_{total}[13] (mg 100 g^{-1})	945	1217
$Al_{extr.}$[14] (mg 100 g^{-1})	76.8	74.7
Al_{total}[15] (mg 100 g^{-1})	548.4	654.8

[1] weight loss on ignition (950°C); [2] extracted in KC1 (1:1.5 (w/v), field-moist soil); [3] extracted in KC1 (1:2.4 (w/v), field-moist soil); [4] extracted in KC1 (1:2.4 (w/v), field-moist soil); [5] modified Kjeldahl; [6] extracted in water (1:60 (v/v), Dutch procedure); [7] extracted in ammonium lactate and acetic acid (pH 4.8); [8] difference between extraction of ignited (550°C) soil and unignited soil in H_2SO_4; [9] digested in $HC10_3$, HNO_3, and H_2SO_4; [10] extracted in HC1 and oxalic acid; [11] digested in $HC10_3$, HNO_3, and H_2SO_4; [12] extracted in ammonium acetate (pH 4.8); [13] digested in $HC10_3$, HNO_3, and H_2SO_4; [14] extracted in ammonium acetate (pH 4.8); [15] digested in $HC10_3$, HNO_3, and H_2SO_4.

deficiency and, to a lesser extent, P-deficiency also reduced shoot growth (Fig. 6). Nutrient supply by the soil itself, therefore appeared to be small because of its restricted volume and/or due to single or multiple deficiencies. The previous management practices imposed for 10 years seemed to bear little effect on soil fertility.

The component species responded differently to the various nutrient solutions. In particular, *Rumex acetosa* and *Holcus lanatus* responded more strongly than *Anthoxanthum odoratum* to complete nutrient solution in the sward that had been cut twice a year for hay (Fig. 7). *Rumex acetosa* was more affected by P-deficiency than *Holcus lanatus*. *Anthoxanthum odoratum* grew little,

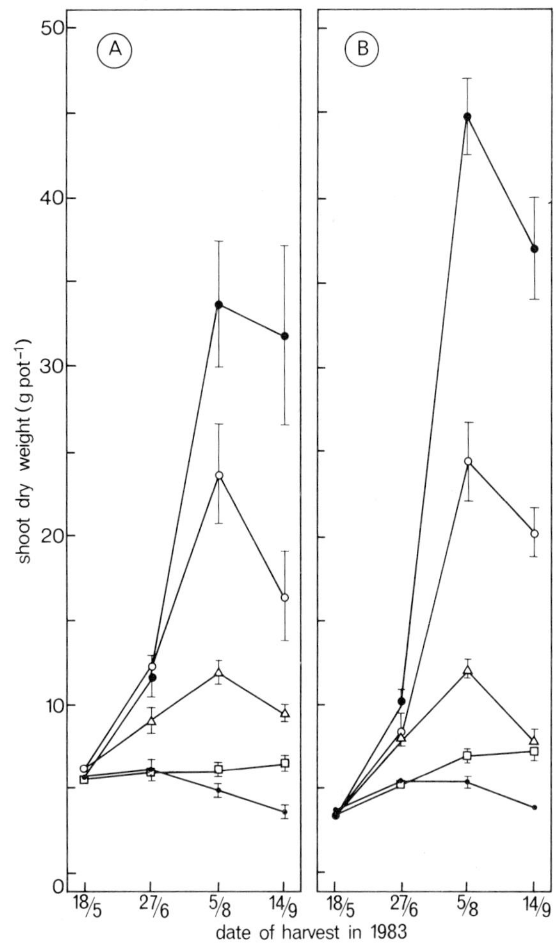

Fig. 6. The shoot yield of turfs (0–10 cm) collected from (a) uncut grassland and (b) grassland cut in July and September 10 years after discontinuation of fertilizer application in 1973, when grown with their roots in water (•) or in nutrient solutions that were N-deficient (□), K-deficient (△), P-deficient (○), or complete (●). Bars denote S.E. (Olde Loohuis & Wemmenhove, 1984).

irrespective of the nutrient solution used, and its weight was adversely reduced by full nutrient solution. This may explain why this species is absent from the uncut sward. This type of experiment largely simulates field conditions, so the results should be ecologically valid. As far as cause and effects are concerned, this type of phytometric approach is unavoidably ineffective since the nature and extent of competition is unknown.

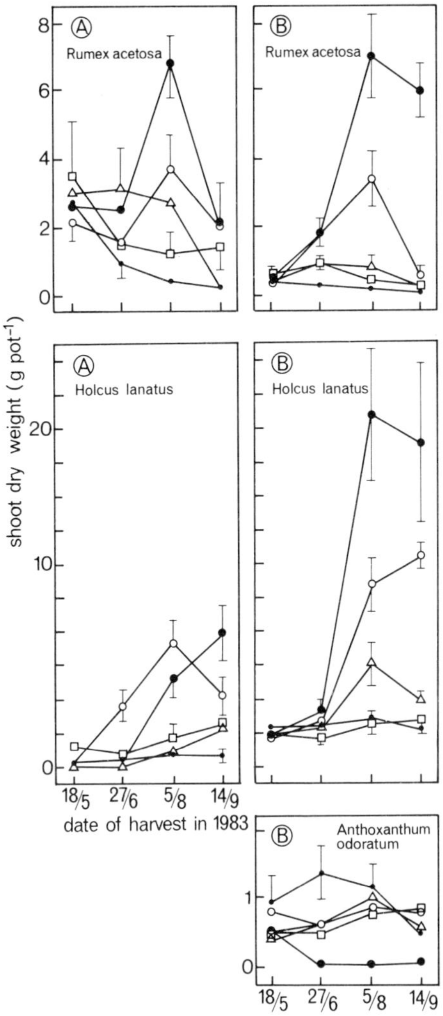

Fig. 7. The shoot yield of *Rumex acetosa, Holcus lanatus,* and *Anthoxanthum odoratum* in turfs (0–10 cm) collected from (a) uncut grassland and (b) grassland cut in July and September 10 years after discontinuation of fertilizer application in 1973, when grown with their roots in water (•), or in nutrient solutions that were N-deficient (□), K-deficient (△), P-deficient (○), or complete (●). Bars denote S.E. (Olde Loohuis & Wemmenhove, 1984).

6.2 Pot experiments with disturbed soil and single species

Pot experiments, using test species and mixed soil samples, were set up to provide supplementary evidence. Seedlings of four species were used to investigate the fertility status of black 'Beek' earth soil.

Desirable features of appropriate indicator species included: (1) potentially rapid growth of seedlings; (2) rapid, flexible growth response of seedlings to varied rates of nutrient availability; (3) wide edaphic tolerance; and (4) occurrence in the sites under investigation. Two grasses and two herbs were selected: *Anthoxanthum odoratum, Holcus lanatus, Rumex acetosa,* and *Urtica dioica* (Table 5). *Anthoxanthum, Holcus* and *Rumex* are typically found in soils of intermediate fertility, but differ in response to N, P and K (Table 7). *Urtica* was included because of its relative high requirement for N and P (Pigott & Taylor, 1964).

Bulk soil samples of the toplayer (0–5 cm) were collected on 8 March 1984 (P-series), 3 May 1984 (N-series), and 24 October 1984 (K-series). The soil samples were thoroughly mixed and placed into 12 cm-pots. Black 'Beek' earth soil likely has a high P-fixing capacity because of high contents of Fe and Al (Table 6). High rates of P were therefore applied to ensure that during the course of the experiments sufficient plant available P was present. Previously presented results (Fig. 6) indicate that multiple deficiencies occur (N, K, and P), so additions of 150 kg N ha^{-1} (as NH_4NO_3), 100 kg K_2O ha^{-1} (as K_2SO_4), and 400 kg P_2O_5 ha^{-1} (as $Ca(H_2PO_4)_2 \cdot H_2O$) were made. Nutrients were added as solution.

Seeds of *Anthoxanthum* and *Holcus*, collected from a sward nearby, and *Rumex* and *Urtica*, from the experimental garden at Haren, were germinated and six (*Urtica* four) matched seedlings were transplanted into each pot. Pots were watered with demineralized water as necessary. After about eight weeks of growth, shoots were clipped and dried.

All four species responded strongly to N, but *Holcus* was most responsive and *Urtica* least responsive (Fig. 8), in spite of the fact that *Urtica* has a reputation of N-response.

When no N-fertilizer was applied, the uncut sward yielded more herbage than the sward cut twice annually. The latter seems to be somewhat N-deficient. Applications of N-fertilizer had a greater effect upon herbage yield when the sward was cut twice annually.

Surprisingly, the herbage yield of *Urtica* was no

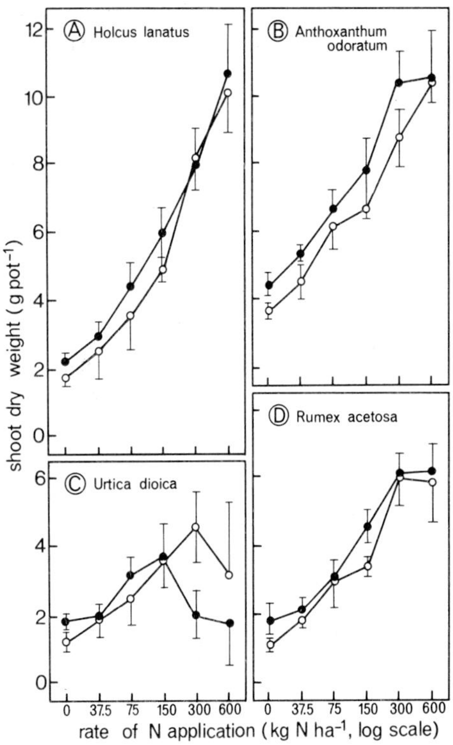

Fig. 8. The effect of added N to the topsoil (0–5 cm) of black 'Beek' earth collected from uncut grassland (●) and grassland cut in July and September (○) 11 years after discontinuation of fertilizer application in 1973, on the shoot yield of young plants of (a) *Holcus lanatus,* (b) *Anthoxanthum odoratum,* (c) *Urtica doica,* and (d) *Rumex acetosa.* Bars denote 95% C.L. (Hoogveld, 1986).

greater with applications of more than 300 kg N ha^{-1} than with no N-fertilizer. However, since ammonium nitrogen is far less beneficial than nitrate to *Urtica* (Ernst, 1978; Rosnitschek-Schimmel, 1982), this may reflect NH_4^+-toxicity.

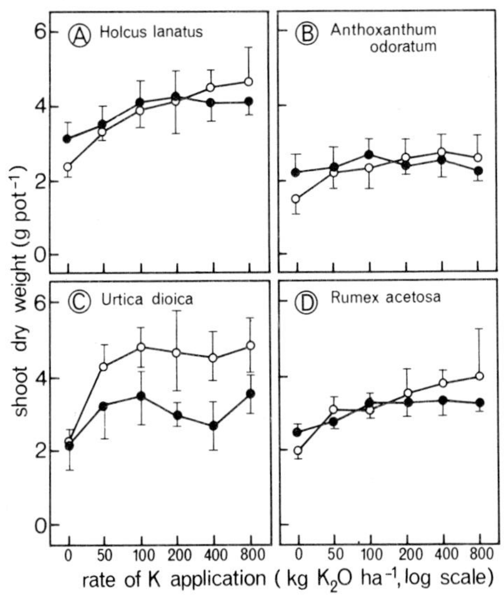

Fig. 9. The effect of added K to the topsoil (0–5 cm) of black 'Beek' earth collected from uncut grassland (●) and grassland cut in July and September (○) 11 years after discontinuation of fertilizer application in 1973, on the shoot yield of young plants of (a) *Holcus lanatus,* (b) *Anthoxanthum odoratum,* (c) *Urtica doica,* and (d) *Rumex acetosa.* Bars denote 95% C.L. (Hoogveld, 1986).

Extra K-supply (Fig. 9) affected herbage yield much less than extra N-supply, indicating that K-deficiency was less than N-deficiency. *Urtica* was the most K-responsive species, while *Anthoxanthum* was the least responsive. This may reflect the fact that *Urtica* has the highest K-content in the shoots and *Anthoxanthum* the least (Table 5).

In the absence of extra K-supply, *Holcus* had the greatest herbage yield and the three other species the least. The uncut sward yielded more herbage of *Holcus*, *Anthoxanthum*, and *Rumex*.

Table 7. Characteristics of *Anthoxanthum odoratum, Holcus lanatus, Rumex acetosa,* and *Urtica dioica* employed to assess relative N, K, and P-supply of black 'Beek' earth soil.

Species	Edaphic range	pH peak	Indicator for soil fertility	RGR	Response to N	P	K
Anthoxanthum odoratum	wide[5]	acid[4]	medium[2,3,11]	low[9]		±[8]	+[7]
Holcus lanatus	wide[5]	acid[4]	medium[2,3,11]	high[12]	+[12]		
Rumex acetosa	wide[5]	acid[4]	medium[2,3,11]	high[5]	+[6]		+[5]
Urtica dioica	wide[1,4,10]	neutral[4]	rich[2,3,11]	high[5]	+[1,10]		+[1,5]

Source: [1] Pigott & Taylor, 1964; [2] Klapp, 1965; [3] Kruijne et al., 1967; [4] Grime & Hodgson, 1969; [5] Rorison, 1969; [6] Gigon & Rorison, 1972; [7] Davies & Snaydon, 1974; [8] Wild et al., 1974; [9] Grime & Hunt, 1975; [10] Ernst, 1978; [11] Ellenberg, 1979; [12] Rorison, 1985.

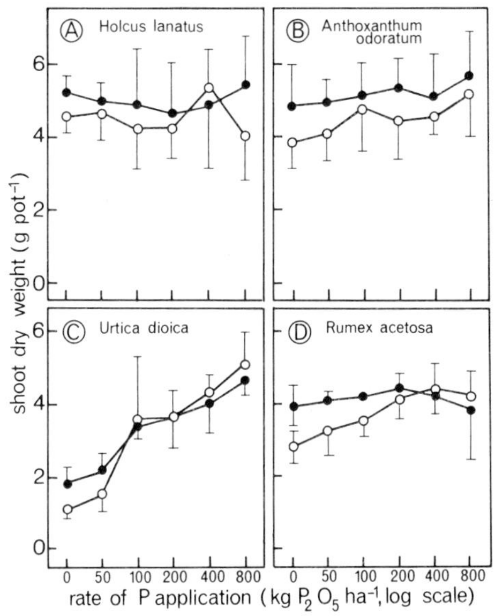

Fig. 10. The effect of added P to the topsoil (0–5 cm) of black 'Beek' earth collected from uncut grassland (●) and grassland cut in July and September (○) 11 years after discontinuation of fertilizer application in 1973, on the shoot yield of young plants of (a) *Holcus lanatus,* (b) *Anthoxanthum odoratum,* (c) *Urtica dioica,* and (d) *Rumex acetosa.* Bars denote 95% C.L. (Hoogveld, 1986).

The herbage yield of the four species responded more to K when the sward was cut twice annually. The latter seems to be somewhat more K-deficient.

Holcus, Anthoxanthum, and *Rumex* hardly responded to added P (Fig. 10), indicating that the P-supply by black 'Beek' earth soil was adequate. However, *Urtica* responded positively to additional P, probably reflecting its relatively high P-requirement (Pigott & Taylor, 1964; Rorison, 1968). *Urtica* generally has a higher P-content in the shoots (Table 5).

The herbage yield of the four species indicated that the P-supplying capacity of the soil beneath the uncut sward was somewhat higher than the soil beneath the sward cut twice annually.

These results resemble those obtained by using undisturbed turfs, though the experiment with the undisturbed turfs did not indicate differences in nutrient status between the soils receiving different cutting management. The apparent differences in nutrient status of the soil, when tested with four test species grown in even-aged monospecific cultures, may be related to the experimental method, and the depth of soil sampling, since undisturbed turfs were taken with a depth of 10 cm, whilst the monocultures of seedlings were grown on the top 5 cm of heavily disturbed soil. The differences may reflect the fact that seedlings used in monocultures are more environment-sensitive.

7. Concluding remarks

(1) When the use of fertilizers was discontinued on grassland that had formerly been used intensively, production declined after 5 years, and an increase in the number of species occurred. Some of the newly invaded and established species indicated nutrient-poor conditions; that was confirmed by bioassay methods.

(2) No direct evidence was obtained of a causal relationship between species diversity and soil nutrient status in semi-natural grassland, but various possibilities are reviewed.

(3) A bioassay method, using undisturbed turfs collected from the field, was applied. In addition, seedling stands of several individual species were grown on soil amended with various nutrients. The results indicated that soil fertility declined somewhat over 11 years when fertilization was discontinued and grassland was cut twice annually for hay. This is linked with the fact that of the total nutrients in the grassland ecosystem, only a small fraction is yearly available to plants (Dickinson, 1984; Bakker, Chapter 12).

(4) Cutting for hay causes openings in the sward which allow a larger number of species to establish. Bakker (Chapter 12) concluded that this was caused by reduced competition for light, but competition for nutrients in the regularly cut sward may be of greater importance (cf. Snaydon & Hope, 1986).

(5) The problem in which way disturbance of soil nutrient status of mixed grassland has to be quantified, is still undissolved.

Acknowledgements

The author acknowledges Mrs. C.W. Biewenga for typing the manuscript and Mr. E. Leeuwinga for preparing the figures.

References

Al-Farraj, M.M., Giller, K.E. & Wheeler B.D. (1984). Phytometric estimation of fertility of waterlogged rich-fen peats using *Epilobium hirsutum* L. Plant and Soil 81: 283–289.

Allen, S.E. (ed.) (1974). Chemical analysis of ecological materials. Blackwell, Oxford.

Armstrong, W. (1964). Oxygen diffusion from the roots of some British bog species. Nature 204: 801–802.

Armstrong, W. & Boatman, D.J. (1967). Some field observations relating to the growth of bog plants to conditions of soil aeration. J. Ecol. 55: 101–110.

Atkinson, D. (1973). Observations on the phosphorus nutrition of two sand dune communities at Ross Links. J. Ecol. 61: 117–133.

Bakker, J.P., Dekker, M. & de Vries, Y. (1980). The effect of different management practices on a grassland community and the resulting fate of seedlings. Acta Bot. Neerl. 29: 469–482.

Bakker, J.P. & De Vries, Y. de (1985). The results of different cutting regimes in grassland taken out of the agricultural system. In: K.H. Schreiber (Ed.), Suksession auf Grünlandbrachen, pp 51–57. Ferdinand Schöningh, Paderborn.

Ball, D.F. & Williams, W.M. (1968). Variability of soil chemical properties in two uncultivated brown earths. J. Soil Sci. 19: 379–391.

Barber, S.A. (1984). Soil nutrient bioavailability. A mechanical approach. Wiley & Sons, New York.

Barber, D.A., Sanderson, J. & Scott-Russel, R. (1968). Influence of micro organisms on the distribution in roots of phosphate labelled with phosphorus-32. Nature 217: 644.

Bates, T.E. (1971). Factors affecting critical nutrient concentrations in plants and their evaluation: a review. Soil Science 112: 116–130.

Beckett, P.H.T. & Webster, R. (1971). Soil variability: a review. Soils and Fertilizers 34: 1–15.

Beddows, A.R. (1961). Biological flora of the British isles. *Holcus lanatus*. J. Ecol. 49: 421–430.

Booy, A.H. (1986). IJzeroer in Drenthe. Ontstaan, voorkomen en winning. Historisch Jaarboek Drenthe 103: 66–88. Van Gorcum, Assen.

Bradshaw, A.D., Chadwick, M.J., Jowett, D. & Snaydon, R.W. (1964). Experimental investigations into the mineral nutrition of several grass species. IV. Nitrogen level. J. Ecol. 52: 665–676.

Brown, J.C. (1978). Mechanism of iron uptake by plants. Plant, Cell & Environment 1: 249–257.

Chapman, H.D. (1966). Diagnostic criteria for plants and soils, pp 730–737. University of California (Division of Agricultural Science), Riverside.

Cornforth, I.S. & Walmsley, D. (1971). Methods of measuring available nutrients in West Indian soils. I. Nitrogen. Plant and Soil 35: 389–399.

Davies, M.S. & Snaydon, R.W. (1974). Physiological differences among populations of *Anthoxanthum odoratum* L. collected from the Park Grass Experiment Rothamsted. III. Response to phosphate. J. Appl. Ecol. 11: 699–707.

Davy, A.J. & Taylor, K. (1974). Seasonal patterns of nitrogen availability in contrasting soils in the Chiltern Hills. J. Ecol. 62: 793–807.

De Bakker, H. (1979). Major soils and soil regions in the Netherlands. Junk, The Hague.

De Both, M.T.J., De Muinck Keizer, M., Wouda, A., Bakker, B. & Drenth, W. (1983). Seizoensfluctuaties in de bodem. Internal report, Dept. of Physical Geography, State University of Groningen (mimeo). 544 pp.

Deist, J., Marais, P.G., Harry, R.B.A. & Heyns, C.F.G. (1971). Relative availability of rock phosphate to different plant species. Agrochemophysica 3: 35–40.

Denudt, G. & Lambert, J. (1976). Étude minerale de quelques plantes de prairies humides. In: J.M. Géhu (Ed.), La végétation des prairies inondables. Colloques phytosociologues 5: 353–357. Cramer, Vaduz.

Dickinson, N.M. (1984). Seasonal dynamics and compartmentation of nutrients in a grassland meadow in lowland England. J. Appl. Ecol. 21: 695–701.

Dirven, J.G.P. (1984). De minerale samenstelling van afzonderlijke graslandplanten. Report 81, Dept. of Field Crops and Grassland Science, Agricultural University, Wageningen. 14 pp.

Dormaar, J.F. (1972). Seasonal pattern of soil organic phosphorus. Can. J. Soil Sci. 52: 107–112.

Ellenberg, H. (1956). Über die Beziehungen zwischen Pflanzengesellschaft, Standort, Bodenprofil und Bodentyp. Bericht über das internationale Symposium Pflanzensoziologie / Bodenkunde. Stolzenau/Weser. 4 pp.

Ellenberg, H. (1979). Indicator values of vascular plants in Central Europe. Scr. Geobot. 9. 122 pp.

Ernst, W.H.O. (1978). Chemical soil factors determining plant growth. In: A.H.J. Freijsen & J.W. Woldendorp (Eds.), Structure and functioning of plant populations, pp 155–187. North Holland Publ. Comp., Amsterdam.

Ernst, W.H.O. & Nelissen, H.J.M. (1979). Growth and mineral nutrition of plant species from clearings on different horizons of an Iron-Humus profile. Oecologia (Berlin) 41: 175–182.

Fitter, A.H. (1982). Influence of soil heterogeneity on the coexistence of grassland species. J. Ecol. 70: 139–148.

Fitter, A.H. (1986). Spatial and temporal patterns of root activity in a species-rich alluvial grassland. Oecologia (Berlin) 69: 594–599.

Foy, C.D., Chaney, R.L. & White, M.C. (1978). The physiology of metal toxicity in plants. Ann. Rev. of Plant Phys. 29: 511–566.

Frankland, J.C., Ovington, J.D. & Macrae, C. (1963). Spatial and seasonal variations in soil, litter and ground vegetation in some lake district woodlands. J. Ecol. 51: 99–112.

Fresco, L.F.M., Hermans, E. & van der Lans, W. (1984). Soil phosphorus as a factor controlling vegetation composition in moist hay fields: a statistical evaluation of some methods for the determination of available phosphate. Plant and Soil 78: 259–269.

Gambrell, R.P. & Patrick, W.H. (1978). Chemical and microbiological properties of anaerobic soils and sediments. In: D.H. Hook & R.M.M. Crawford (Eds.), Plant life in anaerobic environments, pp 375–423. Ann Arbor Science, Ann Arbor.

Gigon, A. & Rorison, I.H. (1972). The response of some ecological distinct plant species to nitrate- and to ammonium-nitrogen. J. Ecol. 60: 93–102.

Godo, G.H. & Reisenauer, H.M. (1980). Plant effects on soil manganese availability. Soil Sci. Soc. Am. Proc. 44: 993–995.

Green, M.S. & Etherington, J.R. (1977). Oxidation of ferrous iron by rice (Oryza sativa L.) roots: a mechanism for waterlogging tolerance. J. Exp. Bot. 28: 678–690.

Grime, J.P. (1979). Plant species and vegetation processes. Wiley, Chichester.

Grime, J.P. & Hodgson, J.G. (1969). An investigation of the ecological significance of lime-chlorosis by means of large-scale comparative experiments. In: I.H. Rorison (Ed.), Ecological aspects of the mineral nutrition of plants, pp 67–99. Blackwell, Oxford.

Grime, J.P. & Hunt, R. (1975). Relative growth rate: its range and adaptive significance in a local flora. J. Ecol. 63: 393–422.

Grootjans, A.P. (1980). Distribution of plant communities along rivulets in relation to management and hydrology. In: O. Wilmanns & R. Tüxen (Eds.), Epharmonie, pp 143–170. Cramer, Vaduz.

Grootjans, A.P. & Ten Klooster, W.Ph. (1980). Changes of ground water regime in wet meadows. Acta Bot. Neerl. 29: 541–554.

Gupta, P.L. & Rorison, I.H. (1975). Seasonal differences in the availability of nutrients down a podzolic profile. J. Ecol. 63: 521–534.

Harrison, A.F. & Helliwell, D.R. (1979). A bioassay for comparing phosphorus availability in soils. J. Ecol. 16: 497–505.

Holst, G. (1974). Über die Stickstoffdisposition in der Nährstoffaufnahme der Pflanzen. Angew. Bot. 48: 77–95.

Hoogveld, J.G.E. (1986). Een testplantexperiment in het kader van onderzoek naar verschraling. Internal report, Dept. of Plant Ecology, State University of Groningen (mimeo). 57 pp.

Hook, D.D. & Scholtens, J.R. (1978). Adaptations and flood tolerance of tree species. In: D.D. Hook & R.M.M. Crawford (Eds.), Plant life in anaerobic environments, pp 299–331. Ann Arbor Science, Ann Arbor.

Islam, A. & Islam, W. (1973). Chemistry of submerged soils and growth and yield of rice. I. Benefits from submergence. Plant and Soil 39: 555–565.

Jansen, H., Pathuis, W., Wemmenhove, A. & Olde Loohuis, J. (1981). De invloed van maaien en afvoeren op de bodemvruchtbaarheid. Internal report, Dept. of Plant Ecology, State University of Groningen (mimeo). 33 pp.

Janssen, B.H. (1974). A double pot technique for rapid soil testing. Trop. Agric. (Trinidad) 51: 161–166.

Jefferies, R.L. & Willis, A.J. (1964). Studies on the calcicole - calcifuge habitat. II. The influence of calcium on the growth and establishment of four species in soil and sand cultures. J. Ecol. 52: 691–707.

Jones, R. (1972). Comparative studies of plant growth and distribution in relation to waterlogging. V. The uptake of iron and manganese by dune and slack plants. J. Ecol. 60: 131–140.

Jones, H.E. & Etherington, J.R. (1970). Comparative studies of plant growth and distribution in relation to waterlogging. I. The survival of *Erica cinerea* L. and *E. tetralix* L. and its apparent relationship to iron and manganese uptake in waterlogged soil. J. Ecol. 58: 487–496.

Klapp, E. (1965). Grünlandvegetation und Standort. Parey, Berlin.

Kruijne, A.A., De Vries, D.M. & Mooi, H. (1967). Bijdrage tot de oecologie van de Nederlandse graslandplanten. Versl. landbouwk. Onderz. 696: 1–65. Wageningen.

Martin, M.H. (1968). Conditions affecting the distribution of *Mercurialis perennis* in certain Cambridgeshire woodlands. J. Ecol. 56: 777–793.

Miller, M.H., Mamaril, C.P. & Blair, G.J. (1970). Ammonium effects on phosphorus absorption through pH changes and phosphorus precipitated at the soil-root interface. Agron. J. 62: 524–527.

Moser, M. & Haselwandter, K. (1983). Ecophysiology of mycorrhizal symbiosis. In: O.L. Lange, P.S. Nobel, C.B. Osmond & H. Ziegler (Eds.), Responses to the chemical and biological environment. Part C., pp 391–421. Encycl. Plant Physiology N.S. vol. 12. Springer, Berlin.

Muller, A. (1979). Deficiency symptoms in cacao seedlings observed in pot experiments with the Bouma-Janssen method. Neth. J. agric. Sci. 27: 211–220.

Olde Loohuis, J. & Wemmenhove, A. (1984). De invloed van maaien en afvoeren op de bodemvruchtbaarheid. Internal report, Dept. of Plant Ecology, State University of Groningen (mimeo). 91 pp.

Olsen, R.A., Clark, R.B. & Bennett, J.H. (1981). The enhancement of soil fertility by plant roots. Am. Sci. 69: 378–384.

Pegtel, D.M. (1976). On the ecology of two varieties of *Sonchus arvensis* L. Ph. D. Thesis, State University of Groningen. 148 pp.

Pegtel, D.M. (1983). Ecological aspects of a nutrient-deficient wet grassland (*Cirsio-Molinietum*). Verh. Ges. Ökol. 10: 217–228.

Pigott, C.D. & Taylor, K. (1964). The distribution of some

woodland herbs in relation to the supply of nitrogen and phosphorus in the soil. J. Ecol. 52 (suppl.): 175–185.

Proctor, J. (1971). The plant ecology of serpentine. II. Plant response to serpentine soils. J. Ecol. 59: 397–410.

Riley, D. & Barber, S.A. (1971). Effect of ammonium and nitrate fertilization on phosphorus uptake as related to root-induced pH changes at the root-soil interface. Soil Sci. Soc. Am. Proc. 35: 301–306.

Rorison, I.H. (1967). A seedling bioassay of some soils in the Sheffield area. J. Ecol. 55: 725–752.

Rorison, I.H. (1968). The response to phosphorus of some ecologically distinct plant species. New Phytol. 67: 913–923.

Rorison, I.H. (1969). Ecological inferences from laboratory experiments on mineral nutrition. In: I.H. Rorison (Ed.), Ecological aspects of the mineral nutrition of plants, pp 155–175. Blackwell, Oxford.

Rorison, I.H. (1985). Nitrogen source and the tolerance of *Deschampsia flexuosa, Holcus lanatus* and *Bromus erectus* to aluminium during seedling growth. J. Ecol. 73: 83–90.

Rosnitschek-Schimmel, I. (1982). Effect of ammonium and nitrate supply on dry matter production and nitrogen distribution in *Urtica dioica*. Z. Pflanzenphysiol. 108: 329–341.

Rouatt, J.W., Katznelson, H. & Payne, T.M.B. (1960). Statistical evaluation of the rhizosphere effect. Soil Sci. Soc. Am. Proc. 24: 271–273.

Rovira, A.D. & Davey, C.B. (1974). Biology of the rhizosphere. In: E.W. Carson (Ed.), The plant root and its environment, pp 153–204. Univ. Press Virginia, Charlottesville.

Safford, L.O. (1982). Correlation of greenhouse bioassay with field response to fertilizer by paper birch. Plant and Soil 64: 167–176.

Saunders, W.M.H. & Metson, A.J. (1971). Seasonal variation of phosphorus in soil and pasture. N.Z. J. agric. Res. 14: 307–328.

Smith, F.W. & Cook, R.L. (1953). A study of the relationship between chemically available phosphorus and plant growth response on several Michigan soils. Soil Sci. Soc. Am. Proc. 17: 26–30.

Snaydon, R.W. & Howe, C.D. (1986). Root and shoot competition between established ryegrass and invading grass seedlings. J. Appl. Ecol. 23: 667–674.

Strigel, A. (1912). Mineralstoffaufnahme verschiedener Pflanzenarten aus ungedüngtem Boden. Über den Einflusz der botanischen Natur, der Herkunft und Erntezeit auf die chemische Zusammensetzung von Wiesenheu. Landw. Jb. 43: 349–371.

Taylor, A.A., De-Felice, J. & Havill, D.C. (1982). Seasonal variation in nitrogen availability and utilization in an acidic and calcareous soil. New Phytol. 92: 141–152.

Van Heuveln, B. (1980). Vegetation and soil in a Drenthian brook valley (the Netherlands). Acta Bot. Neerl. 29: 555–564.

Van Ray, B. & Van Diest, A. (1979). Utilization of phosphate from different sources by six plant species. Plant and Soil 51: 577–589.

Vaughn, C.E., Center, D.M. & Jones, M.B. (1986). Seasonal fluctuations in nutrient availability in some northern California annual range soils. Soil Science 141: 43–51.

Walmsley, D. & Cornforth, I.S. (1973). Methods of measuring available nutrients in West-Indian soils II. Phosphorus. Plant and Soil 39: 93–101.

Wells, M. & Saunders, W.M.H. (1960). Soil studies using sweet vernal to assess element availability. IV. Phosphorus. N.Z.J. agric. Res. 3: 279–299.

Wild, A., Skarlou, V., Clements C.R. & Snaydon, R.W. (1974). Comparison of potassium uptake of four plant species grown in sand and in flowing solution culture. J. Appl. Ecol. 11: 801–812.

Williams, E.C. & Knight, A.H. (1963). Evaluation of soil phosphate status by pot experiments, convential extraction methods, and labile phosphate values estimated with the aid of phosphorus–32. J. Sci. Food Agr. 14: 555–563.

Williams, C.H. & Simpson, J.R. (1965). Some effects of cultivation and waterlogging on the availability of phosphorus in pasture soils. Austr. J. agric. Res. 16: 413–427.

Wöhlbier, W. & Kirchgessner, M. (1957), Der Gehalt von einzelnen Gräser, Leguminosen und Kräutern an Mengen- und Spurenelementen. Landwirtsch. Forsch. 10: 240–251.

Woolhouse, H.W. (1969). Differences in the properties of the acid phophatases of plant roots and their significance in the evolution of edaphic ecotypes. In: I.H. Rorison (Ed.), Ecologicial aspects of the mineral nutrition of plants, pp 357–380. Blackwell, Oxford/Edinburgh.

CHAPTER 5

Hybridization of the habitat in disturbed hay meadows

A.P. Grootjans, G.J.R. Allersma & C. Kik

Detailed contents

1. Introduction

Almost 40 years ago Anderson (1948) drew attention to the importance of human interference in the 'hybridization of the habitat', and its consequent effect on the occurrence of hybrids. Many other examples of the appearance of hybrids have been published since then. The evolutionary significance of hybridization is discussed by Urbanska in chapter 20. We will concentrate, here on the nature of 'hybridized' habitats. Urbanska mainly deals with hybridization in habitats disturbed by man, creating 'intermediate or open habitats that are free to be newly colonized', and which contain 'numerous new microniches'. Both Anderson and Urbanska stress that a hybridized habitat is intermediate between those in which the parental species are found, but is not a mere 'intermingling' of habitats (Stace, 1975). Grant (1971) held the opinion that the creation of open habitats is more important for the success of a hybrid than 'the permissive effects of artificial hybridization by man.' Although many new hybrids occur in open habitats, such as roadcuttings, wood clearings, and glacial morains (Anderson, 1949; Grant, 1971; Briggs & Walters, 1984; Knapp, 1984), hybrids also occur in other man-made environments, such as moderately drained or fertilized grasslands (Bradshaw, 1957; Elveland, 1976; Urbanska, 1977; Kwak, 1979), and in woods (Knapp, 1984).

Our interest in hybridization arose from a nature conservation point of view. 'Hybrid like' populations have been recorded in many Dutch nature reserves, especially after the drainage of surrounding areas (Grootjans, 1980; Corporaal, 1983; Weeda, 1985). Although most of these hybrids are sterile, problems may arise because the protected parent species may be lost while the sterile hybrids may persist through vegetative reproduction, becoming more common than the parental populations (Marchant, 1967; Stace, 1975; Urbanska, 1977; Weeda, 1985).

A generally accepted view is that although hybrids may be formed in an undisturbed closed community, they do not establish succesfully. If this is true, the occurrence of hybrids are indicative of the disturbance in the original habitat. From the point of view of nature conservation, it would be very conveniant to detect disturbances by recording hybrids. We have, therefore, analyzed a number of studies on this subject. In most cases we found that the nature of the 'disturbed habitat' was not properly specified. Indeed, in

J. van Andel et al., (eds.), Disturbance in Grasslands. ISBN 90-6193-640-3.

many cases the presence of a hybridized habitat was inferred from the appearance of hybrids, and there is clearly danger in this circular argument. Discussions on 'hybridization of the habitat', in relation to 'disturbance', will become quite unproductive if the normal functioning of the system is not specified and the degree of disturbance indicated.

In this paper we will deal specifically with 'hybridized habitats' in hay meadow reserves, which have been acquired by the government or private organisations. Efforts have been made to restore the original species-rich vegetation types by infrequent mowing without fertilization. The vegetation, therefore, is in a state of recovery after heavy fertilization. The hay meadows studied are situated in the Drentsche Aa catchment area, which is one of the largest nature reserves on the Drenthian Plateau in the north Netherlands (Barkman & Westhoff, 1969).

2. Undisturbed rivervalley meadows

It is not easy to give a definition of undisturbed river valley systems, because they have all experienced large changes in hydrology and management. Most river valley meadows are man-made, and are the result of initial forest clearance and later management, such as superficial drainage, mowing, grazing, and fertilization (Ellenberg, 1978). Many types of hay meadow still harbour species that are also common in natural woods and mires, e.g. *Caltha palustris, Filipendula ulmaria, Carex aquatilis*, and *Carex lasiocarpa*. The felling of woods in the middle ages was a distinct disturbance in the ancient rivervalleys, as was the drainage of mires that evolved after that period. The constant mowing regime previously applied by the farmers, however, cannot be regarded as a disturbance, but a necessary requirement for the maintenance of the meadows (Ellenberg, 1978).

There is often a large diversity of grassland communities in river valleys, caused by wide variation in soil type and hydrology. The effects of mowing and grazing are superimposed on this abiotic variation. It is important to know which factors give rise to the many different vegetation gradients (ecoclines), despite the attempts by farmers to transform them into productive and uniform pastures. We found the largest differentiation in plant communities in areas where base-rich groundwater was discharged from the main aquifer. Interference with the hydrology (deep drainage, groundwater withdrawal) has generally reduced variation in plant communities. The oligotrophic and mesotrophic communities, in particular, have been affected (Grootjans, 1985) in both agricultural areas and in nature reserves. Brookvalleys with minor interferences with the hydrology and with well developed plant communities are, therefore, considered undisturbed in this study.

3. Hydrological conditions in ecolines

The plant communities in undisturbed river valleys reflect the long term hydrological regime of the site, which not only have direct effects, but indirectly controls the nutrient availability by specific fluctuations in the groundwater table and by the mineral composition of the groundwater (Kemmers, 1986). The hydrological regime, therefore, is an important factor controlling the performance of plant populations at given sites. The effects of hydrological conditions are particularly evident in one type of ecocline, which is found regularly in upstream river valley systems. In this area, a sequence of semi-natural plant communities exists from relic heath-vegetation (Genisto-Callunetum and *Ericetum tetralicis*) on the valley slopes, to stands of Calthion palustris communities on shallow peat layers in the central parts of the valley, and sedge communities, such as the Caricetum curto-echinatae and the Caricetum aquatilis, close to the stream (Fig. 1). This vegetation gradient is influenced by at least three different types of water flow: (i) Flooding occurs close to the stream, leading to silt deposition in the sedge vegetation; (ii) shallow subsurface water entering the valley at the upper part of the gradient; (iii) deep aquifer discharge occurring further down the valley bottom. The presence of

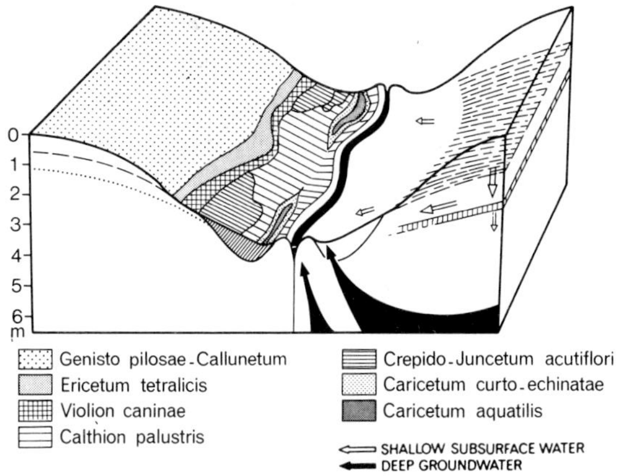

Fig. 1. Distribution of vegetation types and groundwater flows in an upstream river valley system in the northern part of The Netherlands.

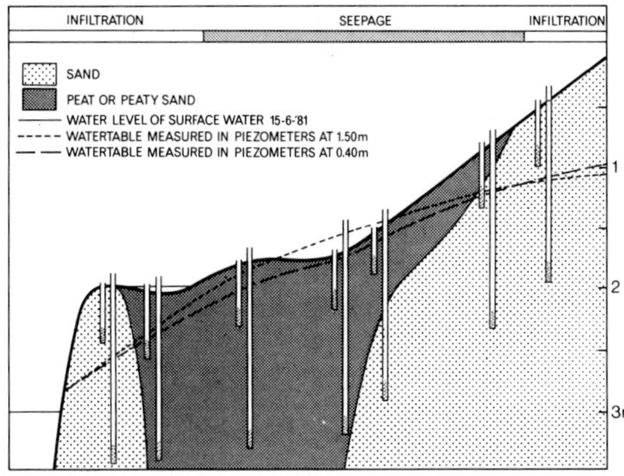

Fig. 2. Cross section of a valley gradient showing seepage and infiltration areas as determined by the water table measurements in piezometers placed at different depths.

these groundwater flows can be detected most easily in winter, when discharging subsurface flow freezes more quickly than the deep groundwater. Subsurface water is white when frozen. Deep aquifer water contains more iron, which gives a brown precipitate when exposed to the air.

Subsurface flows are sustained by local infiltration of rain water, and contain only small amounts of dissolved minerals (Table 1), since it has been in contact with leached substrata for only a relatively short time. Until recently, subsurface flows sustained very base-poor habitats on the valley slopes. By contrast, groundwater from the main aquifer has been in contact with Ca-rich sediments for a long time, and contains large amounts (Table 1) of Ca^{2+} and HCO_3^- ions, and has a high electrical conductivity (EC). We used only a small selection of macro-ions to identify different groundwater flows (see Van Wirdum, 1981); Ca^{2+} concentrations were used to detect deep groundwater and Cl^- concentrations were used to detect pollution.

The study area was a well developed ecocline with *Juncus acutiflorus, Caltha palustris*, and *Carex aquatilis*. Series of groundwater wells (piezometers) were placed at different depths across this vegetation gradient. In the central part of the gradient, the waterlevels in the deeper wells (150 cm) were higher than those in the shallow ones (40 cm), indicating seepage (Fig. 2), but close to

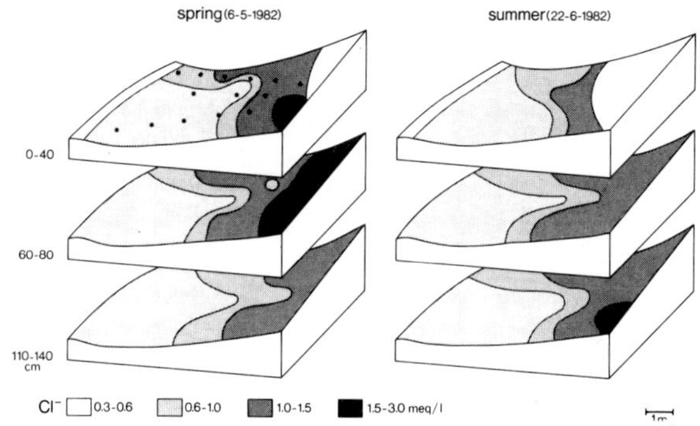

Fig. 3. Cl^--contents of groundwater at three depths in the soil profile. The high Cl-concentrations on the upper slopes are caused by pollution, probably by fertilizers or manure.

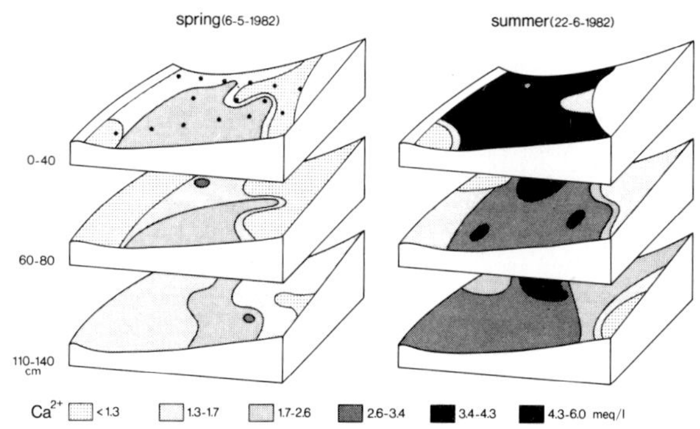

Fig. 4. Ca^{2+}-contents of groundwater at three depths in the soil profile. The Ca^{2+}-concentrations.

the stream and higher up the valley slope the reverse was found, indicating infiltration. This phenomenon was investigated in more detail by analysing water samples in the laboratory. Concentrations of Ca^{2+}, Mg^{2+}, and HCO_3^- proved to be

Table 1. The composition of rainfall (measured in weather station Witteveen), deep groundwater from the main aquifer (measured near Nietap), and subsurface water (measured on the valley slopes at 1–6 meters below the surface).

Water sample	Date	Filter depth (m)	pH	Ca^{2+} meq/l	Mg^{2+} meq/l	Na^+ meq/l	K^+ meq/l	HCO_3^- meq/l	Cl^- meq/l	SO_4^- meq/l	NO_3^- meq/l	E.C.$_{25}$ uS/cm
Rainfall (Witteveen)	1978–1982	–	4.2	0.05	0.04	0.08	0.05	0.01	0.09	0.24	0.06	62
Deep groundwater from the main aquifer (Nietap)	1976	67–130	–	3.55	0.52	0.56	0.04	4.23	0.53	0.09	–	461
Subsurface water from heathlands (Lieveren and Gasteren)	15/7/83	1–1.2	4.1	0.10	0.08	0.27	0.04	0.00	0.31	0.35	0.00	92
	17/10/83	1–1.2	5.9	0.24	0.16	0.17	0.01	0.25	0.20	0.14	0.00	66
	28/7/83	4.5–5.5	6.0	0.12	0.27	0.54	0.07	0.18	0.59	0.58	0.01	151
	21/10/83	4.5–5.5	5.5	0.30	0.31	0.41	0.09	0.15	0.42	0.42	0.00	121
Subsurface water from agricultural areas (Taarlo and Gasteren)	28/7/83	1.3–2.3	4.4	1.40	0.74	3.08	0.51	0.00	3.30	1.75	0.01	700
	21/10/83	1.3–2.3	4.6	0.60	0.39	1.68	0.23	0.00	1.61	1.77	0.01	310
	28/7/83	2.5–3.5	4.2	0.30	1.21	3.58	5.82	0.00	6.49	1.67	1.48	1870
	21/10/83	2.5–3.5	4.4	2.64	1.59	3.22	5.65	0.00	7.05	2.92	1.60	1590

good indicators of water source, but electrical conductivity was less useful. We have relied on the Ca^{2+} and Cl^- contents of the water in order to identify the different groundwater flows (Figs. 3 and 4).

In spring, Cl^- values of 1.5–3 meq/l were measured on the valley slopes; this is much higher than that (0.3–0.6 meq/l) found in natural groundwater (Moore & Bellamy, 1974; Pietsch, 1976). The shallow subsurface flow, with its low Ca^{2+} values is probably polluted by fertilizers or manure. The Cl^- concentrations at the lower end of the gradient were about the same as those found in the unpolluted groundwater from the deep aquifer (Table 1). The discharge of deep aquifer water was apparently enough to isolate most of the polluted shallow subsurface flow to the valley slope.

A similar pattern in Cl-content occurred in the summer, though the highest Cl^- contents were found somewhat deeper in the profile. However, large differences in Ca^{2+} contents were observed between the spring and summer (Fig. 4); the summer values were almost twice as high as those in spring and showed a close resemblance to deep aquifer water (Table 1). The values were especially high in the upper layer (0–40 cm). These high concentrations might be caused by evaporation. Low Ca^{2+} contents were found under the stream bank, where mineral poor infiltration water apparently accumulated.

A model of groundwater flows is presented in Fig. 5. The groundwater from the deep aquifer, which is rich in Ca^{2+}, only reaches the surface in greater amount in summer after a period of low rainfall. The upwelling deep groundwater in spring is diluted with polluted (high Cl-concentrations) subsurface water from the slopes. Dilution with water poor in Ca^{2+} also occurs close to the streambank; this water probably originates from the stream bank itself rather than from the valley slope, since it contains little Cl^-.

Increased concentrations of Cl^- in the shallow subsurface flow from the valley slopes were found on many occasions. In fact, unpolluted subsurface flows are rare in the Netherlands, and restricted to river valleys with heathlands or forests

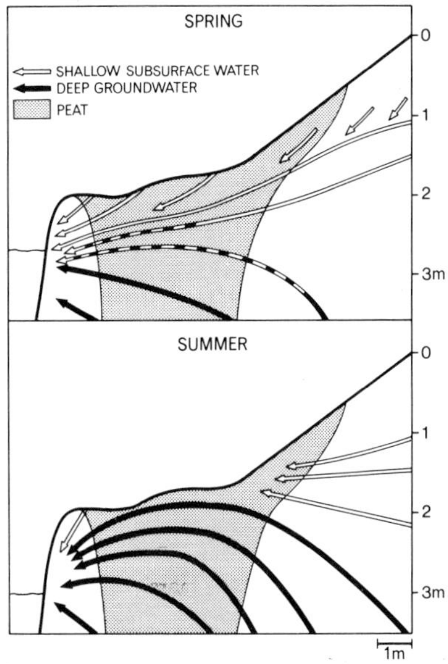

Fig. 5. Model of groundwater flows in a cross section of a small river valley. The centre of the vegetation gradient is influenced by upwelling of groundwater rich in Ca^{2+} in summer. In spring, and to a lesser extent in summer, the vegetation at the stream bank is influenced by infiltration water, which is poor in Ca^{2+}. The valley slope is influenced by a shallow subsurface flow, which is also poor in Ca^{2+}.

on the valley slopes (see e.g. Fig. 2 in Chapter 18). Table 1 shows some results of chemical analyses of groundwater sampled from valley slopes where heathlands and forests are still present, and those from slopes which are in agricultural use (pastures or tillage). The influence of fertilization is clearly expressed in the groundwater composition and its nutrient content.

4. Distribution of plant species

The distribution of some species (Fig. 6), such as *Carex aquatilis, Myosotis palustris*, and *Caltha palustris* was positively associated with areas influenced by groundwater rich in Ca^{2+}; others species, such as *Viola palustris* and *Juncus acutiflorus*, were negatively associated with these areas (Table 2).

The hybrid *Juncus acutiflorus x articulatus* was present in the part of the gradient where deep groundwater and polluted subsurface water mixed, i.e. in a 'hybridized habitat' created by

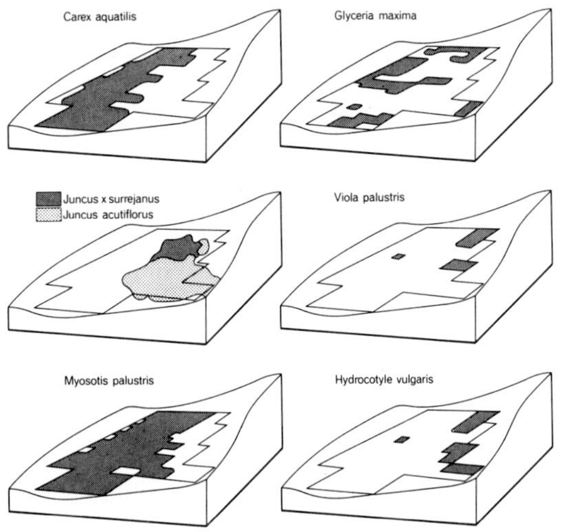

Fig. 6. The distribution of some marsh and meadow species in a river valley.

human disturbances (Anderson, 1948). We should not, however, draw a firm conclusion about hybrids from this single case, since natural hybrids can also be found in intermediate habitats which are not affected by disturbances (Grant, 1971).

5. Habitats of *Juncus* species

In the Drentse Aa catchment area, *Juncus acutiflorus x articulatus* is almost as common as its parent *Juncus acutiflorus*. In contrast, *Juncus articulatus* is quite rare in grasslands, but it is regularly found in open 'muddy' places. In the present study, *Juncus articulatus* was found predominantly in very wet eutrophic vegetation, in association with *Agrostis stolonifera, Carex nigra*, and *Alopecurus geniculatus* (Table 3, group D). By contrast, *Juncus acutiflorus* occurred predominantly in relatively dry and less eutrophic vegetation, with *Agrostis capillaris, Viola palustris*, and *Agrostis canina* (group A), though it was also

Table 2. The relationships between plant species distribution and occurence of Ca^{2+}-rich groundwater, as determined by Chi-square test. +,– = P<0.1; ++,– – = P<0.05; +++,– – – = P<0.001; n.s. = not significant.

N	Species	Spring (6–5) 0–40 cm	Spring (6–5) 60–80	Summer (22–6) 0–40	Summer (22–6) 60–80
90	*Caltha palustris*	+++	+++	+++	+++
57	*Carex aquatilis*	+	++	+++	+++
78	*Myosotis palustris*	+	++	+++	+++
39	*Juncus acutiflorus*	–	–	– – –	n.s.
18	*Viola palustris*	– – –	– – –	n.s.	– – –
10	*Hydrocotyle vulgaris*	n.s.	n.s.	n.s.	n.s.
34	*Glyceria maxima*	n.s.	n.s.	n.s.	n.s.
36	*Juncus x surrejanus*	n.s.	n.s.	n.s.	n.s.

Table 3. Synoptic table of relevees with *Juncus* species from the Drentse Aa catchment area. Clusters (A, B, C, and D) were obtained using a simple centroid clustering. Only differentiating species are shown. I-V = frequency in cluster group; I = 0–20%, V = 81–100%. Superscrypts represent mean cover (%).

Group number	A	B	C	D
Number of relevées	27	7	51	7
Juncus acutiflorus	V^{20}	V^{8}	II^{13}	·
Juncus acutiflorus x articulatus	·	II^{2}	III^{16}	·
Juncus articulatus	·	·	I^{10}	V^{4}
Viola palustris	III^{8}	I^{1}	I^{2}	
Agrostis canina	III^{8}		I^{1}	
Agrostis capillaris	III^{18}	·	I^{5}	I^{1}
Caltha palustris	II^{5}	III^{2}	IV^{4}	III^{8}
Lychnis flos-cuculi	II^{3}	V^{6}	IV^{2}	
Mentha aquatica	I^{1}	V^{4}	·	·
Phragmites australis	·	V^{4}	II^{2}	·
Filipendula ulmaria	II^{6}	IV^{7}	·	II^{2}
Valeriana officinalis	·	III^{3}		
Galium uliginosum	I^{1}	III^{2}	I^{1}	·
Vicia cracca	I^{6}	III^{11}	·	I^{1}
Cardamine pratensis	II^{1}	II^{1}	IV^{1}	V^{1}
Trifolium repens	I^{2}	·	V^{7}	III^{1}
Festuca pratensis	I^{1}	·	IV^{2}	III^{4}
Carex nigra	II^{3}	I^{10}	III^{5}	III^{14}
Cerastium fontanum	II^{1}	I^{1}	V^{1}	II^{1}
Rhinanthus angustifolius	I^{2}	II^{2}	IV^{5}	·
Plantago lanceolata	II^{2}	·	IV^{6}	
Trifolium pratense	·	·	III^{6}	·
Glyceria maxima	I^{1}	·	III^{2}	·
Myosotis palustris	·	I^{1}	III^{1}	I^{1}
Agrostis stolonifera	II^{3}	I^{20}	II^{4}	V^{15}
Alopecurus geniculatus	·	·	I^{3}	III^{2}
Ranunculus flammula	I^{1}	·	I^{3}	III^{2}
Juncus effusus	II^{2}	II^{3}	II^{1}	III^{18}
Leontodon autumnalis	·	·	II^{1}	III^{4}
Eleocharis palustris			II^{1}	III^{1}

found in wet Calthion palustris meadows, associated with tall herbs, such as *Filipendula ulmaria, Valeriana officinalis*, and *Phragmites australis* (group B). The hybrid was most common in wet eutrophic Calthion palustris meadows (group C), in which both parental species are less frequent, and which was intermediate between those of the parental species. The presence of hybrid populations in the lower part of the gradient, with Calthion palustris species such as *Caltha palustris,* and *Myosotis palustris* (Fig. 6), appears to be a regular phenomenon in this type of ecocline.

Juncus acutiflorus x articulatus was most abundant at places where polluted shallow groundwater meets with a base-rich groundwater flow from the main aquifer, thus creating a new series of habitats, in which the hybrid finds opportunities to survive. By contrast, *Juncus acutiflorus* occurs in habitats that probably will become increasingly unfavourable. The original habitat, at the upper part of the gradient, was previously very poor in nutrients. Now, however, these sites have been heavily fertilized, and the relatively fast moving shallow subsurface flow transports nutrients from the hinterland in increasing quantities. Furthermore, *Juncus acutiflorus* is a late flowering species, and in meadows it is mown before seed setting (Allersma, 1979).

Although *Juncus articulatus* is not very common in our study area, it is not an endangered species; it prefers eutrophic habitats and it is able to set seed both before and after mowing (Balátová-Tuláčková, 1971).

Another explanation for the occurrence of the hybrid is, however, that the hay meadows in the Drentse Aa valley are recovering after previous fertilization, thus creating new opportunities for species that have survived in relics of former fen meadow and mire vegetation as 'peripheral populations in ecologically marginal habitats which are subjected to extinction or near extinction' (Lewis, 1962). The populations that are present during the first few years after the new mowing regime, show rapid fluctations in numbers (Chapter 18). Lewis (1962) concluded that 'near extinction followed by regeneration of the population from a few genotypically aberrant survivers, sometimes leads directly to the rapid formation of deviant daughter species'. Furthermore, the species diversity is also increased by the new mowing regime. This alone may create more possibilities for closely related species to form hybrid offspring. The occurrence of hybrids (or other 'deviant species') in nature reserves which are influenced by drainage, could be a temporary success; a part of a changing pattern that has taken place over the course of some decades.

In hydrologically undisturbed river valleys, large differences in soil pH and nutrient availability have always occurred, and have resulted in various ecoclines. Most of this variation was caused by different proportions of deep and shallow groundwater. In seepage areas, base-rich habitats remained as long as the deep groundwater was able to reach the top soil. Superficial drainage applied by the farmers sustained these types of habitats, and accumulation of precipitation water was prevented (Wind, 1986). In most of our newly installed nature reserves, both the subsurface flow and the deep groundwater flow have been affected by man; subsurface flows have been affected by pollution, and deep groundwater flows have been affected by deep drainage and groundwater withdrawal. Decreased discharge of base-rich groundwater leads to increased mineralization of the peat, and to increased infiltration of precipitation in the top soil (Grootjans, 1985; Kemmers, 1986). If shallow groundwater flows are present, their influence would increase in the whole ecocline; the final result would be moist and less alkaline conditions at the upper end of the gradient, and dry, more acid, conditions at the lower end. Pollution of the subsurface flow can create additional changes in almost the whole ecocline.

These various interferences with the hydrology of the river valley lead to changes in the environmental conditions, that are either rare or nonexistant in undisturbed river valleys. These changes can occur over considerable periods: for example, 'pedogenesis' in peat soil may take several decades after drainage (Schothorst, 1977), leading to increased fluxes of nitrogen and phosphorus (Bolt & Bruggenwert, 1976; Kemmers,

1986). The changes will be greatest in the low lying part of the ecocline, which was previously under the influence of baserich groundwater; the nutrient supplies are higher here, and the shift in soil pH is more pronounced after drainage (Grootjans, 1985).

6. Hybrids in 'hybridized' habitats

The new environmental conditions can be considered 'hybridized' habitats, which Anderson (1949) described as 'not only intermediate, but presenting all possible combinations of the contrasting differences of the original habitats'. In our heavily drained nature reserves with peaty soils, these 'hybridized habitats' persist for a long time, giving conditions very favourable for the establisment of hybrid offsprong. Hybrids or hybrid swarms may be able to persist for a long time by vegetative propagation or by polyploidization (Grant, 1971). For example, Bradshaw (1957) found a sterile hybrid between *Agrostis capillaris* and *A. stolonifera* which persisted by vegetative propagation, and seemed to be better adapted to a mowing or grazing regime than either of its parents. Similarly, Urbanska (1977) found a sterile triploid hybrid between *Cardamine amara* and *C. rivularis* which persisted by runners. She also found a fertile hexaploid, which she regarded as a new species *(Cardamine schulzii)*. Other species of fertile hay meadows, which are thought to have a hybrid origin are: *Cardamine pratensis* (Landolt & Grossman, 1968), *Dactylis glomerata* (Grant, 1971) and *Juncus articulatus* (Zandee, 1981). The present regime of hay making without ferrtilization in hydrologically disturbed river valleys does succeed in increasing the species diversity, but at the same time it appears to provide suitable niches for establishment of hybrids.

Using Stace (1975) we have, therefore, compiled a list of possible hybrids between species from different communities in the river valleys of the Drentse Aa and adjoining areas. A sequence of plant species, based on a trophic gradient from oligotrophic to eutrophic, has been derived from proposals by Ellenberg (1978), Dierssen (1982), and Succow (1982). The fertilized hay meadows have been further ranked according to soil water characteristics (Ellenberg, 1978). The species shown represent the first stages of succession in former agricultural grasslands subjected to mowing without fertilization (see Chapter 12). All species can be found in a relatively small area of 100 km^2; many are present in large numbers, though others must be regarded as endangered species, even in nature reserves. In these lists of species there are many closely related species able to form hybrids (Stace, 1975), and at least a dozen hybrids have already been found, with some very common, e.g. *Mentha aquatica x arvensis, Carex nigra x elata, Juncus acutiflorus x articulatus* (Allersma, 1979), *Equisetum fluviatile x arvense, Agrostis stolonifera x capillaris,* and *Rhinanthus angustifolius x minor* (Kwak, 1979). In some drained fen meadows, hybrids or hybrid swarms are more common than their parents, e.g. *Carex hostiana x tumidicarpa,* and *Agrostis stolonifera x capillaris.* (Fig. 7).

The status of hybrids and hybrid swarms in species conservation is unclear. Hybrids may occur as parent species experience increasingly unfavourable conditions, and so may be indicators of species decline. On the other hand, hybridization could help to expand the genetic composition of endangered plant populations, so allowing it to expand and occupy newly created habitats.

Certainly, the paucity of genetic variation within many relic populations may retard their evolutionary expansion (Dobzhansky et al., 1963), while linkage may further retard their ability to respond to selection (Mather, 1953; Falconer, 1981; Hegmann & Dingle, 1982). Hybridization could increase the genetic variation in such populations by introducing new genetic variation or by breaking old linkages. Distribution maps of hybrids and their parental species, presented by Knapp (1984), could indicate that such mechanisms are effective. Knapp found that *Salix alba x fragilis* and *Medicago falcata x sativa* were distributed over a wider area in the Saar region than either parents. On several occasions *Luzula forsteri*

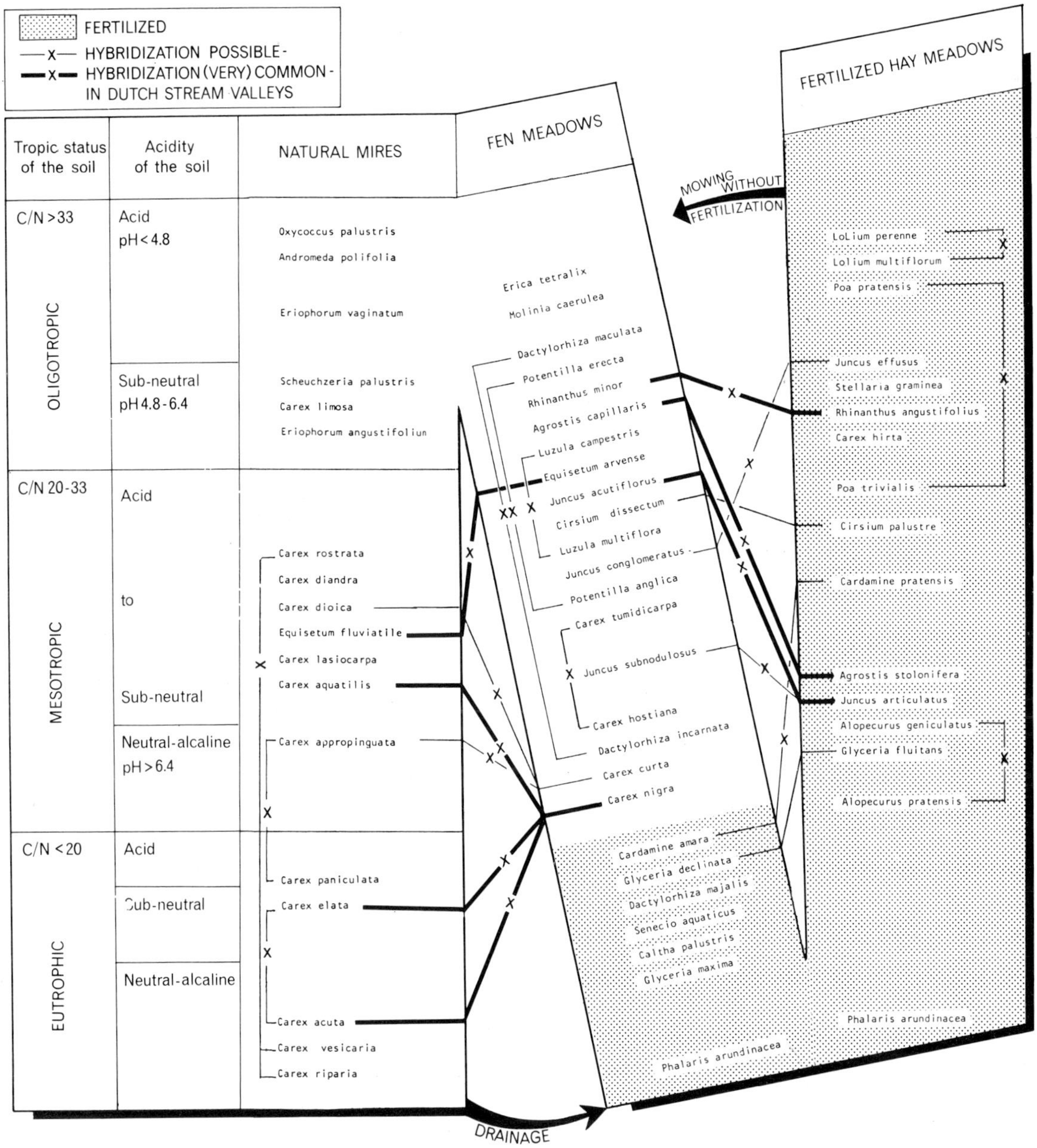

Fig. 7. Possible hybrid formation (according to Stace, 1975) between species from different succession series in a gradient from oligotrophic to eutrophic. Only treeless vegetation types are considered.

forsteri x pilosa was also found outside the distribution areas of the parents.

7. Implications for nature management

The effects of deep drainage on nature reserves can be counterbalanced either by raising the water table, which is sometimes difficult to achieve, or by removing the degrading top soil. The latter strategy can reduce the persistance of endangered species and their hybrids, especially when they do not have a seed bank or propagate exclusively vegetatively. Another technique is to try to artificially increase the species density by sowing or transplanting; the effects of this, depends highly upon the genetic structure and the breeding systems of the populations involved (Frankel & Soule, 1981). Reintroduction of identical populations is not likely to be successful, especially if the environment has been drasticially modified. Recolonization by sowing or transplantation of non-local populations may not prove immediately successful, but natural selection might produce more adapted populations. Even artifically produced hybrids could be used for this purpose. This is, at first glance, not as strange as one may expect, because hybrid swarms are also formed in nature. In some cases the formation of populations or species (compilo species; see Harlan & De Wet, 1963) with new gene complexes occurs, which are better adapted to disturbed habitats, than either parent population or species.

Species with complex and highly diverse breeding systems (often annuals, Grant, 1975) must be treated with caution, especially if they consist of small and ecologically isolated populations with rare exchange of pollen and seeds (James, 1982). In these populations, in particular, interpopulation (or species) gene flow could destabilize the genetic constitution (Mather 1953), which could have marked effects on survival and fecundity. Sowing and transplanting in this sort of populations could, therefore, be harmful.

In conclusion, the input of genes from other populations or the artificial or natural formation of hybrids, may sometimes counterbalance the influences of genetic drift and natural selection. There is, however, a great lack of appropriate data on which to base strategies to preserve endangered species.

References

Allersma, G.J.R. (1979). *Juncus* soorten in het stroomdal van de Drentsc Aa. Danseria 16: 3–5.

Anderson, E. (1948). Hybridization of the habitat. Evolution 2: 1–9.

Anderson, E. (1949). Introgressive hybridization. New York, Wiley.

Barkman, J.J. & Westhoff, V. (1969). Botanical evaluation of the Drenthian District. Vegetatio 19: 330–388.

Bolt, G.H. & Bruggenwert, M.G. (1976). Soil chemistry. A. Basic elements. Elsevier Amsterdam 281 pp.

Bradshaw, A.D. (1957). Natural hybridization of *Agrostis tenuis* Sibth. and *Agrostis stolonifera* L.

Briggs, D. & Walters, S.M. (1984). Plant variation and evolution (2nd edition) Cambridge Univ. Press. 412 pp.

Corporaal, A. (1983). Nieuwe *Carex*-taxa voor Nederland. Gorteria 11: 219–226.

Dierssen, K. (1982). Die wichtigsten Pflanzengesellschaften der Moore N.W. Europas. Conservatoire et jardin botaniques, Geneve, 382 pp.

Dobzhansky, T., Hunter, A.S., Pavlovski, O. Spasski, B. & Wallace, B. (1963). Genetics of natural populations XXXI: Genetics of an isolated population of *Drosophila pseudobscura*, Genetic 48: 91–105.

Ellenberg, H. (1978). Vegetation Mitteleuropas mit den Alpen. Ulmer Verlag.

Falconer, D.S. (1981). Introduction to Quantitative Genetics (2nd. ed.) Longman, 365 pp.

Frankel, O.H. & Soule, M.E. (1981). Conservation and Evolution. London University Press.

Grant, V. (1971). Plant Speciation 1rst Ed. N.York, Columbia Univ. Press.

Grant, V. (1975). Genetics of flowering plants. Columbia Univ. Press. N.Y. and London.

Grootjans, A.P. (1980). Distribution of plant communities along rivulets in relation to hydrology and management. In: Wilmanns O. & R. Tüxen: Epharmony. Berichte uber die internationalen Symposien der I.V.V., 1979, p 143–170. Cramer Verlag.

Grootjans, A.P. (1985). Changes of groundwater regime in wet meadows. Thesis University of Groningen.

Harlan J.R. & de Wet, J.M.J. (1963). The compilo species concept. Evolution 17: 497–501.

Hegmann, J.P. & Dingle, H. (1982). Phenotypic and genetic covariance structure in Milkweed bug life history thraits. In: H. Dingle & J.P. Hegman (Eds.), Evolution and genetics of life histories: 177–185, Springer Verlag.

Kemmers, R.H. (1986). Calcium as hydrochemical characteristic for ecological states. Proc. 7th Int. Symp. on Problems of Landscape Ecol. Research, Bratislawa, 21—26 Oct. 1985.

Knapp, R. (1984). Perspectives on species differentiation and hybridization in vegetation analysis and in phytosociological releves. In: Knapp, R. (Ed.) Sampling methods and taxon analysis in vegetation science 23–24. Junk, The Hague.

Kwak, M.M. (1979). The role of bumblebees in the pollination and variation of some *Rhinanthoideae (Scrophilariaceae)*. Thesis University of Groningen, 114 pp.

Landolt, E. & Grossmann, F. (1968). Zur vermutlich hybridogenen Entstehung von einigen Wiesenpflanzen des schweizerischen Mittellandes. Verh. Schweiz. naturf. Ges. 1968: 114–117.

Lewis, H. (1962). Catastrophic selection as a factor in speciation. Evolution 16: 257–271.

Marchant, C.J. (1967). Evolution in *Spartina (Gramineae)* 1. The history and morphology of the genus in Britain. Journal of the Linnean Society (Botany) 60: 1–24.

Mather, K. (1953). The genetical structure of populations. Symp. Soc. Exp. Biol. 7: 66–95.

Moore, P.D. & Bellamy, D.J. (1974). The Peatlands. Springer Verlag, London-New York.

Pietsch, W. (1976). On the relation between the vegetation and the absolute and relative ion content of mire waters in Middle Europe. Proc. 5th. Int. Peat Congres, Vol. 2. Poznan, Poland, Sept. 21–25, p 62–72.

Schothorst, C.J. (1977). Subsidence of low moor peat soils in the Western Netherlands. Geoderma 17: 265–291.

Stace, C.A. (1975). Hybridization and the flora of the British Isles, Ac. Press, London.

Streefkerk, A.J. (1985). Hydrologische ingrepen in het stroomgebied van de Drentse Aa en de gevolgen voor het landschapsreservaat 'Stroomdallandschap Drentse Aa'. Report SBB, Utrecht, 52 pp.

Succow, M. (1971). Die Talmoore des Nordostdeutschen Flachlandes, ein Beitrag zur Characterisierung des Moortypes Niedermoor. Arch. Naturschutz. u. Landschaftsforch. 11, 133–168.

Succow, M. (1982). Topische und chorische Naturraumtypen der Moore. In: Kopp et al. Naturraumliche Grundlagen der Landnutzung. Akad. Verlag, Berlin.

Urbanska, K.M. (1977). Reproduction in natural triploid hybrids (2n=24) between *Cardamine rivularis* Schur and *C. amara* L. Ber. Geobot. Inst. ETH Stiftung Rubel 44, 42–85.

Weeda, E.J. (1985). Veranderingen in het voorkomen van vaatplanten in Nederland. In: Mennema e.a. Atlas van de Nederlandse flora 2. 10–47, Bohn, Scheltema & Holkema, Utrecht.

Wind, G.P. (1986). Slootpeilverlaging en grondwaterstandsdaling in veenweidegebieden. Cultuurtechnisch Tijdschrift: 321–330.

Wirdum, van. G. (1981). Linking up the natec subsystem in models for the watermanagement. Committee for Hydrological Research TNO. Proc. & Inf. 27: 108–128.

Zandee, M. (1981). Contribution to the biosystematics of two cyanogenic polyploid complexes (*Juncus articulatus* and *Holcus mollus*). Thesis, University of Leiden, 154 pp.

CHAPTER 6

Disturbance in transplanted grasslands and wetlands

F. Klötzli

Detailed contents

1. Introduction

Attempts have been made to transplant grassland communities in steppe (Skripchinsky et al., 1975) and prairie areas (e.g. Curtis prairie, Cottam & Kline pers.comm.), but transplantation of wetland communities had so far never succeeded on a greater scale. In this study, we have tried to transplant communities of high biological value in the vicinity of Zürich-Airport. The area which contained a number of very rare plant communities and species, would have been destroyed or heavily disturbed if it was not moved to a safe area.

The airport lies on a flat alluvial plain, forming the ancient bed of a post-glacial shallow lake dotted by small morainic hills. Because of these variable conditions a great diversity of wet and dry sites had developed; these sites had been further modified by man, creating an intense mosaic of open surfaces, meadows, fertilized and unfertilized, fields, peathags, interspersed by semi-natural woodlands. Some types of straw-meadows and mires occurred only in this area on a greater scale. Consequently, species diversity was high with 500 species in about 10 km^2; about 10% of these species could be considered rare or very rare (Klötzli, 1969, 1975, 1981; Klötzli & Keel, 1976).

'Transplantation' of at least some of the more important wetland communities was undertaken by the department of puplic constructions of the Canton Zürich in collaboration with planning authorities, engineering firms and scientists.

This paper describes and evaluates methods and results of community transplantations, with particular reference to disturbance, recovery, and implications for nature management. The mainly qualitative analysis, presented here, is based on a lot of quantitative information, partly published previously (e.g. Klötzli 1980, 1981).

J. van Andel et al., (eds.), Disturbance in Grasslands. ISBN 90-6193-640-3.

2. Description of transplanted vegetation

2.1 Molinion, straw meadows s.str.

Straw meadows are part of the anthropogeneous grassland growing on soils influenced by groundwater or waterlogging and lack of nutrients (especielly N,P), which are usually cut in autumn and only once a year. They all show one common combination of species groups, which are listed in Table 1 (i.e. groups I, G, H, K, Z, Ω acc. to Klötzli, 1969, see also for references and more details on their ecology). If they are cut too early then low growing hemicryptophytes (rosette plants) and quickly regenerating meadow plants spread to the detriment of true straw meadow species. To the contrary tall massive plants ('macrophorbs') are sponsored if cutting is neglected and nutrients are accumulating at the surface of the soil (development of Filipendulion meadows, groups Q, R). Transformation into hay meadows (Arrhenatherion, Calthion resp. on wetter sites) is possible in one to two years by more mowing and fertilising at least twice a year. Consequently, Molinia meadows show a rather sensitive combination of plants which is quickly altered if management is not appropriate.

Normally, *Molinia* straw meadows on base rich subsoil are prevailing (Stachyo-Molinietum). Their typical species are found in groups H, I′, I″, J, K, indicating intermittently wet or dry base rich site consitions (pH 5–10 cm below surface usually around 6.0 to 7.5). In flowering seasons the following species may be conspicuous: *Molinia coerulea* and *Selinum carvifolia*, on wetter sites also *Gentiana pneumonanthe, Parnassia palustris, Epipactis palustris* (St.-Mol.caricetosum hostianae, incorporating groups K, L, M, N), on drier sites also *Silaum silaus, Betonica officinalis, Chrysanthemum leucanthemum, Succisa pratensis* (St.-Mol.caricetosum tomentosae, incorporating groups C, D, E, G). All details on groundwater conditions may be taken from Table 2a.

Table 1. Species groups in straw meadows and fens of alluvial plains.

Group	Species	Group	Species
A	*Salvia pratensis*	E	*Plantago lanceolata*
	Pimpinella saxifraga		*Hypericum perforatum*
	Medicago lupulina		*Primula veris*
	Trisetum flavescens		*Luzula campestris*
	Arrhenatherum elatius		*Rhytidiadelphus triquetrus*
A′	*Daucus carota*	E′	*Koeleria cristata*
	Cerastium caespitosum		*Dactylis glomerata*
B	*Knautia arvensis*		*Trifolium pratense*
	Plantago media		*Senecio erucifolius*
	Galium pumilum		*Silaum silaus*
C	*Carex caryophyllea*		*Colchicum autumnale*
	Brachypodium pinnatum (>1)		*Viola hirta*
	Hippocrepis comosa		*Chrysanthemum leucanthemum*
	Sanguisorba minor	F	*Calamagrostis epigeios* (>1)
	Campanula glomerata	G	*Carex flacca*
	Rhinanthus minor		*Festuca ovina*
	Veronica chamaedrys		*Trifolium montanum*
	Carex tomentosa		*Rhytidiadelphus squarrosus*
C′	*Orchis morio*	H	*Stachys officinalis*
	Equisetum arvense		*Centaurea jacea*
D	*Bromus erectus*		*Prunella vulgaris*
	Avena pubescens		*Linum catharticum*
	Thymus serpyllum[1]		*Ranunculus nemorosus*
	Polygala vulgaris		*Briza media*
	Scabiosa columbaria		*Polygala amarella*
	Achilea millefolium		*Ctenidium molluscum*

	Thuidium philibertii[2]
I	*Festuca rubra*
	Holcus lanatus
	Anthoxanthum odoratum
I′	*Galium mollugo*
	Galium verum
	Ajuga reptans
I″	*Festuca arundinacea*
	Vicia cracca
	Lotus corniculatus
	Scleropodium purum
J	*Carex pulicaris*
	Orchis latifolia
	Serratula tinctoria
K	*Deschampsia caespitosa*
	Cirsium oleraceum
	Angelica silvestris
	Climacium dendroides
K′	*Gymnadenia conopea*
	Mnium seligeri[3]
L	*Fissidens adiantoides*
	Carex hostiana (>1)
	Carex davalliana
	Carex lepidocarpa
	Epipactis palustris
	Orchis incarnata
M	*Lythrum salicaria*
	Galium palustre
	Cirsium palustre
	Lotus uliginosus
N	*Chrysohypnum stellatum*[4]
	Drepanocladus revolvens
	Juncus alpinus
	Gentiana pneumonanthe
Q	*Geranium palustre*
	Symphytum officinale
	Convolculus sepium
	Ranunculus repens
	Primula elatior
R	*Carex acutiformis*
	Carex gracilis
	Thalictrum flavum
	Carex riparia
	Lathyrus palustris (>+)
S	*Juncus subnodulosus*
	Schoenus ferrugineus
S′	*Parnassia palustris*
	Schoenus nigricans
	Dicranum scoparium paludosum[7]
	Primula farinosa
	Tofieldia calyculata
	Aneura pinguis
T	*Pinguicula vulgaris*
	Drosera anglica
	Liparis loeselii
U	*Rhynchospora alba*
	Sphagnum contortum[5]
	Sphagnum cymbifolium
V	*Carex elata*
	Carex lasiocarpa
	Peucedanum palustre
	Hydrocotyle vulgaris
	Scorpidium scorpioides
V′	*Iris pseudacorus*
	Senecio paludosus
	Carex appropinquata
	Caltha palustris
	Epilobium palustre[6]
	Pedicularis palustris
	Utricularia neglecta
	Utricularia minor
W	*Eriophorum latifolium*
	Eriophorum angustifolium
	Carex oederi
W′	*Menyanthes trifoliata*
	Ranunculus flammula
	Equisetum fluviatile
X	*Carex fusca (>2)*
	Carex rostrata
	Carex diandra
	Carex vesicaria
	Comarum palustre
Y	*Solidago gigantea var. leiophylla*
	Cirsium arvense
	Rubus caesius
Z	*Molinia coerulea*
	Carex panicea
	Acrocladium cuspidatum
	Succisa pratensis
	Galium uliginosum
	Potentilla erecta
	Lathyrus pratensis
Ω	*Filipendula ulmaria*
	Lysimachia vulgaris
	Eupatorium cannabinum
	Agrostis alba
	Mentha aquatica
	Equisetum palustre
	Selinum carvifolia
	Valeriana dioeca
	Phragmites communis

---

[1] partly *Th. pulegïoides*
[2] partly *Th. delicatulum*
[3] *Mnium affine* Group
[4] syn. *Campylium* (partly *C. proteusum)*
[5,6] & related species
[7] partly *D. benjeani* + D. *bergeri*

Table 2a. Duration line group, soil type and plant community (from Klötzli, 1969).

DL-group	groundwater table, median cm	max. yearly fluctuation cm	max. weekly fluctuation cm	groundwater table, median growing season cm	inundation weeks	plant community (most important)	soil type	typical DL
2	100-140	>170	r. .13100->143 s. 33- 59	95-120	–	Stachyo-Brometum typicum	Cambisol-(calcic) Gleysol (eutric), Fluvisol (gley spots from 35 cm under the surface	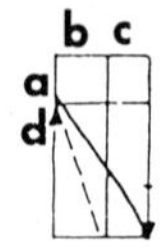
3	75-90	–>150 (-158)	St.>75-114 Si. 37- 48	70-90	–	Stachyo-Molinietum caricetosum tomentosae	Gleysol, Cambisol)	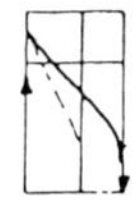
4	50-65	–>150 120	aSt.>88,Si.31 bSt.>35,Si.26	50-65	–	[a])Stachyo-Molinietum typicum	a loam b sand	
5	25-40	[a])132- >150	St. 47-82 Si. 21-38	30-40	some days	[a])Stachyo-Molinietum caricetosum hostianae	wet Gleysol	
		[b])120-132	St. 67-81 Si. 21-36		-3	[c])Stachyo-Molinietum juncetosum and Schoenetosum, Valeriano-Filipenduletum, moister types	wet Gleysol (partly dystric) or 'Anmoor'	
		[c])116	St.30, Si.40			[d])Primulo-Schoeneteum stachyetosum	'Anmoor'	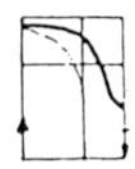
6	15-25	89	St./Si.20	7-25	13(9)	Carici-Agrostietum	Histosol (partly dystric)	
7	65	St.30, Si.12			7½(3)	Caricetum diandrae	Histosol on Gyttja	
	40	St.8, Si.7			9(½)	Schoenetum wet types	Gleysol, very wet	

Legend DL year; —— DL, growing season; r (St.) raising water table; s (Si.) sinking water table; [a] 50 cm; [b] 28 weeks (growing season); [c] 52 weeks; [d] max. weekly fluctuations (arrows).

2.2 Mesobromion, semi-dry brome meadows (partly managed as straw meadows)

The driest flat and small elevations (around 50 cm higher than Molinion) in alluvial plains carrying straw meadows may be dominated by (now exceedingly rare) brome meadows, as a rule in the 20th century replaced by Arrhenatherion or cultivated fields.

Conspicuous are species of the alliances Mesobromion and Molinion, but also Molinietalia species. *Bromus erectus* is dominant, often accompanied by *Betonica officinalis* and *Chrysanthemum leucanthemum* in summer. But also many dryness indicators are to be seen (groups A to G): amongst others e.g. *Salvia pratensis, Daucus carota, Pimpinella saxifraga, Plantago media, Hippocrepis comosa, Carex caryophyllea* (= *verna*). (More on plantsociological peculiarities and ecological characteristics in Klötzli, 1969).

2.3 Caricion davallianae (= Eriophorion latifolii), basiphilous small sedge meadows

Small sedge meadows are normally found on very wet gley soils – in spring fens and the like –,

Table 2b. Duration line group and species groups (from Klötzli, 1969).

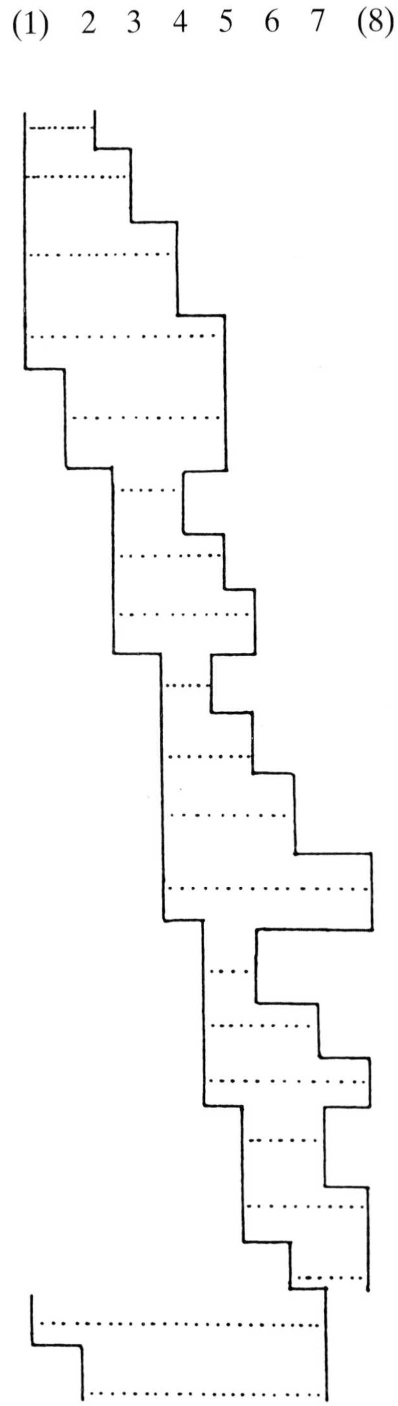

Indicators for	Species groups
Dryness	A B
Dryness	C D
Drynes(*)	E E'
Partly dryness(*)	
Common meadow species	G H I-I''
Dryness(*)	E'''
Partly dryness(*)	F O Y
Humidity(*)	J P
Humidity(*)	K
Partly humidity(*)	O'
Humidity(*)	K'
Wetness(*)	L
Wetness(*)	M
Humidity	Q
Wetness (spring)(*)	R S
Wetness	N V
Wetness (spring)	S'
Wetness (extreme)	U W
Wetness (extreme)	T V' W' X
Humidity(*)	Z
Humidity(*)	Ω

* intermittent

where groundwater is influencing the root horizons permanently and which may be flooded for a couple of weeks even in the growing season. These 'parvocariceta' are generally rich in species of the alliance Caricion davallianae and of the higher hierarchic units Tofieldietalia and Scheuchzerio-Caricetea fuscae (groups L, N, S, S′, T). Physiognomy is accentuated by the name-giving species of the associations (e.g. Primulo-Schoenetum, Valeriano-Caricetum davallianae) *Schoenus nigricans* and *Schoenus ferrugineus, Carex davalliana* and *Carex hostiana,* and also *Primula farinosa, Parnassia palustris, Eriophorum latifolium*.

In flat alluvial plains *Schoenus* meadows (also managed as straw meadows) are only found in the area of former or existing oligotrophic base rich lakes, partly on sandy fen peat, partly on lake marl, Anmoor or other base rich gleyic soils. Wetter types are differentiated by groups, T, U, V, W; the driest type is floristically very near to *Molinia* meadows, but has no species of groups C, D, E, E′. On the other hand *Schoenus* meadows may border fens dominated by *Carex elata*, especially under the influence of spring water coming from adjacent slopes.

2.4 Caricion (canescenti-) fuscae, acidophilous small sedge meadows

Plant communities on acid fen peat, lying in depressions of alluvial plains with permanent high groundwater level in peat gyttja or peat anmoor are generally dominated by *Carex fusca*, They mostly form complexes with other acidophilous fens with e.g. dominating *Carex rostrata* as a pioneer stage (Caricion rostratae sensu Balátová-Tulačková, cited in Klötzli, 1969) or then dominating *Carex diandra* (Caricetum diandrae sensu Jonas, 1933, lit.cit.) showing a similar basic composition as the slightly drier *Carex fusca* fens (Carici-Agrostietum, medium groundwater table only 10 cm higher). All three plant communities may have the common dominance of *Comarum palustre* and sometimes of *Menyanthes trifoliata* (groups W, W′, X).

The driest parts in such a complex may already show a considerable amount of species of Molinion and Molinietalia, but usually the Caricion fuscae species prevail. The *Carex diandra* fen is often flooded (as also the *Carex rostrata* fen on less thick peat, often bordering this complex) and growing on 80–350 cm of peat and gyttja, having developed of *Carex rostrata* stages. The general hydrarch succession, also to be seen in the transplanted fen, is going from *Menyanthes/Carex rostrata* quaking mats, often flooded by 20 cm of water, to *Carex diandra* and sometimes *Carex appropinquata* fens and finally to the *Carex fusca/Agrostis canina* fen with transitions to acid *Molinia* meadows.

2.5 Surrounding vegetation

Plant communities surrounding the transplants are chiefly of the same types as the former vegetation taken away before the process of transplantation. As a rule these places were chosen in herds of *Solidago serotina* and/or *Calamagrostis epigeios*, or in *Phalaris* herds (special groups F, Y), in order not to destroy any other valuable surface with straw meadows. Only the *Carex diandra*-complex was placed in the basin of a former *Carex elata* fen surrounded by oligotrophic meadows of different types (Belonging to Molinion, Filipendulion and also Mesobromion).

3. Methods of transplantation

3.1 Historical background

Previous attempts have been made to save wetlands by manipulating groundwater tables. The first experiment to transplant a fen was made in 1968. The procedure was quite primitive: the fen basin was enlarged and the tussocks of the sedges (Caricetum paniculatae, a rare fen type in Central Europe) were put on to the newly prepared site, where the water table was kept at a suitable hight.

3.2 Preparations

Before undertaking any transplantation, the decisive site factors must be assessed and the conditions of the old and new site matched. The following conditions are likely to be most important:

(a) *water regime* (Table 2a, b)
  (i) mean depth & fluctuations of the groundwater
  (ii) duration of inundations in the main root horizon
  (iii) soil texture within the range of water table fluctuations and in the main root horizon, if intermittently dry

(b) *nutrient regime*
  (i) concentration of nutrients, e.g. available N, P within the rooting zone
  (ii) base status, e.g. pH and exchangeable cations
  (iii) buffering capacity of peat

The main problems relate to transport and weight of the soil profiles to be transported. The 'transplantation units' should be as large as possible to maintain homogeneity. In most cases the turf blocks were transported on pallets (90 × 130 cm). If the whole rooting depth is included (30–50 cm deep), weights of up to ¾ to 1 metric ton can be expected, so machines must be heavily built and have a special shovel. Trax or 'Gradall' were found to be suitable if working on a platform in wet areas.

For transport lorries (5–10 tons loading capacity) may be used, though in rougher areas 'snow cats' (2–3 t) may be needed. They are loaded by the same or an other trax, under favourable subsoil conditions also by a fork-elevator.

The transplantation units may be unloaded at the new site by elevators or by trax using ropes around the pallets, and reset by the same or, in the more difficult cases, by hand (8–10 men/pallet) or dragline.

Depending on the textures, uniform resetting of the turf may be difficult. It can be very tedious or impossible to obtain a level surface, without cracks or fissures. Reconstruction is made even more difficult because the original site may not have been level. As a result, a rather undulating new surface is created which needs immediate smoothing. The units should not lie too closely, because tilting up of the borders would change aeration of the top soil and stimulate mineralisation processes. Consequently, N-indicators would finally spread and give a tendency to change the whole new surface (Fig. 1).

Obviously, all work should be done outside of the growing season, i.e. from November to February, or perhaps mid-March. During this season, peat is frozen and more easily moved and hardly any mineralisation will take place.

For technical and time reasons the surfaces had to be kept small. Normally 100 m^2 were moved, but for the wettest sites 200 to 5000 m^2 had to be reset.

3.3 Management and control

The following site variables should be controlled:

1. Composition of vegetation
2. Water level and water fluctuations
3. Influence of nutrients

Weeds tend to invade the cracks between transplant units. This invasion can be controlled, to a certain extent, by putting asbestos tiles in the soil at the edges of the outer units.

At the end of the first growing season new surfaces are analyzed phytosociologically, giving special attention to the presence and vitality of certain indicator plants, e.g. N-indicators. So far, apart from invasions by weeds at the edges of units (see above) only minor changes in the original status have been detected.

It is important to control the water level of each community; any additional water that is applied should pass through a buffering substrate of oligotrophic sand and peat.

In the case of drier straw- or litter-meadows (Molinion, Filipendulion etc.) the water regime is less important and no regulating devices are required.

Influx of nutrients from the surrounding areas are monitored by analyzing both groundwater

Fig. 1. Reaction in the labile phase of transplanted straw meadows and fens (Klötzli, 1980).

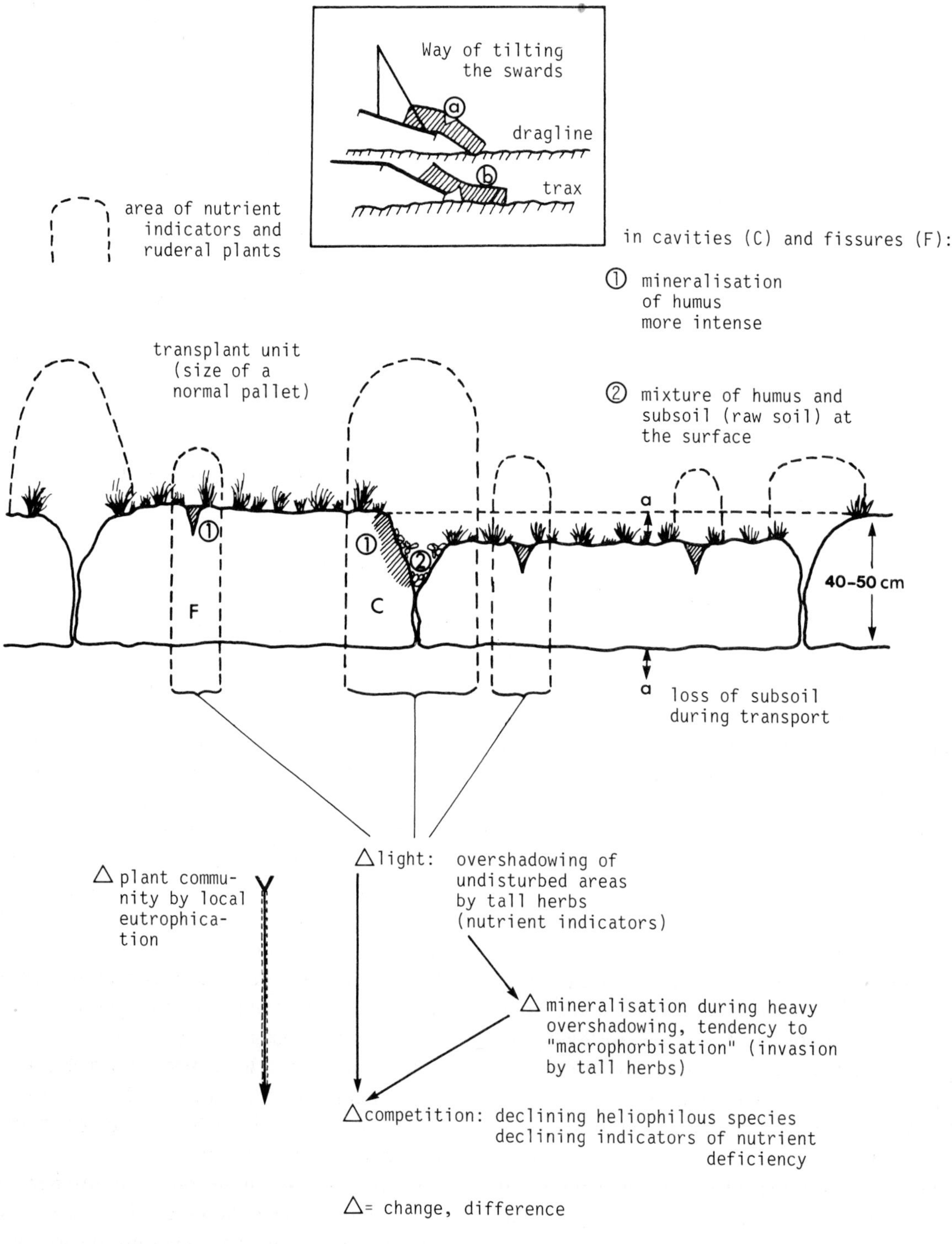

and substrate, and by the presence of indicator plants. These data are then compared with those of the old sites.

4. Community reponses

4.1 'Transplantation shock'

Most communities suffered from a 'transplantation shock', as a result some species expanded, some species disappeared (or fluctuated strongly) and some species invaded the communities (Fig. 2, Table 6). These changes were mostly due to disturbance during transplantation, e.g. changes in the topsoil, loss of subsoil or consequent change in the nutrient and water regimes. The most obvious changes occurred in cracks and fissures between the units, where greater N-mineralisation and aeration takes place and where weed species were able to invade, causing local 'macrophorbisation', i.e. spreading of high herbs (Fig. 1). These changes generally initiate a higher production of the transplanted units and a retreat of the heliophilous species. However, with appropriate management, such disturbances are quickly reduced and the new surface approaches its original state.

4.2 Changes

A survey of changes in species composition in different transplants is shown in Tables 3a,b,c,d. Certain changes are apparent in the composition of all the transplants (Tables 4,5); these include:

a) increase in the abundance of larger tussock-grasses, and immigration of stoloniferous grasses and mostly some bushes producing root tillers (compare Briemle, 1985)
b) slight increase in abundance of smaller tussock-grasses, and the partial disappearance (or even elimination) of low growing species and rosettes
c) no changes in the proportion of legumes and sedges
d) an increase in species indicating 'disturbance', e.g. *Festuca rubra, Rubus caesius*, pioneer species of bare clay soils, and stoloniferous pioneers of intermittently dry sandy soils
e) moisture indicators on locally more humid (micro-) sites (small depressions)
f) a decline in the abundance of heliophilous species and indicators of poor soils.

Some species disappeared almost completely (Orchidaceae, Apiaceae, *Equisetum*), or temporarily declined because of damage on roots or rhizomes.

Most of these changes were obviously due to disturbance of the sward during transplantation (fissures, undulating surface), which affected the availability of water or nutrients (indicated by typical species), but in part, it may reflect the influx of nutrients from surrounding areas or intensity of management (low mowing intensity provokes accumulation of nutrients, Klötzli, 1980, especially Table 5).

Another factor that may have induced change was the closer proximity of contrasting communities, leading to migration of new species, either by seed, or by immigration of stoloniferous species. Obviously the appearance of new species largely depends on the composition of the surrounding vegetation and on the probability of seed dispersal to the transplant site. According to the Island Theory, the chance of successful installation of species depends on the site of the nearest similar site and its distance from the new site (Gorman, 1979, more references). The success of immigration will depend on the suitability of conditions for new species and the competitive ability of the existing species. In any case invasion is likely to be a slower process than the disappearence of existing species, and most gaps are likely to be filled by pre-existing species, mostly disturbance indicators. But the combination of indicator species groups allows to assess the general trends in the development of the sward accompanied by occasional (each 3—5 years) soil and water analysis. Even more pronounced changes occur when the groundwater level is changed or fertilizers are applied (van den Bergh, 1981; Boller-Elmer, 1977; Egloff, 1986; Egloff & Näf, 1982; Klötzli, 1980; Näf, 1980; Fig. 2b). Obviously, the sponsoring of macrophorbs leads to a restratifica-

Table 3a. Development of transplanted straw meadows and fens: e.g. Mesobromion normal development.

Surface	St.-Brom.typ 'Pflanzgarten'*								
Year of control	71	74	75	76	77	78	79	80	82	85
originally dominating species:										
Bromus erectus										
Brachypodium pinnatum										
Koeleria cristata										
Festuca rubra										
Agrostis tenuis										
Anthoxanthum odoratum										
Dactylis glomerata										
Festuca arundinacea										
newly dominating species:										
Molinia litoralis										
Poa pratensis angustif.										
Molinia coerulea										
Deschampsia caespitosa										
Holcus lanatus										
further species:										
Avena pubescens										
Briza media										
Phleum pratense										
Arrhenatherum elatius										
Trisetum flavescens										
Sieglingia decumbens										
originally dominating species:										
Carex flacca										
Carex verna										
Carex pallescens										
new species:										
Carex panicea										
Carex hirta										
Juncus effusus										
originally dominating species:										
Lathyrus pratensis										
Lotus corniculatus										
Trifolium pratense										
Vicia sepium										
Hippocrepis comosa										
further species:										
Trifolium montanum										
Trifolium dubium										
Lotus uliginosus										
most important mosses:										
Scleropodium purum										
Rhytidiadelphus squarr.										
Rh. triquetrus										
Thuidium philibertii										
Ctenidium molluscum										
Acrocladium cuspidatum										

Surface	St.-Brom.typ 'Pflanzgarten'*								
Year of control	71	74	75	76	77	78	79	80	82	85
originally dominating species:										
Stachys officinalis										
Galium verum										
Sanguisorba minor										
Galium album										
Plantago lanceolata										
Hypericum perforatum										
Potentilla erecta										
Viola hirta										
Primula veris										
Scabiosa columbaria										
Veronica chamaedrys										
Silaum silaus										
Plantago media										
Chrysanthemum leucanth.										
Linum catharticum										
Galium pumilum										
Succisa pratensis										
Knautia arvensis										
Achillea millefolium										
newly dominating species:										
Rubus caesius										
Valeriana officinalis					o		o			
further species:										
Ajuga reptans										
Ranunculus nemorosus										
Prunella vulgaris										
Rumex acetosa										
Hypericum maculatum										
Pastinaca sativa										
Polygala vulgaris										
Campanula glomerata										
Campanula patula										
Pimpinella saxifraga										
new species:										
Senecio erucifolius										
Geranium palustre										
Potentilla sterilis										
Filipendula ulmaria										
Lythrum salicaria										
Equisetum arvense										
Euphorbia cyparissias										
weeds:										
Epilobium parviflorum										
Cirsium arvense										
Cirsium lanceolatum										
Galeopsis bifida										
Verbena officinalis										
most important woody plants:										
Alnus glutinosa										
Frangula alnus										
Crataegus monogyna										
Viburnum opulus										
Quercus robur S.										
Rhamnus cathartica										

Legend (table 3a-3d: ---- + 1–5 abundance-dominance-scale of Braun-Blanquet; o reduced vitality; √ dispatched; ⋆ 'Pflanzgarten' = special place; S = Sapling.

Table 3b. Development of transplanted straw meadows and fens: e.g. Development of Molinion from Mesobromion.

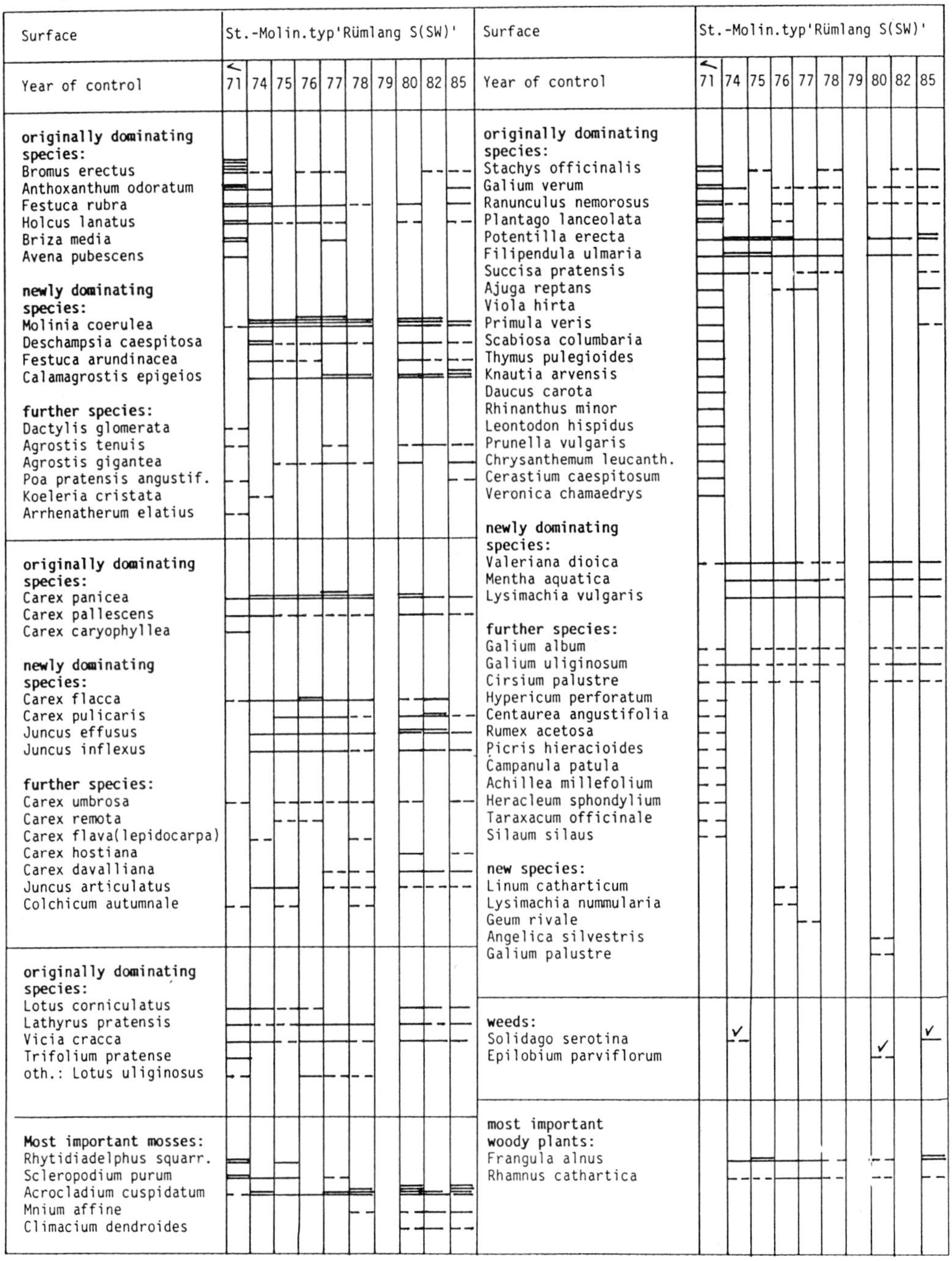

tion of nutrients (plant-available) in the soil due to changing light and water conditions on the top soil and consequently to hardly reversible changes (compare changes under different managements in Briemle, 1985, and also in Klötzli, 1978, and Kienzle, 1979). Specific management techniques are then necessary to steer an optimum development, i.e. to get near to the typical composition of a given plant community (e.g. early mowing, retention of ground or ditch water, cut-down of vole activity).

Tabel 3c. Development of transplanted straw meadows and fens: e.g. Eriophorion latifolii.

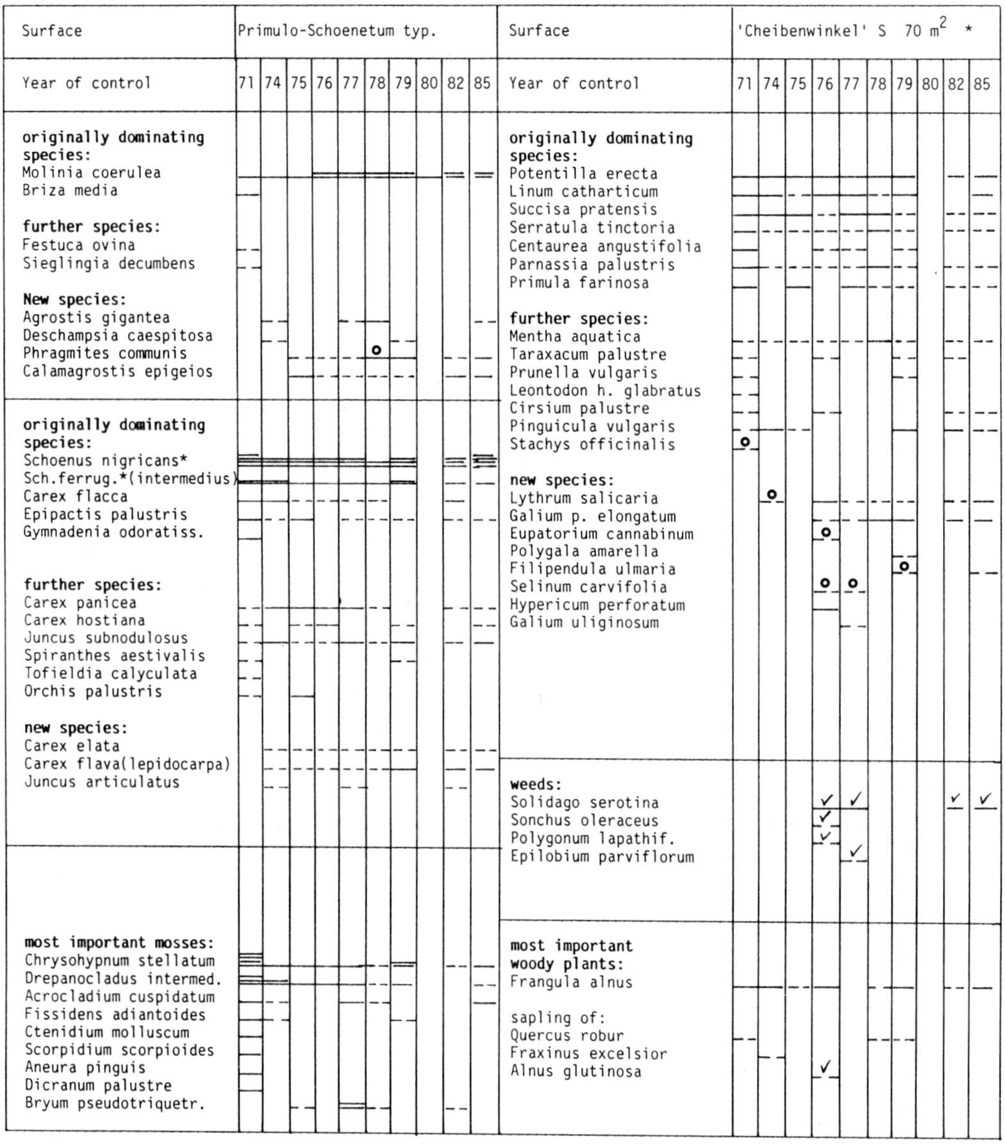

* 'Cheibenwinkel' = special place.

4.3 Causes of change (Table 5)

The main causes of the changes in vegetation composition and structure initiated by transplantation were:

a) higher mineralisation rates of nutrients in the upper soil horizons, leading to an increase in tall herbs ('macrophorbs'), rushes and some sedges (e.g. *Carex acutiformis*), similar to the changes occurring when meadows are fertilized (Briemle, 1985; Boller-Elmer, 1977).
b) disturbance of root system during transplanting, leading to the loss of Apiaceae *(Silaum, Selinum, Daucus, Peucedanum palustre)*, Orchidaceae and *Equisetum* spec.
c) changes in light conditions in the sward, leading to loss of heliophilous species, especially indicators of nutrient deficiency *(Linum catharticum, Briza media)* and some rosette forming plants (e.g. *Chrysanthemum leucanthemum*).
d) invasion of stoloniferous species (e.g. *Phrag-*

Table 3d. Development of transplanted straw meadows and fens: e.g. *Caricion fuscae* 'Tägerloh' * (only innermost permanent plot (Klötzli, 1980).

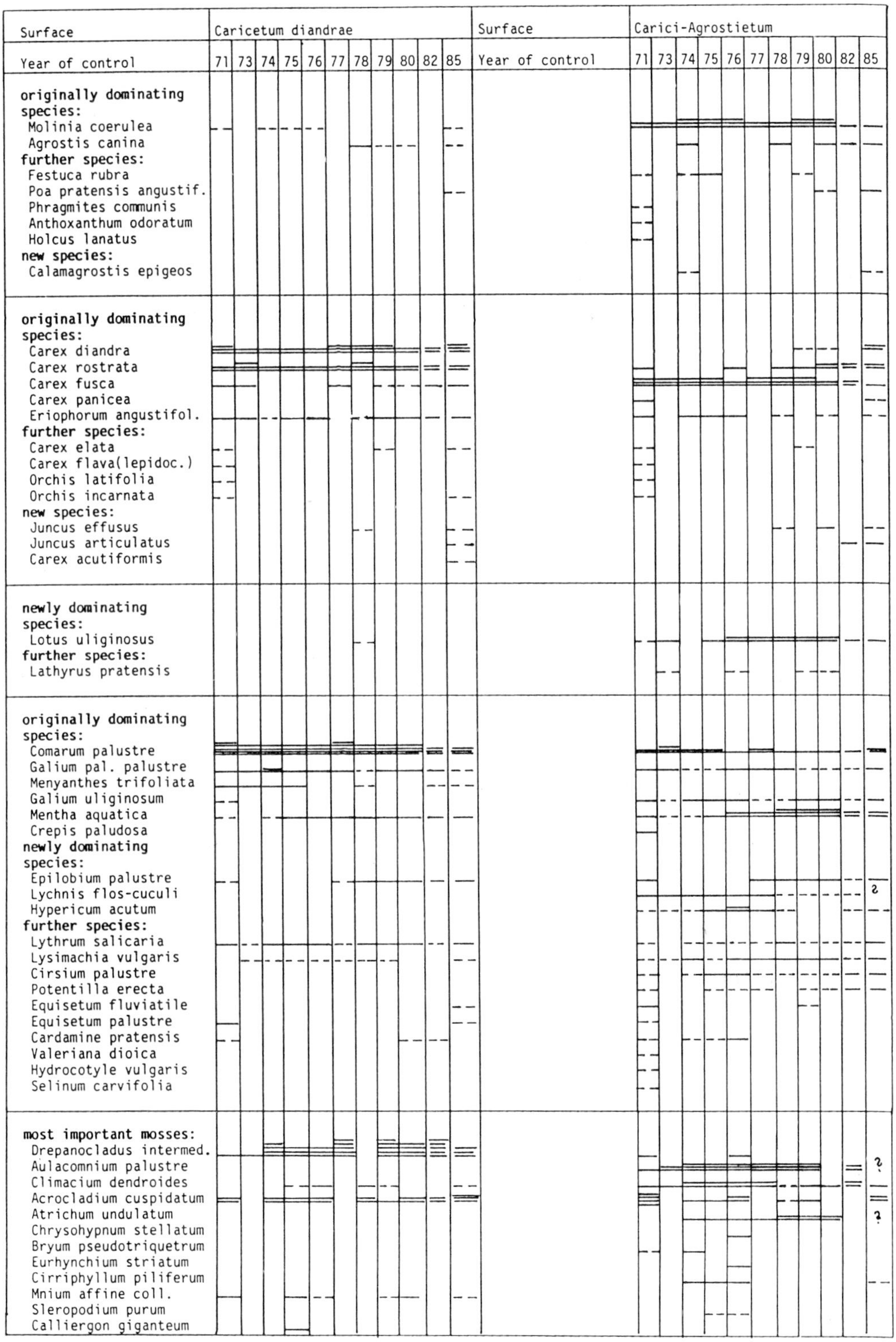

* 'Tägerloh' = special place.

Table 4. Plant Groups in the 'Succession Rain' on transplanted swards.

persistent species	*Molinia coerulea*	*Carex flacca*
	(Festuca rubra)	*- panicea*
	Brachypodium pinnatum	*- pallescens*
	Galium verum	*- diandra*
	Potentilla erecta	*- rostrata*
	(Filipendula ulmaria)	*- fusca*
	(Hypericum perforatum)	*Schoenus* spec.
	(Mentha aquatica)	*Acrocladium cuspidatum*
	Comarum palustre	*Chrysohypnum stellatum*
	Lathyrus pratensis	*Drepanocladus intermed.*
retreating species	*(Briza media)*	*Gymnadenia conopea*
	* *Selinum carvifolia*	*- odoratissima*
	* *Silaum silaus*	*Colchicum autumnale*
	* *Stachys officinalis*	
	Chrysanthemum leucanthemum	*Equisetum* spec.
	Crepis paludosa	
	Linum catharticum	
	Trifolium pratense	
penetrating species	*(Phragmites communis* A	*Juncus spec.* H
	(Calamagrostis epigejos A	*Carex pulicaris* A
	Deschampsia caespitosa	
	* *Molinia litoralis*	*Cirsium palustre*
	Lythrum salicaria A	* *Cirsium oleraceum*
	* *Galium elongatum* A	*Angelica sylvestris*
	Epilobium palustre	* *Valeriana officinalis*
	* *Lotus uliginosus*	*(Mentha aquatica)*
invading species	*Phragmites communis* A	*Juncus* spec. A U
	Calamagrostis epigejos A	
	(Festuca rubra A	*(Ajuga reptans)* A
	Solidago serotina A	*(Mentha aquatica)* A
	Cirsium arvense U	*(Cirsium oleraceum)*
	Epilobium parviflora U	*(Hypericum perforatum)*
	(Lotus uliginosus)	
fluctuating species	*Agrostis canina* A	*Orchis* spec.
	(Briza media	*Epipactis palustris*
	Viola hirta	*Spiranthes aestivalis*
	Galium uliginosum	*Eriophorum angustifolium*
	Galium palustre	*(Carex rostrata)*
	Pinguicula vulgaris	
	Primula farinosa	*Lychnis flos-cuculi*
	Cardamine pratensis	*Ranunculus nemorosus*
	Hydrocotyle vulgaris	*Centaurea angustifolia*
	Serratula tinctoria	*(Stachys officinalis)*

A stolones; H tussock; U weed; ( ) more or (less; also in other groups.
* diagnostically relevant, partly invading.

mites, Calamagrostis epigeios, Solidago serotina, Mentha aquatica).

e) general disturbance of sward microclimate (presumably to account for the loss of such common species as *Trifolium pratense, Chrysanthemum leucanthemum, Crepis paludosa*).

These changes are to some extent comparable with changes in fallow land where similar changes

Table 5. Changes of plant groups in transplanted swards (Klötzli, 1980).

	Plant species	Meso	Mol	Schoe	Cc dia fu	Remarks	
⊕	*Molinia litoralis*	**				Advantage on raw loam	
New or sponsored species	*Festuca rubra*	*	*			Local invading species after disturbance	
	Poa pratensis angustifolia	*	*			*N- & disturb. indicators* also (partly) in reconstitution phase	
	Carex hirta	*	*				
	Hypericum perforatum	*	*				
	Rubus caesius	*	*				
With equal site conditions	*Deschampsia caespitosa*	*	**	*		*N- & disturb. indicators* mostly in labile phase	
	Juncus inflexus/effusus/articulatus	*	**	*	*		
Mostly nutrient indicators and pioneers of raw soils	*Carex pulicaris*		**			Mostly *N-indicators* tendency to Filipendulion	
	Filipendula ulmaria/(Lysimachia		*				
	vulg.)		*				
	Cirsium oleraceum		*				
	Valeriana officinalis		*				
	Lotus uliginosus		*				
	Cirsium palustre		*				
	Angelica silvestris		**				
	Climacium dendroides						
	Agrostis alba gigantea		*	*		Mostly *pioneer species* on raw soil	
	○*Carex flava (lepidocarpa)/*		*	*			
	davalliana		*	*			
	○*Lythrum salicaria/Valeriana dioica*						
	Phragmites communis		*	*	*	Species penetrating *with stolons*	
	Calamagrostis epigeios		**	*	*		
	Mentha aquatica/(Carex acutiformis)		**	*	*		
	○*Carex elata / Scutellaria*			*		*Species of locally wetter sites, 'spontane' wetness indicators* ⊛	
	galericulata			*	*		
	○*Galium palustre*						
⊖	*Bromus erectus/Dactylis glomerata*	**	**			L	Inhibition of
Disadvantaged species	*Colchicum autumnale*	*	*			G	L heliophilous species
	Orchis morio / mascula	*	*			L G	G geophytes (mostly orchids)
	Trifolium pratense / montanum	**	**			?	
	Silaum silaus	**	**			Pf	
	Chrysanthemum leucanthemum	**	**			(L) ?	
	Prunella vulgaris / Plantago lanceolata	*	*			L	
	Ranuculus nemorosus						
With equal site conditions	*Briza media*	*	*	*		*L*	Pf species with tap roots & other deep-rooting species (Apiaceae, *Equisetum)*
	Equisetum arvense / palustre	*	*	*		Pf	
	Stachys officinalis / Linum catharticum	*	**	*		*(L) ?*	
	Gymnadenia conopea / odoratissima	*	*	*		*G*	
Mostly heliophilous species, geophytes and deep-rooting species	*Selinum carvifolia*		**	*	*	Pf	
	Polygala amarella		*	*		L	
	Orchis palustris/Spiranthes aestivalis			*		*G*	
	Orchis incarnata				*	G	? unknown causes, mostly lack of light, ⊕ !)
	Equisetum fluviatile				*	Pf	
	Crepis paludosa				*	?	
⊖	① *Festuca ovina capillata*	**				⓪ systematically important	
Dryness indicators	⓪ *Koeleria cristata / Avena*	**					
	pubescens	**					
Sensitive species mostly heliophilous, dryness indicators	⓪ *Hippocrepis comosa*	**				① tendency from Mesobromion to Molinion	
	Thymus serpyllum pulegioides	**					
	⓪ *Polygala vulgaris/Rhinanthus*	**					
	minor	**					
	⓪ *Pimpinella saxifraga/Daucus*	**				② tendency from Molinion to Eriophorion latifolii	
	carota						
	⓪ *Campanula glomerata/Achillea*						
	millefolium						
	Knautia arvensis						
	② *Sanguisorba minor / Viola hirta*	*	**				
	Primula veris / Stachys officinalis	*	**				
	Scabiosa columb./Centaurea j. angustif.	*	**				

** heavy change; * minor change.

regime (incl. absorption of nutrients, etc.). In a way, transplanted mires may serve as models to determine carrying capacity (i.e. the tolerance towards changes in the original environmental conditions) of parts of our environment. (For further detail see Klötzli, 1981).

Transplantations are only applicable for rare plant communities since the costs are high (300–500 SFr/m^2). The probability of successful reconstitution of a wetland community is high if the sites are carefully matched (homogeneity of the surface, buffering of disturbances) and the subsequent management is appropriate. However, slight disturbances may lead to weed invasions and the spread of certain indigenous species.

To conclude, transplantations may occasionally be useful for the conservation of vegetation, but are no substitute for conservation *in situ*. It is an expensive emergency solution to be used only when other forms of conservation are impossible. A transplanted and reconstituted or renaturalized surface should not be considered as good as a naturally grown and original surface. A good copy of a Mona Lisa is still not *the* Mona Lisa.

References

Boller-Elmer, K. (1977). Düngungseinflüsse von Intensivgrünland auf Streu- und Moorwiesen. Veröff. Geobot. Inst. ETH, Stiftung Rübel, Zürich 63: 103 pp.

Briemle, G. (1985). Vegetations- und Standortsentwicklung auf Niedermoor unter dem Einfluss verschiedener Pflegemassnahmen. Telma 15: 197–221.

Egloff, Th.B. (1986). Auswirkungen und Beseitigung von Düngungseinflüssen auf Streuwiesen. Eutrophierungssimulation und Regenerationsexperimente im nördlichen Schweizer Mittelland. Veröff. Geobot. Inst. ETH, Stiftung Rübel, Zürich 89: 183 pp.

Egloff, Th. & Näf, E. (1982). Grundwassermessung in Streuwiesen des unteren Reuss-Tales. Ber. Geobot. Inst. ETH, Stiftung Rübel, Zürich 49: 154–194.

Gorman, M. (1979). Island ecology. In: Outline studies in ecology. Chapman & Hall, London. 64 pp.

Kienzle, U. (1979). Sukzessionen in brachliegenden Magerwiesen des Jura und des Napfgebietes. Sarnen: von Ah. 104 pp.

Klein, A. (1980). Die Vegetation an Nationalstrassenböschungen der Nordschweiz und ihre Eignung für den Naturschutz. Veröff. Geobot. Inst. ETH, Stiftung Rübel, Zürich 72: 75 pp.

Klötzli, F (1969). Die Grundwasserbeziehungen der Streu- und Moorwiesen im nördlichen Schweizer Mittelland. Beitr. Geobot. Landesaufn. 52: 296 pp.

Klötzli, F. (1975). Naturschutz im Flughafengebiet – Konflikte und Symbiose. FLughafen-Information 3: 3–13,21.

Klötzli, F. (1978). Zur Bewaldungsfähigkeit von Mooren der Schweiz. Telma 8: 183–192.

Klötzli, F. (1980). Zur Verpflanzung von Streu- und Moorwiesen. Erfahrungen von 1969 bis 1980. Ber. ANL 5: 41–50.

Klötzli, F. (1981). Zur Reaktion verpflanzter Oekosysteme der Feuchtgebiete. Dat. u. Dokum. Umweltschutz. Stuttgart, Univ. Hohenheim, 31: 107–117.

Klötzli, F. & Keel, A. (1976). Verpflanzung von Streu- und Moorwiesen im Raume des Klotener Riets bei Zürich. Exk. fhr. Geobot. Inst. ETH, Stiftung Rübel, Zürich. 17 pp.

Miles, J. (1979). Vegetation dynamics. In: Outline studies in ecology. Chapman & Hall, London. 80 pp.

Näf, E. (1980). Zur Entstehung und Erhaltung von Mooren und Streuwiesen im Reusstal. Diploma thesis. Geobot. Inst. ETH, Zürich. 134 pp.

Schreiber, K.F. (1980). Brachflächen in der Kulturlandschaft. Dat. Dok. Umw. schutz 30: 61–93.

Skripchinsky, V.V., sr.; Tanfiljen, V.G., Dzibov, D.S.; Dubar, J.A. & Skripchinsky, V.V., jr. (1975). Some ways to recreate certain phytocoenosis. Abstr. 12th Intern. Bot. Congr. p. 545.

Van den Bergh, J.P. (1981). Interactions between plants and population dynamics. Verh. Ges. Oekol. 9: 155–163.

Van der Maarel, E. & Werger, M.J.A. (1978). On the treatment of succession data. Phytocoenosis 7: 257–278.

Van der Valk, A.G. (1982). Succession in temperate north american wetlands. In: Gopal, B., Turner, R.E., Wetzel, R.G. & Wigham, D.F. (Eds.). Wetlands. Ecology & management. pp 169–179. NIE/Int. Sci. Publ., New Dehli.

Wegelin, Th. (1984). Schaffung artenreicher Magerwiesen auf Strassenböschungen. Veröff. Geobot. Inst. ETH, Stiftung Rübel, Zürich 82: 104 pp.

CHAPTER 7

Vegetation responses to changes in tidal inundation of salt marshes

W.G. Beeftink

Detailed contents

1. Introduction

About five centuries ago the estuaries of western Europe became important as overseas trade routes; this led to large urban and industrial settlements and started an ever increasing human pressure on the estuarine environment. Estuaries and their fringing coastal marshes were, and are still, subjected not only to an array of man-made detritus, litter and chemical substances, but are also modified by embankment of marsh lands, the building of dams, barrages and other stream-conducting structures, and by dredging activities.

The increased occurrence of environmental disasters were first considered acts of God, but later it was recognized that some of them were the result of reclamation works. However, these activities and their effects cannot be clearly separated from socio-economic factors with which they were associated (O'Keefe et al., 1976). For example open-cast mining of peat for salt-winning and fuel, and associated political quarrels in the estuarine region of the south-west Netherlands led to such disasters (Beeftink, 1975), while the flood disaster of February 1st 1953 was partly due to inadequate maintenance of the sea-defending dikes during World War II (De Haan, pers. comm.).

As a result of the 1953 disaster, the Dutch Government decided to construct a series of barrages across estuaries (Fig. 1) to prevent further catastrophes. This Delta Plan has a number of environmental side-effects, causing a complicated series of large-scale changes in the tides and salinity of the estuarine waterbodies. For the salt marshes bordering these estuarine arms, the most important environmental changes are:

1. an abrupt increase of the tidal amplitude at the seaward side of the barrages, combined with a rise in salinity of that waterbody (Volkerak-Oosterschelde system, foredelta Haringvliet).
2. an abrupt disappearance of the tides at the landward side of the barrages, by cutting off all tidal influence and transforming the water-

J. van Andel et al., (eds.), Disturbance in Grasslands. ISBN 90-6193-640-3.

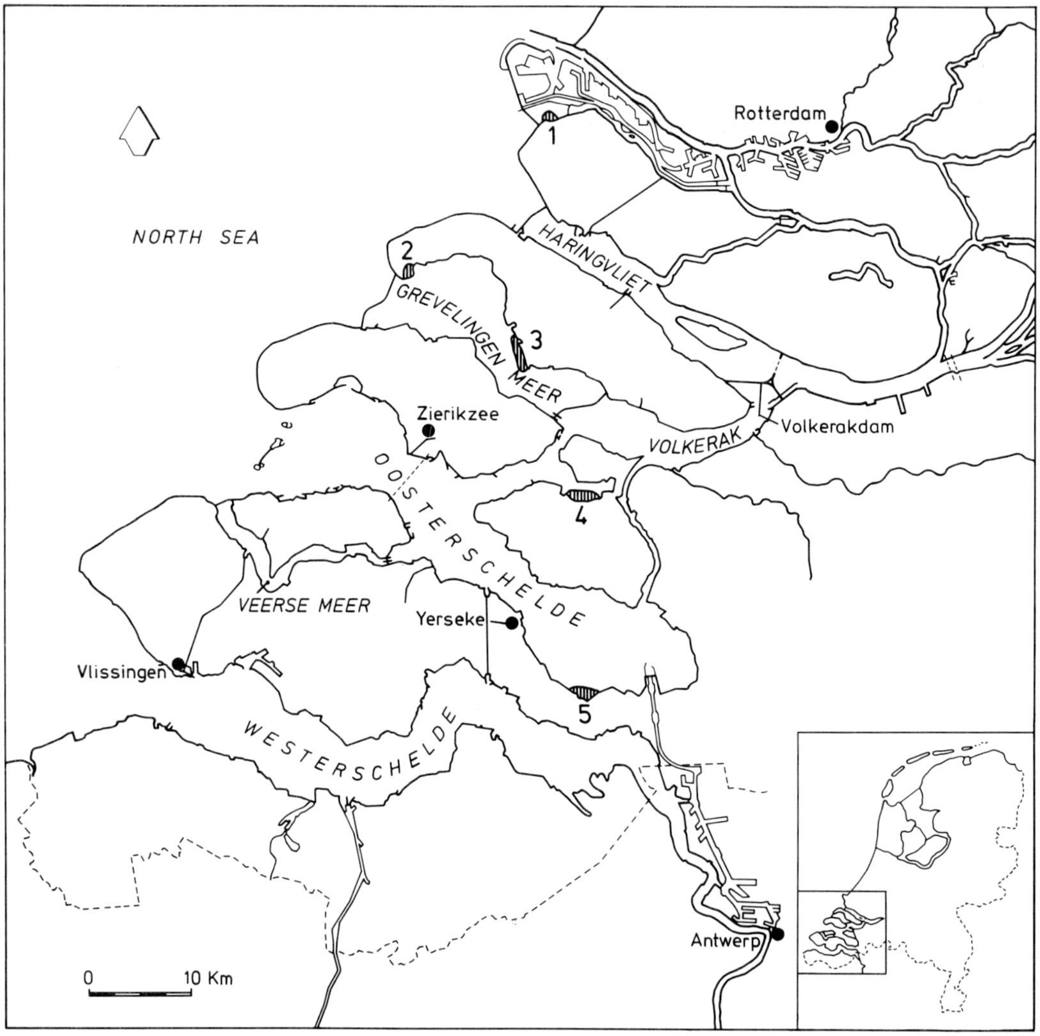

Fig. 1. Map of the south-west Netherlands showing the sites of barrage constructions as part of the Delta Plan. The main salt marshes considered in this paper are: 1 = Groene Strand, 2 = Springersgors, 3 = Slikken van Flakkee, 4 = St. Annaland, 5 = Krabbendijke salt marsh.

body into a lake (Veerse Meer, Grevelingen Meer), so also decreasing the salinity.

3. a decrease of the tidal amplitude after the construction of a storm-surge barrier (Oosterschelde) with little change in the salinity regime.

These changes in the tidal rhythm and salinity regimes give reason for the estuarine ecologist to reflect on the concept of disturbance. At the outset we need to make a clear distinction between natural and human-induced disturbances and disasters (O'Keefe et al., 1976). Natural disturbances, and their concomitant recovery mechanisms, are now acknowledged as integral parts of normal ecosystem behaviour (Reiners, 1983). Estuarine salt marshes, in particular, are characterized by natural disturbance (Odum, 1971), but they have also been subjected to man-made disturbances of various kind. Reclamation and improvement for agricultural use and open-cast mining for clay and peat are among the oldest human activities (Beeftink, 1975).

Most early human-induced disturbances in the salt marsh closely mimiced natural disturbances (Beeftink, 1975). Most modern civil-engineering works, however, go far beyond natural disturbances in scale and intensity. The ecological side-effects of the Delta Plan are examples of the lat-

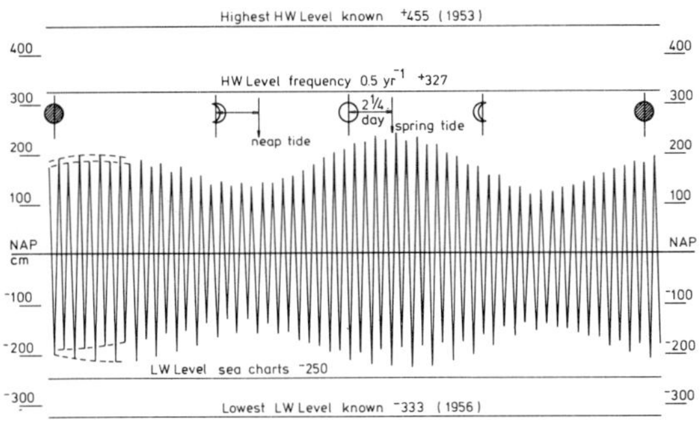

Fig. 2. Tidal patterns at Vlissingen during a moon cycle (after Rijkswaterstaat, 1985).

ter, and make prediction of ecological effects very difficult.

Any definition of disturbance should be sound and unequivocal with respect to a broad range of environmental impacts. The definition accepted at the Groningen Workshop does not wholly meet these requirements: 'interfering with a normal functioning of a given biological system' remains vague with respect to what is meant with 'interfering' and 'normal functioning', particularly if 'normal functioning' includes natural disturbances and its concomitant recovery mechanisms (Reiners, 1983). In the context of salt marshes as decribed in this paper, disturbance can be defined as *'the mechanisms which limit the plant biomass by causing its partial or total destruction'* (Grime, 1979), with the caveat that senescence of individuals is excluded (Grubb, 1985). In explaining this definition, Grubb (1985) stated that disturbance may involve killing plant parts *in situ,* severance of plant parts followed by decomposition *in situ,* loss by combustion, severance and removal from the site, or burial followed by death and decay (if any) *in situ.* Disturbance must always be initiated from outside the plant community in question.

In this context two reservations have to be made. First, disturbance can only be recognized in the field by its harmful effects on organisms. In other words, the cause is defined by its effects, leaving their relation questionable.

The second reservation is that different levels of biological organization require different interpretations of disturbance; for example, a harmful effect on the individuals of a certain species does not mean that disturbance will be recognized at the ecosystem level. Grazing of a salt marsh, for instance, will generally be considered system-inherent, but may be at the same time the cause of disturbing individuals of some vulnerable species in this system.

In this paper, I consider some examples of vegetation changes, caused by both natural and man-induced variations in the vertical component of the tides. How far these vegetation changes can be considered disturbances will be discussed.

2. Tidal variation

The tides are mainly the result of the movement of the moon and sun with respect tot the earth, causing the circulation of tidal waves from East to West, mainly through the Antarctic Ocean. This mechanism explains the regularity of tides.

Most variations in the tidal pattern reflect changes in both lunar and solar orbits and positions. The tidal variation with the highest frequency (24 hours and 50 minutes) is the semi-diurnal cycle with differences in high-water levels of about 50 cm at Vlissingen (Fig. 2). The spring-neap tide cycle has a period of 27.55 days and a mean difference in high-water level of 35 cm at Vlissingen (Fig. 2). The declination of the sun gives a third cycle, causing a deviation of about 5 cm, with a frequency of one year. The moon's declination also results in a slight deviation of about 3–5 cm from mean high-water level. A fifth (nodal) cycle is the result of the variation in the moon's orbit, with a tidal variation of about 6 cm at Vlissingen and frequency of 18.6 years. There were maxima in 1923, 1941, 1960 and 1978 (Fig. 3). Finally, long-term cycles, caused by precession of the equinoxes, nutation of the earth's axis, and excentricity of the earthian orbit, which have periodicities of about 20, 40 and 100.10^3 years respectively, affect either the tidal pattern or the overall sea level (Hays et al., 1976). Tides are also subjected to tectonic, glacio- and hydro-isostatic movements, and are affected by the wind. The effect of these factors on tide levels are less predictable. Postglacial isostatic and tectonic

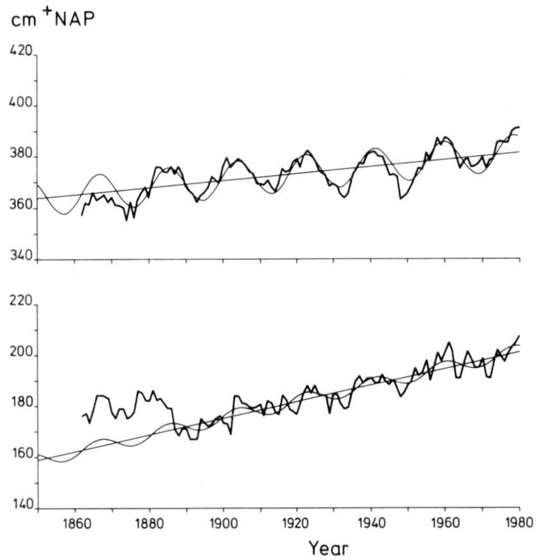

Fig. 3. Yearly averages of (a) tidal difference and (b) high-water level at Vlissingen from 1850 to 1980. The general rise in sea level (straight line), the nodal cycle of 18.6 years (sinusoidal line) and deviations from this pattern, largely caused by wind effects, are all apparent. NAP = Dutch Ordnance Level (after de Ronde, 1982).

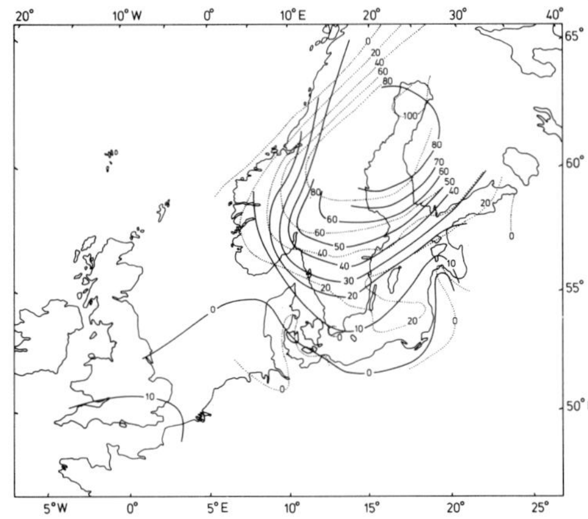

Fig. 4. Postglacial crust movements in Fennoscandia (isobases in cm per century); positive values denote land uplift. Solid lines according to Rossiter (1967), dashed lines after Gutenberg (1941). See also Fig. 6 in Ericson & Wallentinus (1979).

movements occurred in north-west Europe (Fig. 4). The centre of land uplift is in the Gulf of Bothnia, with a maximum of 1 m per century; this is compensated by land subsidence south-west of a line through Jutland and the southern Baltic coast. Land subsidence in the Netherlands is about 4–5 cm per century (Van der Plassche, 1983). However, in addition to this crust movement, the sea-level is rising at a rate of about 15 cm per century through climatic change, so that the relative sea-level in the Netherlands is rising by about 15–20 cm per century (Fig. 3, Table 1). Dredging activities in the river mouths may increase the high-water level even further. These slow changes in tide levels may not be considered by some as disturbing salt-marsh ecosystems.

Winds blowing up and off the sea-water masses, can have a considerable effect on tide levels, especially in the funnel-shaped North Sea, and in areas (e.g. the Baltic) where the tidal amplitude is small. In these cases, the effects of storm floods can extend for several days (Table 2). On a year-basis the mean deviation of the high-tide level and the tidal amplitude caused by storms show an irregular pattern, with a tendency to clustering, either above or below the normal nodal cycle, for successive years (Fig. 3).

Table 1. Relative sea-level rise at Zierikzee and Vlissingen over 1933–80 (cm per century) and, calculated from these figures, for the period 1965–1984 (cm per 20 years). Unpublished data delivered by De Ronde (pers. comm.).

Period	Zierikzee 1933–80	Zierikzee 1965–84	Vlissingen 1933–80	Vlissingen 1965–84
Av. sea level	20	4.0	15	3.0
Av. high-water level	38	7.6	34	6.8
Av. low-water level	4	0.8	7	1.4
Av. tidal difference	34	6.8	27	5.4

Table 2. The effect of wind on tide levels on the continental North Sea coast (tidal difference 1.3–5.0 m) and the Baltic coast (tidal difference 0.1–0.3 m).

	North Sea coasts	Baltic coasts
Storm flood raise	max. 1.50–3.00 m	max. 1.70–2.50 m
Duration of inundation	2–4 high tides (max. some hours per tide)	1–2 days

Table 3. Estimated cover percentages of plant species and numbers of molluscs in four permanent plots in the Springersgors salt marsh during 1962–1970. Deviations (cm) from long-term averages calculated on the basis of monthly averages over the periods September-August. Cover estimates for the plant species according to Doing Kraft (1954), and transformed into cover percentages.

Year Deviation from av. high-water level (cm)	1962 +8.1	63 −4.3	64 −2.4	65 −0.3	66 +5.0	67 +4.0	68 +1.8	69 +1.0	70 +2.8
Plot Nr. 12 (1.41 m +NAP, 2×2 m).									
Puccinellia maritima	80	80	90	80	90	60	10	15	10
Aster tripolium	2	5	10	4	4	2	2	1	1
Glaux maritima	–	–	1	2	2	–	–	–	–
Plantago maritima	–	1	1	1	1	1	–	–	–
Limonium vulgare	10	20	30	20	20	20	30	20	20
Suaeda maritima	3	3	2	3	2	3	3	3	10
Halimione portulacoides	3	3	3	3	2	1	3	–	1
Spartina anglica	2	2	2	1	1	1	2	3	2
Salicornia europaea	5	3	2	3	4	30	30	15	20
Triglochin maritima	–	1	2	3	3	10	20	15	10
Hydrobia ulvae	a	p	p	p	r	r	m	a	–
Plot Nr. 5 (1.49 m +NAP, 5×5 m).									
Festuca rubra	3	5	10	10	10	2	2	3	
Elymus pycnanthus	–	–	–	1	1	–	–	–	
Artemisia maritima	–	2	2	2	2	3	1	–	
Parapholis strigosa	2	–	–	–	–	–	–	–	
Plantago maritima	5	15	10	10	10	10	10	10	
Aster tripolium	3	15	10	30	20	4	4	20	
Glaux maritima	10	10	10	10	10	4	4	4	
Halimione portulacoides	30	25	30	20	10	10	4	4	
Salicornia europaea	4	3	–	1	4	20	10	5	
Suaeda maritima	4	3	–	3	5	20	15	4	
Puccinellia maritima	1	–	1	–	1	1	2	2	
Spartina anglica	–	–	–	–	–	1	1	2	
Limonium vulgare	50	40	40	30	50	60	70	70	
Triglochin maritima	2	1	1	2	3	2	3	2	
Spergularia maritima	2	3	3	1	2	2	2	2	
Littorina saxatilis	–	–	–	r	r	r	a	a	
Assiminea grayana	p	–	p	–	r	–	p	a	
Ovatella moysotis	p	–	–	–	–	–	–	–	
Plot Nr. 10 (1.54 m +NAP, 4½×5½ m).									
Elymus pycnanthus	100	100	100	100	100	70	80	70	
Halimione portulacoides	30	2	1	–	2	2	10	10	
Aster tripolium	10	3	3	–	2	4	5	10	
Suaeda maritima	4	–	2	1	20	45	20	30	
Artemisia maritima	2	1	1	–	1	1	–	–	
Limonium vulgare	–	–	–	–	–	1	–	–	
Salicornia europaea	–	–	–	–	–	–	–	1	
Assiminea grayana	p	p	a	m	r	–	–	–	
Ovatella myosotis	m	p	p	p	m	m	m	p	
Plot Nr. 1 (1.70 m +NAP, 2×3 m).									
Elymus pycnanthus	5	30	40	60	70	70	30	50	
Festuca rubra	70	80	100	90	80	40	55	60	
Artemisia maritima	40	30	20	10	10	4	3	3	
Salicornia europaea	–	–	–	–	–	2	4	1	
Suaeda maritima	2	–	–	–	–	10	30	3	
Aster tripolium	3	2	1	2	1	4	4	4	
Puccinellia maritima	2	–	–	–	–	–	–	–	
Spergularia maritima	1	–	–	–	–	–	–	–	
Plantago maritima	4	3	2	3	2	2	2	1	
Limonium vulgare	2	1	2	1	1	1	1	1	
Halimione portulacoides	3	2	1	2	2	3	3	2	
Assiminea grayana	a	–	–	–	r	p	–	p	
Ovatella myosotis	p	–	–	–	a	a	m	p	

3. A case study

3.1 The Delta Works in the south-west Netherlands

The effects of changes in the tidal amplitude, caused by the Delta Works, have been studied on several hundred permanent plots laid out in the salt marshes in the early 1960s. The plots were sited in parts of the Delta area where distinct changes in the tides were expected, or were not expected (Fig. 1). The plots were distributed as regularly as possible over the phytosociological units (Beeftink, 1977a) present in the marsh. The size of the plots exceeded the minimal area, so that they can be considered representative in a phytosociological sense.

Each permanent plot was analysed each year, and, as far as possible, on the same date. The method of Doing Kraft (1954) was used for the floristic and structural analysis. For the sake of uniformity all estimates of cover and abundance were transformed into cover percentages.

The height of the soil surface, relative to N.A.P. (Dutch Ordnance Level), was measured at the start and the end of the period. Differences in height between the start and end of the period indicate net accretion or net shrinkage of soil.

Relevant environmental factors were monitored, in case of tidal attributes by the Rijkswaterstaat (1961–1982, 1964, and unpublished data), rainfall by the Royal Netherlands Meteorological Institute, KNMI (unpublished data), and temperature and salinity in the Oosterschelde by the Delta Institute (Van Boven, 1981, and unpublished data).

Table 4. Cover percentages of *Aster tripolium, Halimione portulacoides* and *Salicornia europaea* in four plots in the Springersgors salt marsh during 1962–1970. For further explanation see Table 3.

Plot nr.	+NAP m	1962	63	64	65	66	67	68	69	70
		+8.1	−4.3	−2.4	−0.3	+5.0	+4.0	+1.8	+1.0	+2.8
Aster tripolium										
1	1,70	3	2	1	2	1	4	4	4	
10	1,54	10	3	3	0	2	4	5	10	
5	1,49	3	15	10	30	20	4	4	20	
12	1,41	2	5	10	4	4	2	2	1	1
Halimione portulacoides										
1	1,70	3	2	1	2	2	3	3	2	
10	1,54	30	2	1	0	2	2	10	10	
5	1,49	30	25	30	20	10	10	4	4	
12	1,41	3	3	3	3	2	1	3	0	1
Salicornia europaea										
1	1,70	0	0	0	0	0	2	4	1	
10	1,54	0	0	0	0	0	0	0	1	
5	1,49	4	3	0	1	4	20	10	5	
12	1,41	5	3	2	3	4	30	30	15	20

3.2 Effects of natural short-term tidal variations

Fig. 3 indicates that wind effects on the high-tide level usually occur in clusters of some years with deviations of either an increased or a decreased level, with respect to the 'normal' high-tide level. Table 3 illustrates the vegetation response to such a cluster of decreased high-tide levels. The years 1962–1963, 1963–1964 and 1964–1965 had lower tide levels than normal (Fig. 3), and the years before and after had higher than normal tides, mainly due to the effects of wind.

The period of relatively low tides in 1962–65 coincides with changes in the vegetation composition in the middle and upper marshes. Some plant species moved downward or upward in the marsh in response to changes in the mean high-water levels. For example lower cover densities of *Aster tripolium* were found in the two higher plots, and higher densities in the two lower plots, during the period of low tide levels (Table 4). The same occurred in *Halimione portulacoides,* but this species is generally much less flexible than *Aster tripolium* in its response, and found a lower limit of response between plots 12 and 5.

The cover of *Salicornia europaea* and *Suaeda maritima* was low during 1962–65 in the lower marsh where the frequency of storm floods decreased. In part this may reflect the fact that seeds of these species are dispersed over the whole area by wind-raised high tides (Koutstaal, Markusse & De Munck, unpublished). *Spartina anglica, Triglochin maritima* and *Limonium vulgare* had higher cover percentages under those ti-

Table 3. Transformed cover percentages of plant species in five permanent plots in the Springersgors salt marsh, since Spring 1970 (plot nr. 22) or Spring 1971 (nrs. 3 and 33) isolated from tidal influence.

A. Plot Nr. 3. Puccinellietum maritimae (1.39 m −NAP. $1\frac{1}{2} \times 4\frac{1}{2}$ m).

Year	1970	71	72	73	74	75	76	77	78	79	80	81	82	83	84
Total cover (%)	98	90	99	99	99	35	80	85	85	85	80	80	70	65	90
Spartina anglica	3	2	–	–	–	–	–	–	–	–	–	–	–	–	–
Aster tripolium	2	–	–	–	–	–	–	–	–	–	–	–	–	–	–
Salicornia europaea	3	4	–	–	–	–	–	–	–	–	–	–	–	–	–
Suaeda maritima	10	30	4	2	–	–	–	–	–	–	–	–	–	–	–
Triglochin maritima	40	50	40	1	–	–	–	–	–	–	–	–	–	–	–
Puccinellia maritima	40	5	30	90	80	2	–	–	–	–	–	–	–	–	–
Limonium vulgare	20	20	30	15	2	–	–	–	–	–	–	–	–	–	–
Halimione portucaloides	5	4	3	10	3	–	–	–	–	–	–	–	–	–	–
Urtica dioica	–	–	1	2	5	20	70	55	30	30	4	4	4	4	4
Solanum dulcamara	–	–	–	1	1	2	2	2	2	–	–	–	–	–	–
Cirsium arvense	–	–	–	–	2	3	5	25	40	10	10	4	20	10	4
Senecio jacobaea	–	–	–	–	2	5	4	25	30	35	35	3	3	2	15
Poa trivialis	–	–	–	–	–	2	2	3	3	25	30	50	25	20	60
Epilobium adenocaulon	–	–	–	1	1	2	2	2	4	20	20	30	25	10	4
Cardamine hirsuta	–	–	–	–	–	–	–	4	4	4	4	5	5	4	4
Cynoglossum officinale	–	–	–	–	–	–	–	1	2	1	2	2	5	4	4
Rubus spp.	–	–	1	2	2	2	1	1	3	1	2	1	2	10	3
Sonchus arvensis	–	–	–	–	–	–	–	–	–	–	1	2	2	3	3
Prunella vulgaris	–	–	–	–	–	–	–	–	–	–	2	1	2	10	10
Calamagrostis epigejos	–	–	–	–	–	–	–	–	–	–	–	–	2	2	2

B. Plot Nr. 33. Halimionetum portulacoidis (1.42 m −NAP. 4 × 4 m).

Year	1970	71	72	73	74	75	76	77	78	79	80	81	82	83	84
Total cover (%)	98	80	95	95	85	60	75	90	90	80	70	70	80	98	98
Spartina anglica	2	1	1	–	–	–	–	–	–	–	–	–	–	–	–
Aster tripolium	2	1	–	–	–	–	–	–	–	–	–	–	–	–	–
Salicornia europaea	4	4	–	–	–	–	–	–	–	–	–	–	–	–	–
Suaeda maritima	10	30	60	20	–	–	–	–	–	–	–	–	–	–	–
Triglochin maritima	2	2	3	2	1	–	–	–	–	–	–	–	–	–	–
Puccinellia maritima	1	–	–	–	–	–	–	–	–	–	–	–	–	–	–
Limonium vulgare	25	20	20	30	25	20	10	20	10	4	4	2	–	–	–
Halimione portulacoides	70	30	10	3	1	–	–	–	–	–	–	–	–	–	–
Sambucus nigra	–	–	4	2	2	3	2	2	2	1	1	–	–	–	–
Glaux maritima	–	1	3	30	10	10	3	4	5	4	10	15	15	30	10
Cirsium arvense	–	–	1	5	20	10	4	4	15	5	4	4	3	4	2
Urtica diocia	–	–	3	10	10	10	50	30	10	4	4	4	2	3	2
Sclanum dulcamara	–	–	2	2	2	2	2	4	2	–	–	–	–	–	–
Senecio jacobaea	–	–	1	1	3	15	3	30	25	20	10	5	4	4	5
Epilobium adenocaulon	–	–	–	–	2	3	1	3	4	20	4	4	2	2	1
Poa pratensis	–	–	1	1	2	2	2	10	10	20	20	20	30	30	30
Poa trivialis	–	–	–	–	–	–	–	–	2	2	5	4	–	–	–
Cynoglossum officinale	–	–	–	–	–	1	1	1	2	1	3	3	4	3	–
Rubus spp.	–	–	–	2	2	2	1	2	2	3	2	2	5	10	25
Festuca rubra	–	–	–	–	1	1	1	4	3	3	5	5	10	10	15
Agrostis stolonifera	–	–	–	–	2	2	2	10	10	15	15	20	20	35	40
Sonchus arvensis	–	–	–	1	3	2	1	2	2	3	3	1	2	2	4

C. Plot Nr. 22. Artemisietum maritimae (1.60 m −NAP. 2 × 6 m).

Year	1970	71	72	73	74	75	76	77	78	79	80	81	82	83	84
Total cover (%)	98	99	99	95	99	99	75	99	99	99	99	99	99	99	99
Aster tripolium	10	2	–	–	–	–	–	–	–	–	–	–	–	–	–
Suaeda maritima	4	–	–	–	–	–	–	–	–	–	–	–	–	–	–
Limonium vulgare	1	–	–	–	–	–	–	–	–	–	–	–	–	–	–
Halimione portulacoides	10	2	2	2	–	–	–	–	–	–	–	–	–	–	–
Plantago maritima	2	1	1	–	–	–	–	–	–	–	–	–	–	–	–
Artemisia maritima	10	10	5	4	3	–	–	–	–	–	–	–	–	–	–
Festuca rubra	90	90	60	40	40	10	3	3	3	2	2	2	2	4	10
Elymus pyonanthus	5	20	50	60	70	95	70	99	99	95	95	95	80	95	95
Epilobium adenocaulon	–	–	–	1	2	1	–	–	–	–	–	–	–	–	–
Cirsium vulgare	–	–	–	–	2	2	2	1	1	1	–	–	–	–	1
Galium aparine	–	–	–	–	–	–	–	2	1	2	10	3	4	2	2
Cirsium arvense	–	–	–	–	–	–	–	15	25	30	20	40	30	15	20
Urtica dioica	–	–	–	–	–	–	–	–	1	2	3	3	3	3	2
Poa pratensis	–	–	–	–	–	–	–	–	–	–	–	–	3	2	2
Rubus spp.	–	–	–	–	–	–	–	–	–	–	–	–	1	1	10

dal conditions, mainly as a consequence of vegetative reproduction. *Glaux maritima, Festuca rubra, Artemisia maritima* and *Elymus pycnanthus* showed the reverse effect, being sensitive to higher flooding frequences.

In the marsh levels where these four permanent plots were laid out, species of the lower and middle marsh increased, and those which have their ecological optimum in the upper marsh decreased, when tidal influence increases. Species with a wide ecological amplitude, such as *Aster tripolium,* moved along the marsh gradient parallel with the tidal influence. The severe winter of 1962–1963 caused local damage to *Halimione portulacoides,* in places where the snow cover had blown away (Beeftink et al., 1978). Little damage was observed in the four plots in the Springersgors salt marsh (Table 3), though the sharp decline of *Halimione* in plot 10 may partly be explained by frost damage.

3.3 Effects of disappearance of the tidal factor

Removal of tidal influence can have very marked effects on salt marshes, accelerating soil processes which are normally delayed by tidal activity. These soil processes include drying and salinization of soil, shrinkage in soil and the formation of cracks which accelerate oxidation processes and the decomposition of organic matter. Most salt-marsh species soon disappear because of high salinity, water deficit, oxidation processes, high nutrient levels, and damage of the roots by horizontal shrinkage.

Table 5 gives examples of the fate of changes in population assemblages after the construction of a barrier in spring 1971, and the subsequent loss of tides. Within three years after the closure of the barrier, the abundance of some species increased considerably, while others declined. *Suaeda maritima,* for instance, increased to high abundance after mass-death of *Spartina anglica* in the lower marsh (not in Table 5), and also of *Puccinellia maritima* (Table 5A) and *Halimione portulacoides* (Table 5B) in the middle marsh. *Puccinellia* and *Halimione* also increased initially, while *Triglochin maritima* declined, but all became extinct sooner or later. Other salt-marsh

Table 6. The average survival time (years) of halophyte species in permanent plots of salt marshes after the building of barrages had removed tidal influence.

Area	Grevelingen Springersgors	Slikken van Flakkee Flakkee	Brielse Gat Groene Strand	Salt marsh level
Salinity	saline	saline	brackish	
Management	no grazing	grazed	no grazing	
Number (n)	7–21	5–11	1–8	
Salicornia europaea	1.8	2.2	–	
Spartina anglica	2.4	3.2	2.0	lower
Aster tripolium	2.6	4.5	2.2	
Suaeda maritima	2.8	3.1	–	
Triglochin maritima	3.2	3.4	2.6	
Halimione portulacoides	3.8	3.8	–	
Plantago maritima	3.7	> 5.8	2.6	
Puccinellia maritima	4.2	> 4.7	2.5	middle
Artemisia maritima	4.1	–	3.0	
Spergularia maritima	4.9	> 5.5	1.0	
Limonium vulgare	5.3	4.2	2.0	
Glaux maritima	> 11.2	> 5.6	> 11.1	
Parapholis strigosa	9.0	4.0	–	
Juncus gerardii	9.0	> 4.5	> 14.0	upper
Festuca rubra	> 15.0	> 7.6	> 3.4	
Elymus pycnanthus	> 15.0	> 8.0	> 10.6	

species, however, survived many years, such as *Glaux maritima, Festuca rubra* and *Elymus pycnanthus,* or even establish after loss of tides.

The average survival time for the salt-marsh species at the various salt marsh levels is given in Table 6. The results are:

1. Populations of species characteristic of the lower marsh, such as *Spartina anglica* and *Salicornia europaea,* remain only for a very short time (1–3 years), those of the upper marsh remain for much longer (9–> 15 years).
2. Species growing in the lower and middle areas of an originally brackish marsh survive for a shorter time than the same species growing in an originally saline marsh.
3. Grazing extends the survival of populations growing in the lower marsh, and some populations from the middle marsh, but shortens the survival of species in the upper marsh.

Ruderal species of fresh habitats, mostly tall herbs, soon invaded the salt marsh after closure of the barrier, and rapidly replace the halophytes (Table 5). The species composition of these ruderal invasions does not differ very much between plots in the course of succession. A similar fairly consistent, succession pattern was also found in the plots during the first 8–9 years after former tidal sand flats were isolated from the tides (Hogeweg et al., 1985).

3.4 Effects of disturbance regimes

During 1965–1984 a number of environmental changes, both natural and man-induces, occurred in the salt marshes of the Oosterschelde-Volkerak system. Their impact on the vegetation cannot be considered apart from each other, because the after-effects overlap and intervene. The changes were caused by various factors:

(i) The prolonged and severe winters 1962/63, 1978/79, 1984/85 (Table 7). *Halimione portulacoides,* if not covered with snow and ice, and to a lesser extent *Spartina anglica,* die if the temperature falls below –15°C for a series of subsequent nights (Beeftink, 1979, Beeftink et al., 1978). Low air temperatures are reflected in the temperature of the Oosterschelde water, which shows a minimum in the winter of 1978/79 (Table 7).

(ii) Low rainfall and high summer temperatures increased the salinity of soil and water so that in 1976, when rainfall was well below average, chlorinity reached a maximum of 19‰ Cl^- in spring 1977. Such dry and hot summers promote soil development processes, resulting in vertical shrinkage of some centimeters and a higher flooding frequency.

(iii) Apart from these natural fluctuations in chlorinity, caused by climatic conditions, the Delta Works influenced the relation between sea and river water in the estuaries. After the Grevelingendam was built in summer 1964, and until May 1969 when the Volkerakdam was finished, more river water reached the Oosterschelde than before, so that the chlorinity of Oosterschelde water decreased (Table 7) with minima of 8–9‰ Cl^- in winter and early spring. After May 1969, most river discharge was cut off and salinity increased again. Such changes in water salinity could lead to changes in the density or growth of existing species, or to changes in the location of species along gradients of soil salinity and flooding (Beeftink, 1979).

(iv) Siltation and vertical shrinkage of soil have caused changes in the soil level on the permanent plots varying from −10 to + 18 cm in the St. Annaland salt marsh and from +3 tot +22 cm in the Krabbendijke salt marsh over a 20 year period. This net accretion during 1965–85 would be affected by both deposition (siltation),

Table 7. Annual variation in rainfall (mm, March-August) at St. Annaland, and average chlorinity (‰, March-August) and temperature (°C, November-April) based on monthly averages in the Oosterschelde at Yerseke during 1965–1984.

Year	1965	66	67	68	69	70	71	72	73	74
Rainfall	512	466	390	450	432	282	322	395	319	329
Chlorinity	*14.1*	*14.4*	*15.2*	*14.8*	*14.9*	15.3	16.6	16.8	16.8	16.9
Temperature	5.0	5.4	6.0	5.6	4.5	4.3	5.7	6.1	6.1	6.8
Year	**1975**	**76**	**77**	**78**	**79**	**80**	**81**	**82**	**83**	**84**
Rainfall	408	*124*	284	324	430	437	438	301	426	303
Chlorinity	15.8	*17.0*	*17.2*	16.0	15.8	15.0	15.2	15.7	15.7	15.7
Temperature	6.8	5.5	5.8	5.6	*3.9*	5.7	5.6	4.6	5.8	5.5

partly as a consequence of the Delta Works, and shrinkage caused by summer drought in 1976.

(v) The mean high-water level in the Oosterschelde (Zierikzee) increased about 7.5 cm (Table 1) over the period of observation (De Ronde, pers. comm.). A moderate net accretion rate can neutralize this tidal rise. The construction of the Volkerakdam in 1969 caused an increase in the high-water level on the sea side of the dam; this effect ranged from about 1 m in the vicinity of the dam to about 8 cm in the eastern part of the Oosterschelde. This rise is achieved in subsequent periods as appears from Table 8; for the St. Annaland salt marsh the main increase (24 cm on an average) occurred soon after the dam had been finished. Further increases of about 11 cm and 7 cm occurred in 1973–74 and 1978–79 respectively. The equivalent increases for the Krabbendijke salt marsh were 4, 10 and 4 cm respectively. The potential storage capacity of the estuary was not fully used after the Volkerakdam was built, because the tidal channels limited the flow of water into the estuary. However, the channels were later eroded. It took at least nine years for a new hydraulic equilibrium to be achieved.

Both the natural sea-level rise and the net accretion rate have to be taken into account when estimating the ecological effect of building barrages.

The effects of these climatic and hydraulic changes will now be discussed in relation to changes in the vegetation of permanent plots in the St. Annaland and Krabbendijke salt marshes.

St. Annaland salt marsh

Plot nr. 11, *Spartinetum anglicae* (Fig. 5). Height of the soil surface 1.62 m + NAP (1966). Effective high-water level rise 42–15 = 27 cm.

Although severe winter frost in 1962, 1978 and 1984 and summer drought in 1976 had no effect on the *Spartina* canopy, changes in the tidal regime did have an effect: cover decreased in 1969, 1975 and 1981, following the stepwise increases in the high-water level of 24, 11 and 7 cm respectively (Table 8). These temporary decreases in

Table 8. Variations in high-water levels during the period 1964–1984 at St. Annaland and Krabbendijke salt marshes, calculated from monthly averages of tides. All values are expressed as deviations from long-term averages till ultimo 1961.

Period	1964/65	65/66	66/67	67/68	68/69	69/70	70/71	71/72	72/73	73/74
St. Annaland salt marsh (long-term av. HWL 1.73 m +NAP = Bruinisse + 22 cm)										
year (Sept.-Aug.)	–4.1	2.0	– 0.4	– 0.6	+ 6.8	23.8	21.1	20.8	23.2	28.9
winter half year (Sept.-Febr.)	–2.7	4.0	+4.5	+3.3	– 1.0	24.2	25.0	17.2	23.2	35.5
summer half year (March-Aug.)	–5.5	0.0	– 5.3	– 4.5	+14.5	23.3	17.2	24.5	23.2	22.3
Krabbendijke salt marsh (long-term av. HWL 1,82 m +NAP)										
year (Sept.-Aug.)	+0.1	3.0	+2.2	+0.2	1.3	4.0	1.8	–0.2	3.5	7.5
winter half year (Sept.-Febr.)	–0.8	3.2	+6.5	+2.8	1.0	4.3	4.3	–3.7	3.5	13.2
summer half year (March-Aug.)	+0.8	2.8	–2.2	– 2.5	1.7	3.7	–0.8	+3.3	3.7	1.8

Period	1974/75	75/76	76/77	77/78	78/79	79/80	80/81	81/82	82/83	83/84
St. Annaland										
year (Sept.-Aug.)	34.1	27.7	33.7	34.0	37.7	40.8	39.3	41.0	42.1	42.3
winter half year (Sept.-Febr.)	40.0	32.3	35.5	38.2	37.0	38.7	40.5	45.7	42.0	47.8
summer half year (March-Aug.)	28.2	23.0	31.8	29.8	38.3	42.8	38.2	36.3	42.2	36.8
Krabbendijke										
year (Sept.-Aug.)	13.4	6.2	9.6	13.3	14.0	16.9	15.8	15.3	16.1	18.0
winter half year (Sept.-Febr.)	20.8	10.8	10.5	14.5	13.2	13.3	18.0	20.5	14.9	22.4
summer half year (March-Aug.)	6.2	1.5	8.7	12.0	14.8	20.5	13.7	10.0	17.2	13.5

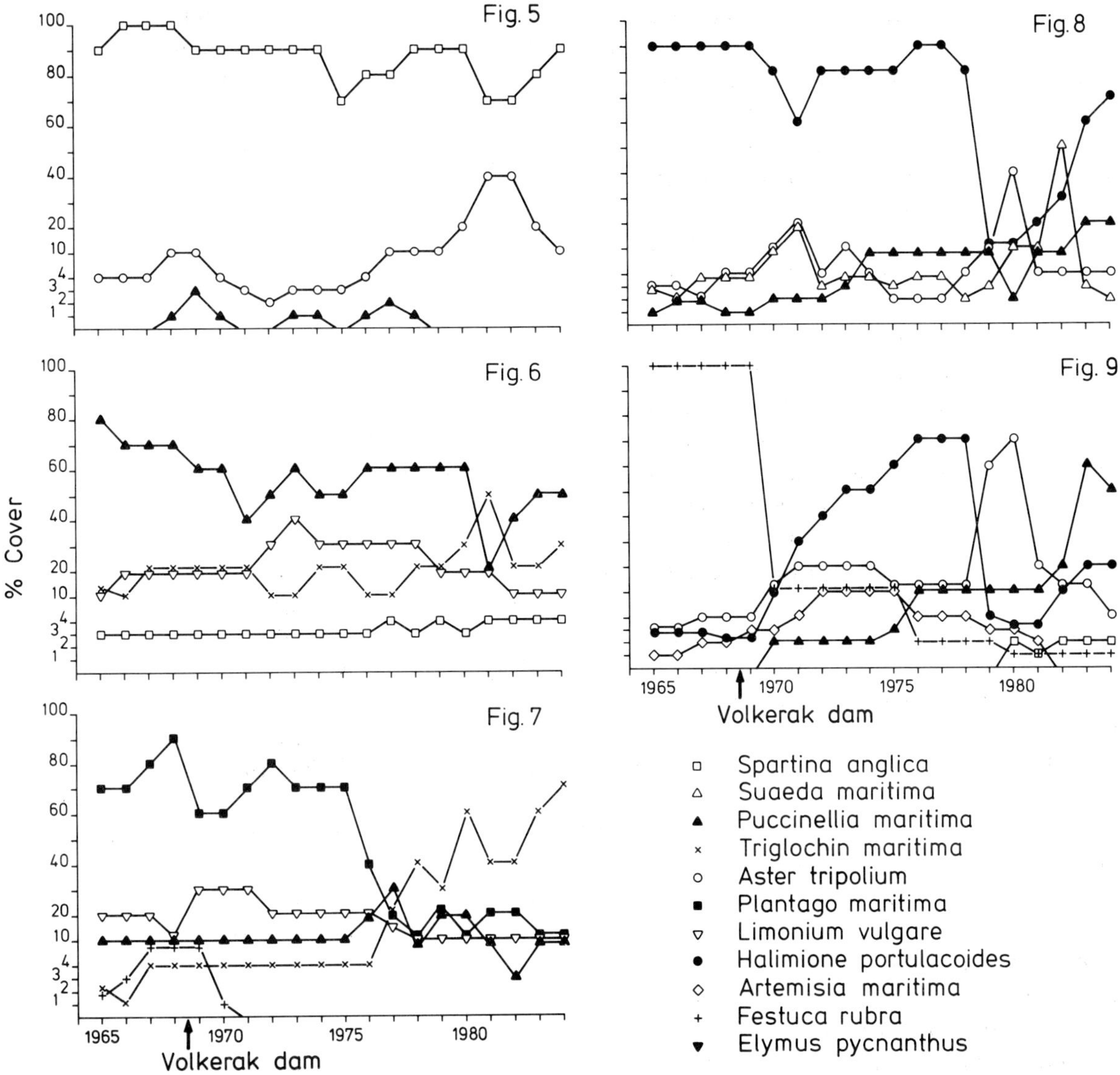

Figs. 5–9. Changes in the abundance (cover %) of the main species in 5 permanent plots of the St. Annaland salt marsh during 1965–1984.

5. Plot Nr. 11 Spartinetum anglicae. Surface 5 × 6 m.
6. Plot Nr. 7 Puccinellietum maritimae. Surface 2 × 6 m.
7. Plot Nr. 1 Plantagini-Limonietum. Surface 3 × 4 m.
8. Plot Nr. 13 Halimionetum portulacoidis. Surface 1×5 m.
9. Plot Nr. 3 Artemisietum maritimae. Surface 3×6 m.

Spartina anglica were balanced by an increase in *Aster tripolium* and, to a lesser extent, *Puccinellia maritima.* An overall effective high-water level rise of 36 cm did not affect the community as a whole.

Plot nr. 7, *Puccinellietum maritimae* (Fig. 6). Height of the soil surface 1.69 m+NAP (1966). Effective high-water level rise 42–4 = 38 cm.

The cover of *Puccinellia maritima* was least in 1971 and 1981, probably again related to increases in the high water level (Table 8), but later recovered, as in the case of *Spartina anglica* on plot nr. 11. *Triglochin maritima* generally increased when *Puccinellia* decreased. *Limonium vulgare* reached a maximum of 40% cover in 1973, probably because the flood rise cause that this species went over its optimal growth conditions with respect to the tidal factor. The cover of

Spartina anglica increased very gradually over the last 10 years. On the whole, the plant community has not yet changed structurally and syntaxonomically, but the tendencies in vegetational change show that it may do so in the long run.

Plot nr. 1, *Plantagini-Limonietum* (Fig. 7). Height of the soil surface 1.78 m+NAP (1966). Effective high-water level rise 42–(–6) = 48 cm.

The sudden flood rise in 1969 seems to have caused a decrease in *Plantago maritima* and an increase in *Limonium vulgare* in the following four years. The very dry summer of 1976 led to a drastic fall in cover/abundance of *Plantago maritima*, after which the species dit not recover, and a dramatic rise of *Triglochin maritima*. The plant community underwent an essential change in species composition giving reason to syntaxonomical transition from the Plantigini-Limonietum to a Puccinellietum maritimae rich in *Triglochin*.

Plot nr. 13. *Halimionetum portulacoidis* (Fig. 8). Height of the soil surface 1.84 +NAP (1966). Effective high-water level rise 42–13 = 29 cm.

The flood rise of 1969, and perhaps the winter frost in 1969/70, caused damage in *Halimione* with a minimal cover in 1971. Severe winter frost in 1978/79 brought about a dramatic fall from 80 to 10% cover, but the species recovered gradually in the following years. *Aster tripolium* and *Suaeda maritima* filled the 'clearings' in the *Halimione* canopy. In the meantime *Puccinellia maritima* increased gradually from 1 to 20%, probably indicating a syntaxonomical transition to a Puccinellietum maritimae. In another permanent plot bearing a *Halimione* community, situated 3 cm higher up the marsh but less silted up, *Puccinellia* had outstripped *Halimione* in cover percentage during the last years.

Plot nr. 3, *Artemisietum maritimae* (Fig. 9). Height of the soil surface 1.90 +NAP (1966). Effective high-water level rise 42−(−2) = 44 cm.

The flood rise caused *Festuca rubra* to decline dramatically, from 100 to 1% cover. *Artemisia maritima* had a maximum density of 10% in 1972–75, similar to that of *Limonium vulgare* (Fig. 6). *Halimione* took the place of *Festuca*, but was damaged by severe winter frost in 1976/77. *Puccinellia* seems to be the perennial species replacing *Halimione* at the end of the series. *Aster tripolium* filled the canopy gaps of perennial species. *Spartina anglica* invaded the plot unobtrusively. On the whole, the plant community changed from an Artemisietum maritimae to a Halimionetum portulacoidis, and will perhaps, under the stress of frost, change to a Puccinellietum maritimae.

Krabbendijke salt marsh

Plot nr. 3, *Puccinellietum maritimae* (Table 9). Height of the soil surface 2.29 m +NAP (1965). Effective high-water level rise 18–20 = −2 cm.

The increase in cover of *Aster tripolium* in 1967–1970 could be ascribed to the lowered salinity concentrations of the tidal water in 1965–69 (Table 7). From 1971 onwards *Triglochin* increased, and *Puccinellia* decreased, in cover percentages. After 1976, this trend was accelerated for some years, during which the cover of *Puccinellia* was minimal or even zero. These changes in cover of both species seem to be connected with changes in inundation frequencies, first by a rise in high-water level, and second by shrinkage of the soil. The increase in cover of *Limonium vulgare* during 1977–1983 may be the result of an accelerated decomposition rate of organic matter in 1976. Since the increase in flood level was balanced by siltation, this had little effect on the floristic composition of the community.

Plot nr 5, *Plantagini-Limonietum* (Table 10). Height of the soil surface 2.36 m +NAP (1965). Effective high-water level rise 18–13 = 5 cm.

In this plot *Puccinellia* increased, and *Plantago maritima* decreased after the closure of the Volkerakdam but, after 1976, their cover declined to about the original percentages. Over the whole period, *Triglochin* and *Limonium* increased, while *Plantago* decreased and *Armeria maritima* disappeared. The effective rise of the high-water level by 5 cm caused a general decline in species that are characteristic of the upper salt marsh

Table 9–12. Estimated cover percentages of plant species in four permanent plots in the Krabbendijke salt marsh during 1964–1984. For further explanations see Table 3.

Table 9. Puccinellietum maritimae (Surface 3 × 4 m).

	1964	65	66	67	68	69	70	71	72	73	74	75	76	77	78	79	80	81	82	83	84
Salicornia europaea	1	2		2	3	4	5	4	4	3	4	4	30	4	10	4	4	3	10	1	1
Suaeda maritima	10	4		1	3	5	5	20	5	4	10	4	20	4	4	2	4	3	4	1	2
Spartina anglica	1	1		1	1	2	2	2	2	2	3	2	2	3	3	2	2	3	3	3	4
Aster tripolium	4	3		10	20	30	10	3	4	3	3	2	2	3	3	2	2	1	3	4	3
Triglochin maritima	15	3		4	4	4	4	10	20	10	10	10	30	60	60	60	40	20	10	15	30
Puccinellia maritima	80	90		90	80	90	85	60	65	80	75	75	35	3	–	2	10	35	45	45	45
Limonium vulgare	30	30		20	20	30	30	30	30	30	25	25	25	45	45	55	55	55	50	40	25
Halimione portulacoides	2	2		2	2	1	2	1	2	1	1	1	2	2	2	1	2	2	2	2	2
Spergularia maritima	–	–		–	1	1	1	–	–	–	–	1	–	–	–	–	–	–	–	–	1

Table 10. Plantagini-Limonietum (Surface 2×6 m).

	1964	65	66	67	68	69	70	71	72	73	74	75	76	77	78	79	80	81	82	83	84
Salicornia europaea	2	1		2	2	2	2	4	–	3	4	1	4	2	4	1	3	1	4	1	2
Suaeda maritima	4	2		1	1	1	2	20	2	2	4	–	4	3	4	4	4	1	4	1	2
Spartina anglica	–	–		–	–	–	–	–	–	1	–	–	–	–	–	–	–	–	–	–	–
Aster tripolium	4	3		4	4	10	10	4	10	4	4	3	3	4	10	10	3	2	3	4	10
Triglochin maritima	3	3		4	4	3	3	3	3	3	2	4	4	4	4	10	5	10	40	50	35
Puccinellia maritima	3	5		10	10	10	10	30	25	30	40	20	35	3	5	10	15	4	5	10	20
Limonium vulgare	30	30		20	25	30	30	35	30	30	35	30	30	45	45	40	55	25	35	40	45
Halimione portulacoides	3	3		3	3	2	2	2	2	2	3	2	4	3	2	2	1	1	2	2	1
Spergularia maritima	1	2		3	2	3	1	4	4	4	2	3	3	2	2	2	3	3	3	2	3
Plantago maritima	40	50		70	70	70	70	35	40	40	40	55	45	55	60	65	55	70	30	3	3
Glaux maritima	–	1		–	3	3	3	3	2	2	3	2	4	2	2	2	2	3	3	4	4
Festuca rubra	2	3		20	2	10	10	3	2	1	1	1	1	2	1	1	1	1	2	1	1
Juncus gerardii	–	1		2	2	2	3	2	1	2	3	3	–	3	–	–	–	3	1	1	1
Artemisia maritima	30	10		10	10	15	15	4	2	1	–	3	–	–	2	–	–	–	–	–	–
Elymus pycnanthus	–	–		–	–	1	1	–	2	2	2	3	2	1	–	–	–	–	–	–	–
Atriplex hastata	–	–		–	–	–	–	–	2	–	–	–	–	–	–	–	1	–	–	–	–

Table 11. Halimionetum portulacoidis (Surface 2 × 6 m).

	1964	65	66	67	68	69	70	71	72	73	74	75	76	77	78	79	80	81	82	83	84
Salicornia europaea	–	–		–	–	–	1	–	–	–	–	–	–	–	–	–	–	–	–	–	–
Suaeda maritima	2	4		4	2	2	10	10	1	1	3	–	3	10	4	4	3	3	3	3	–
Spartina anglica	3	3		3	2	1	1	–	–	–	–	–	–	–	–	–	1	2	2	2	–
Aster tripolium	–	–		4	3	3	3	3	3	1	3	2	2	1	1	1	3	2	2	3	2
Triglochin maritima	–	1		–	–	–	–	–	–	–	–	–	–	–	–	–	–	–	–	–	–
Puccinellia maritima	–	1		20	15	10	2	2	1	2	1	–	–	–	–	–	–	–	–	–	2
Limonium vulgare	1	2		2	–	–	–	–	–	–	–	–	–	–	–	–	–	–	–	–	–
Halimione portulacoides	99	90		85	80	60	60	50	20	40	30	40	60	60	60	30	30	10	35	35	65
Festuca rubra	–	1		5	5	10	1	1	1	–	–	–	–	–	–	–	–	–	–	–	–
Elymus pycnanthus	–	–		3	20	40	50	60	90	60	80	65	40	45	50	75	75	90	75	75	40
Atriplex hastata	–	–		–	1	1	5	2	2	2	3	3	15	10	2	2	2	2	10	3	–
Altriplex littoralis	–	–		–	–	–	1	–	–	–	–	–	–	–	–	–	–	–	–	–	–

Table 12. Artemisietum maritimae (Surface 2 × 6 m).

	1964	65	66	67	68	69	70	71	72	73	74	75	76	77	78	79	80	81	82	83	84
Salicornia europaea	–	–		–	–	–	2	5	–	3	3	4	2	4	10	2	2	–	30	2	2
Aster tripolium	2	1		2	1	–	–	1	1	–	2	2	–	1	–	2	2	2	2	5	2
Puccinellia maritima	–	1		–	–	–	1	1	1	2	1	–	–	–	–	–	1	–	–	–	–
Limonium vulgare	–	–		–	–	–	–	–	–	–	–	–	–	–	–	–	–	–	–	–	1
Halimione portulacoides	1	1		1	1	–	–	1	–	–	–	–	–	–	–	–	1	1	1	2	5
Spergularia maritima	–	–		–	–	–	–	–	–	–	–	–	–	–	–	–	–	–	–	–	2
Plantago maritima	–	1		–	–	–	–	–	–	–	–	–	–	–	–	–	–	–	–	–	–
Festuca rubra	40	70		70	90	90	10	–	–	–	–	–	–	–	–	–	–	–	–	–	–
Artemisia maritima	70	40		45	25	15	3	3	3	3	3	1	–	1	2	5	2	2	2	1	2
Elymus pycnanthus	–	1		2	5	20	10	10	40	60	99	75	15	4	45	90	99	99	65	80	90
Atriplex hastata	2	3		5	20	40	90	95	60	50	15	30	90	99	75	55	5	2	25	2	1
Atriplex littoralis	3	1		1	–	–	1	1	1	1	–	1	2	2	1	–	–	–	1	1	–

(Armerion maritimae), and an extension of the species with optima in the middel marsh (Puccinellion maritimae). The community therefore changed from the Plantagini-Limonietum to the Puccinellietum maritimae.

Plot nr. 7, *Halimionetum portulacoidis* (Table 11). Height of the soil surface 2.39 m +NAP (1965). Effective high-water level rise 18–20 = −2 cm.

The vegetation shows a dynamic relation between *Halimione portulacoides* and *Elymus pycnanthus*. The latter entered the plot two or three years before 1969, perhaps as a result of the lowered salinity of the tidal water (Table 7). Another explanation is that the species established by supply of nutrients from algae washed ashore in the plot. In any case yearly variations in the amount of these algae are the cause of reversed reactions of both species. The chamaephyte *Halimione* is easily choked by these algae, while the geophyte *Elymus* is not damaged. *Puccinellia* and *Festuca rubra* disappeared, either because of the algal mats, or as a result of the heavy growth and litter production of *Elymus*. The effect of the flood-level rise was minimal or masked by the influence of the decaying organic matter.

Plot nr. 9, *Artemisietum maritimae* (Table 12). Height of the soil surface 2.49 m +NAP (1965). Effective high-water level rise 18–14 = 4 cm.

This plot shows changes similar to those on plot 7. Large amounts of organic matter choked *Festuca rubra* and *Artemisia maritima*, and promoted the expansion of *Elymus pycnanthus*, or even choked this species in favour of annual *Atriplex* species. It is not known whether the change of this plant community from Artemisietum maritimae to Atriplici-Elytrigietum pungentis was caused primarily by a lowered salinity of the substrate in the years before 1969, or by an increased supply of organic matter in which the rise of the high-water level may be a driving force.

4. Discussion

4.1 Cosmic variations in the tidal cycle

It is well-known that the vertical distribution of tidal marsh biota is closely related to the vertical movements of the tides (Johnson & York, 1915; Chapman, 1960; Ranwell, 1972). This zonation is the result of cyclic patterns of salinity, water potential and osmolarity of the substrate (Jefferies et al., 1979).

Tidal cycles are highly predictable, though modified by climatic deviations and isostatic changes in the earth's crust (Fig. 3); these deviations can result in vegetational changes which may be considered natural disturbances. However, the existing biota are generally adapted to the naturally occurring cycles, and the semi-diurnal cycle of alternating high-water levels does not seem to 'cause partial or total destruction of plant biomass'. For example, many arthropods living in the salt marsh have structures or mechanisms enabling them to avoid or endure a limited submergence with salt water (Foster & Treherne, 1976a). Some other species have periods in their life cycle during which they are vulnerable to tidal submergence. Newly born apterae of the root-aphid *Pemphigus trehernei* Foster, for instance, are killed by a 2–h submergence in the first subsequent 80 min. of their life, probably by the absence of an adequately waterproofed cuticle (Foster & Treherne, 1976b). In the lower marsh, the response of arthropods to the frequency and duration of submergence differ according to their feeding habit (Heydemann, 1979). Only a few species show a rhythm that is in phase with the 25–h tidal cycle: the intertidal collembolan *Anurida maritima* (Guérin), for instance, emerges from its underground refuges for foraging shortly after the retreat of the tide, and retire underground at least 1 h before the return of the tide (Foster & Moreton, 1981).

In some cases the life cycle is related to the spring-neap tide cycle (Fig. 2). For example, young seedlings of *Salicornia* and *Spartina* establish in the mud at the marsh's edge during neap tide, before being uprooted by mechanical action of waves and tidal currents at spring tide (Wiehe, 1935; Groenendijk, 1985). Wiehe (1935) observed a sharp rise in the mortality rate of *Salicornia* at the level where the marsh is subjected to daily flooding. It appeared that seedlings of this species need to be left uncovered for at least 2–3 days after germination for successful establishment, with the note that both germination and root penetration are temperature dependent.

Chapman (1960) reported that, in the case of *Aster tripolium,* normally five days must elapse before seedlings are sufficiently rooted to prevent being swept away by tidal action. Hubbard (1969) found that germination of seed of *Spartina anglica* is inhibited by light and suggested that, under the sheltered positions of Poole Harbour, tidal immersion would favour germination by a reduction in the intensity and duration of light. However, in non-exposed sites, as at bridgwater Bay, Somerset, *Spartina* can establish when its immersion time is less than ⅓ of that at Poole Harbour (Morley, 1973). Groenendijk (1985) has shown that seed of *Spartina* can be lost through local short-term erosion/sedimentation processes in a 3 cm thick top layer of the mud. This seed loss decreased with soil depth, but increased significantly on lower levels of the mud flat.

The spring-neap tide cycle also influences physiological activities, such as photosynthesis, respiration, and salt secretion of plants in the lower marsh. Ranwell et al. (1964) pointed out, that plants of *Spartina* in the lower marsh of Poole Harbour were more susceptible to small but consistent changes in sea level than in those growing in the upper reaches of the estuary. Hubbard (1969) attributed the more limited seaward spread of *Spartina* to periods of immersion during neap tides rather than spring tides, though Groenendijk (1985) stressed hydrodynamic (wave) action as the dominant factor. Areas higher up the estuary are submerged with water of lower salinities and are less exposed to mechanical action of waves and currents. Various salt-marsh species and aquatic fauna elements of supralittoral biotopes show therefore the brackish-water submergence or diving effect (Beeftink, 1977a) along this gradient.

The spring-neap tide cycle also causes variations in soil aeration in the salt marsh. Armstrong et al. (1985) found that neap tides caused brief intervals of surface oxidation even in the lower marsh, while spring tides caused short intervals of surface reduction even in the upper marsh. The oxidation-reduction fluctuations showed different patterns in different community types. However, as Armstrong et al. (1985) stressed, the relationship between vegetation types and frequency of tidal flooding does not neccessarily indicate a causal relationship between soil aeration and vegetation type. Nevertheless, our results have shown that change in the tidal pattern does result in vegetational change. Whether such a vegetational change will be interpreted as succession or regression is a matter of syntaxonomical evaluation of the species in question.

Although tidal variations caused by the declination of both sun and moon may be too small (3–5 cm on a year basis each) to affect vegetation, nodal cycles (Fig. 3) may cause accretion of the salt marsh during the period or rising tides (about 9 years), and may thus promote a stepwise succession, though with alternating successions and regressions. Studying these vegetation dynamics requires a long series of vegetation analyses, with little variation in other factors.

4.2 Other natural factors affecting the water level

Wind is a major factor causing deviations in the tidal pattern of the south-west Netherlands (Fig. 3). This may then influence the cover/abundance of some plant species. Even one year of tides deviating, on average, by more than 5–10 cm over the whole year, can lead to changes in the density of some plant populations (Tables 3 and 4)). Such changes seem to be more rapid when the tides are lower than normal compared with those occurring when the tides are higher than normal. Succession trends may thus be more speedy than regression trends, so that the tidal mechanism in itself would promote succession rather than regression. This tendency may be associated with an overall gradual emergence of the soil substrate from tidal influence, by progressive and irreversible step-by-step changes in drainage conditions and soil ripening.

In general, net accretion of soil on the salt marshes investigated have kept pace with the rise in sea-level (15–20 cm century^{-1}), or even the rise in high-water level (34–38 cm century^{-1}). Redfield & Rubin (1962) and Redfield (1967) studying the Atlantic Barnstable estuary (Massachusetts), and Phillips (1981) studying estuaries along the southern California coast, have also found accretion of tide lands keeping up the rise in sea level over periods of about 2000 years. However, over periods of less than the last 100 years, Stevenson et al. (1985), working in Chesapeake Bay, and Baumann et al., (1984), working in the Mississippi delta, found that accretion had not kept pace with sea-level rise. The ocean's level is expected to rise rapidly as a consequence of the increasing CO_2-emissions through burning of fossil fuels; estimates of between 0.3 and 4.5 cm yr^{-1} have been made for the coming two centuries (Delft Hydraulics Laboratory, 1980; Prins, 1986). Accretion is supposed to match such a sharp rise in only very few localities, leading to catastrophic degradation and deterioration of tidal wetlands (see also Orson et al., 1985).

Seismic activity can give rise to large variations in the relative levels of sea and land (Steers, 1977), often causing complex sequences of marsh formation and erosion (Chapman & Ronaldson, 1958). These movements are often slow, though long continued, and changes in the sea level are generally compensated for by marsh accretion.

Fennoscandia, Scotland and much of the northern part of America are noticeable areas with postglacial and tectonic land uplift (Ericson & Wallentinus, 1979). Especially the coasts around the Gulf of Bothnia are marked by an upward movement to about 100 cm century^{-1} (Fig. 4), causing considerable sea-shore displacements. Coastal marsh thickness is minimal, and accretion is mainly creating newly emerging land areas. Those marshes, fine-grained shores and bedrock outcrops, are subjected to different kinds of isolation: topographic, hydrological and biological (Ingmar, 1982). Apart from these external isolations, long-lived perennial plants growing on these sites may get isolated from their environments, even within their life-span. Succession is markedly accelerated by land uplift, but does not show elements of disturbance except in cases of sudden isolation (Ericson & Wallentinus, 1979). As quoted by Cramer (1980), Schwanck (1974) found however, that the influence of land uplift predominates the internal dynamics of the vegetation; species from higher zones invade before the vegetation has become fully developed.

4.3 Effects of coastal engineering works

The closure of estuarine arms with barrages generally results in a sudden increase of the tidal difference at the sea-side, and a decrease or total disappearance of the tides at the river-side of the barrage. The rise in high-water levels decreases from the barrage to the sea. Salinity has usually been increased at the sea-side of the barrage. There, bigger water masses erode the tidal gullies and creeks, and throw sand and silt upon the marsh surface, neutralizing partly the effect of flood rise.

From studies in the Volkerak-Oosterschelde system it appeared that the vegetation of the upper marsh is most vulnerable to an abrupt rise of the high-water level, and the vegetation of the lower marsh least (Table 13). This is related to changes in submergence: In low marsh levels with daily flooding only the duration of submergence has slightly been increased. On higher marsh levels, however, the frequency of submergence increased progressively. Each dominant plant spe-

Table 13. Maximal abrupt increase in high-water level (cm) at which the vegetation recovers after disturbance through increase in tidal difference and salinity (from oligo/meso- to polyhalinity) as a result of barrage building in an estuary (modified after Beeftink, 1979).

Community	HW increase	Salt-marsh level
Agropyro-Rumicion crispi	1–3	upper edge
Armerion maritimae	3–5 (10)	upper
Puccinellion maritimae	10–15 (40)	middle
Spartinion	30–50	low

cies characteristic of a particular salt-marsh level is supposed to have a threshold value in frequency and duration of submergence above which a flood rise causes die-back. The mechanism of flood rise suggests that this die-back is caused by an increased number of exposures to submergence, changing the redox potential of the rooted substrate to more reduced conditions, as Armstrong et al. (1985) found.

Clearings or gaps in the canopy are occupied by mostly annual and short-lived perennial species *(Salicornia, Suaeda, Aster, Puccinellia maritima)*. These species start a secondary succession series which ends either in the same developmental vegetation stage as before (favoured by an accelerated accretion of the soil surface), or in an earlier developmental stage related to the magnitude of the tidal rise (see also Beeftink, 1979; Beeftink et al., 1978).

Removal of tidal circulation at the riverside of the barrage causes dramatical changes in halophyte populations. Several species start with an increased density *(Salicornia)* and productivity *(Puccinellia, Triglochin, Limonium, Halimione)* before dying (see also Zedler et at., 1980; Groenendijk et al., 1987). They are replaced by luxuriously growing plant species of fresh habitats, mostly ruderals and grassland species. Only some halophyte species from the upper marsh, such as *Glaux maritima, Festuca rubra* and *Elymus pycnanthus,* persist many years, or even establish (Table 5).

Contrary to the effect of flood rise, removal of the tides has its highest impact in the lower marsh, and its lowest in the upper marsh (Table 6). This gradient is the reflexion of a gradient of accelerated processes of soil ripening (desiccation, shrinkage, aeration, decomposition of organic materials, desalinization), which is most intense in the lower muddy salt marshes of brackish origin. Grazing may interfere in the lower marsh, in so far as trampling impedes the process of soil ripening, and grazing keeps the vegetation open. In the upper marsh, however, the soil underwent already some ripening under tidal conditions. There, early desalinization and grazing promote the establishment and vegetative reproduction of some grassland spiecies, such as *Poa trivialis, P. pratensis, Agrostis stolonifera* and *Trifolium repens.* These species may compete with the halophyte species more than under ungrazed conditions.

The construction of a storm-surge barrier such as in the mouth of the Oosterschelde, will result in a permanent reduction of the tides. Groenendijk et al. (1987) estimate that 50–60 cm tidal reduction will not cause a dramatic vegetational change. Although the proportional share of the species in the upper marsh will change detrimental to the species with their optimum in the lower marsh, the authors found no evidence that the *Spartina* belt which fringes most salt marshes, will deteriorate, so that little space will be left for the species growing originally in the middle and upper marsh areas.

Long-term investigations in the salt marsh reveal that environmental changes of different kind succeed each other, or even appear in the same season. The examples showed that relative sea-level rise, soil accretion, fluctuation in salinity of the flooding water and in air temperature, drought, and changes in tidal sumbergence as a consequence of coastal engineering occur within a period of 20 years. Each of these environmental changes influence the vegetation more or less. Some of them are neutralizing such as the relative sea-level rise and soil accretion. Others intensify each other, such as the relative sea-level rise, shrinkage through drought, and the increased flood level caused by the construction of a barrage across an estuarine arm. Also severe winter frost can accelerate vegetation regression. This illustrates that the impact of a single environmental change on the vegetation is difficult to establish in the field, because different concomitant regimes of disturbance are possible (see also Beeftink, 1977b).

4.4 Types of disturbance and their effects on biota

A basic distinction can be made between continuous and periodic disturbance. Continuous disturbance produces a long-persistent type of vegetation distinct from non-disturbed communities in floristic and structural sense (for instance, grazing will simultaneously suppress vulnerable species, and favour others (Denslow, 1980); this can find expression syntaxonomically: a grazed Puccinellietum maritimae versus an ungrazed Plantagini-Limonietum. Periodic disturbance produces transient gaps in the canopy in which regeneration occurs along secondary succession series (Beeftink, 1979). The abrupt rise of the high-water level after the construction of a barrage accross an estuarine arm is an illustration of this type of disturbance.

Periodic disturbance differs in frequency, extent and intensity. Frequencies of less than 0.25 yr^{-1} are found to be maximal for recovery of early-developmental plant communities in the salt marsh, and 0.08 yr^{-1} for mature plant communities, premised that the substrate remained undamaged (Beeftink, 1979). Large areas covering different plant communities might need a much longer period of recovery, as Dicks (1977) has shown for the recovery of a 60 ha salt marsh denuded by repeated oilings over a period of 20 years. The impact of disturbances of different intensities is illustrated by the influence of flood rises of different magnitude, but also by different grazing intensities, removal of the vegetation with or without the underlying soil by erosion or open-cast mining, etc. (Beeftink, 1977b).

Steele (1985), discussing the difference between temporal variability of marine and terrestrial ecological systems distinghuished between white and red noise (by analogy of the spectrum of light). Red noise (for time scales > 50 yr) would be typical for marine environments, because atmospheric short-term variability is damped off by the very large heat dissolving capacity of the oceans. Terrestrial environments should be more subject to external forces by white noise including also frequencies of < 0.02 yr^{-1}. The internal dynamics and structures of marine and terrestrial systems are expected to differ in significant ways in response to the temporal character of each physical environment leading to significantly different types of systems. The tidal salt-marsh ecosystem might take an intermediate position: In the lower parts marine characteristics would be preponderant; biota higher up might be gradually more subject to disturbance of high frequencies. Such a gradient is suggested in the vegetation response to an abrupt rise in the high-water level (Table 13).

It is evident that an abrupt rise or fall in the tidal difference as a consequence of coastal engineering has all properties of a disturbance *sensu* Grime (1979) with respect to the salt-marsh ecosystem. On the other hand, also for most gradual and very predictable changes in the tidal circulation initiated by cosmic influences and isostatic or tectonic earth movements disturbance may be a very precarious concept. Plant biomass is mostly not affected by these environmental changes, but among the invertebrates disturbance may occur. This argues that different groups of organisms need – or have the opportunity to avoid – disturbance of different kind and different time scales.

The effect of disturbance increases with its intensity. Tidal deviations caused by wind effects only result in partial die-back in distinct plant species growing on distinct levels of the marsh. Very large abrupt changes in the tidal circulation may effect the vegetation of large parts of the marsh. (Partial) mortality for some individuals or (sub)populations causing gaps in the canopy is at the same time a source of establishment of others. Natural salt-marsh vegetation can largely be considered a mosaic of disturbance patches of different kinds and intensities of impact and different size-frequencies, all together composing the disturbance regime (Godron & Forman, 1984).

This view closely corresponds to the hypotheses of both Connell (1978) and Huston (1979) who postulate that variations in species diversity of plant communities are the result of variations in the rates of within-community disturbance (Connell, 1978), or of variations on the 'dynamic equilibrium' between disturbance rates and the rates of competitive displacements of species (Huston, 1979). Both argue that a competitive equilibrium is probably not common in nature. In mosaics like the salt-marsh vegetation a competitive equilibrium would only occur in an environment in which the size-frequency distribution of component species have evolved (Denslow, 1980). The historic size-frequency distribution of disturbances determined the conditions of the mortality and regeneration patterns of the existing species, and therefore influenced within- and between-community variations in diversity. Denslow (1980) concluded that this model offers an evolutionary reason why levels of disturbance lower or higher than an 'intermediate level' are likely to result in lower species diversity; she also suggested that the so-called 'intermediate level' of disturbance with highest total species diversity (Connell, 1978; Huston, 1979) may be largely different from community to community. It throws new light not only on the mechanisms shaping plant communities, but also on the diagnostic value of this shaping in the classification of vegetation for evaluating man-induced environmental forces.

Acknowledgements

The author is much indebted to Messrs. Tj. de Haan and J.G. de Ronde (Rijkswaterstaat, The Hague) for technical information on tide circulations and relative sea-level rise, to the Royal Netherlands Meteorological Institute (De Bilt) for delivering climatic data, and to Dr. J. van Andel (Groningen) and Dr. R.W. Snaydon (Reading) for critical remarks on the text. Most vegetation analyses were carried out by my collaborator Mr. W. de Munck; the levelling excercises were executed by Messrs. M.C. Daane and W. de Munck. Mrs. M.J. van Leerdam and Mr. A.A. Bolsius prepared typework and figures.

References

Armstrong, W., Wright, E.J., Lythe, S. & Gaynard, T.J. (1985). Plant zonation and the effect of the spring-neap tidal cycle on soil aeration in a Humber salt marsh. J. Ecol. 73: 323–339.

Baumann, R.H., Day, J.W. & Miller, C.A. (1984). Mississippi deltaic wetland survival: sedimentation versus coastal submergence. Science 224: 1093–1095.

Beeftink, W.G. (1975). The ecological significance of embankment and drainage with respect to the vegetation of the South-West Netherlands. J. Ecop. 63: 423–458.

Beeftink, W.G. (1977a). The coastal salt marshes of Western and Northern Europe: an ecological and phytosociological approach. In: V.J. Chapman (Ed.), Wet coastral ecosystems, pp 109–155. Elsevier, Amsterdam.

Beeftink, W.G. (1977b). Salt-marshes. In: R.S.K. Barnes (Ed.), The coastline, pp 93–121. Wiley, London.

Beeftink, W.G. (1979). The structure of salt marsh communities in relation to environmental disturbances. In: R.L. Jefferies & A.J. Davy (Eds.), Ecological processes in coastal environments, pp 77–93. Blackwell, Oxford.

Beeftink, W.G., Daane, M.C., de Munck, W. & Nieuwenhuize, J. (1978). Aspects of population dynamics in *Halimione portulacoides* communities. Vegetatio 36: 31–43.

Chapman, V.J. (1960). Salt marshes and salt deserts of the world. Leonard Hill, London, 392 pp.

Chapman, V.J. & Ronaldson, J.W. (1958). The mangrove and salt marsh flats of the Auckland Isthmus. N.Z. Dep. Sci. Res. Bull. 125: 1–79.

Connell, J.H. (1978). Diversity in tropical rain forests and coral reefs. Science 199: 1302–1310.

Cramer, W. (1980). Strandverschiebung und Strandvegetation auf Häverö Prästäng. Acta Phytogeogr. Suecica 68: 42–50.

De Ronde, J.G. (1982),. Changes of relative mean sea-level and of mean tidal amplitude along the Dutch coast. Proc. NATO Advanced Research Workshop, Utrecht, June 1–4, 1982.

Delft Hydraulics Laboratory (1980). Effect of a mean ocean level rise on the tidal system of the North Sea. Note prepared for the IFIAS Seminar 'Scanning our Changing Planet', Delft, October 1980.

Denslow, J.S. (1980). Patterns of plant species diversity during succession under different disturbance regimes. Oecologia 46: 18–21.

Dicks, B. (1977). Changes in the vegetation of an oiled Southampton Water salt marsh. In : J. Cairns, K.L. Dickson & E.E. Herricks (Eds.), Recovery and restoration of damaged ecosystems, pp 208–240. University Press of Virginia, Charlottesville.

Doing Kraft, H. (1954). L'Analyse des carrés permanents. Acta Bot. Neerl. 3: 421–424.

Ericson, L. & Wallentinus, H.G. (1979). Sea-shore vegetation around the Gulf of Bothnia. Wahlenbergia 5: 1–142.

Foster, W.A. & Moreton, R.B. (1981). Synchronization of activity rhythms with the tide in a salt marsh collebolan *Anurida maritima*. Oecologica 50: 265–270.

Foster, W.A. & Treherne, J.E. (1976a). Insects of marine salt marshes: problems and adaptations. In: L. Cheng (Ed.), Marine insects, pp 5–42. North-Holland Publ. Cy. Amsterdam.

Foster, W.A. & Treherne, J.E. (1976b). The effects of tidal submergence on an intertidal aphid, *Pemphigus treherni* Foster. J. Animal Ecol. 45: 291–301.

Godron, M. & Forman, R.T.T. (1984). Landscape modification and changing ecological characteristics. In: Mooney, H.A. & Godron, M. (Eds.), Disturbance and ecosystems. Components of response, pp 12–28. Springer-Verlag, Berlin.

Grime, J.P. (1979). Plant strategies and vegetation processes. Wiley, Chichester, 222 pp.

Groenendijk, A.M. (1985). Establishment of a *Spartina anglica* population on a tidal mudflat: a field experiment J. Environm. Manag. (accepted).

Groenendijk, A.M., Spieksma, J.G.J. & Vink-Lievaart, M.A. (1987). Effects of different flooding regimes on growth of and interactions between salt-marsh species. Vegetatis (accepted).

Grubb, P.J. (1985). Plant populations and vegetation in relation to habitat, disturbances and competition: problems of generalization. In: J. White (Ed.), The population structure of vegetation, pp 595–621. Dr. W. Junk Publishers, Dordrecht.

Gutenberg, B. (1941). Changes in sea level, postglacial uplift, and mobility of the earth's interior. Bull. geol. Soc. Amer. 52: 721–772.

Hays, J.D., Imbrie, J. & Shackleton, N.J. (1976). Variations in the earth's orbit: pacemaker of the ice ages. Science 194: 1121–1132.

Heydemann, B. (1979). Responses of animals to spatial and temporal environmental heterogeneity within salt marshes. In: R.L. Jefferies & A.J. Davy (Eds.), Ecological processes in coastal environments, pp 145–163. Blackwell, Oxford.

Hogeweg, P., Hesper, B., van Schaik, C. & Beeftink, W.G. (1985). Patterns in vegetation succession. An ecomorphological study. In: J. White (Ed.), The population structure of vegetation, pp 637–666. Junk, Den Haag.

Hubbard, J.C.E. (1969). Light in relation to tidal immersion and the growth of *Spartina townsendii* (s.l.). J. Ecol. 57: 795–804.

Huston, M. (1979). A general hypothesis of species diversity. Amer. Nat. 113: 81–101.

Ingmar, T. (1982). Isolation of brackish water basins due to land uplift. Medd. Växtbiol. Inst. Uppsala 1982 (3): 121–125.

Jefferies, R.L., Davy, A.J. & Rudmik, T. (1979). The growth strategies of coastal halophytes. In: R.L. Jefferies & A.J. Davy (Eds.), Ecological processes in coastal environments pp 243–268. Blackwell, Oxford.

Johnson, D.S. & York, H.H. (1915). The relation of plants to tide levels. Washington, Carnegie Inst., 162 pp.

Morley, J.V. (1973). Tidal immersion of *Spartina* marsh at Bridgwater Bay, Somerset. J. Ecol. 61: 383–386.

Odum, E.P. (1971). Fundamentals of ecology. Third ed. Saunders Company, Philadelphia, 574 pp.

O'keefe, Ph., Westgate, K. & Wisner, B. (1976). Taking naturalness out of natural disasters. Nature 260: 566–567.

Orson, R., Panageotou, W. & Leatherman, S.P. (1985). Response of tidal salt marshes of the U.S. Atlantic and Gulf coasts to raising sea levels. J. Coast. Res. 1: 29–37.

Phillips, R.P. (1981). The long term natural stability of the estuarine marshlands along the southern California coast. Bull. Marine Sci. 31: 814.

Prins, J.E. (Ed.) (1986). Impact of sea level rise on society. Part 1. Proc. Workshop 27-29 August 1986, Delft Hydraulics Lab., Delft, 34 pp.

Ranwell, D.S. (1972). Ecology of salt marshes and sand dunes. Chapman & Hall, London, 258 pp.

Ranwell, D.S., Bird. E.C.F., Hubbard, J.C.E. & Stebbings, R.E. (1964). *Spartina* salt marshes in Southern England. V. Tidal submergence and chlorinity in Poole Harbour. J. Ecol. 52: 627–641.

Redfield, A.C. (1967). The ontogeny of a salt marsh estuary. In : G.H. Lauff (Ed.), Estuaries, pp 108–114. Am. Assoc. Adv. Sci. Publ. 83.

Redfield, A.C. & Rubin, M. (1962). The age of saltmarsh peat and its relation to recent changes in sea level at Barnstable, Mass. Proc. Natl. Acad. Sci. U.S. 48: 1728–1735.

Reiners, W.A. (1983). Disturbance and basic porperties of ecosystem energetics. In: Mooney, H.A. & Godron, M. (Eds.), Disturbance and ecosystems. Components of response, pp 83–98. Springer-Verlag, Berlin.

Rijkswaterstaat (1961–1982). Jaarboeken der waterhoogten, afvoeren, watertemperaturen en windgegevens. Directie Waterhuishouding en Waterbeweging, Den Haag.

Rijkswaterstaat (1964). Tienjarig overzicht der waterhoogten en afvoeren 1951–1960. Staatsuitgeverij, Den Haag, 441 pp.

Rijkswaterstaat (1985). Getijtafels voor Nederlands 1986. Staatsuitgeverij, Den Haag, 151 pp.

Rossiter, J.R. (1967). An analysis of annual sea level variations in European waters. Geophys. J.R. astr. Soc. 12: 259–299.

Schwanck, B. (1974). Vegetationens förändring på de framväxande skären. In: Nordenskiöldsamf. i Finland (Ed.), Skärgård i omvandling, pp. 79–85. Helsingfors.

Steele, J.H. (1985). A comparison of terrestrial and marine ecological systems. Nature 313: 355–358.

Steers, J.A. (1977). Physiography. In: V.J. Chapman (Ed.), Wet coastal ecosystems, pp 31–60. Elsevier, Amsterdam.

Stevenson, J.C., Kearney, M.S. & Pendleton, E.C. (1985). Sedimentation and erosion in a Chesapeake Bay brackish marsh system. Marine Geol. 67: 213-235.

Van Boven, P.J. (1981). Chloridegehalte en watertemperatuur van de Oosterschelde bij Yerseke over de periode 1964–1980. Rapp. Versl. DIHO, Yerseke, 70 pp.

Van der Plassche, O. (1983). Zeespiegelbewegingen in Nederland gedurende de laatste 7000 jaar. Jaarboek 1982 Neth. Organization for Pure Scient. Res., Versl. & Beschouwingen, pp. 149–186.

Wiehe, P.O. (1935). A quantitative study of the influence of tide upon populations of *Salicornia europaea.* J. Ecol. 23: 232–333.

Zedler, J.B., Winfield, T. & Williams, P. (1985). Salt marsh productivity with natural and altered tidal circulation. Oecologia 44; 236–240.

CHAPTER 8

Vegetation development from denuded ground to grassland

A. Fischer

Detailed contents

1. Introduction

Most European grasslands, i.e. meadows and pastures, arose as a result of lifestock production. Nowadays, however, much uncultivated grassland occurs on land between areas intensively used for agriculture, traffic, habitation, and recreation. Much of this grassland is situated on embankments of roads, railways, and agricultural terraces. Its main function is to prevent soil erosion. Many of these grasslands arise as a result of the total destruction of the previous relief, soil, and vegetation, followed by the reconstruction of a new system, which normally differs substantially from the previous one.

To create an efficient protection against soil erosion within a short time, a seed mixture is usually sown on the steep slopes, immediately after the construction of the new areas. The mixture commonly consists mainly of various grass and legume species. The species differ in speed of establishment and the grassland vegetation develops during the following years.

Sometimes the vegetation on embankments and terrace slopes is managed by mowing, either once a year or irregularly every two or three years, usually in late summer. This grassland management serves for safety on road and railway. As a consequence of this management, a floristic composition arises which is similar to that of a meadow community under cultivation, although it is not used as such. This vegetation type, regularly dominated by *Arrhenatherum elatius*, is known as 'ruderal meadow'. It is a very common grassland community in town boroughs of Central Europe, especially on road and railway embankments (Fischer, 1985a). It is classified as a community of the alliance Arrhenatherion (tall oatgrass meadows), and closely related to the cultivated meadow community Arrhenatheretum elatioris Br.-Bl. ex Scherr. 25, from which it differs by the occurrence of such species as *Artemisia vulgaris, Tanacetum vulgare, Linaria vulgaris, Urtica dioica,* and *Solidago canadensis.*

Where the terrace slopes are very steep, mowing management is neither practicable nor necessary, and the resulting unmown grasslands are not very similar to cultivated grasslands.

Disturbance of these unmown slopes, therefore, initially takes place by (i) ground denudation and (ii) sowing of a seed mixture (Fig. 1). Thereafter, however, there is often little distur-

J. van Andel et al., (eds.), Disturbance in Grasslands. ISBN 90-6193-640-3.

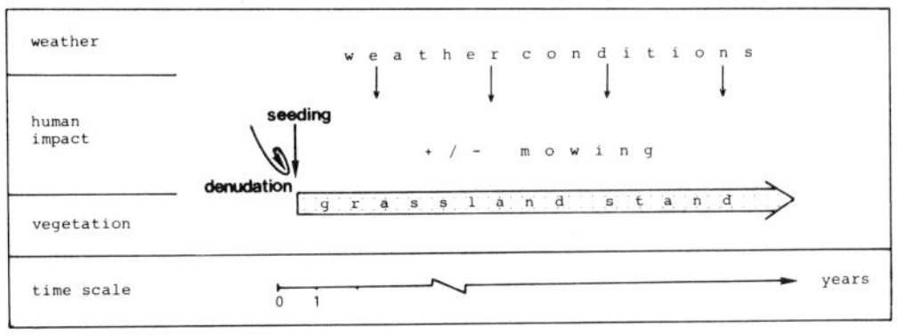

Fig. 1. Diagrammatic representation of the factors affecting the revegetation of reconstructed slopes.

bance, though changing weather conditions and plant-induced soil changes (Chapter 3) constitute environmental variation; this has little influence on community structure during the first decades of community development.

This embankment vegetation provides an excellent opportunity to study secondary grassland succession after severe disturbance.

2. Study area

The study area, the little mountain range 'Kaiserstuhl' (92 square kilometers), lies within the climatically warm and dry Upper Rhine Plain, between Schwarzwald (Black Forest) and Vogesen, a few kilometers west of Freiburg, W.-Germany. The area is covered by a thick loess layer, usually 5 to 15 m deep. During the last 1000 years the loess covered hills had been terraced, using primitive techniques, for the purpose of agriculture and viticulture. More recently, in the process of consolidation of vineyards, modern techniques have been used to transform the small terrace landscape into one of large-scale terracing, so that high-intensity wine production can be undertaken. Between 1968 and 1980 about four square kilometers of new terrace slopes have been produced. The large-scale terracing is carried out in units of about 100 ha, each unit being denuded and reconstructed within a few months. The species normally used for sowing are: *Festuca ovina* agg., *F. rubra* agg., *Lolium perenne, L. multiflorum, Poa pratensis, Bromus inermis, Trifolium dubium, Trifolium repens, Lotus corniculatus, Medicago lupulina*, and *Sanguisorba minor*. As a result of this procedure, extensive stands of unmanaged grassland vegetation have been developed on terrace slopes in the Kaiserstuhl (Fig. 2) during the last twenty years.

3. Vegetation development following ground denudation and seeding

Two methods were used to study succession on terrace slopes: (a) phytosociological analysis and (b) permanent plot analysis.

a) Analysis of phytosociological relevés (Braun-Blanquet, 1964) were carried out on vegetation stands of different age but similar site conditions, i.e. similar exposure and similar loess type (see below). Floristical and structural differences between different aged stands were interpreted as vegetation changes in succession series (Fischer, 1982).
b) Detailed analyses were carried out on a set of permanent plots (Fischer, 1985b, 1986).

Methods (a) and (b) lead to similar descriptions of succession (Fischer, 1985b). In the present paper, studies of vegetation development over an 8 year period on a separate set of permanent plots are presented.

The results are firstly presented in a condensed form (Fig. 3) by grouping species into sown and wild species; the wild species are further grouped into annuals, biennials, and perennials.

i) *Sown grassland species:* Sown species reached a peak in number and abundance two to three years after slope construction and seeding (Fig. 3). On S-facing slopes both species number and abundance then decreased, however; the decline was less marked on N-facing slopes. Most of the sown species disappeared during the first decade with a loss of about one to two species per year (Fischer, 1986); *Festuca ovina* and *F. rubra* remained for a longer period, and often dominated the stands (Fig. 6).
ii) *Wild species:* The number and abundance of annual species was initially high but then decreased rapidly. The number and abundance of biennial species was quite variable; they tended to increase after four years, but never became dominants in the vegetation. The

Fig. 2. Photograph of an area of vineyards in the Kaiserstuhl taken in June 1978, a few years after large-scale terracing and seeding. Different types of grass-dominated vegetation developed on the terraces influenced mainly by different loess types (see Fig. 4) and exposure. Photo Fischer.

number of perennial species increading during the first two years, then remained constant. The abundance of perennial species increased rapidly during the second half of the first decade (Fig. 3).

The general pattern of vegetation succession, therefore, was an initial dominance of annual species, followed by an increase in biennial and perennial species, and a final dominance by perennial wild species, which developed under a cover of sown species. The sequence is similar to that observed by Schmidt (1981) and Pickett (1982) on abandoned arable fields, and by Knapp & Knapp (1955) and Meisel & Hübschmann (1973).

The abundance of annual species was largely unaffected by the presence of sown species, because they reached their peak abundance early. In contrast, the abundance of both biennial (including few winter annual) and perennial species was heavily suppressed by the sown species (Fischer, 1986).

The three groups of 'wild' species appear to be rather heterogeneous conglomerations of subgroups differing in population longevity and ecological requirements:

Annuals constituted the first group which in-

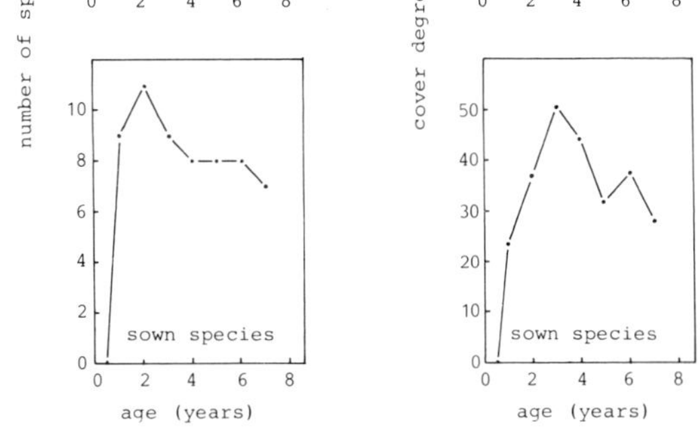

Fig. 3. Changes in the number of species and their abundance (cover degree) in a permanent plot on a reconstracted SE-facing slope of the Kaiserstuhl. A = annuals, B = biennials, P = perennials.

cluded short-lived summer annuals (e.g. *Polygonum persicaria, Euphorbia helioscopia, Atriplex patula*), occuring during the first months after slope construction; a second subgroup of annuals also occurred during the first few months of succession, but persisted into the following year (e.g. *Mercurialis annua, Stellaria media, Veronica persica, Sonchus oleraceus*); a third subgroup consists of facultative annuals or biennials/short-living perennials, which persisted about five years (e.g. *Poa annua, Capsella bursa-pastoris*); the final subgroup contains overwintering annuals and facultative annuals or biennials, which appeared while the other annuals began to decrease but before the biennials increased (e.g. *Conyza canadensis, Lactuca serriola*).

The biennials fall into two subgroups: biennials which prefer the warmer and dryer S-facing slopes (e.g. *Isatis tinctoria*), and biennials which are more or less restricted to the colder and moister N-facing slopes (e.g. *Daucus carota*).

The perennials can be divided into four subgroups: (i) species of ruderal dry meadows (e.g. *Elymus repens, Saponaria officinalis, Chondrilla juncea*), which occurred frequently, even at the beginning of succession, but were limited to the dry S-facing slopes after a few years (most of these species posess rhizomateous organs); (ii) species of dry meadow and tall-forb communities (e.g. *Euphorbia cyparissias, Artemisia campestris, Origanum vulgare, Hypericum perforatum*), restricted to the S-facing slopes and most abundand after 5 to 10 years; (iii) species of Arrhenatherion communities, which were most abundant between years 5 to 10 and occurred predominantly on N-facing slopes, some of them were restricted to those slopes (e.g. *Arrhenatherum elatius, Lathyrus pratensis*), though others also occurred an S-facing slopes (e.g. *Taraxacum officinale* agg., *Achillea millefolium, Galium mollugo* agg.); (iv) rhizomatous species with high requirements for nutients and soil water (e.g. *Urtica dioica, Cirsium arvense*), which were more or less restricted to the moister N-facing slopes.

A vegetation mosaic, therefore, evolved in both space (site) and time, depending largely on water regime, loess type, and developmental stage of vegetation.

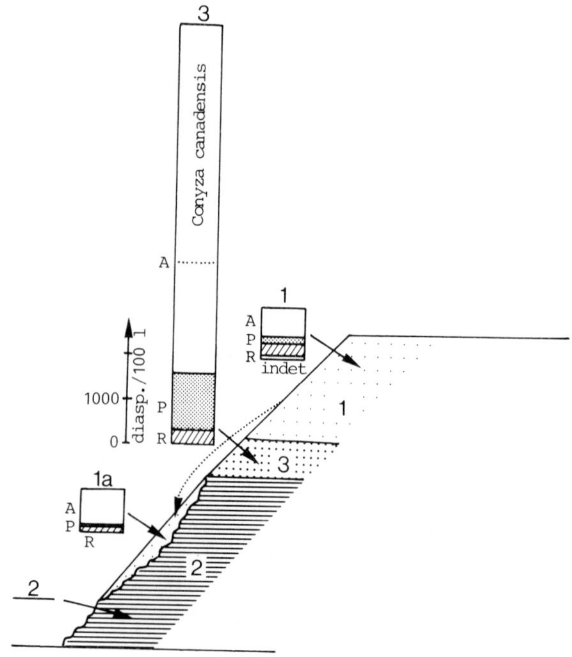

Fig. 4. The abundance of seeds (A = annuals, P = perennials) and of viable vegetative fragments of perennials (R) in the loess material immideately after terrace slope construction. 1 = disturbed loess material, heaped up during terrace construction; 1a = material slipped down from 1; 2 = primary loess; 3 = rests of previous soil. (Based on data from Fischer, 1982).

4. Factors controlling succession

Significant numbers of annuals and perennials occurred on the slopes within a year of their formation. These plants must have survived, probably as seed or vegetative fragments, in the loess material during the terrace construction. Subsequently the perennials spread, perhaps largely by vegetative means, to become the dominant component of the vegetation (Fig. 3). The factors determining these processes will now be considered.

4.1 Seed bank

The reconstructed terrace slopes consisted of three types of stratum:

(i) primary loess (2 in Fig. 4), (ii) loess material which had been heaped up during terrace construction (1 in Fig. 4; 'transported loess'), and (iii) top soil remaining from the previous surface, some of which was left after the removal of most of the previous soil.

Primary loss (2 in Fig. 4) was free of any living plant material. Transported loess (1 in Fig. 4) had become contaminated during terracing: 100 l of that material hold about 600 viable seeds of annuals/biennials and about 200 seeds of perennials. This material partly slips down the slope and contaminates the surface of primary loess below (1a in Fig. 4). Remnants of the previous top soil (3 in Fig. 4) contained about ten times as many viable seeds of annuals/biennials, and perennials than the contaminated loess (1). The most common annual/biennial species of the seed bank were *Conyza canadensis, Lactuca serriola, Chenopodium album, Stellaria media, Mercurialis annua, Capsella bursa-pastoris, Atriplex patula, Veronica persica,* and *Lamium purpureum*; these species dominated the initial phase of succession. The most frequent perennials found as viable seeds were *Urtica dioica, Artemisia campestris, Origanum vulgare, Hypericum perforatum*, and *Chondrilla juncea*; these species reached prominance in the slope vegetation 5 to 10 years after disturbance.

By counting of the annual rings of *Artemisia campestris*, a woody undershrub, which is dominant on dry, S-facing slopes, it was possible to determine when plants became established. The oldest specimens were as old as the terrace but, on terrace sections with steep slopes consisting of primary loess, the oldest specimens were one or two years younger than the terrace, reflecting the later arrival of seed (Fischer, 1982).

Many perennial species (e.g. *Taraxacum officinale* agg., *Achillea millefolium, Euphorbia cyparissias, Elymus repens, Convolvulus arvensis, Saponaria officinalis, Urtica dioica, Cirsium arvense*, and *Rubus caesius*) regenerated from vegetative organs such as rhizomes, tubers, root buds, and root-stocks.

It seems, therefore, that most of the species which colonise the slopes during the first ten to fifteen years after their formation, are already present in the loess material when the terraces are under construction. The remaining top soil is especially rich in such seed and vegetative fragments.

Following initial studies in Germany (Putensen, 1882, near Lüneburg; Peter, 1893/94, near Göttingen), studies of seed banks have been made all over the world, from Alaska (Leck, 1980) to Australia (Carroll & Ashton, 1965). The studies have shown that soils possess a large number of viable seeds, and that soils beneath arable fields are especially rich in buried viable seeds, while meadows, pastures and (temperate) forests have smaller seed banks (Table 1).

The values in Table 1 are based on different methods and different soil depths (Fischer, 1987). Even with a standardised technique there is still considerable variation in the seed bank content, depending on different ecological conditions (Table 2).

4.2 *Species immigration*

Immigration of species by seed rain was studied using 'wet seed traps' (i.e. water-filled bassins) and 'dry seed traps' (i.e. funnels of different size with controlled water discharge). During the initial establishment period seeds of 14 non phane-

Table 1. Typical values for the seed content of soils beneath different vegetation types.

vegetation type	seeds/fruits (1 m²)
arable fields	10.000–50.000
meadows, pastures	2.000–15.000
temperature forests	1.000– 5.000

Table 2. The seed content of soil beneath different grassland communities near Giessen (BRD).

vegetation type	soil depth (cm)	seeds/fruits (1 m²)
wet meadow (Molinietalia community)	0– 6 0–20	40.000 75.000
dry meadow (Arrhenatheretum ranunculetosum bulb.)	0– 6	7.000
dry rough pasture (Gentiano-Koelerietum)	0– 6	5.700

rophytic species were caught. Four of them (*Senecio vulgaris, Cirsium arvense, Lactuca serriola*, and *Sonchus oleraceus*) had morphological adaptations to wind-dispersal, but only 1.2% of the total seed input belonged to these species, and then only if at least one individual of the species was fruiting within about 5 m of the trap. Immigration of wind-dispersed species was therefore not an important feature of the early stages of succession on the Kaiserstuhl loess slopes (Fischer, 1982).

Similar results have been reported by Ryvarden (1971, p. 215) in Norway; he concluded that 'a high and constant percentage of the diaspores was not dispersed beyond 5 meters and no really long-distance-dispersed diaspore was represented'. In different grassland communities near Giessen (BRD) these species growing in vegetation patches of 1 square meter, with the seed trap in the centre, accounted to 97 to 99 percent of the total seed input; seeds originating from outside the grassland stands made maximally half a percent of the total (Fischer, 1987).

This is confirmed by the fact that wind-dispersed species made only a small contribution to the initial slope vegetation (Fig. 5). The increase of the number of species with special morphological adaptations of the seed to wind-dispersal ('wind-dispersed species') was not greater than that of species without such morphological properties; indeed, it was often lower and tended to decrease in older stands (Fig. 5e). In a few cases, the abundance of species with wind-dispersed seeds increased rapidly between years 5 to 15 but, as in the case of Bötzingen (Fig. 5b), this increase was caused by *Calamagrostis epigejos* which, although it possesses wind-dispersed propagules, invaded the plots exclusively by vegetative spread.

During nearly a decade of observations, the largest invasion by wind-dispersed species occured on a WNW-facing slope (Fig. 5f), which was covered by a dense turf of sown fescue when first observed in 1977, five years after terrace construction. The permanent plot (72 square meters) initially contained a few, small specimens of six wild species, two of which had morphological adaptations to wind-dispersal (*Crepis capillaris* and *Taraxacum officinale* agg.). During the next three years, the abundance of sown species increased and wild species (whether wind-dispersed or not) remained constant. After the dry summer of 1982 and 1983, sown species decreased rapidly and wild species increased, so that in 1985 32 wild species were present, 18 of them with wind-dispersed propagules. All of these invading species, however, were common in the surrounding vegetation. No such invasion by wind-dispersed species occured on the SE-facing slope only 40 m away (Fig. 5g).

Immigration by seed from outside, therefore, seems to be a fairly rare event and not the main cause of recolonisation.

Small scale mapping of selected species on large terrace slopes, with remnants of undisturbed grassland and brush vegetation in its centre, demonstrated that many species did not spread from the patches of old vegetation into directly adjacent, newly established slopes during the first decade. Even pioneer species with wind-dispersal adaptations (e.g. *Anemone sylvestris*) were not invasive. Those species which did spread (e.g. *Valeriana officinalis* ssp. *collina, Hieracium pilosella, Hieracium umbellatum, Clinopodium vulgare,* and *Origanum vulgare*) moved a maximum of 5 m within the first decade (Fischer, 1982).

This slow spread can be seen by eye on permanent transects (Fig. 6): meter by meter, populations of plants, which had survived the terrace construction, spread over the new slopes, the slower-spreading species include *Salvia glutinosa* and *Solidago gigantea*, while the quicker-spreading species include *Artemisia campestris* and *Achillea millefolium.*

It is therefore apparent that vegetation succession on large denuded areas during the first decades, is primarily determined by short distance dispersal, starting from remaining plant and vegetation fragments. Long distance dispersal is a rare event. Such rare long distance immigrations, however, may be very successful over longer time periods (i.e. centuries).

These results are different from those of

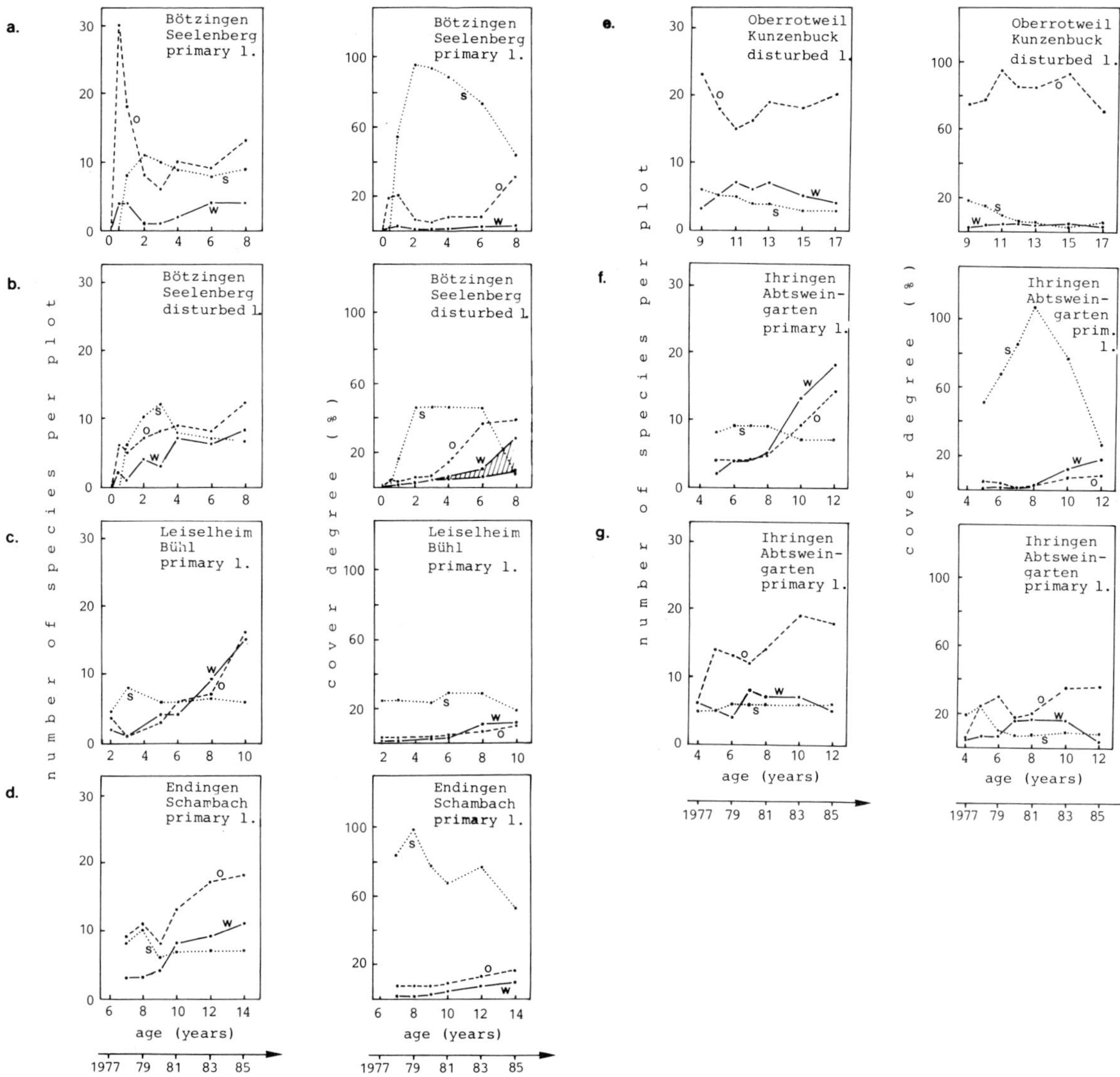

Fig. 5. Vegetation succession on permanent plots: Changes in the number of species and their abundance (cover degree) during an eight year observation period. S (dotted curve) = sown species; O (dashed curve) = wild species without wind-dispersed propagules; W (soild curve) = wild species with wind-dispersed propagules; shaded in 5b: portion of Calamagrostis epigejos. The name of village, the name of vineyard area, and the type of parent loess material are inserted.

Schmidt (1981) who worked on areas of only a few hundred square meters, the soil of which had been sterilised by heat treatment. These relatively small areas of bare ground were invaded by many species in a short time (months to few years).

4.3 Influence of extreme weather conditions

Extreme weather conditions heavily influence the progress of vegetation succession. During wet periods, established mesophytes with high productivity may suppress smaller and less competitive species, especially their seedlings. The result is to halt succession. In contrast, during drought periods, mesophytes may die leaving open spaces in which new species, especially xerophytic ones,

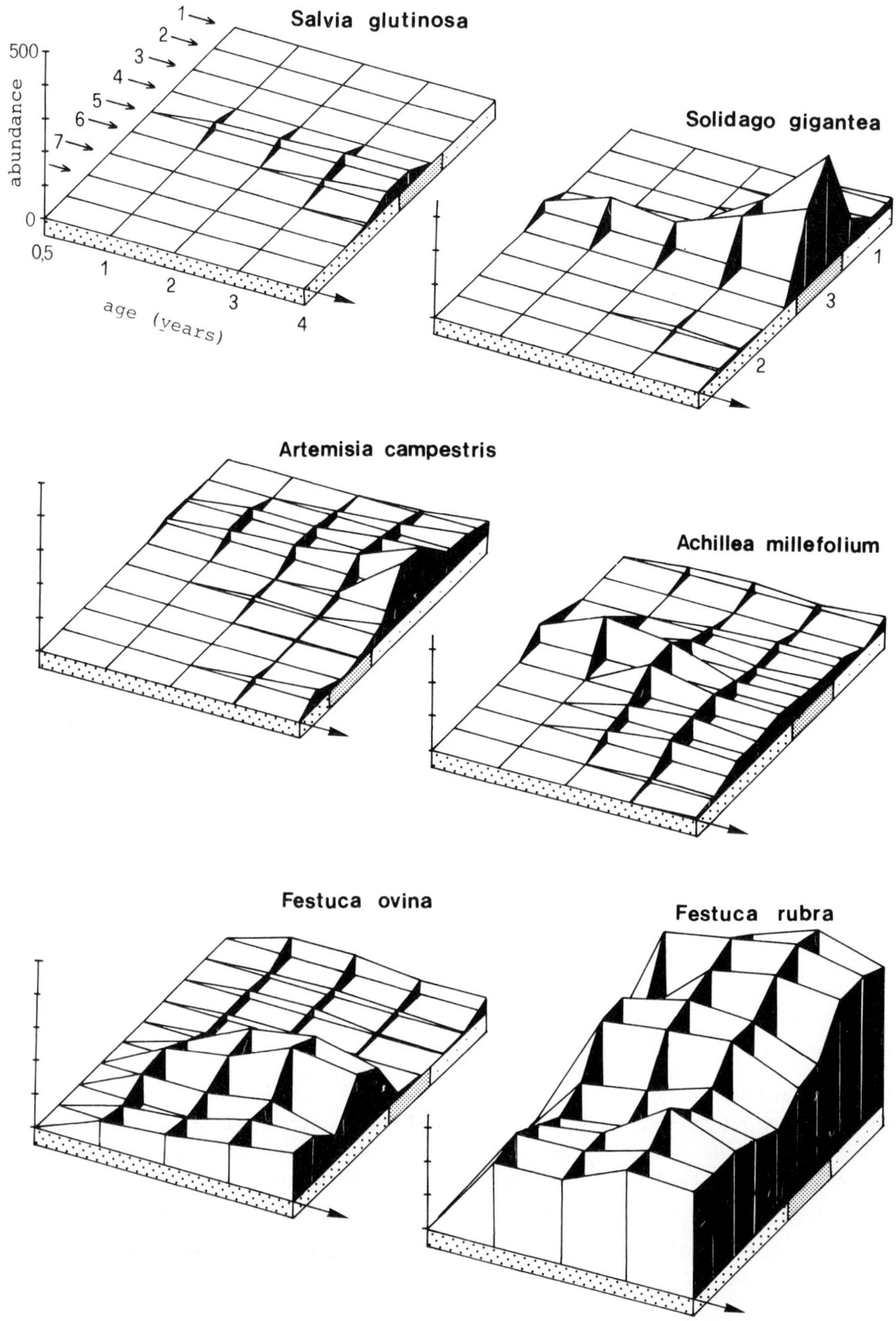

Fig. 6. Changes in the abundance of two sown species (*Festuca ovina* agg. and *F. rubra* agg.) and four indigenous species in permanent transects. All transects (each with 10 quadrats of one square meter) were on e ENE-facing slope except for *Artemisia campestris,* which was on a SE-facing slope. The parent material was disturbed loess material (1), primary loess (2), and rest of previous top soil. A special method had been used for determining species quantity (Fischer, 1985b).

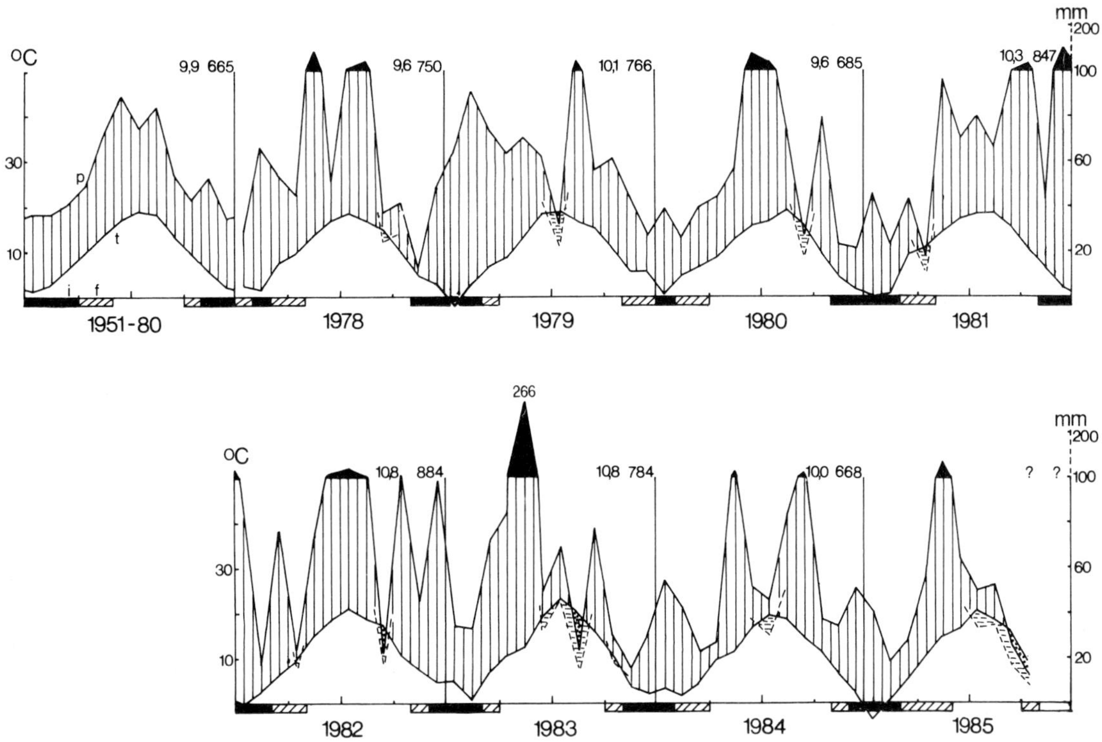

Fig. 7. Weather conditions at Oberrotweil/Kaiserstuhl during study period 1978 to 85 and the mean values of climatical parameters 1951 to 1980. Data from 'Deutscher Wetterdienst, Wetteramt Freiburg/BRD'). Monthly mean temperature (t, lower curve); precipitation per month (p, upper curve; scale reduced and in black if more than 100 mm); months with icedays and frostdays (i, f; bars along the bottom); mean temperature of the year and precipitation per year (values in the top). Vertically shaded: humid phase; dotted: arid phase (strong drought); horizontally dashed: period of moderate drought.

may establish, and succession may be accelerated.

On the Kaiserstuhl terrace slopes, the abundance of sown species, which are mostly mesophytes, decreased (Fig. 5) after the drought of 1982 and 1983 (Fig. 7). Only the permanent plot at Ihringen (Fig. 5g) showed little effect of drought, but this exposed, S-facing slope was already populated by xeric vegetation with incomplete coverage, before the period of dryness started.

5. Discussion

The recolonisation of large, bare loess slopes in this study originated mainly from seed or vegetative fragments of species already existing at the beginning of the succession, and by short distance dispersal of these species. Initial populations developed from viable seeds or vegetative organs of the original vegetation remaining in the soil.

Extreme weather conditons can inhibit or accelerate vegetation succession. In contrast to the disturbance of ground denudation extreme weather conditions represent inherent environmental variation which probably cannot be regarded as disturbance.

Although the general sequence of succession can be charaterised, the particular floristic composition of succession stages cannot be accurately predicted; it is subject to stochastic factors, such as species composition of the seed bank, the sequence of weather conditions during succession and, over longer time periods, the immigration of species from outside ('Nahverbreitung', 'Distanzverbreitung'; Luftensteiner, 1982).

The ability of species to create a permanent seed bank or to regenerate from vegetative fragments is a 'pre-adaptation' to such disturbance ('dispersal in time', Harper, 1977). Not all the species of the seed bank, however, can pass the 'sieve of site conditions' after germination. For example, many biennials and perennials are restricted during succession to either the moister N-facing or the drier S-facing slopes.

Grasslands on terrace slopes and embankments, in areas of intensive agriculture, are often the last uncultivated and unused patches of vegetation within this landscape, and are therefore important in nature conservation. The reservoir of viable plant parts in the soil is essential to the subsequent species composition of vegetation, and must therefore be taken into consideration during the planning of landscape projects.

Acknowledgements

I would like to thank Prof. Dr. O. Wilmanns/Freiburg i.Br. (BRD) and N. Crompton M.Sc./Giessen (BRD) for helpful discussion and revision of the manuscript.

References

Braun-Blanquet, J. (1964). Pflanzensoziologie. 3. ed. Wien, New York.

Carroll, E.J. & Ashton, D.H. (1965). Seed storage in soils of several Victorian plant communities. Vict. Nat. 82: 102–110.

Fischer, A. (1982). Mosaik und Syndynamik der Pflanzengesellschaften von Lößböschungen im Kaiserstuhl (Südbaden). Phytocoenologia 10: 73–256.

Fischer, A. (1985a). 'Ruderale Wiesen' – Ein Beitrag zur Kenntnis des Arrhenatherion-Verbandes. Tuexenia 5: 237–248.

Fischer, A. (1985b). Feinanalytische Sukzessionsuntersuchungen in Grünlandbrachen – Methode und Methodenvergleich. Münstersche Geograph. Arb. 20: 213–223.

Fischer, A. (1986): Feinanalytische Sukzessionsuntersuchungen in Grünlandbrachen – Vegetationsentwicklung ungelenkt und nach Begrünung. Veröff. Natursch. Landschaftspfl. Bad.-Württ. 61: 349–390.

Fischer, A. (1987). Untersuchungen zur Populationsdynamik am Beginn von Sekundärsukzessionen. Masch.schr. Habil.schr. Giessen, 230 S.

Harper, J.L. (1977): Population Biology of Plants. London, New York, Toronto, Sydney, San Francisco: Academic Press.

Knapp, G. & Knapp, R. (1955). Über Möglichkeiten der Durchsetzung und Ausbreitung von Pflanzenindividuen auf Grund verschiedener Wuchsformen. Ber. Dtsch. Bot. Ges. 67: 410–419.

Leck, M.A. (1980). Germination in barrow, Alaska, tundra soil cores. Arct. Alp. Res. 12: 343–349.

Luftensteiner, H.W. (1982). Untersuchungen zur Ausbreitungsbiologie von Pflanzengemeinschaften an vier Standorten in Niederösterreich. Bibl. Bot. 135: 1–68.

Meisel, K. & Hübschmann, A. von (1973). Grundzüge der Vegetationsentwicklung auf Brachflächen. Natur u. Landsch. 48: 70–74.

Peter, A. (1893/94). Culturversuche mit 'ruhenden' Samen. Nachr. Ges. Wiss. Gött., Mitt. 17/1893: 673–691, Mitt. 4/1894: 372–393.

Pickett, S.T.A. (1982). Population patterns through twenty years of oldfield succession. Vegetatio 49: 45–59.

Putensen, H. (1882). Untersuchungen über die im Ackerboden enthaltenen Unkrautsämereien. Hann. Land. u. Forstwirstsch. Vereinsbl. 21: 514–524.

Ryvarden, L. (1971). Studies in seed dispersal I: Trapping of diaspores in the alpine zone at Finse, Norway. Norw. J. Bot. 18: 215–226.

Schmidt, W. (1981). Ungestörte und gelenkte Sukzession auf Brachäckern. Scripta Geobot. 15: 199 p.

PART 3

Vegetation response to defoliation and pathogens

Since regular defoliation (cutting or grazing) is a necessary requirement for the continued existence of most grasslands, defoliation per se will not result in disturbance; however, changes in the frequency or intensity of defoliation can be regarded as disturbance (Chapter 1). These changes are considered in Part 3. Many chapters in this book in fact deal with the combined effects of changing soil conditions and changing defoliation practices. Miles (Chapter 3) discussed the mutual relationship between soil changes and vegetational response following reduced grazing intensities. The studies of Pegtel (Chapter 4), Grootjans (Chapter 5) and Bakker (Chapter 12) were carried out in a nature reserve in which cessation of fertilizer applications coincided with later and less frequent cutting than under former agricultural practice. To some extent, changes in soil conditions and defoliation practices go hand in hand in grazing regimes, because grazing is accompanied by trampling and the voiding of dung. The simultaneous occurrence of various kinds of disturbance is discussed in Part 4.

In the considerations of vegetation response to changing soil conditions in Part 2, it became apparant that the canopy structure can be an important factor in explaining vegetational changes (Pegtel in Chapter 4). The effect of changing soil conditions, which in turn affects canopy structure and leads to vegetational response, is worked out by Bakker (Chapter 12). He concludes that the vegetational response to hay-making and grazing without fertilizer applications is governed by removal of above-ground biomass in the short term, and by nutrient removal in the long term. Nutrient removal is unimportant in the short-term because annual removal is small, compared to the total size of the nutrient pool in the soil.

Fresco et al. (Chapter 10) model the effect of increased grazing intensity in a desalinating polder. The predicted result is an increase in the abundance of halophytes after a few years. This is in agreement with exclosure experiments which show that halophytes decrease in the absence of grazing (Joenje, 1985). Thalen et al. (Chapter 11) introduce different grazing intensities or different kinds of grazers in their model of vegetational structure. The validity of their model is difficult to ascertain, however, because the changes take hundreds of years to occur.

At first sight the study by Drent & Prins (Chapter 9) of herbivores grazing in Lake Manyara National Park, Tanzania, does not include any disturbance. However, the buffalo population (one of the two dominant grazers) was depleted by Rinderpest in 1959. This depletion, however, seems to have had little effect on the overall densities of large herbivores or on the vegetation composition. This apparant anomaly is explained by seasonal habitat shifts in grazing, in which a 'gap' left by one herbivore species is immediately 'filled' by another species, resulting in an overall removal of 80% of the annual herbage accumulation. These features imply a close fit between herbivore numbers and the food supply in a given grazing system and a situation in which grazers seem to manipulate their own food quality (McNaughton, 1979). Herbivory is commonly patchy (Crawley, 1983), especially if only one herbivore species is present. However, if several herbivores are present the herbage untouched by one species is usually eaten by another species (Pratt et al., 1986).

Zadoks (Chapter 13) stresses the importance of mutual relationship between pathogenic fungi and their hosts. Pathogens are an essential part of every biocoenosis but, because of coevolution of host and pathogen, the impact of pathogens on the vegetation is usually quite small. Disturbance, caused by pathogenic attack, i.e. resulting in marked vegetational response, is quite rare in nature (as opposed to man-made monocultures). Only when host and pathogen meet for the first time or meet again after long separation, real disturbance can occur.

There is some analogy between the short-term grazing model of Fresco et al. (Chapter 10) dealing with species and the long-term model of Thalen et al. (Chapter 11) dealing with structural classes of the vegetation at the one hand, and the effects of plant pathogenic fungi in natural communities on the population and the vegetation level put forward by Zadoks (Chapter 13) at the other hand. He indicates the short-term effects of population control and patchiness in the vegetation, and the long-term effects on evolution and succession at the population and vegetation level, respectively. The effects of disturbances at the population level, i.e. on genetic structure will be dealt with in Part 4.

It seems that not only the effects of changes in defoliation and pathogens, i.e. the vegetational response, are well described, but that more insight exists into their causes than to those concerning soil conditions (Part 2). This seems obvious since it is often easy to see how the frequency and intensity of defoliation changed under new management practices. More knowledge, however, is needed on the effects of selectivity on patchiness, and also on interactions between different herbivore species, and this should be taken into account in models concerning the vegetational response to grazing. Another step in modelling should consider the level of plant communities and not only vegetational structure.

References

Crawley, M.J. (1983). Herbivory. Blackwell, Oxford.

Joenje, W. (1985): The significance of waterfowl grazing in the primary succession on embanked sandflats. Vegetatio 62: 399–406.

McNaughton, S.J. (1979). Grazing as an optimization process: Grass ungulate relationships in the Serengeti. Am. Nat. 113: 691–707.

Pratt, R.M., Putman, R.J., Ekins, J.R. & Edwards, P.J. (1986). Use of habitat by free-ranging cattle and ponies in the New Forest, southern England. J. Appl. Ecol. 23: 539–557.

CHAPTER 9

The herbivore as prisoner of its food supply

R.H. Drent & H.H.T. Prins

Detailed contents

1. Introduction

In this review we consider the asymmetry in the relation between the herbivore and its food supply: successively we will touch upon the limits to feasible exploitation as the food supply is depleted, the repercussions for the reproductive performance of the herbivore when a mismatch between requirement and food on offer occurs, and finally the universal feature of seasonal habitat shifts in grazing systems, with the herbivore tracking the availability of forage of adequate quality. All of these features, taken together, imply a close fit between herbivore numbers and the food supply in a given grazing system; we will provide an example of this from a region where climatic vagaries are sufficiently restricted to allow a review of animal biomass statistics to be undertaken against a backdrop of assumed constant forage production.

Our examples will be restricted to bird and mammal herbivores, and we consider only the direct link of phytophagy 'the cow eating the grass', the simplest of the many interactions between plants and animals (Southwood, 1985). Southwood rightly emphasizes the crucial importance of biochemical evolution, the need to break the intractable bonds of plant compounds during the digestive process, and the key role of micro-organisms in filling this need. In our discussion we lean heavily on our experience with geese, that rely to only a moderate extent on energy by the symbiosis with micro-organisms in the gut, and as a result show an unmistakable dependence on foods with relatively low content of structural plant carbohydrates. Geese show a strong preference for young growing shoots whenever available and rely primarily on soluble carbohydrates and proteins contained in the cell sap to satisfy their demands. Precisely on account of these limitations geese are excellent subjects to explore the limitations of herbivory as a way of life.

2. Grazing thresholds in relation to resource depletion

Our first theme is that grazing frequently entails high rates of utilisation, leading to resource depletion sufficient to drastically reduce intake, so

J. van Andel et al., (eds.), Disturbance in Grasslands. ISBN 90–6193–640–3.

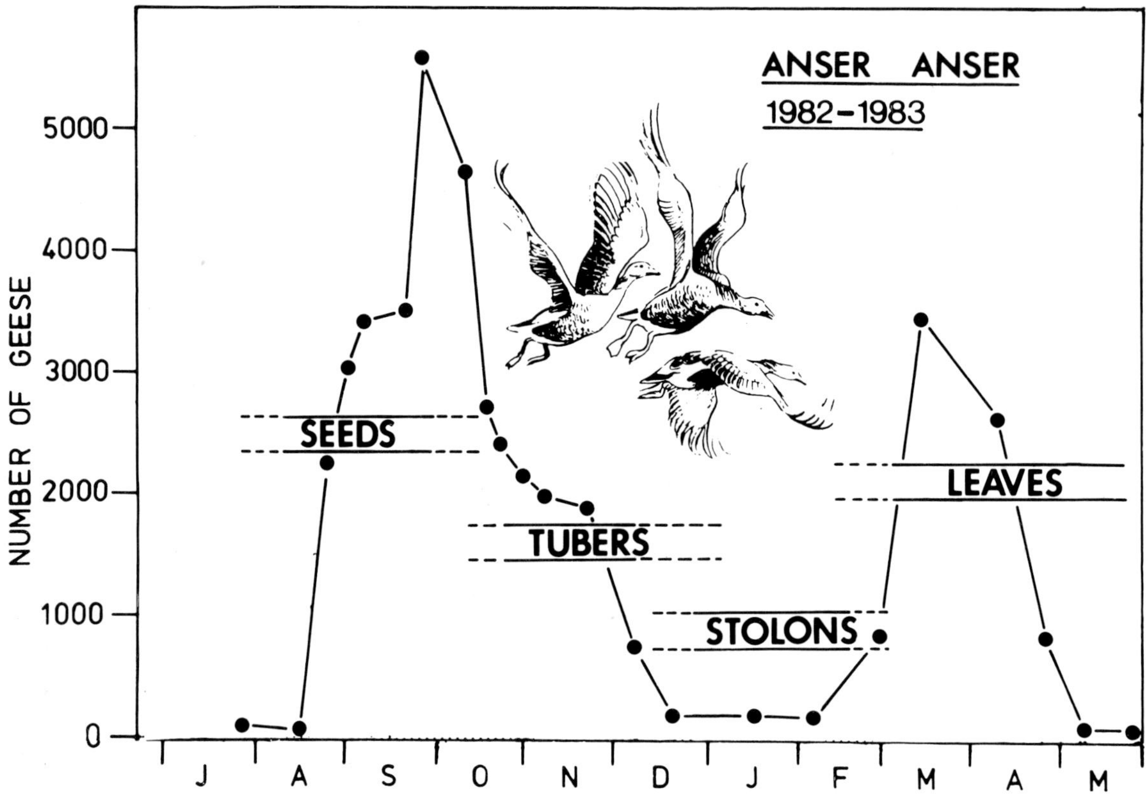

Fig. 1. Numbers of Greylag Geese utilizing the Lauwersmeer area in the northern Netherlands in the winter of 1982–1983 when detailed observations were made on diet. *Agrostis* shows two periods of utilisation: late summer when seeds are taken (further considered in Fig. 2) and spring when fresh leaves are utilized (from van Eerden et al., in prep.).

that the herbivore must shift its diet, feeding site, or both. Resource depletion is by no means limited to herbivory, but because grazers are typically group-living animals with limited ability to deal with the various counter strategies deployed by the plants (secondary compounds etc., see Crawley, 1983 for review), the net result is to focus heavy grazing pressures on specific plants during restricted growth phases. We will consider two examples from local goose studies.

In late summer Greylag (*Anser anser*) abound in the northern Netherlands, and the enclosure of the estuarine Lauwerszee in 1969 resulted in over 5000 ha of former sea bottom being converted to extensive natural vegetation allowing quantitative study of the dependence of the geese on their food supply. Fig. 1 depicts goose numbers in the study area, with an indication of the main diet components. We will concern ourselves here with the utilisation of seed heads of *Agrostis stolonifera*, the prime food in August shortly after the geese arrive in large numbers. We have investigated if the numbers of geese using the former estuary area in the late summer period can be predicted on the basis of the seed crop, and will show that the duration of stay of the goose flocks is determined by the rate of depletion of the food stock. This project, carried out by our department in conjunction with the reclamation authority (RIJP) will be reported in detail elsewhere (van Eerden, Slager & Soldaat, in prep.) but can be summarized as follows. To determine the lower cut-off point of goose utilisation of *Agrostis* seed, wild-caught captive Greylags were restricted to a pen of movable netting erected in the natural vegetation. During the day the birds were observed from a hide, and at days' end the geese were moved to a holding cage, the droppings col-

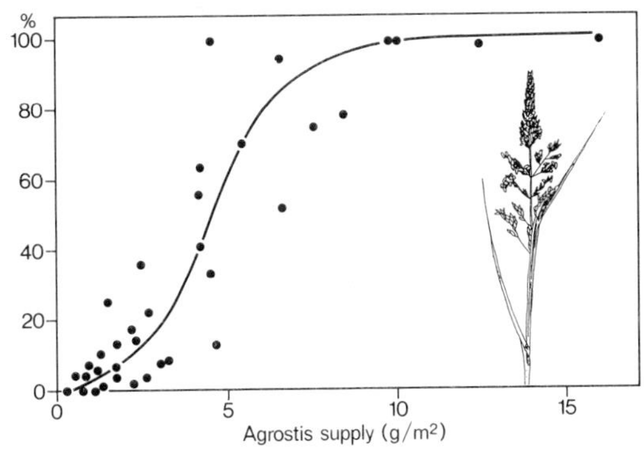

Fig. 2. Feeding trials with captive Greylag maintained on the natural vegetation in late summer. Depicted is the dependence of diet (expressed as % contribution by *Agrostis stolonifera* seed) on food supply (the seed stocks, depleted by the birds in the course of the observations). Each dot represents mean daily diet for an individual Greylag, and the data are used to define the exploitation threshold (taken as 5 g.m^{-2}, see text) as explored in Fig. 3 (from van Eerden et al., in prep.).

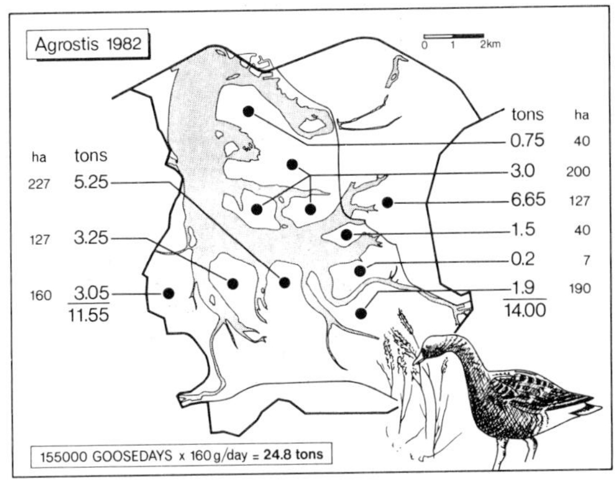

Fig. 3. Greylag Goose usage of the Lauwersmeer area in the northern Netherlands, in relation to seed stocks of *Agrostis stolonifera*, as estimated at the outset of goose utilisation in the late summer of 1982. Seed stocks in the central area were measured in the sampling units, depicted, and converted by means of the relevant vegetation area (given in hectare per sampling unit). Seed stocks (in metric tons) are given in terms of food availability to the geese (i.e., biomass in excess of the threshold value if 5 g.m^{-2} adduced in Fig. 2). The total exploitable seed supply (25.55 tons) is compared to the estimated consumption at the bottom of the figure (goose the utilisation from detailed counts, daily ration from trials with wild-caught individuals) as explained in the text. Shaded is the discharge lake (from van Eerden et al., in prep.).

lected and the seed supply on offer determined by counting seed heads along a transect through the pen and scoring each head for seed content (full, medium, light). Analysis of the droppings revealed the diet of the birds (see Owen, 1975 for technique) as confirmed by direct observation. The captives showed successive use of food plants in the same sequence as their wild counterparts, feeding first on seed heads of *Agrostis*, and shifting later to *Aster tripolium* and leaves of *Agrostis*. When the ground was waterlogged, the birds extracted the starchy nodules of *Triglochin palustris*, a food source used by several species of geese throughout its range at this time of year (Thomas, 1980).

The data on diet are arranged in relation to the *Agrostis* seed supply (the preferred food type during this period) in Fig. 2. A sharp decrease in seed use was apparent when the seed stock dwindled to below 10 g.m^{-2}, and only 50% of the diet was composed of seed when the seed stock had dropped to 5 g.m^{-2}. Observations on unrestrained Greylags confirmed that the flocks abandoned *Agrostis* seed as the dominant food source below the 5 g.m^{-2} level, and provisionally we can accept this as a rough indication of the cut-off point. The captive trials shed some light on what determines this lower threshold of acceptance, in that the mean bite size clearly declined as the seed stock was exhausted. We presume that dwindling intake rates trigger the switch to other foods.

We used this threshold value to quantify goose usage of the entire Lauwerszee. By means of a massive sampling programme the supply of *Agrostis* seeds on offer over the entire area was estimated. Assuming the 5 g.m^{-2} threshold value, and correcting for wind loss (data from exclosures not presented here) the total available food supply can be calculated from the point of view of the geese (see Fig. 3). If the total seed supply is divided by the daily requirement of seed (160 g dry weight) the number of goose-days the area support in the year of sampling can be estimated. In comparison, we can make a first approximation of the actual 'Greylag seed days' in this season on the basis of our weekly goose counts (where each flock was assigned to a vegetation type, i.e. we

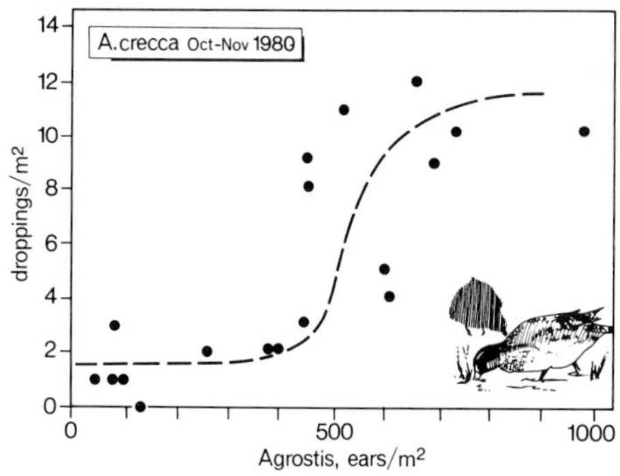

Fig. 4. Exploitation of *Agrostis* seed bu Teal: utilisation threshold as revealed by droppings counts in relation to seed stock on offer (van Eerden, 1984).

can estimate the number of seed-eaters). The figure on observed Greylag seed-days is within 5% of the number expected on the basis of the quantification of the seed supply, so the method certainly holds promise. Once the *Agrostis* seed has fallen to the ground it is safe from further utilisation by the Greylag, but Teal (*Anas crecca*) can seive the seed from the substrate if covered by a shallow film of water. In one season it was possible to quantify this second phase of seed exploitation (van Eerden, 1984), and again a threshold value emerges from the data (Fig. 4). From the point of view of the seed crop, therefore, refuge from consumption is only attained after passing through a family of threshold curves, acting as a successive series of sieves.

In an earlier stage of succession following enclosure, the former tidal flats were invaded by *Salicornia europaea, Sueda maritima,* and *Spergularia maritima*, and the seed of these plants was utilized by waterfowl in the fall (primarily October-November). Work similar to that described for the Greylag was carried out in that period (van Eerden, 1984), directed towards the chief consumers Wigeon (*Anas penelope*), Barnacle Goose (*Branta leucopsis*) and Teal (*Anas crecca*). For *Salicornia* alone, seed production of the area was estimated at 200 tons dry weight for the 1975 season, and in that year 46% of the seed crop was consumed by Wigeon, 24% by Barnacle Geese and 7% by Teal, so that the waterfowl removed more than three-quarters of the *Salicornia* seed. In other years this pattern held as well, the Wigeon, being the first on the scene, claiming the lion's share of the crop. The *Salicornia* crop in the former Lauwerszee declined through the seventies (the peak year was 1972) and concomitant with this declining food base the birds were forced to shift to other foods to an increasing extent. The most spectacular effect of the enormous but temporary seed supply was to attract virtually the entire population of the Barnacle Goose wintering in continental Europe (the Novaya Zembla breeders as distinct from the Greenland and Spitsbergen populations wintering in Ireland and Scotland) at that time some 40 thousand birds. This population advanced its time of arrival to the area by three weeks during the years of heavy seed crops, but since 1978 with the dwindling seed supply this trend has been reversed and the birds have protracted their stay in more northerly haunts (primarily coastal Germany adjoining the Danish border).

The utilisation of seed crops constitutes a finite food supply that is not replenished until the following year, and the fact that waterfowl deplete the food supply in a matter of weeks underlines how important this food concentrate (high quality, high density) is in the diet at that time of year. It may come as a surprise to realize that a similar pattern of heavy usage leading to local depletion also typifies waterfowl grazing at other times of the year. The limited ability of these birds to digest plant material makes them heavily dependent on just a few plant species at any locality at a given moment, and the rate at which food can be collected determines a lower acceptance threshold for plant density. These relations can be illustrated by quantification of spring grazing, taking as our example Brent (*Branta bernicla*) on the saltmarshes of the Wadden Sea islands. The nominate or dark-bellied form *B. bernicla bernicla* has a high-arctic breeding range along the coastline of Siberia, and virtually the entire population stages in the Wadden Sea of Denmark, Germany and the Netherlands from the latter part of March through to the end of May when they migrate to the breeding grounds (Ebbinge *et al.*, 1981). Of vital importance at this time of year are the saltmarshes of the Wadden

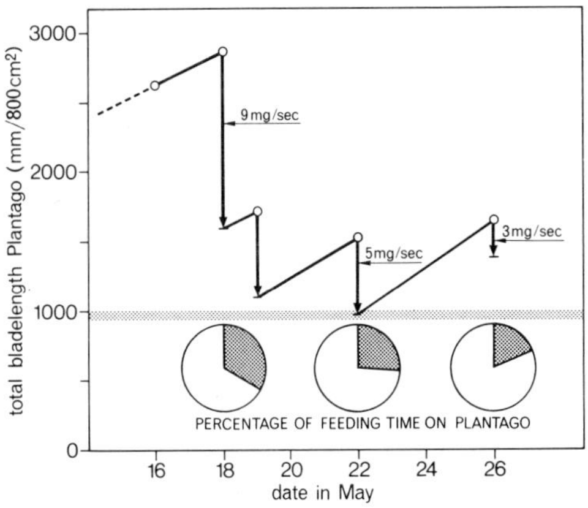

Fig. 5. Grazing impact of Brent on successive visits to the saltmarsh study plot. Measurements of the preferred food, *Plantago maritima*, are given as cumulative bladelength (determinded by stereoscopic photography before and after each goose visit). Mean intake rate of geese utilizing *Plantago* is shown for each episode, as well as the proportion of feeding time devoted to the dwindling food stocks (lower panel). The stippled bar indicates the approximate lower threshold for *Plantago* utilisation (from Prop & Loonen, in press).

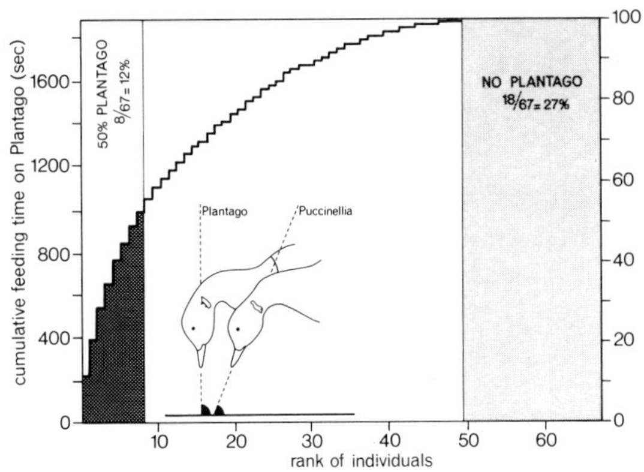

Fig. 6. Dietary choice of individual Brent in the flock visiting the saltmarsh plot depicted as a cumulative curve of *Plantago* feeding time when the 67 individuals observed are ranked in decreasing order of *Plantago* utilisation. Half of all *Plantago* feeding time is accounted for by only 8 birds, and 18 obtained no *Plantago* at all. Dietary choice was assessed from analysis of film, head stance (inset) revealing food intake (from Prop & Loonen, in press).

Sea islands and coastline, tidal foreshore habitats where early spring growth of *Festuca rubra, Puccinellia maritima, Plantago maritima* and *Triglochin maritima* is the staple food.

Studies on the saltmarsh of the island Schiermonnikoog opposite the estuary area of the former Lauwerszee, where approximately three thousand Brent gather in spring, showed that goose visitations followed a pulsating pattern with massive visits every 3–5 days. The entire harvestable fraction was removed during each flock visit, and indeed there was evidence that the rear birds in the flock already faced a sharply depleted food supply (Prins *et al.*, 1980; Drent & van Eerden, 1980). These points have recently been followed up in more detail employing ciné-film records from which the tracks of individual geese and their grazing dynamics can be traced, in combination with stereo photographs of small plots in the vegetation taken before and after goose visitation and enabling quantification of the exact amount of plant material removed (Prop & Loonen, *in press*). A summary of events for the observation quadrangle (5 × 20 m) is presented in Fig. 5, and shows that the supply of *Plantago maritima*, the preferred food plant in this zone was reduced dramatically during each visit, and that subsequent regrowth was repeatedly removed. Comparing the diet assembled by the birds using the plot shows a declining reliance on *Plantago* in the course of the season, with a compensating switch to *Puccinellia* leaf.

Geese were filmed at close range and details of the head stance and the angle between bill and substrate could be discerned on the film, making it possible to discriminate between *Plantago* bites and *Puccinellia* bites. This allows reconstruction of the diet within the flock vist, and an example is presented in Fig. 6. Only 12% of the individuals accounted for half of the *Plantago* cropped during the group visit, i.e. there is a very unequal feeding opportunity for the individuals within the flock. In fact, 25% of all geese traversing the observation quadrats failed to collect any *Plantago*. The reason for the strong preference of this plant with respect to the alternative (*Pucinella*) can be sought in the much higher harvesting rates (coupled with a slightly greater digestibility). From the film analysis in conjunction with the stereo photos (i.e., relating total blade length cropped to total feeding time devoted to the plot in question) *Plantago* intake rates of up to 10 mg.sec^{-1} were calibrated, in sharp contrast to the 2–3 mg.sec^{-1} typical for *Puccinellia* feeding. When

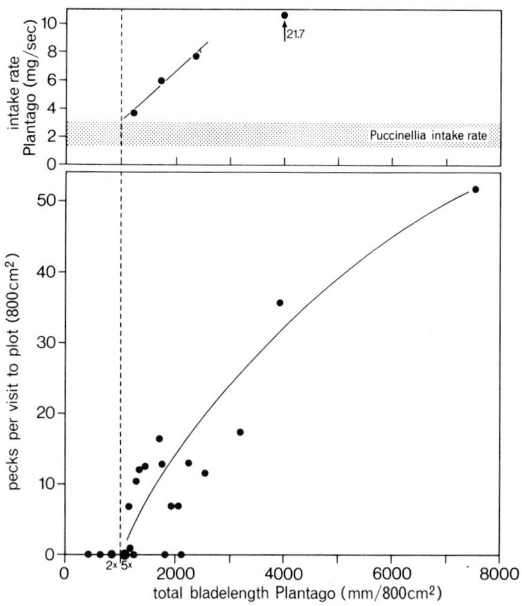

Fig. 7. Grazing threshold for Brent utilizing *Plantago maritima* in relation to a background vegetation of *Puccinellia maritima*. Data are from photoplots (800 cm^2) in the natural vegetation, food on offer given as cumulative blade length (from stereoscopic photography before each goose visit) and goose grazing as pecks per visit directed to the plot in question (from ciné-film) for the grazing episodes summarized in Fig. 5. Lower limit for *Plantago* utilisation is shown by broken line, and as shown in the upper panel this coincides with the point at which *Plantago* no longer confers an advantage in intake rate (from Prop & Loonen, in press).

intake of *Plantago* fell by depletion of the food stock to the range experienced with *Puccinellia* the birds moved on, their strategy apparently being directed towards discovering another patch of *Plantago* (Fig. 7). During the wave of exploitation, between one-third and one-quarter of the living biomass of *Plantago* leaf is removed, and for any one site the main period of use is restricted to a period of one to two weeks.

Summing up so far, we find that in exploiting either seed stocks or the early spring growth, waterfowl bring such heavy grazing pressures to bear that the food source is rapidly depleted to a level beyond which exploitation is no longer profitable and the waterfowl move to another food source. In the spring situation, food quality declines rapidly as the plants age and accumulate structural components resistent to digestion and inhibiting the digestibility of the soluble carbohydrates and proteins that are vital to the geese. Hence geese must exploit the plants in the narrow window delimited by minimal bite size and minimal nutritive value. As has been emphasized for other grazing animals (McNaughton, 1979) heavy exploitation of early growth stages prolongs the period of peak digestibility, as illustrated by data on rates of decline in nitrogen content in grazed and ungrazed plants (see Fig. 8 for two examples from goose studies) and in this sense the herbivores can be said to 'manipulate' their food supply (Ydenberg & Prins, 1981).

Utilisation curves can be given in the form of 'functional response' curves depicting dependence of diet on food density. For example, sheep, kangaroos and rabbits switch from pasture herbs to the Bluebush shrub (*Maireane pyramidata*) as the herbs are depleted (Fig. 9 from Short, 1985). Note that the sheep for the situation depicted (where Bluebush made up 13% of the initial biomass) had depleted this alternate food source by the time the pasture biomass had declined to 125 kg.ha^{-1}, while exploitation of Bluebush by kangaroos and rabbits increased steadily as herb biomass declined.

In general, food intake rates fall as plant biomass declines. Short found that intake rate had been depressed by one-third of plateau values when herb biomass had fallen to 111 kg.ha^{-1} for the sheep, but this value was reached by kangaroo only at 84 kg.ha^{-1} hinting at a difference in harvesting efficiency. When faced with a declining supply and restricted to a monoculture, herbivores attempt to cope with declining bite size by increasing bite rate, but this compensatory mechanism has its limits as illustrated for sheep in Fig. 10 (from Allden & Whittaker, 1970).

In all of these studies, lower limits to profitable exploitation have been determined, and the task of modelling the optimal response of the grazing animal has begun (Noy-Meir, 1975). Our concern here has not been with the choice of the individual harvester, but rather on the impact of the cumulative grazing pressure on the food stocks, with the general conclusion that the rate of food depletion in many cases defines the period of utilisation and thus impinges on site selection and

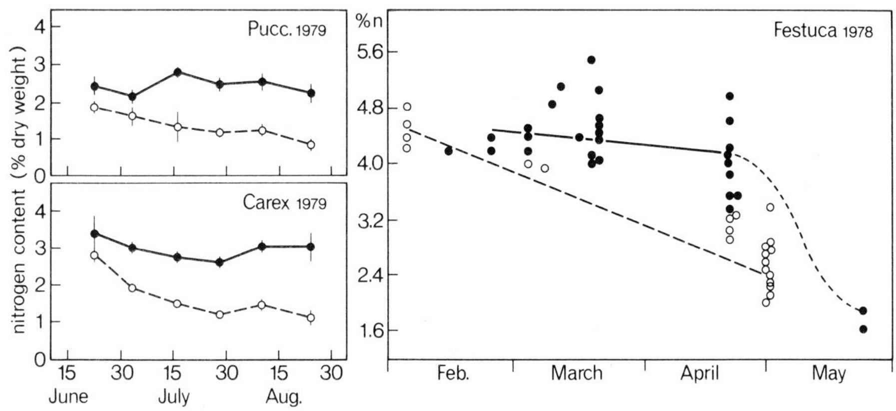

Fig. 8. Total nitrogen content (% dry weight) of shoots of *Puccinellia phryganoides* (upper left), *Carex subspathacea* (lower left) and *Festuca rubra* (right) from grazed (solid dots) and ungrazed (open circles) sites. The 1979 data refer to Snow Goose grazing on the Hudson Bay (Cargill & Jefferies, 1984) the 1978 data to Barnacle Goose grazing on the Schiermonnikoog saltmarsh (recomputed from Ydenberg & Prins, 1981).

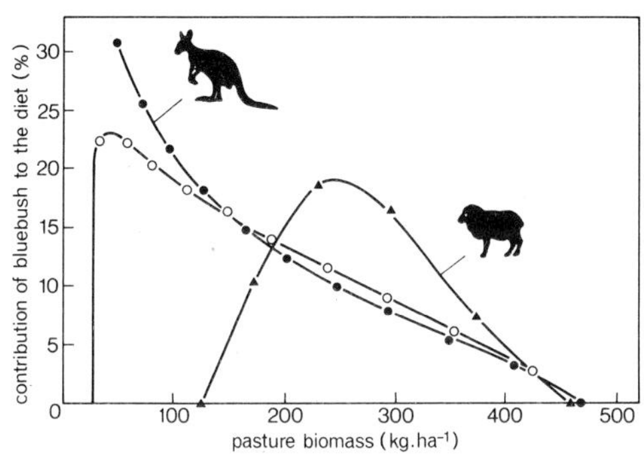

Fig. 9. Bluebush (*Maireana pyramidata*) exploitation by sheep, Kangaroos and rabbits (open circles) as alternative food source as the preferred pasture species are depleted. Results for pastureland with low initial incidence of Bluebush (Short, 1985).

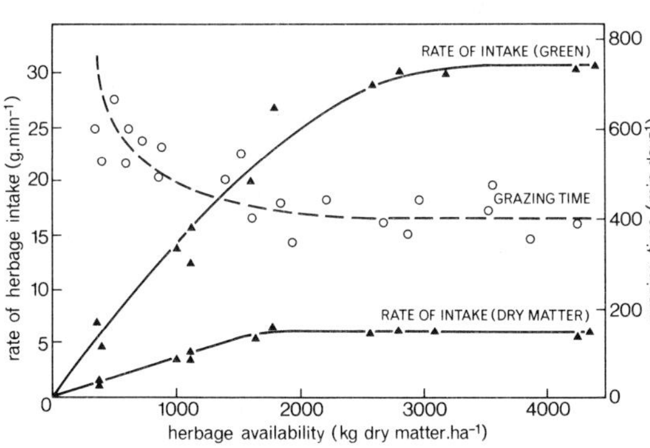

Fig. 10. Grazing parameters in sheep in relation to herbage biomass (Allden & Whittaker, 1970). Note that intake rate (triangles) per unit grazing time declines below about 2000 kg.ha^{-1}, and that sheep attempt to compensate by increasing feeding time (open circles).

the evolution of the migratory strategy, as will be taken up in the next section.

3. Repercussions for the consumer: mismatch between food supply and food demand

Reproduction in herbivores entails negotiating nutritional bottlenecks. Maternal budgets are strained in birds during egg laying and incubation, and in mammals during lactation. Study of year-to-year variations in reproductive success of resident herbivores implicates the crucial role of spring feeding conditions for reproduction, and an extensive study on Red Grouse (*Lagopus lagopus*) demontrated a positive correlation between growth conditions for the major food source for the hens (heather) and subsequent chick production in that season (Moss *et al.*, 1975). When taken in conjunction with the feeding trials carried out on the related Ruff Grouse, *Bonasa umbellus* (Beckerton & Middleton, 1982) these results lead to the notion that local feeding conditions in the pre-egg phase determine maternal condition (i.e. state of body reserves) which in turn determines reproductive performance.

An added complication is that seasonal shifts in range occur in many species to allow the young to take advantage of the spring flush of vegetative growth. However, this may entail the parents entering these habitats early to anticipate developments, at a time of low and sometimes even minimal food supply. If this occurs, achieving adequate maternal condition depends on feeding conditions prior to departure for the breeding grounds. For example, feeding conditions in spring staging areas are of paramount importance for enabling breeding in high arctic nesting birds. For the Brent Goose considered in the previous section with respect to resource depletion, there are two avenues of evidence bearing on the relation between spring condition and subsequent breeding success. In this species, staging in the Wadden Sea and breeding along the Siberian coast, juveniles remain with their parents throughout their first winter of life, and are recognizable on the basis of plumage characters. Hence families can be recognized as units and, providing the adults are individually marked their reproductive performance can be assessed when they return to their northwestern European wintering haunts in the fall. By capturing and weighing females in the spring, and determining subsequent success, Ebbinge *et al.* (1982) were able to demonstrate a weight differential between successful and unsuccessful parents, females accompanied by young in the fall having attained higher body mass before departure in the spring. Another approach to this problem is to consider the year-to-year variation in reproductive success in the population as a whole, and to relate this to mean female body mass determined from mass catches by cannon netting samples in May. When the available data are collected together (Fig. 11) it will be seen that juvenile percentage determined in the fall is unmistakably related to spring body mass, success in the intervening summer only being high when the females exceeded 1600 g, i.e. when they were presumably able to lay down considerable amounts of body fat (see inset in Fig. 11) and body protein (analyses from H. Korte, work in progress). Similar correlations between spring parental mass and breeding success have been adduced for several high arctic nesting waders (ten Have *et al.*, in progress) although of course the spring diet is quite different.

It is not clear why certain spring seasons result in larger body reserves than others. In general,

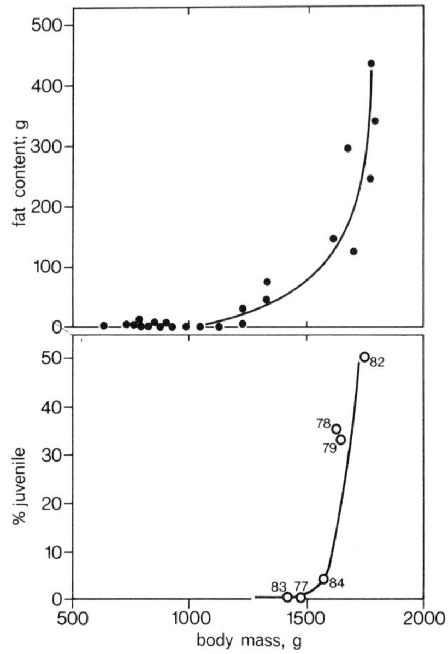

Fig. 11. Dependence of reproductive success in Brent on spring body condition: season means of percent juveniles in the flock in fall in relation to female body mass at the end of May for six years where both are available (from Ebbinge et al., in progress). The critical importance of the fat depot is suggested by the upper panel, where individual values from carcass analysis are plotted on the same body mass scale (from Korte, in prep.).

cool, 'late' springs in the staging areas correlate with high body mass at departure, and we suggest that in cold springs the phenology of the major food plants is more suitably timed to allow rapid accumulation of body reserves. In our study area the period of rapid growth of *Plantago maritima* and *Triglochin maritima* may well be critical, against the background of the staple food *Puccinellia maritima*. If spring comes too early, the growth burst of *Plantago* is over too soon, and the geese are left with no alternative providing the combination of high nutritive content, high digestibility, and high intake rate. Current work is designed to test this general idea. A subsidary effect is that in a cold spring, as a result of the low temperatures, both the digestibility and nutrient content (sugars, protein) of growing grasses are enhanced (reviewed by Deinum, 1985).

As the NW European-wintering population of Brent has expanded following cessation of hunting activities, it has become increasingly difficult to achieve breeding success: a lower proportion of birds of appropriate age, the potential parents, return with young in the fall (Ebbinge, 1985). This finding is in line with a more general phenomenon, a negative density dependence in reproductive performance in birds as well as in mammals (Fowler, 1981) and it is in filling in the links between the food resource on the one hand and parental condition on the other that the study of herbivore systems is liable to make an important contribution to general theory.

4. Herbivore numbers and food supply: Manyara

As a case study on how food supply dictates herbivore numbers, we will consider Lake Manyara National Park, Tanzania. The basic data concern nine complete reconstructions of the populations of large herbivores within present park boundaries (100 km^2) during the last 25 years (We use the term 'reconstructions' as in some cases extrapolations were made, the original census material showing gaps). The basic message from the material is that total food consumption within the system has varied much less through the years than the species composition of the herbivore assemblage. In the absence of clear-cut trends in climatological variables (in particular, rainfall and its distribution through the years) we assume the production rate of forage in the system to be roughly constant, and total food on offer will depend largely on the area under grass. Variations in the extent of Lake Manyara are thus seen as the main source of variation in food supply and when this is taken into account, grazing pressures per hectare can be arrived at. From this analysis it can be concluded that the only consistent trend involves an increasing Cape buffalo grazing pres-

Table 1. Numbers of large herbivores in Lake Manyara National Park since 1959, and Park size.

	1959	1961	1965	1967	1970	1975	1976	1981	1984
Elephant	380[a]	421	421	300	427	386	453	483	373
Buffalo	1750	1507	1507	1500	1548	1823	2032	1800	2097
Hippopotamus	150[b]	150[b]	150[b]	150[b]	150[b]	151	118	150	179
Burchell's zebra	180[c]	115	115	100	75	156	157[d]	162	255
Impala	600[c]	451	285	380	451	985	1071[d]	1500[e]	150[e]
Black rhinoceros	30[c]	23[f]	20[f]	20[f]	23[f]	35[f]	20[f]	20[f]	25[f]
Giraffe	60[a]	60	31	50	60	60	60	72	66
Wildebeest	250[c]	250[a]	7[a]	2[a]	0	80[a]	80	169	675
Waterbuck	20[a]	20[a]	20	20[a]	20	20[a]	20[d]	16	17
Bushbuck	20[a]	20[a]	20	20[d]	20	50[b]	50[b]	50[g]	50[g]
Reedbuck	220[a]	220[a]	220	120[a]	20	20[a]	20[d]	10[g]	10[g]
Baboon	550[c]	450[a]	400	400[a]	400	450[a]	450	500[g]	500[g]
Warthog	0[c]	2[c]	6[c]	6[a]	35[c]	50[d]	50[c]	65	95
Size of Park (km^2)	100	100	86	86	84	98	100	100	100

Sources: 1961 – Morgan-Davies, cit. in Watson & Turner, 1965; 1965 – Watson & Turner, 1965; 1967 – Vesey-FitzGerald, 1969; 1970 – Douglas-Hamilton, 1972; 1975 – Mwalyosi, 1977; 1976 – Douglas-Hamilton unpubl.; 1981 – Douglas-Hamilton unpubl.; 1984 – Douglas-Hamilton unpubl.

Due to Lake level changes, the real size of the area in the Park which can be utilized by herbivores is fluctuating. The Park size figures in this Table refer to the area of Park which is covered by vegetation, and the acreage estimates based on aerial photographs, Warden's Reports, and information from Parks Rangers.

[a] = Unpublished estimates based on counts of other years and inferences from Warden's Reports.

[b] = Unpublished estimates; the published counts are certainly too low (oral information from Parks Ragers) (1961: 10; 1965: 10; 1967: 20; 1970: 10).

[c] = Unpublished estimates mentioned in Warden's Reports.

[d] = Interpolation.

[e] = Estimates based on road sampling (Prins & Weyerhauser, 1987).

[f] = These data are published rough estimates and very uncertain.

[g] = Estimates on personal information from Parks Rangers and personal observations.

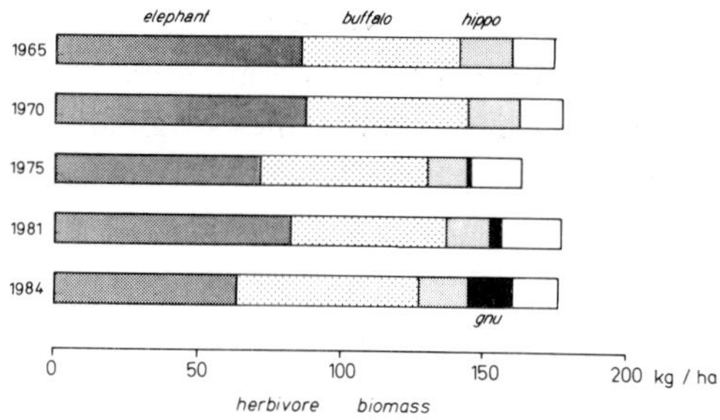

Fig. 12. Large herbivores in the Manyara census area (from left to right elephant, buffalo, hippopotamus, gnu or wildebeest, and other = open bloc) presented as biomass (live weights) ha^{-1}. Over a twenty year period total biomass has varied less than the constituent species (see text).

sure, and an explanation for this apparent contradiction will be offered.

A reconstruction of total herbivore numbers is attempted in Table 1. Estimates for 'bush species' (Black Rhinoceros, Bushbuck, Impala, Waterbuck and Baboon) are less reliable than those for 'plain species' (Burchell's Zebra, Wildebeest, Warthog); the most reliable estimates are for the large and conspicuous species (African Elephant, Cape Buffalo, Giraffe and perhaps Reedbuck). Gross consumption was calculated from the empirical relationship that large herbivores require herbage dry matter equivalent to 2.5% of their live body mass per day (van Soest, 1984; van Wijngaarden, 1985). Average live body mass, and an indication of diet, is given in Table 2 (we have ignored the effects of trampling, although we realize that these may be quantitatively important under certain conditions, Sinclair & Norton-Griffiths, 1979). Lake Manyara shows wide fluctuations, sometimes inundating parts of the adjoining grasslands and woodland habitats after periods of heavy rainfall, and at other periods receding and thus enabling the shorelines zone to be colonized by grasses. The area actually available for grazing was determined from aerial photographs (see Loth & Prins, 1986) supplemented by information from the Park Rangers (see Table 1).

Grazing pressure (kg biomass herbivores consumed per ha grazing land per year) is depicted in Fig 12. The Manyara system is dominated by two large herbivores, namely the buffalo and the elephant. These two species account for 35% and 43% respectively of the total offtake, while the eleven remaining species collectively consume only 22%. Most of the food consumed is grass (82%) while browse consumption is quantitatively less important. For further analysis, herbivores are categorized in five functional groupings (see Table 2) buffalo, grazing elephant, other grazers (hippo, zebra, impala, wildebeest, waterbuck, bushbuck, reedbuck, baboon and warthog, to the extent they are grass eaters) and the browsers (elephant, impala, rhino, giraffe, water-, bush- and reedbuck, baboon and warthog, to the extent browse occurs in the diet). When the grazing pressure exerted by each of these categeories was regressed on time, only one significant deviation from a zero slope was obtained. With the exception of the grazing pressure of the

Table 2. Manyara herbivores: nomenclature, body mass and diet composition.

Species	body mass (kg)	percent grass in diet	percent browse in diet
Elephant *Loxodonta africana*	1.700	70	30
Buffalo *Syncerus caffer*	325	100	0
Hippopotamus *Hippopotamus amphibius*	1.000	100	0
Burchell's zebra *Equus burchelli*	250	100	0
Impala *Aepyceros melampus*	50	95	5
Black rhinoceros *Diceros bicornis*	1.000	0	100
Giraffe *Giraffe camelopardalis*	765	0	100
Wildebeest *Connochaetus taurinus*	225	100	0
Waterbuck *Kobus ellipsiprymus*	200	80	20
Bushbuck *Tragelaphus scriptus*	60	10	90
Reedbuck *Redunca redunca*	40	95	5
Baboon *Papio anubis*	20	20	70
Warthog *Phacochoerus aethiopicus*	50	90	10

buffalo (which shows an increasing trend) all categories are stable with time (although the component species show fluctuations). There were also no significant correlates between grazing pressures of the five categories and annual rainfall either in the year of census, or the year of census plus previous year, nor was there a correlation with the cumulative rainfall over a three year period closing with the census. It appears that grazing pressures, as here defined, have fluctuated around equilibrium values for the past twenty-five years, except for buffalo. Before considering the implications of this apparent stability, the exceptional case of the buffalo requires comment.

A first possibility is to examine the grazing components of the system for possible competitive interactions. There are strong negative correlations between grass consumption by elephant and that by all other grazers ($r = -.81$, $p<.01$) and between elephant and buffalo alone ($r = -.73$, $p<.02$) since these two species were the major consumers in this system. Similarly, there was a significant negative relation between elephant browse consumption, and that by all other browsers combined ($r = -.45$, $p<0.01$). Since there were no significant trends in elephant numbers with time, elephant numbers cannot be the cause of changes in buffalo numbers. Past outbreaks of Rinderpest are the most likely explanation of changes in buffalo numbers. In the Manyara area, the last epidemic occurred at the end of 1959, eliminating some 250 buffalo (15% of the population) within a few weeks (Prins & Weyerhaeuser, 1987). It is unfortunate that no pre-Rinderpest census figures for buffalo are available, but our interpretation of the counts is that recovery was slow until 1970, and hence the trend discerned can be considered a recovery following a severe demographic disturbance.

The overall densities of large herbivores in Manyara are higher than expected in view of the comparative aridity of the area (Coe *et al.*, 1976). If consumption by the wild herbivores is translated into a stocking rate for beef steers of 400 kg, then Manyara supports 0.42 steer units $.ha^{-1}.yr^{-1}$. This is comparable to legume-based tropical pastures in an area with the same rainfal regime, in Queensland, Australia, where cattle gain weight under these range conditions (Whiteman, 1980; Table 5.15). There are reports of exceptionally high densities of game in Manyara in the 1940's (Watermeyer & Elliot, 1943) so we assume that the current herbivore density and offtake level is sustainable, since the vegetation of Manyara appears to be virtually unchanged since the fifties (Lamprey, pers.comm., Loth & Prins, 1986; Greenway & Vesey-Fitzgerald, 1969), although some change has taken place since the turn of the century (Baumann, 1894). We incline to the view that the high density of herbivores can be sustained due to the high level of soil fertility in combination with high ground water tables (further discussed in Loth & Prins, 1986).

The Manyara census area is undoubtedly too restricted in extent to provide an understanding of numerical fluctuations of the component species. The elephant counts are a case in point, as these represent totals and not extrapolations from sample areas, and result from the intensive programma of Douglas-Hamilton (1972). Nevertheless, the resulting series can only be interpreted as revealing extensive interchange with the surroundings, as the fluctuations are on too short a time base to have been generated internally (see also Croze *et al.*, 1981). We favour the interpretation that Manyara functions as a local optimum and in particular as a refugium during periods of dry season drought, and feel that Douglas-Hamilton (1972) has laid too much emphasis on effects of changing human pressures. The overall stability in Manyara contrasts with the neighbouring Serengeti ecosystem, where herbivore populations have increased spectacularly in the same time period. Sinclair (1985, and in Sinclair & Norton Griffiths, 1979) has provided convincing evidence that this increase can be explained by an expanding food base caused by a trend of increasing dry season rainfall over the years. By contrast, dry season rainfall in Manyara has, if anything, decreased over the period 1959–84.

Whatever interpretation is given to the data, the historic perspective makes abundantly clear that in the Manyara area current rates of offtake have been in effect for some decades at least. This makes the determination of total consumption in relation to production (as was undertaken in the 1983–84 period) of salient interest. By following standing crop both within exclosures and without, it was possible to measure empirically the cumulative consumption by all herbivores as well as the local production for the most important grasses. Overall, 80% of the annual production is cropped by herbivores, as will be shown in the following section.

5. Seasonal habitat shifts in grazing systems: Manyara

The most important grazing areas in Manyara can be grouped in six vegetation units, characterized by the following food plants:

A_2 *Sporobolus spicatus* dominated alkaline grasslands
A_3 *Cynodon dactylon* dominated deltas
G_1 *C. dactylon* dominated forest glades
A_1 *Cyperus laevigatus* dominated swamps
W woodlands with *Cynodon plectostachyus, Sporobolus helvolus, S. cordofanus* and other grasses
G_2 *Chloris gayana* dominated wooded (bush) grasslands

(code refers to the key adopted by Loth & Prins, 1986, where more detailed information can be found; *Cyperus* is a sedge, the other species listed are all grasses).

Monospecies stands of the grassland types listed were selected for quantitative study (types A_3 through G_2). For the woodlands type (W) the three food species have been considered together.

Eight elephant and buffalo-proof exclosures (made form steel angle-iron with iron bars at intervals of 20 cm, area 2 × 2 m and depending on the vegetation 1 or 2 m high), which did not necessarily exclude small herbivores such as rodents and insects, were placed in the vegetation and shifted within the stand every ten weeks. Two types of measurements were collected. Every time the exclosures were shifted herbage was clipped in 50 × 50 cm plots to ground level, both inside and outside the exclosure, hand sorted into four fractions (live leaf, dead leaf, live stalk, dead stalk) and dried in the sun on metal sheets (approx. 60° C) for subsequent analysis.

Since large herbivores in Manyara, as in other East African grassland areas (Sinclair, 1977) utilize almost exclusively the green leaves, the standing crop of living leaf was determined at approximately fortnightly intervals using spectral reflectance (at 840 nm = infrared, 660 nm = red, and 550 nm = green), allowing a rapid comparison to be made between incoming and reflected light. Readings were integrated by coupling the instrument to a Sharp programmable calculator (for further description of this unit, developed by the Agricultural University Wageningen, see van Kasteren, 1982, van Wirdum, 1983). The instrument was calibrated using clipped samples and the ratio of infrared to red reflectance (corrected for quality of incoming light) gave satisfactory estimates of live leaf standing crop (r^2 ranging from 0.62 to 0.91).

Annual production of leaf material was estimated from the sum of leaf accumulation inside exclosure plus a correction for dead material inside exlosure (empirical relationship 0.8 g dead material = 1.0 g live dry weight for *Sporobolus spicatus, Cyperus laevigatus* and *Chloris gayana*: other species 0.5 g dead = 1.0 g live dry weight). This is a minimal estimate since consumption by small herbivores was not included, nor are trampling losses included. We further assume that under the prevailing grazing regime growth stimulation by grazing can be ignored.

For comparison with these date on production, consumption by large herbivores was calculated as the difference between accumulation of live leafy material inside and outside the exclosures.

Both sets of estimates are collected in Table 3, and lead to the conclusion that the large herbivores remove the greater part of the leafy material available to them, i.e. most leaves in this grazing system will be consumed by herbivores instead of senescing. On the basis of clipping, total aboveground biomass production (to include stalk and flower production) is estimated as three times the leaf biomass production. On this basis, annual production would be on the order of 30 tons $.ha^{-1}.yr^{-1}$, which is well within the range given for semi-arid regions in the tropics with a reasonable water supply as collated by Snaydon (1981).

The different grass species showed a different seasonal pattern of production (and consumption) (Fig. 13), which for most grasses, except *Chloris gayana*, was clearly geared to the rainfall pattern. This is most clearly seen in the case of *Sporobolus spicatus*, a perennial species which grows on sandy soils (Loth & Prins, 1986). Growth started as soon as the 'short rains' start falling at the end of October, stopped during the short dry season of January, and started again at the beginning of the 'long rains' in February. The other grasses and the sedge *Cyperus laevigatus* showed a similar pattern but reacted dramatically to the first rains. In part, this response to rainfall was probably due to the mineralisation of nitrogen in the soil (Dradu & Harrington, 1972), while the second rains result in a slightly lower production peak (smaller nitrogen flux in the soil at the end of the short dry season, see Dradu & Harrington, 1972). *Cynodon dactylon* and the woodland species grow on more loamy/clayey soils than *Sporobolus spicatus*, so their growth peaks are more extended, while *Cyperus laevigatus* grows on areas with upwelling groundwater, which presumably has been replenished already after the short rains. It appears likely that the production of *Cynodon* dactylon is much higher than that of the woodland species (including the

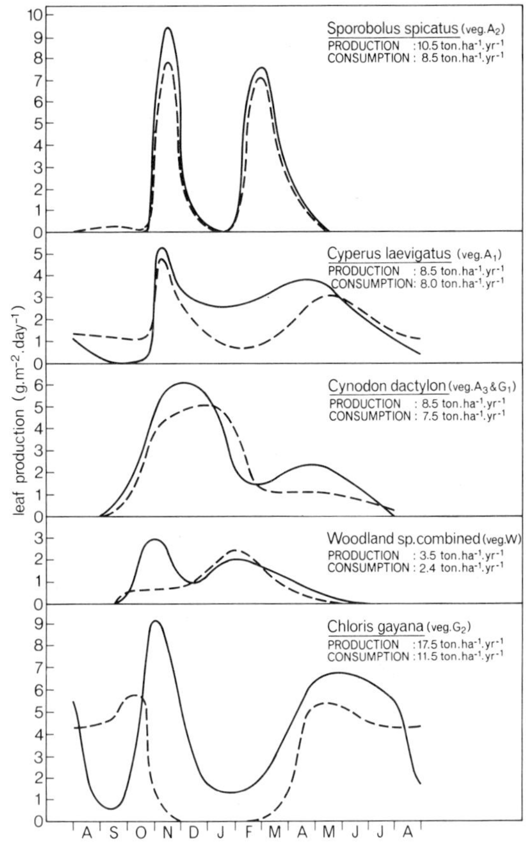

Fig. 13. Production (solid line) and consumption (broken line) curves for Manyara grasses in the course of the annual cycle (depicted from August to August) determined by sampling insiede and outside exclosures (see text). Note that consumption is staggered (compare with fig. 15 showing seasonal shifts for one of the herbivores in relation to vegetation types identified here by letter code).

Table 3. Consumption by large herbivores in relation to leaf production in Manyara grasslands (see text for restrictions).

Species	Annual production of leaves (metric tons.$ha^{-1}.yr^{-1}$)	Annual consumption as % of annual production
S. spicatus	10,5	82
C. dactylon	8,5	88
C. laevigatus	8,5	94
Woodland Species	3,5	71
C. gayana	17,5	65
Mean	9,5	80

closely related *C. plectostachyus*), because it grows in areas without tree cover. However, the second peak of *C. dactylon* is lower, perhaps because most nitrogen was already 'captured' in the sward due to the vigorous growth during the first rains. The production pattern of *Chloris gayana* is very different from the other species: although the first production peak occurs immediately during the short rains, the second peak occurs at the end of the long rains. It is not easy to explain this, but a major difference between this species and the others considered is that *C. gayana* forms very dense tall swards (100% ground cover in several layers and a sward height of an average 110 cm) while the other species rarely attain a

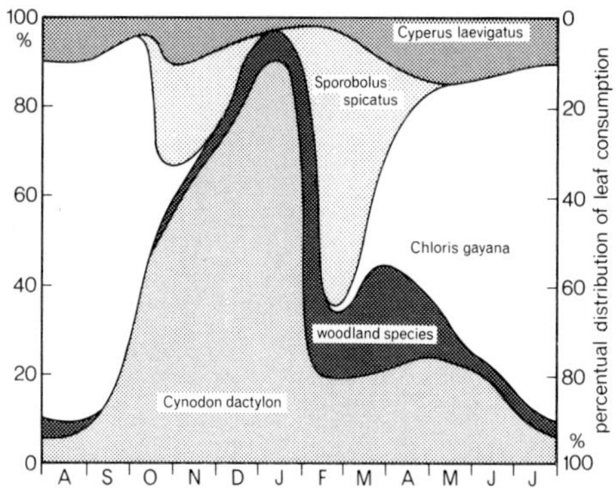

Fig. 14. Leaf consumption in Manyara: total offtake by all large herbivores in the course of the year (August through July, compiled form fig. 13) expressed as per cent contribution of each species or species group. Note that the bulk of the consumption at any one time is usually accounted for by only 2 species.

ground cover of 100%, are normally single-layered, with a sward height of at most 40 cm. *C. gayana* appears to be ignored by the grazers during the wet season, and is grazed only during the dry deason. It is possible, then, that herbage removal during these dry periods stimulates the growth of new leaves, causing the very late production peak in this species.

The salient feature of the species patterns presented in Fig. 13 is that the peaks in production of leafy material are staggered in relation to one another. During most months of the year fresh growth is accumulating at some point in the Manyara system, an expression of the diversity of edaphic and hydrological conditions over a restricted geographic area. The total offtake of all large herbivores is given in Fig. 14, and this summation reveals that at any one time the bulk of the food consumption in the Manyara grassland is accounted for by only two or three plant species. The period of *C. gayana* exploitation may well represent the bottleneck phase in this system, as far as our preliminary analyses of nutritive quality can be taken in conjunction with biomass on offer. For the present, it will suffice to emphasize that the wide spread of months during which the extensive grasslands are in a phase of rapid vegetative growth, provides the underpinning of the high biomass of the Manyara herbivore community.

The diversity of species and their growth pattern form a mozaic in space and time, leading to extensive movements of the large herbivores, shifting their focus to track the changing pattern of food availability. The annual cycle of migration brings most of the large herbivores (elephant, zebra, wildebeest) beyond the park boundaries but buffalo movement is restricted to the 100 km^2 park area. Fig. 15 displays the seasonal shifts of buffalo in relation to the main grassland vegetation types described earlier, and the year-round summary shows how completely the system is utilized.

6. Herbivory and disturbance

Disturbance has been defined in this book as 'change in conditions interfering with a normal functioning of a given biological system'. The question here, is whether herbivory can be said to constitute a disturbance from the point of view of the continuity of the vegetation and hence of the herbivores themselves. In the present chapter we have described the utilization of the vegetation by two different groups of herbivores (waterfowl and large mammals) in two contrasting areas (the northern Netherlands and a part of East Africa, respectively). Looking back on the information presented, we can draw a series of conclusions which appear to hold for both systems, and hence may be of more general validity.

a) Grazing pressures (both on seed stocks as well as growing leaves) are so heavy that the food stock is effectively depleted to a level beyond which exploitation is no longer profitable for the individual herbivore; this level we have termed the grazing or exploitation threshold.
b) Freedom from herbivory is only attained after the vegetation had passed through a family of utilization curves defined by the attributes of the different herbivore species exploiting the resource ('successive depletion').
c) In natural systems vertebrate herbivores are not restricted to limited areas, and after a given target plant has been grazed down to the threshold value the herbivore switches to an-

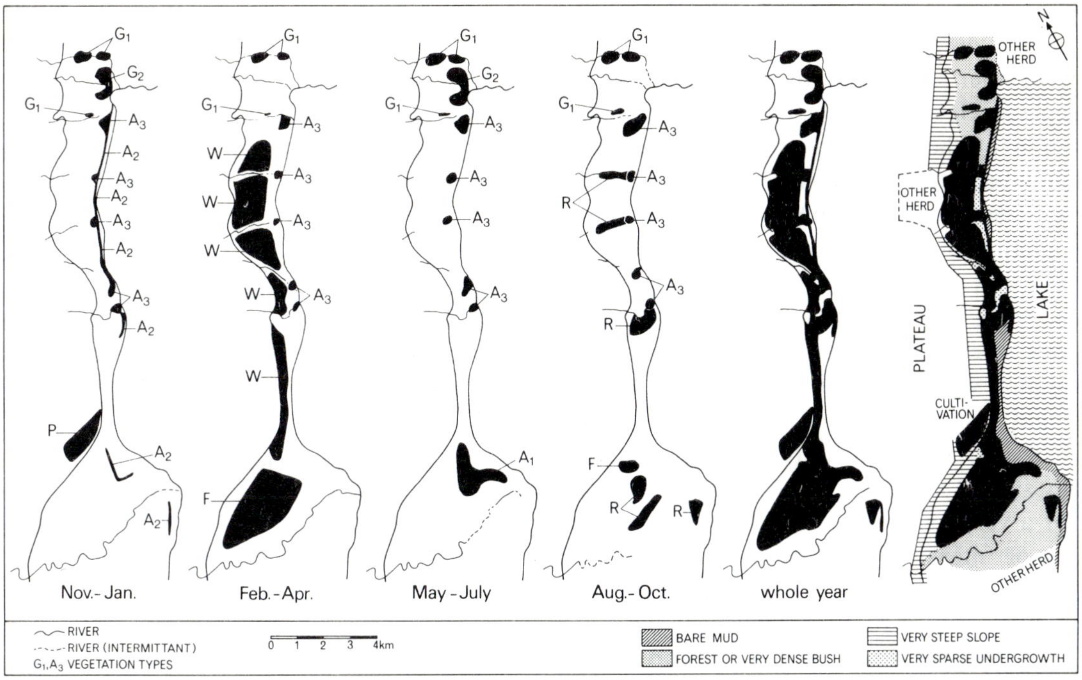

Fig. 15. Seasonal distribution of Buffalo in the main study area, Manyara (for abbreviations of vegetation types, see text) in three-month periods and summed throughout the year. At far right, general physiography is given (legend below). Note that the open habitat types are all used in the course of the year.

other food source, which may involve a shift into another habitat or even extensive seasonal movements. Such shifts in habitat use can be quantified in relation to the grazing threshold if intake rates are known in relation to energy and nutrient requirements, since in our experience only a restricted number of alternative food sources are at a suitable growth stage for exploitation at any given moment.

d) For Manyara, an area of reasonably consistent rainfall and hence primary production, census data on large mammalian herbivores depending on grassland indicate relative stability of the total herbivore biomass over at least twenty years, even though constituent species may fluctuate. This supports the contention that food supply sets the population limits of the herbivores.

Our major conclusion, then, is that because the herbivore is prisoner of its food supply, the vegetation (under natural conditions at least) is free from disturbance due to the herbivore. We believe that this balance will change dramatically if the herbivores are restricted in their movements, and disturbance in the form of local overgrazing can be expected. The importance of spatial and temporal heterogeneity, so striking in natural systems, is that the herbivore can shift readily to alternative food sources. Current efforts to model effects of herbivores on the vegetation (Thalen *et al.*, chapter 11) do not yet incorporate this basic feature adequately, and there is a challenging task for animal ecologists in the years ahead.

Having concluded that under natural conditions grazing is unlikely to have destructive effects. i.e. constitute a disturbance to the system, we feel that we must end on a cautionary note. Local overgrazing in a completely natural setting has been documented for the bog vegetation of Thjorsarver in central Iceland, where heavy utilization by the breeding Pinkfooted Geese, *Anser brachyrhynchus*, has led to degradation and serious loss of foraging habitat (Gardarsson & Sigurdsson, 1972). A causal explanation for this mismatch between geese and their food supply has not yet been forthcoming, but extensive bo-

CHAPTER 10

Ecological modeling of short-term plant community dynamics under grazing with and without disturbance

L.F.M. Fresco, H.P.M. van Laarhoven, M.J.J.E. Loonen & T. Moesker

Detailed contents

1. Introduction

1.1 Models of plant community- and ecosystem-dynamics

A simulation model is a simplified presentation of a system in the real world around us, where this system is a perceived -limited- part of reality. The user determines the system with respect to his own purposes and to the natural structure of the real system (Penning de Vries, 1984). The identification of the system involves (i) the scale, including patch size (spatial size), the time steps plus duration and organization level, (ii) the variables to be explained (e.g. vegetation dynamics, animal behaviour, meat production, forage availability), (iii) the 'independent' variables and parameters, the use of which is determined by organization level and a prior insight of the user. It is important that the model should be 'open', i.e. there is opportunity to include variables which later appear to be important.

The decision as to which type of model is to be built depends mainly on the aims of the project. In the context of vegetation and landscape research these aims can either be to predict the changes to be expected or to understand the mechanisms causing these changes (= explanation). In the first case it is a primary requirement that the model and the simulation programme are widely applicable. In the second case, i.e. causal-analytical models, this generalization follows from the spatial and temporal dimensions of the real system that was initially chosen to collect data from. This model starts with a theoretical consideration of the internal structure of the system. In such models a sequence of steps, i.e. building the model, writing the programme, executing the simulation, validating the parameters, analyzing the sensitivity of the model, may be re-

J. van Andel et al., (eds.), Disturbance in Grasslands. ISBN 90-6193-640-3.

peated, leading to a different, improved structure than the one the process started with. A model can either be stochastic or deterministic or a combination of both. A general rule is: a deterministic model is contructed first and stochastic parts are added only if it proves essential.

The use of discrete time steps instead of differential equations has it advantages when describing a complex ecological system: 1. Ecosystems are rarely in a state of static equilibrium and, even under constant environmental conditions, are continuously changing (Grubb, 1979). Such dynamic change is better described with a discrete model. 2. Natural discontinuities in time exist (diurnal, yearly). The parameters in the system, having 'time' in their dimensionality, can be measured or estimated in a direct way with a constant time step from experiments and field observations. Climatological and other information, which is supplied in yearly or daily observations, can be applied directly. 3. A simulation can have three approaches: 'event', 'activity' or 'process' (Graybeal & Pooch, 1980). An event occurs at a point in time, an activity occurs over a period of time, a process is a continuing change. The scale on which one looks at the system should be chosen in such a way that changes can be described in terms of 'events' and 'activities', rather than 'processes'.

The discrete model obviously has also its disadvantages. The most important of these is the fact that the outcome of continuous processes is systematically underestimated (De Wit & Goudriaan, 1978). If parameters are used (by means of adjusting and 'tuning') to fit the simulated results with observations, these estimated parameters will have systematic errors. This should be taken into account with parameter-evaluation.

1.2 The use of an explanatory model

The necessity for one or other type of model is largely determined by the number of variables concerned and the complexity of their interrelations. Purely descriptive models, such as those underlying multivariate techniques (e.g. transition matrices) have the limitation of not allowing for mutual interrelations between independent variables. Another disadvantage of this type of model is that mutual effects of abiotic and biotic factors cannot be described or will not be revealed. In most cases the results of such analyses can only be formulated in terms of hypotheses. Frequently these hypotheses only have a local validity even after having been tested by means of an experimental approach.

Do explanatory models really give an 'explanation' or do they remain at the level of 'theory' or 'hypothesis'? Taking into account that a field-phenomenon is practically never reproducable, and that any explanation of such a phenomenon is therefore different to an explanation of a physical event, simulation can best be considered as a link in a closed chain of observation, theory, hypothesis, experiment and simulation. The most important function of the explanatory model seems to be to bring information (characteristics of variables, interrelations, existing submodels) together into a manageable form.

In general, the separation of studies of biocoenoses into descriptive or causal-analytical is subjective and artificial and therefore irrelevant. However, studies such as the present one, which is aimed at understanding patterns and dynamics on a certain level of organization from properties on the next lower level, can perhaps be classed as 'causal-analytical'.

A serious problem in simulating plant community dynamics is the great number of parameters involved in such a simulation. It has been said that: 'with five parameters I can draw you an elephant, with two more I can let it dance the mazurka'. The number of parameters needed to make the results fit observations must therefore be kept as low as possible, and the parameters should only be varied within the limits of observation and of scientific probability. It is essential that all parameters are measurable, and are expressed in imaginable dimensions.

The question of whether sufficient environmental and population/life-history data have been collected to explain the dynamics of a system can only be answered by means of a set of mathematical equations. These equations have to present ecological reality as well as possible, given the necessary simplifications. The set of equations must be structured usually in a systems-flow-diagram. This constitutes the model. This model must be translated into a computer language and can than be used for a simulation. The initial state of a simulation is of primary importance, it has to consist of real (observed) values. If the results of the simulation are consistent with observed dynamics, then it may represent reality.

1.3 The modeling of disturbance

Disturbance can be simulated by constructing a mathematical model of a system, and then subjecting the model to events, that can be regarded as a disturbance, with respect to that system, taking into account the relevant scale (spatial, temporal, integration level). The word 'disturbance' should only be used with reference to a given system (or family of systems) and a given scale. For example, grazing is certainly a disturbance for an individual leaf, or in most cases an inidividual plant. It is also a disturbance for a grassland community just taken into grazing, but not for a community after a couple of years of grazing.

The problems and possibilities of modeling disturbance can best be illustrated by a simple and well-known example (Fig. 1). A model can be constructed to describe the effects of disturbance on organisms in a petri dish, based upon the assumptions: (i) reproduction is positively proportional to the population size and (ii) there is a maximum population size, determined by available resources and space. If we further assume that (iii) death does not play a significant role in the chosen time span, and (iv) any increase in the speed of growth, due to an increasing population size, and decrease, caused by approaching the carrying capacity, are equally important, then the logistic equation can be used. If growth is temporarily disturbed by adding a toxin to the petri dish on the fourth day of the experiment, the growth formula can then be modified using a simple, descriptive equation to quantify the results. This class of models to describe the results of disturbance, will be called 'type A'. If, however, the

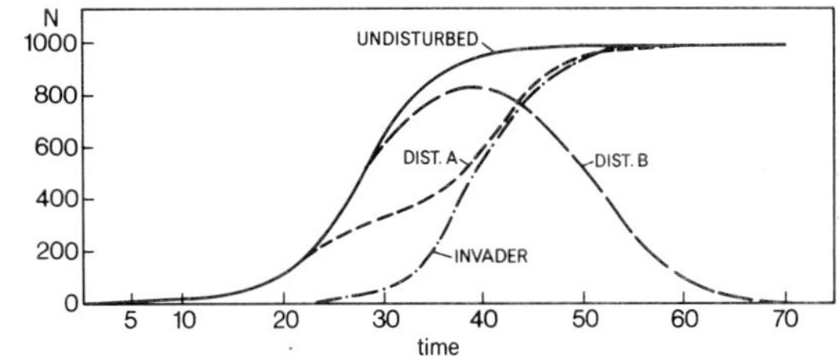

Fig. 1. (Logistic) growth curves of hypothetical populations. Undisturbed growth rate (AX) = 0.25; carrying capacity (KX) = 1000. Equation: DX/DT = AX ⋆ X ⋆ (1 − X/KX). Disturbance 'A' reduces the growth rate to minimally 0.025 during T(20–40). Than: AX = ABS(T-30) ⋆ 0.025. Disturbance 'B' is an invasion of a second organism at time = 20. Equations:

DX/DT = 0.25 ⋆ X ⋆ (1 − X/1000) ⋆ (1 − 2.00⋆Y/1000) and
DY/DT = 0.50 ⋆ Y ⋆ (1 − Y/1000) ⋆ (1 − 0.50⋆X/1000).

For further explanation see text.

culture is infected with a bacteria, which eleminates the organism in a few days, a serious modification of the mathematical model is needed. Since two populations are mutually affecting each other, an extension of the scale of the model is necessary to encompass a higher level of organization. This class of modeling of disturbance will be called 'type B'. Examples of the B-type modeling of disturbance are those of Doyle (1981), simulating the effects of hurricanes on the structure of tropical rain forest, West et al. (1980), simulating the influence of air pollution on temperate forest and Miller et al. (1984) simulating disturbance in a tundra ecosystem with the model 'ARTUS'.

The validity of a model of a system has its limits, which should be defined so that they equal the limits in general of the environmental (vector-) space to which the system is adapted. Disturbance occurs when these boundaries are exceeded, i.e. conditions transcend the limits of adaptation (or flexibility) of the system. Hence the modeling of disturbance needs a model on a more extended scale (spatial and/or temporal and/or level of organization) than the undisturbed system itself.

The following definition of disturbance might be useful in this context: 'any event exceeding the environmental conditions to which the system is is adapted'. Disturbance is not identical to unpredictable variation of a normally variable factor to which variation a system is adapted. Every factor in a system has its own 'predictable unpredictability'. Terms like 'adaptation to disturbance' or 'continuous disturbance' are not useful. Each system has its own autogenic dynamics: short-term and long-term, random, cyclic and/or directed (succession). After a system has been disturbed, it no longer exists in its previous form as it has been replaced by a different system with its own dynamics but, after a certain period of time, convergent succession may cause the disturbed and the undisturbed systems to become identical. The classification 'disturbance-dependent system' can only be applied to the instable, temporarily existing situation, following a disturbance.

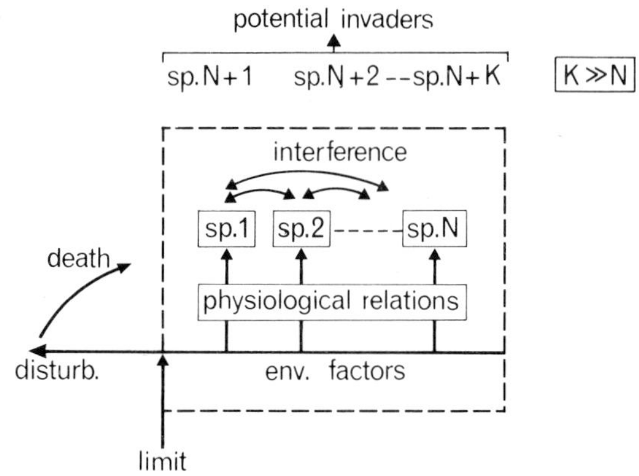

Fig. 2. Generalized design of an explanatory model of plant community dynamics. Disturbance is defined as an event exceeding the limits of validity of the model.

1.4 The aims of modeling short-term plant community dynamics

The understanding and causal explanation of an ecosystem or plant community requires knowledge of 'the species life-history and physiological characteristics which determine, to a large extent, potential population responses to the changing competitive environment' (Peet & Christensen, 1980). Ellenberg (1954) argued that: '...questions came up with respect to the causes of these strict relations (between the mosaic of plant communities and environmental conditions). These questions can be answered by three supplementary means: (i) quantitative comparison of numerous communities and their habitats, (ii) quantitative analysis of relevant environmental variables and of characteristics and performance of plants in their own habitats, (iii) experimental studies under more or less simplified conditions'. The second and third of these can be achieved more succesfully with the help of an explanatory model.

An explanatory model of plant community dynamics must contain characteristics of the species present, including their repsonses to the relevant environmental factors. Interference between species populations has to be included in such a model, which has to separate competition for different resources (Fig. 2). One can regard the system as 'a mozaic of phytocoenoses' consisting of subsystems namely the vegetation patches, or homogeneous or heterogeneous pieces of vegetation, depending on how the investigator wishes to define them. Each of these subsystems has its own flexibility with respect to changes in environmental conditions. Any change wich exceeds these limits is a disturbance of the subsystem. The limits of validity of the ideal model should equal the limits of adaptation (flexibility) of the system as a whole. In this case the disturbance may cause the replacement of one subsystem by another, which is already part of the main system, i.e. within the limits of validity of the model, hence being of 'type A'. If an environmental factor exceeds these limits, death of individuals and/or extinction of populations starts to play a role, open space may be generated, and invasion of new genotypes or species may occur (Gremmen et al., 1984). To quantify these changes a 'type B' model is needed. Many species can be potential invaders, depending on the composition of the seedbanks, their presence in adjacent areas, the time of the year, the weather, etc. Since potential invaders can be of various life forms and strategies, it is a very complex task to model these changes.

Disturbance of a plant community, or any other system, can be described in two dimensions, namely intensity and duration or frequency. The time dimension of disturbance has not been covered in this study. Although, in most cases, disturbance will cause destruction of living phytomass or accumulated detritus (Grime, 1979; Reiners, 1983), it is better that this effect is not included in a definition.

In spite of the problems of simulating disturbance, such an excercise is useful, since it may give the user an idea of the trends and general patterns, that are likely to occur. It is also an excellent method to test the basic model and to find its weak spots and errors. This kind of simulation will only be predictive in the very short term. Inductive models, based on numerous observations, will be more effective for long-term prediction.

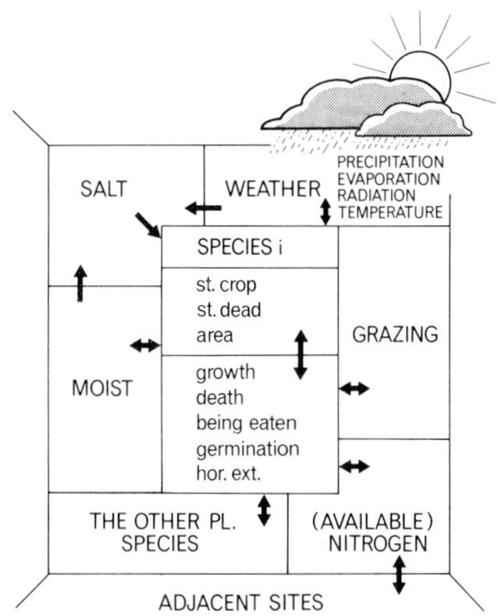

Fig. 3. State variables and driving factors influencing the growth of a species in a site.

2. Description of the model 'EGRAS'

2.1. General

The model 'EGRAS' ('Ecological Grassland Simulator') is concerned with a section of the former Lauwers Sea, which was a part of the Wadden Sea (in the north of the Netherlands), which was embanked in 1969. The actual area involved ('Schildhoek en Pampusplaat') is approximately 200 ha and was designated a feeding area for wintering and transmigrating wild fowl in 1979. However, this aim was endangered with the increasing amount of litter and standing dead material and increasing dominance of *Agrostis stolonifera*, especially in the dryer parts of the area. Therefore, since 1982, the area has been grazed by young cattle at a stocking rate of about 1 animal per hectare. The cows enter the area annually in early June and leave it in October. A large exclosure has been established for the purpose of scientific research. Data concerning vegetational changes, herbage accumulation and herbage removal by cattle and water fowl and the changing (desalinating) abiotic environment have been published (Joenje, 1978, 1979, 1982). Other information concerning growth characteristics and mutual competitive ability of some plant species, was collected in the cause of this study.

The aim of this project was to develop a model and a simulation computer programme, explaining short-term fluctuations and changes of vegetation composition from physiological characteristics of plant species and populations. The decision as to which abiotic factors were to be considered was initially left open. The initial version of the programme only contained total nitrogen, chloride and cattle grazing. This proved to be insufficient. The abiotic part of the model was therefore extended to include soil moisture, weather characteristics and the partitioning of nitrogen into soil available N, soil non-available N and biomass N (Fig. 3).

The earlier model 'ARTUS' (Miller et al., 1984) dealt with one day steps on one square meter sites and described the growth, above and below ground, of ten plant species, six of which were mosses. It differed in spatial scale and specification from the present study. Instead of describing a complete nutrient budget and explaining the phytomass production and allocation using independent driving factors, this study attempts to discover which abiotic factors, and which inter-population interference parameters, are minimally required to understand observed spatial and temporal variation in quantitative species composition. This is a synecological problem rather than merely an investigation of an ecosystem. A number of solutions and assump-

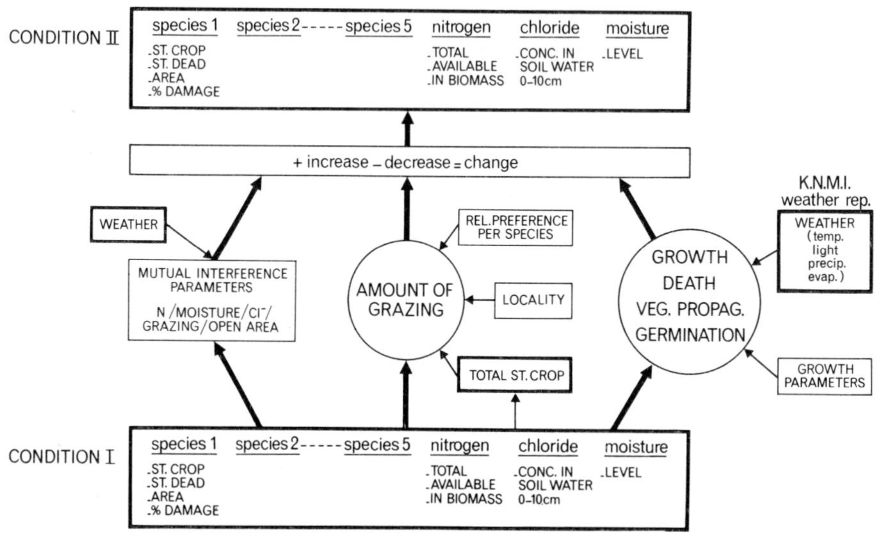

Fig. 4. Flow diagram of a single step (1 day) of the simulation programme 'EGRAS'. 'Condition I' represents the state of the system at the end of the previous day (day 1), 'conditions II' represents the state at the end of day 2. For further explanation see text.

tions from 'ARTUS' have been used in 'EGRAS'.

The area was subdivided into a grid of hexagonal plots, each sized 400 m^2. These sites were each classified, with the help of a vegetation map of the area (Reitsma, 1981). All operations were carried out on a summarizing 'map', containing 25 different quadrats (total size 1 ha.) representing the present vegetation typology. The programme used one day time-steps. The duration of a complete simulation was one grazing season. A simulation of the changes in the late autumn, winter and early spring (with water fowl grazing) has not yet been accomplished. Under a set of assumptions, the most important of which was that all relevant changes in vegetation composition took place between May and October, it was possible to simulate the dynamics in subsequent years. The initial state in all simulations was the actual (described) state in May 1982, at the start of the new management, which caused a disturbance, according to the definition.

The model uses five plant species namely *Agrostis stolonifera, Aster tripolium, Phragmites australis, Puccinellia maritima* and *Salix repens*, which constitute at least 50% of the vegetation coverage in about 80% of the area.

An outline of the model is presented in Fig. 4. The three main processes, represented by thick arrows in the figure, are mainly independent. All three use the state at the beginning of the step (day) and are used in the sequence: 1. grazing, 2. growth & death, 3. abiotic interference. Since there are a few operational links between the lines, the programme is not completely indifferent to this sequence. However, a change of sequence of operational steps never resulted in a difference of more than 0.44% over an entire growing season. Listings of the most important equations and most parameters are given in the appendix. The simulation programme was written in PASCAL. This programme was not meant to be applicable for other areas than the one just described.

The simulated system in this study is a 'landscape' rather than an 'ecosystem', since it deals with an ecological mosaic of ecosystems which is always highly heterogeneous (Godron & Forman, 1983).

2.2 Growth & death

The growth equations (appendix) use discrete versions of the logistic equation, based on relative growth rate of plant species per unit area (RGRMAX) and with the maximum possible phytomass per species per unit area occupied (SCMAX). The maximum growth rate is diminished by several effects. In the equations these effects are represented by factors which can have values between zero (growth stops) and one (no effect) (Fresco, 1982). The environmental factors considered to influence growth are:

1. C: competition with the other species populations for resources other than nitrogen
2. ND: nitrogen as a deficient factor
3. NT: nitrogen as a toxic factor (factors ND and NT represent a complete response curve with respect to nitrogen)
4. ST: sodium chloride as a toxic factor
4. G: grazing intensity
5. T: temperature
6. L: light
7. W: soil water.

Most of the parameters (see appendix) have been obtained either experimentally or from literature.

The total of environmental influences which restrict the relative growth is a factor C × ND × NT × ST × G × T × L × W = TGLF ('Total of Growth Limiting Factors'). The total biomass of a species in a site is divided into living (SCLI) ('Standing Crop - Living') and dead (SCDE) standing phytomass. The below ground biomass had to be neglected because of inadequate information. The equations, describing growth of a species, the transitions from living to dead standing material and from standing dead crop to litter are:

$$RGR(t) = RGRMAX \times TGLF(t) \quad (1a)$$

$$DSCLI(t) = RGR(t) \times (1 - SCLI(t)/SCMAX) \times SCLI(t) \quad (1b)$$

$$SCLI(t) = SCLI(t) - DEATH \times SCLI(t-1) \quad (1c)$$

$$SCDE(t) = SCDE(t-1) - DECOD \times SCDE(t-1) + DEATH \times SCLI(t-1) \quad (1d)$$

DECOD (g/g/day) represents the coefficient of transformation from standing dead material to litter, which values were obtained from personal observations in the area mentioned and from literature. DEATH and DECOD are calculated within each one-day-run. DEATH (g/g/day) is directly proportional to the total growth limitation by causes other than by approaching the carrying capacity (factor TGLF). If this growth limitation exceeds a value that is given per species (GDLI, 'Growth and Death Limitation'), the relative amount of dying remains at a constant value (MINDE).

Each species occupies a certain amount of area, e.g. AREA(j) for species j and the growth state parameters concern this covered area. For the area of a species in a site to expand, there has to be open, uncovered, surface (OPAREA), i.e. a species has to establish in the bare soil. This can be done either by germination or by vegetative propagation. Germination per unit open area is a function of: (i) chloride conditions in the upper soil layer, (ii) the air temperature, (iii) the amount of living seeds present. Variable (iii) was large for the five species considered, though variable both in space and time, therefore seed was not assumed to be a limiting factor. The vegetation propagation is a linear function of: (i) available open area (OPAREA), (ii) the speed of growth, though Jaquard & Heim (1983) found a non-linear asymptotic relationship between sprout and tiller production in *Dactylis glomerata*, (iii) the length of the boundary between open area and vegetation, both in the site and in adjacent sites, (iv) chloride in the upper soil layer, (v) potential horizontal extension of other species.

2.3 Grazing

The grazing has been simulated as the grazing of one single 'supercow'. For the test-map of 25 × 400 m^2 = 1 ha, this 'supercow' was equal to one single cow (except for the disturbance 'overgrazing', see later). The total amount of biomass consumed by a cow increases from 5.2 kg dry matter per day in early June to 10.8 kg in med-August, than remains constant for the rest of the grazing season. The relative preference of the cows for the various species was expressed in parameters PRSP(j), obtained by means of observations in the area. The total amount of phytomass grazed in a site is divided over the species proportionally with

$$((SCLI(i,j) + SCDE(i,j)) \times PRSP(j) \quad (2)$$

The minimum amount of above-ground phytomass remaining after grazing was estimated for each species from the results of field measurements (RESIDU(j)). The relative grazing preference values for a given site i are calculated next (every day):

$$RPSI(i) = PRLO(i) \times PRFO(i) \times PRAD(i) \quad (3)$$

where PRLO is the 'locality preference', determined by factors such as the regular routes of the cattle to sleeping sites, the presence of fences, etc. PRFO is the 'forage preference' (Goodall, 1969), which is proportional to the logarithm of the standing crop and to the preference for the species. PRAD is determined by the preference-values of adjacent sites.

After the calculation of PRSI-values, each of the sites is classified into one out of five equidistant preference-classes. The grazing starts in a randomly chosen site of the highest class. This site is grazed, until its classification drops down one class. Next the grazing continues in the nearest site (if necessary chosen at random) of the highest class.

2.4 Abiotic factors

Since no evidence is available that nutrients other than nitrogen are limiting or controling factors in the system, the nutrient-conditions are represented by nitrogen only. The amount of total nitrogen remains constant in each site (400 m^2). The total amount of nitrogen (NITOT) is divided into: (i) a fraction in the live standing crop of each of the species j (NISCLI(j)), (ii) a fraction in the dead standing crop (NISCDE(j)), (iii) the fraction in the litter and in other forms not directly available to the plants (NINAV) and (iv) the fraction which becomes available during the time-step (NIAV).

Amounts of nitrogen are expressed in the model relative to standard units (u.N. = units nitrogen). This dimension has been adopted because of a transmission of the results of experiments. The plant responses to nitrogen in these experiments were considerably more exact and reproducible than a chemical re-analysis of the amount of nitrogen. Therefore nitrogen values derived from simulation are not easy to compare with those from chemical soil analysis. There is a significant non-linear, but monotonous relationship between values derived from the model 'EGRAS' and those measured in the field.

The amount of nitrogen in the newly-grown biomass depends on parameters of the species and the amount of available N:

$$NIFPL(i,j) = P1(j) + P2(j) \times \ln(NIAV(i)) \quad (4)$$

where NIFPL(i,j) is the relative amount of nitrogen in the new biomass of species j in site i and Pl(j) and P2(j) are constants, obtained from simulation, using the results of growth experiments with varying nitrogen concentrations.

The models applied here for the simulation of decomposition and availability of nitrogen are those of Olson (1963), Davidson et al. (1978), Runge (1983) and Kemmers & Jansen (1985). Olson's decay coefficient is a function of temperature and soil moisture:

$$NIIN(t) = NIIN(t\text{-}1) + f(sum(NISCDE),DECOD - DECOL \times NIIN(t\text{-}1) \quad (5a)$$

$$NINAV(t) = NINAV(t\text{-}1) + DECOL \times NIIN(t\text{-}1) - DEMIN \times NINAV(t\text{-}1) \quad (5b)$$

$$NIAV = NIAV(t\text{-}1) + DEMIN \times NINAV(t\text{-}1) - f(growth) \quad (5c)$$

where NIIN is the detritus input (expressed in u.N.) and DECOL is the decay coefficient (a function of temperature and soil moisture) and DEMIN is the demineralisation constant. Each site was treated as a closed system with respect to soil fertility.

The biomass which is removed by the cattle (including the trampled biomass and the consume-loss) immediately loses its nitrogen to NINAV. The model would be improved using time delays for the decay of urine, faeces and lost litter, but so far no adequate information is available.

The salinity of the majority of the area has been declining slowly since the embankment was finished in 1969, but there is still considerable temporal and spatial variation which controls vegetation composition. So far no suitable model, describing the relationships between salt, moisture, soil texture and topography has been found in literature. A simple empirical model has been developed. Because of the yearly fluctuation of the salt values desalinisation has not been taken into account in the five-year runs of the simulation. The model assumes a negative linear relationship between soil moisture condition and the salt concentration, if there is no water above the soil surface. The maximum salt concentration under extremely dry conditions (SAMAX) has been shown to be a function of the percentage of the soil surface which is not covered with vegetation. However, there is some circular reasoning in this part of the model. Therefore it is better regarded descriptive, rather than explanatory.

Weather conditions have been taken from monthly reports of the Royal Netherlands Meterological Institute (K.M.N.I.) at Leeuwarden Air Force base (ca. 40 km from the area). The average temperature and the total amount of radiation were used for the simulation of the plant growth, whereas precipitation and evaporation were used for the estimation of soil moisture conditions.

2.5 Simulation of a sequence of seasons

During the winter season (mid-October until May), the area is at first intensively grazed by transmigrating water fowl (Joenje, 1985). Later (December to February) there is more extensive grazing by overwintering wigeons. In the spring grazing by migrating geese and ducks can again be observed. Data concerning winter-grazing have recently been collected, they have not been incorporated into the model.

Most of the parameters in the model reach their validity-limits at minimum average diurnal temperatures, such as those in the middle of October. An extension of the model to that of a whole year is being worked on, but has not yet been accomplished.

Sequences of not more than five years were simulated; in each case the simulation started in early June 1982. The weather conditions of 1982 were repeated five times, so as to facilitate interpretation; the spring and summer of 1982 were average, except for a rather dry early summer. If we assume that all changes in relative vegetation composition occur during the summer season, that the biomass removed by winter grazing and death during winter is replaced by spring regrowth, and that the percentage standing dead material remains at the value it had after the first (1982) grazing season, simulation of the period October till May can be replaced by a data operation and a restart (Fig. 5).

Although the programme simulated vegetation dynamics in all 25 sites at each simulation run, for simplicity only six sites have been chosen for a final interpretation (Table 1).

Table 1. The six 400 m^2 sites (out of 25), selected for presentation and interpretation. The species names in italics are those of species with an initial (1982) standing crop of more than 100 g/m^2.

Site nr.	Rel. salt cond.	Rel. moist. cond.	Soil texture	Vegetation type
[1.2]	fresh	dry	clay	*A.stolonifera*/*P.australis*/S.repens
[1.3]	fresh	dry	sand	*P.australis*/*S.repens*
[2.2]	s/f	d/w	s/cl	*A.stolinifera*/*A.tripolium*/*P.australis*/P.maritima
[3.3]	s/f	wet	clay	*A.stolonifera*/S.repens/A.tripolium/P.maritima
[5.3]	salt	wet	clay	*A.stolonifera*/A.tripolium/P.maritima
[5.4]	salt	wet	clay	*P.maritima*/A.stolonifera/A.tripolium

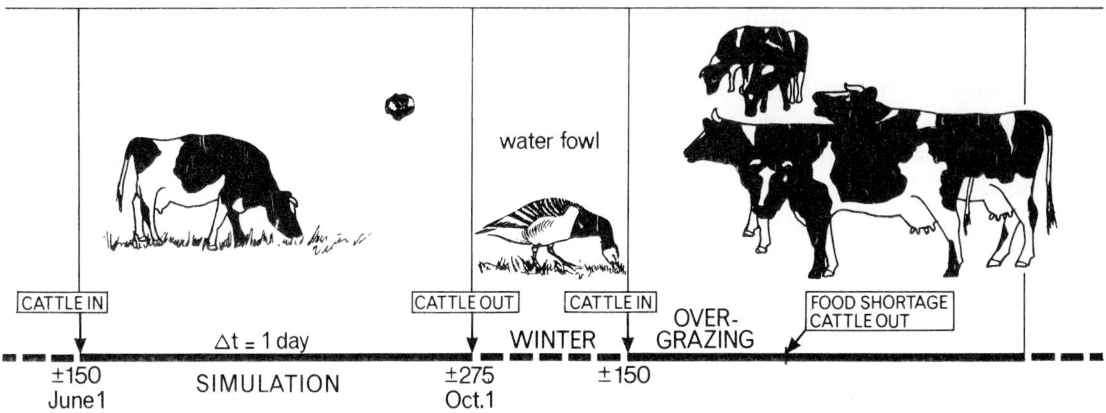

Fig. 5. A scheme of the simulation of a sequence of years. The events in the winter period are not simulated, but approached as one single event by means of an estimation procedure (see text). The programme has been run without grazing, with grazing and with overgrazing in the second year. Where a shortage of forage occurs, the cattle are removed and no grazing occurs for the remainder of the season.

2.6 Disturbance

Three types of disturbance were simulated: (i) an extremely dry season, simulated by lowering the soil water level by 30 cm over the entire area in the second year of the simulation and returning it to normal in the third year, (ii) raising the soil water level by 30 cm in the second year and returning it to normal in the third year, (iii) quadrupling the stocking rate in the second year and returning it to normal in the third year.

Under high grazing pressure a food shortage could occur in the simulation. If the amount of consumable forage was less than 20% of that needed by the cattle (5.2 – 10.8 kg of dry matter per cow per day) during ten sequential days, the simulated cattle was removed from the area, which than remained ungrazed during the rest of the season.

3. Results

Simulated values for living phytomass did not depart significantly from the observed values over the growing season (Table 2, Fig. 6). Simulation tended to underestimate *P.maritima* and *A.stolonifera B* (with abundant flowering and little tillering), but overestimates *A.stolonifera A* (with abundant tillering and not much flowering) in general. The results of the simulation of living and total phytomass are, when compared with observed values, satisfactory, but can be improved. However, the computer programme failed to simulate changes in the percentage standing dead phytomass correctly, especially in the ungrazed simulation of 1982 (Fig. 7). The cause of this seasonal pattern is still unknown, therefore cannot be simulated. The simulated patterns in the following year, 1983, are according to expectation; a decrease of the percentage standing dead matter until July/August and an increase in the late summer and autumn. The peculiar pattern of *A. stolinifera B* in 1983 underlines the necessity to investigate the relationship between weather conditions and the dying of the phytomass.

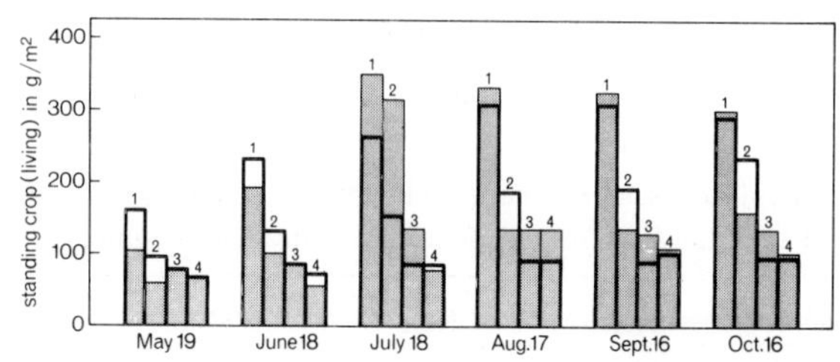

Fig. 6. A comparison of simulated values of living aboveground phytomass with observed values in the growing season of 1982. Meaning of the columns in each histogram from left to right: 1. *A.stolonifera B* (with rapid tillering and no or hardly any flowering); 2. *A.stolonifera A* (with relatively littlle tillering and abundant flowering); 3. *P.maritima*; 4. *A.tripolium*. The bold-line histograms represent the simulated values, the thin-line histograms observed values.

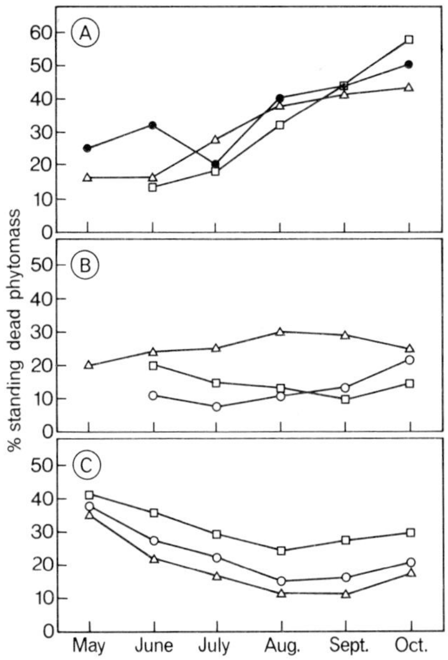

Fig. 7. Standing dead matter as a percentage of total (living + dead) phyto-mass. Observed values in 1982 (A) and 1983 (B), compared with simulated values (C). Squares: *A.stolonifera*A*; triangles: *A.stolonifera*B*; open circles; *P.maritima*; closed circles: *P.australis*.

Simulations of 'wet' and 'dry' years indicated that: 1. Bare ground (up to 91% in the second year) was recolonised slowly, taking at least three years; 2. *A.stolonifera* was least influenced by either 'dry' and 'wet' years, although its standing crop was lower after the disturbance, it recovered quickly; 3. *P.australis* was most affected by the 'dry' second years, when it failed to re-establish, but was benefited by the 'wet' years. 4. *S.repens* was hardly affected by the 'dry' disturbance, but was greatly benefited by the 'wet' second year; 5. *P.maritima* was not affected by 'wet' years but was benefited from the 'dry' disturbance, especially in the sites where the species was absent: there it established; 6. *A.tripolium* invaded, resp. increased after either 'dry' and 'wet' disturbance; all events causing an increase of bare ground were obviously beneficial for the establishment and competitive situation of *A.tripolium*.

Table 2. The depature (% difference) of the simulated living phytomass from observed values of four species in five months during the 1982 growing season. The *A.stolonifera A* form with rapid vegetative propagation and hardly any flowering; *A.stolonifera B* has abundant flowering and relatively little vegetative tillering. The analysis of variance shows a significant difference between the species, but no difference between months. The data are the same as those used in Fig. 6.

	June	July	Aug.	Sept.	Oct.	Total
*A.stolonifera*A*	+30	−52	+36	+43	+47	+104
*A.stolonifera*B*	−20	−22	− 6	− 6	− 3	− 57
P.maritima	0	−36	−29	−29	−29	−123
A.tripolium	+17	+13	−29	0	0	+ 0
Total	+27	−97	−28	+ 8	+15	− 75

Item	D.F.	M.Sq.	F
Species	3	2339.17	4.855*
Month	4	490.24	1.018 n.s.
Residual	12	481.78	

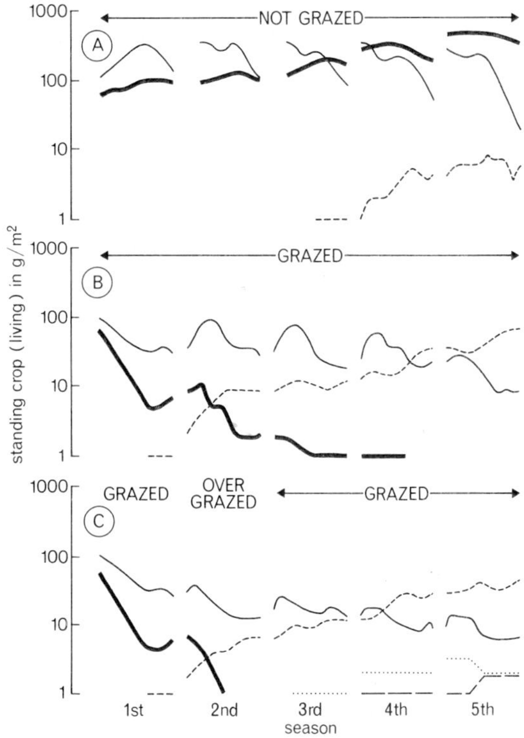

Fig. 8. An example of a graphical presentation of the results of a five years simulation (site [1.3]). The winter period between two subsequent years causes a discontinuity! From top to bottom: not grazed, 'normally' grazed and overgrazed. Legenda: bold lines: *S.repens*; thin lines: *P.australis*; 'long' interrupted lines: *P.maritima*; 'short' interrupted lines: *A.stolonifera*; dotted lines: *A.tripolium*.

For the graphical presentation of a five-year simulation (not grazed, grazed and overgrazed) of a site see Fig. 8. The results of these more-year

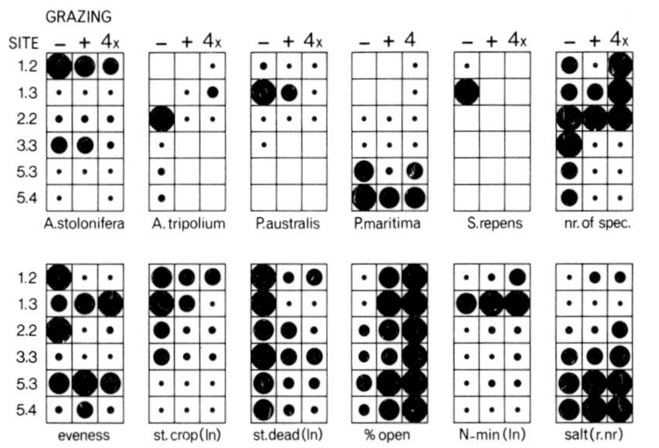

Fig. 9. The state of twelve biotic and abiotic variables on day nr. 210 (August 28.) of the fifth simulated year. All variables were independently divided into 5 equidistant classes, represented by the five different dot sizes. The colums represent the simulated regimes 'not grazed' grazed and 'overgrazed', the rows stand for the six sites (Table 2). The intervals of the outcomes of each variable, used for the classification are: *A.stolonifera* 0–149 g/m²; *A.tripolium* 0–21 g/m²; *P.australis* 0–244 g/m²; *P.maritima* 0–28 g/m²; *S.repens* 0–449 g/m²; nr. of species per site 2–4; eveness (rel. Shannon measure) .118–.998; standing crop 17–698 g/m²; standing dead 0–379 g/m²; %% open area 23–93%; mineral nitrogen 59–1000 u.N.; salt (rank numbers) 1–18.

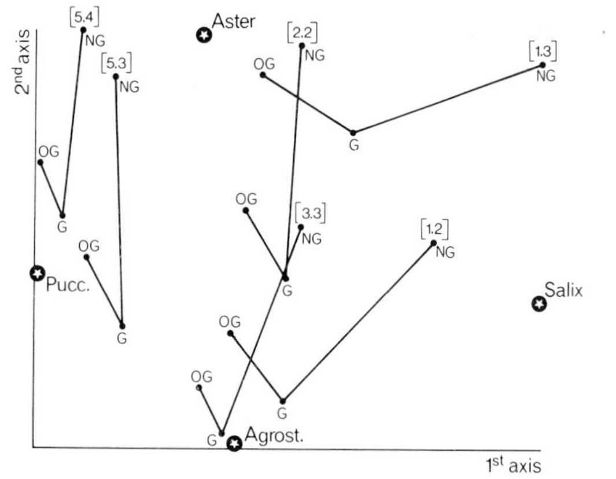

Fig. 10. The results of a DECORANA-ordination with the untransformed phytomass-data of the 210-th day of the fifth year. The data of the five species in the six sites under the regimes 'not grazed' ('NG'), 'grazed' ('OG') and 'overgrazed' ('O') were used.

simulations (1982–1987) could, of course, not be compared with actual values. However, many trends, such as the disappearance of willow under grazing, the establisment of Sea Aster in disturbed places and the advantage of the halophytic species under grazing, are consistent with observations. The simulated effects of the grazing and overgrazing on twelve biotic and abiotic variables are presented graphically in Fig. 9. The effects of grazing on the phytomass of the species was further subjected to multivariate ordination, using 'Detrended Correspondance Analysis' (Gauch, 1982). For that purpose simulated values on day nr. 210 of the fifth year were used. The results (Fig. 10) show that grazing changes vegetation containing a great deal of *S. repens* and *P.australis* into vegetation characterized by *A. tripolium* and *A. stolonifera*. Vegetation originally dominated by *A. stolonifera, A. tripolium* and *P. australis* was not greatly affected by grazing. The overgrazing in the second year causes a species replacement towards *A. tripolium* and *P. maritima* in all cases.

Finally, the dynamics of the plant species populations under the three simulated treatments ('not grazed', 'grazed' and 'overgrazed') were compared (Table 3). This comparison makes clear that the halophytes *P. maritima* and *A. tripolium* benefit from overgrazing when compared with 'normal' grazing. For *P. maritima* an extremely high grazing density even seems to be more beneficial than a normal density, at least in relatively dry habitats. *A.tripolium* remained stable in the salt sites, whereas it increased strongly in phytomass in the fresh and dry sites. An exception is the intermediate site 2.2, where it was replaced by *P.maritima*.

4. Discussion

The questions to be answered with the help of the results of this simulation model are: 1. Simulation models are simplifications of actual processes; to what extent do they improve our knowledge of mutual interference of biotic and abiotic factors and predict changes in species composition? 2. Is THIS model a sufficient description of THIS system · in other wordt: do we or don't we need other variables and/or other relationships to understand the dynamics of this grassland-system? 3. Does the extension of the model – perhaps into circumstances which will never be observed – supply us with information with respect to the

Table 3. Dynamics of the species populations in each site for each of the grazing simulations over five years. Explanation of the figures: +3 = establishment, +2 = increase of phytomass of at least 100%, +1 = increase of <100% but >=20%, 0 = stable (change of phytomass <20%) or absent, −1 = decrease <100% but >=20%, −2 = decrease >3=100%, −3 = disappearance.

site	Grazing −	+	4×	−	+	4×	−	+	4×	−	+	4×	−	+	4×
1.2	+2	+2	+2	0	0	+ 3	0	0	+3	+1	−2	−2	+2	−3	−3
1.3	+3	+3	+3	0	0	+ 3	0	+3	+3	+2	+1	−2	+2	−3	−3
2.2	+2	+2	+2	+2	0	+ 2	+2	−2	−2	+2	−1	−2	0	0	0
3.3	+2	+2	+2	+1	0	0	+1	0	0	+1	−3	−3	0	0	0
5.3	−1	+1	−1	+2	+2	+ 2	+2	0	0	0	0	0	0	0	0
5.4	−1	−3	−1	+2	+2	+ 2	+1	0	0	0	0	0	0	0	0
sum	+7	+7	+7	+7	+4	+12	+6	+1	+4	+6	−5	−9	+4	−6	−6
	A.stolonifera			*P.maritima*			*A.tripolium*			*P.australis*			*S.repens*		

general rules concerning disturbance and, if so, what degree of generalization do these results justify?

4.1 Evaluation of the simulation approach

The problem 'finding a combination of species properties such that the actual vegetation dynamics can be explained adequately', can be approached in a more simple way. Assuming that three environmental gradients affect species composition: nutrient poor (sand) vs. rich (clay), fresh vs. salt and not grazed vs. grazed, the species parameters can be looked for which can be expected to describe the species' ability to maintain or to establish in a given habitat. These parameters could be: 1. concerning the nutrient poor environment, the minimum amount of nitrogen required for growth (MINN); 2. concerning the rich environment, the speed of growth (RGRMAX); 3. concerning the fresh environment, unknown; 4. concerning the salt environment, the maximum tolerated amount of salt in the soil (LETC); 5. concerning the not-grazed situation, unknown; 6. concerning grazing, the vulnerability of the species with respect to grazing (LETG). Values for all these parameters were determined experimentally.

Since the relative weight of each of those parameters was not known, they were replaced by their rank numbers (Table 4). Average rank numbers for each of the habitat types give an indication of the rank order of abundance of the species in a habitat type. When compared with either observed and simulated values, these predictions turned out to be a rather close approach. The parameters RGRMAX, MINN, LETC and LETG therefore provide a good description of the behaviour of these five species in the area studied.

Table 4. Prediction of the species composition, based on rank numbers of the most important simulation parameters. The Fig. I, II and III indicate relative abundance, derived from average rank numbers. For further explanation see text.

	poor	rich	fresh	salt	not-g.	grazed
A.stolonifera	2.5	5	3	3	2.5	4.5
P.maritima	2.5	4	3	4.5	2.5	4.5
A.tripolium	2.5	3	3	4.5	2.5	2
P.australis	2.5	2	3	1.5	5*	3
S.repens	5	1	3	1.5	2.5	0**

grazing	+	+	+	+	−	−	−	−
nutrient	R	P	R	P	R	R	P	P
salt	F	F	S	S	S	F	S	F
A.stolonifera	III	III	III	II	II	III		
P.maritima	II	III	III	III	III	I	III	
A.tripolium			II	I	I		III	
P.australis						II	II	III
S.repens							II	III
compare with site(s)	3.3	1.2 1.3	5.3 5.4	5.3 5.4	5.3 5.4	3.3	5.3 5.4	1.2 1.3

The simulation does, however, provide more information than such an analysis and the simple relationships given above. Effects, such as disturbance and the influence of varying weather conditions can only be investigated with the help of a simulation programme, where the relationships between the variables are expressed as mathematical equations. Changes in the model can be evaluated whether or not they provide an improvement by means of comparison with actual values. However, this type of simulation approach is yet only possible for simple species-poor communities.

4.2 The sufficiency of the model

The model described here has developed by gradual extension. It is still being extended, for example the species *Calamagrostis epigeios* and *Alopecurus geniculatus* have recently been included. The edaphic factors were selected gradually. Climatic variables have been introduced after simulations showed them to be necessary.

The addition of externally developed sub-models and the development of sub-models as separate projects constitutes a danger in that they can introduce a large number of variables and parameters. The programme would then acquire a pretention which it cannot prove. In this context: the nomenclature and the origin of the parameters concerning the nitrogen-cycle could suggest that this model simulates a nitrogen cycle, but this is not so: in fact this has been emphasised by the use of an exclusive relative dimension (u.N). The aim of the procedures used in this model is to describe the relevant variations in soil fertility, taking into account factors such as changes of the nitrogen contents of newly formed phytomass, and competition for nutrient as a resource. Having studied the results of greenhouse experiments and field observations, we concluded that those phenomena were indispensable for the understanding of vegetation dynamics.

Sensitivity analysis of the parameters used showed the simulation model to be not too sensitive with respect to variation in any parameter. An independent variation of each of the parameters, with plus and minus ten percent, resulted in a maximum deviation of two percent. Initial values do have a clear effect. In the simulations demonstrated observed field values were used as starting values. The influence of (properties of) plant species that were not included in the model could not be determined.

Considering that the within-season dynamics did not depart systematically from the values that were measured in 1982 and 1983 (Table 2, Fig. 6), the conclusion can be drawn that the variables chosen and the values given to them, closely approach reality in this arithmetical arrangement.

If the model is used for further prediction, the system must be observed for a longer period of time, and more experimental disturbance must be imposed: more theory-building may also be needed. Even then it is not sure that simulation will be more successful than an inductive prediction. In order to formulate hypotheses, and to design experimentation, simulation is a time-saving alternative to experiments, and as such an attractive approach.

4.3 The use of simulating disturbance

In the introduction of this paper we tried to explain that building an explanatory model of disturbance in plant communities is a task which cannot, in general, be fulfilled. Through the occurrence of unexpected events the reliability of a prediction with such a model would be low. To obtain insight in the course of processes to be expected after disturbance, the plant species, involved in the simulation should represent sufficient variation of life and growth forms. With respect to strategies the dispersion mechanisms are of particular importance.

The effect of every major disturbance is that it causes an increase in bare ground. In the area studied this means that through stronger evaporation at the soil surface this is to cause an increase of the concentration of sodium-chloride in the upper soil layer, though this may be reversed by 'wet' disturbance. The saltier environment is beneficious to halophytic plant species as has been observed in salt marsh systems (Hansen, 1982), although Bakker (1985) found that grazing benefitted halophytic plant species, where no increase of the salt concentration in the top soil could be measured.

Simulation cannot replace numerical analysis (direct or indirect) of the biocoenoses in question. Explanatory simulation and description by means of numerical techniques rather supplement each other in the process of getting to understand vegetation dynamics (Weinstein & Shugart, 1983).

The results of this simulation confirm theories about the recovery of biocoenoses after disturbance with some time delay (Reiners, 1983). The speed with which this recovery takes place in the simulation may not entirely reproduce reality.

Application of the model to disturbances has led to a better insight in its limitations, so that a number of improvements could be carried out. We conclude that causal/logical models of systems-dynamics are adequate tools, helping us to improve our understanding of processes leading to structural changes in vegetation. The main value of a model is probably the experience gained during its construction. A model which is not yet finished, or perhaps never will be finished, might therefore be among the most valuable.

Acknowledgements

The authors wish to thank all those who placed the results of their field and experimental studies at our disposal (Nico Beemster, Mieleke van Deursen, Cor Dijkstra, Jan van Esch, Bart Friso, Eric Quené, Willy Schulten, Maarten Smit, Willy Strik, Hanneke Terpstra, everyone we forget to mention here, and Dr. Wouter Joenje and Ir. Hans Drost, who coordinated the field work). We are also grateful to those who helped to improve the manuscript by making remarks on language, style and contents, especially Dr. Roy Snaydon. Finally our thanks are due to E. Leeuwinga and D. Visser for drawing the graphs.

References

Bakker, J.P. (1985). The impact of grazing on plant communities, plant populations and soil conditions on salt marshes. Vegetatio 62: 391–398.

Boersma, S.K.T. & Hoenderkamp, T. (1985). Simulatie, een moderne methode van onderzoek. Academic Service, The Hague. 315 pp.

Davidson, J.M., Graetz, D.A., Suresh, P.m Rao, C. & Magdi Selim, H. (1978). Simulation of nitrogen movement, transformation and uptake in plant root zones. Ecological Research Series 600/3–78–029. Environmental Protection Agency, Athens, 105 pp.

Doyle, T.W. (1981), A simulation model of forest succession in the Puerto Rican tropical montane forest. In: D.C. West, H.H. Shugart & D. Botkin (Eds.) Forest succession: patterns and applications. Springer, Berlin, Heidelberg, New York.

Ellenberg, H. (1954). Ueber einige Fortschritte der kausalen Vegetationskunde. Vegetatio 5/6: 199–211.

Fresco, L.F.M. (1982). An analysis of species response curves and of competition from field data: some results from heath vegetation. Vegetatio 48: 175–185.

Gauch, H.G.Jr. (1982). Multivariate analysis in community ecology. Cambridge University Press, Cambridge. 298 pp.

Godron, M. & Forman, R.T.T. (1983). Landscape modification and changing ecological characteristics. In: H. Mooney & M. Godron (Eds.) Disturbance and ecosystems, pp 12–28. Springer, Berlin, Heidelberg, New York.

Goodall, D.W. (1969). Simulating the grazing situation. In: F. Heinmets (Ed.): Concepts and models of biomathematical simulation techniques and methods 1: 211—236. Marcel Dekker Inc., New York.

Graybeal, W. & Pooch, U.W. (1980). Simulation: principles and methods. Cambridge University Press, Cambridge.

Gremmen, N.J.M., Reijnen, M.J.S.M., Wiertz, J. & van Wirdum, G. (1984). Modelling for the effects of groundwater withdrawal on the species composition of the vegetation in the Pleistocene areas of The Netherlands. In: Annual Report 1984 of the State Institute for Nature Management: 89–111.

Grime, J.P. (1979). Plant Strategies and Vegetation Processes. Wiley, London. 222 pp.

Grubb, P.J. (1979). The maintenance of species-richness in plant communities: the importance of the regeneration niche. Biological Review 52: 107–145.

Hansen, D. (1982). Entwicklung und Beeinflussung der Nettoprimaerproduktion auf Vorlandflaechen und im Vogelschutzgebiet Hauke-Haien-Koog. Schriftenreihe Inst. Wasserw. und Landschaftsoekologie Christian-Albr.-Universitaet Kiel I: 1–273.

Jaquard, P. & Heim. G. (1983). Demographic strategies and originating environment. In: H. Mooney & Godron, M. (Eds.) Disturbance and ecosystems, pp 226–239. Springer, Berlin, Heidelberg, New York.

Joenje, W. (1978). Plant colonization and succession on embanked sandflats. Thesis, Groningen. 160 pp.

Joenje, W. (1979). Plant succession and nature conservation of newly embanked flats in the Lauwerszeepolder. In: R.L. Jefferies & A.J. Davy (Eds.) Ecological processes in coastal environments, pp 617–634. Blackwell, Oxford.

Joenje, W. (1982). Nature in new Wadden-polders; conservation by exploitation. In: Polders of the world. Vol. 2. pp 68–82. Pudoc, Wageningen.

Joenje, W. (1985). The significance of waterfowl grazing in the primary vegetation succession on embanked sandflats. Vegetatio 62: 399–406.

Kemmers, R.H. & Jansen, P.C. (1985). Stikstofmineralisatie in onbemest half-natuurlijke graslanden. Rapport 14, Instituut voor Cultuurtechniek en Waterhuishouding, Wageningen. 15 pp.

Miller, P.C., Miller, P.M., Blake-Jacobson, M., Stuart Chapin III, F., Everett, K.R., Hilbert, D.W., Kummerow, J., Linkins, A.E., Marion, G.M., Oechel, W.C., Roberts, S.W. & Stuart, L. (1984). Plant-soil processes in Eriophorum vaginatum tussock tundra in Alaska: a systems modeling approach. Ecological Monographs 54: 361–405.

Olson, J.S. (1963). Energy storage and the balance of producers and decomposers in ecological systems. Ecology 44: 322–331.

Peet, R.K. & Christensen, N.L. (1980). Succession: a population process. Vegetatio 43: 131–140.

Penning de Vries, F.W.T. (1984). Practical uses of dynamic and explanatory models for simulating crop growth. Wissenschaftliche Zeitschrift der Humboldt-Universität zu Berlin, Math.-Nat,. 33: 343–345.

Reiners, W.A. (1983). Disturbance and basic properties of ecosystem energetics. In: H. Mooney & M. Godron (Eds). Disturbance and ecosystems, pp 83–98. Springer, Berlin, Heidelberg, New York.

Reitsma, J.H. (1981). De vegetatie in het Lauwersmeergebied in 1980. Internal report State Authority of the IJssellake Polders, Lelystad.

Runge, M. (1983). Physiology and ecology of nitrogen nutrition. In: O.L. Lange, P.S. Nobel, C.B. Osmond & H. Ziegen (Eds.): Encyclopedia of plant physiology. New series. Vol. 12C: 163–200. Springer, New York, Berlin, Heidelberg.

Weinstein, D.A. & Shugart, H.H. (1983). Ecological modeling of landscape dynamics. In: H. Mooney & M. Godron (Eds.): Disturbance and ecosystems, pp 29–45. Springer, Berlin, Heidelberg, New York.

West, D.C., McLaughlin, S.B., Shugart, H.H. (1980). Simulated forest response to chronic air pollution stress. J. Environ. Qual. 9: 43–49.

Wit, C.T. de & Goudriaan, J. (1978). Simulation of ecological processes. Pudoc, Wageningen. 167 pp.

Appendix. Information concerning the most important functions and parameters in 'EGRAS'.

Parameters	explanation	Value				
		Agr.	Aster	Phrgm.	Pucc.	Salix
RGRMAX	max. rel. growth g/g/day	.31	.20	.13	.21	.08
SCMAX	max. carr. cap. g/m^2	884	884	2000	884	2000
INTIJ	coeff. of comp. 0–1	all values 0, except Phrgm. and Salix, excluding (1) all species.				
MINN	min. req. N-tot for growth dimensionality of N: text	300	300	300	300	150
LETN	lethal N-tot in soil	all values VERY large.				
LETC	lethal NaCl in top soil	97.5	162.5	19.8	162.5	13.5
GROPT	optimal grazing press. 0–1	all values 0				
GRLE	lethal grazing press. 0–1	1.00	1.00	.27	1.00	.05
TEMIN	min. air temp. for growth C.	2	2	2	2	2
TEOPT	opt. air temp. for growth C.	20	20	20	20	20
TEMAX	max. air temp. for growth C.	30	30	30	30	30
LIMIN	min. light for growth J/cm^2	100	100	100	100	100
LIMAX	max. light for growth J/cm^2	2500	2500	2500	2500	2500
WAMIN	min. water level cm	−20	−20	−50	−20	−50
WAMAX	max. water level cm	+30	+10	+30	+10	?

1. Procedures concerning growth and death

$$PLGR(i) = RGRMAX \times (1-SCLI/SCMAX) \times C \times ND \times NT \times ST \times G \times T \times L \times W$$

All factors except RGRMAX: $0 <= \text{factor} <= 1$;
For explanation of the factors: see text.

2. Procedures concerning grazing

The total consumable biomass in the area = SCTOT.
The ungrazed residual (varying per species) = SCRES.
The speed of consumption increases linearly with (SCTOT-SCRES), if (SCTOT-SCRES) is less than a constant SCSAT (Noy-Meir 1978).
The angle of increase depends of the time of the year.
If (SCTOT-SCRES) > = SCSAT then
GRSP (speed of grazing in g/day) equals GRSPMAX (a constant).

For calculation of the preference-class: see text.

The amount of living and dead biomass, removed from the site by trampling:
For each species:
SCLITR = (ARTR × SCLI)/AREA, where
ARTR = AREA × C3 × 32/400 and
SCDETR = (ARTR × SCDE)/AREA.
The coefficient C3 is a function of soil moisture and the growth form of the species. The figure '32' is a consequence of an assumption that each cow-step affects 1 dm^2 of the surface.

The loss of biomass by grazing (not-consumed, not-trampled) has been estimated at 1%.

All the nitrogen in the removed biomass returns to the system.

Parameter explanation		Value				
		Agr.	Aster	Phrgm.	Pucc.	Salix
SCRES	residual s.cr. after grazing g dr.m./m^2	25	50	200	25	250
PRSP	relative graz. pref.	.17	.26	.28	.26	.03
C3	a constant, depend. of growth form (s. above)	.015	.050	.024	.015	.050

3. Procedures concerning abiotic factors

Parameter explanation		Value				
		Agr.	Aster	Phrgm.	Pucc.	Salix
P1	min. N-intake u.N./g	.0254	.0339	.0014	.0536	.0141
P2	N-intake u.N./g per ln u.N. available N	.0028	.0028	.0099	.0085	.0000

Salt as a function of vegetation coverage:
SAMAX = −0.021 × SUMAREA + 2.329; SAMAX is an absolute value (kg NaCl).

CHAPTER 11

Modelling the structural changes in vegetation under different grazing regimes

D.C.P. Thalen, H. Poorter, L.A.P. Lotz & P. Oosterveld

Detailed contents

1. Introduction

1.1 Grazing as a management practice for conservation

Most grasslands and other areas with a short herbaceous vegetation in The Netherlands will, without grazing or cutting, follow a succession towards woodland. This fact is of crucial importance to all managers of conservation areas who wish to maintain or create diversity in short vegetation, and is now usually recognised in the management of nature reserves. Most of the grasslands currently managed by conservation agencies in The Netherlands were previously grazed or mown for hay. Under these practices, sometimes continued for centuries, a semi-natural landscape developed with a great variety of habitats, rich in plant and animal species.

J. van Andel et al., (eds.), Disturbance in Grasslands. ISBN 90-6193-640-3.

The situation has changed in the past decades with the introduction of artificial fertilizers and the mechanical means to lower the groundwater table. That has allowed more intensive grazing and mechanical harvesting, resulting in a drastical reduction in the floristic and micro-structural diversity of the Dutch grasslands.

The nature conservation agencies, both governmental and private, are now attempting to restore the diversity of grasslands under their control. Grazing animals have proved to be promising tools in the conservation or even creation of diverse habitats (Thalen, 1984a; Oosterveld, 1985). Experience has shown that two conditions should be met in the management of nature reserves (e.g. Westhoff, 1971, 1977):

1. the management should lead to a spatially varied but (in time) constant influence, and
2. the type of influence should not be alien to the system.

Three, more practical, aspects of management might also be taken into account:
3. it should be applicable to areas of different size,
4. it should be economically feasible,
5. the 'tools' to be used should be fitting in the landscape (scenery).

There is ample evidence that large grazing animals, both domestic and wild, if properly chosen in terms of species and densities, can be useful management tools (Oosterveld, 1975, 1985; Van de Veen & Van Wieren, 1980; Bakker et al., 1983; Thalen, 1984a). This is now widely recognised in The Netherlands. Some 35 conservation areas were grazed by domestic animals (cattle, horses, ponies, sheep and goats) in 1971. This number has increased by more than tenfold since then.

1.2 Pattern development under grazing

In their search for food, water, and shelter, herbivores select areas in their daily pattern of activities. From the larger to the smaller units choices are made between landscapes, plant communities and, especially where food is concerned, between plant species and parts of individual plants (e.g. Pratt et al., 1986). Each individual animal acts according to its physiological and social status, but there are also fairly constant differences between species. For example 'grazers' (bulk and roughage feeders), 'browsers' (concentrate selectors) and 'intermediate feeders' (versatile feeders able to adapt their diet) have been recognised (Hofmann, 1973, 1976; Van de Veen, 1979; Van de Veen & Van Wieren, 1980). Depending on whether an animal is a 'grazer' or a 'browser' there is a preference for different types of vegetation resulting in a different impact of the foraging animal on the vegetation and the terrain (e.g. Putman, 1986).

The behaviour of cattle, ponies, sheep, and goats has been studied in a number of nature reserves in The Netherlands. It was found that a more or less fixed utilization pattern is soon established. Moderate to low stocking rates are usually applied, but a relatively small part of the terrain is always used very heavily, while in more than half of the area the animals may rarely be seen (Table 1). The daily routine of terrain use is partly governed by the availability of drinking water and shelter or of favoured resting places, partly also by the locations where preferred forage is available. Areas previously grazed will be revisited regularly to harvest the regrowth of young shoots of high palatability and nutritive value; this leads to a mosaic pattern of heavily used and virtually unused vegetation. A pattern of the levels 'land unit', 'between community' and 'within community' can be distinguished (Oosterveld, 1976; Bakker et al., 1984; Hillegers, 1984; Putman, 1986). If the grazing management is stable, this pattern will be enhanced; areas of short herbaceous cover will be maintained, while the ungrazed areas follow a succession via tall grasses and forbs and shrub and tree establishment to woodland and forest. The first phases of such successions can now be observed in many nature reserves in The Netherlands that have been grazed by large herbivores (Oosterveld, 1985; Fig. 1).

Table 1. Intensity of terrain use (%) in two nature reserves, where grazing with large domestic animals is used as a management tool for conservation. The figures were calculated from presence scores, carried out continuously two days a month over a full year (from: Oosterveld, 1979). Grazing pressure (in number of cattle or horses per hectare): intensive - 1.0 or more; normal - 0.3 to 1.0; extensive - 0.1 to 0.3; very extensive - 0.1 or less.

Type of terrain and type of grazing	Grazing pressure Intensive	Normal	Extensive	Very extensive
Cattle and horses in a saltmarsh, low dune area, 360 ha, 1 animal per 3 ha, seasonal grazing	6	12	57	25
Icelandic ponies in abandoned fields, heathland, pine plantations, area 100 ha, 1 animal per 5 ha, permanent grazing	3	10	60	27

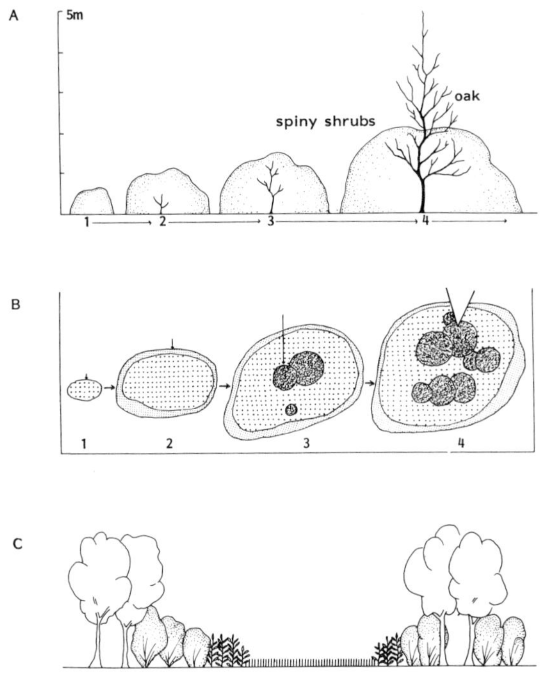

Fig. 2. The establishment of trees within the protection of unpalatable species (a, b), and the resulting physiognomy (c) of the vegetation as observed in the Borkener Paradies. The vegetation has been extensively grazed for many years by cattle (see 1.2 and 5.2). From Burrichter et al. (1980).

Convincing examples of all the phases of succession and the final dynamic equilibrium can only be seen in areas with a very long history of stable and not too intensive grazing, for example in the New Forest in England (Putman, 1986) and on the 'Alvar' (calcareous heath) areas of the Swedish islands Oland and Gotland (Petterson, 1965; Sjögren, 1970; Rosen & Sjögren, 1973). Probably the most thoroughly analysed and well described area is the 'Borkener Paradies' in north-west Germany (Burrichter et al., 1980; Jaspers, 1983; see also 5.2). *Quercus robur* develops here under the protection of such unpalatable species as *Prunus spinosa, Rhamnus catharticus, Crataegus* spp., *Rosa canina* and *Rubus* spp. (Fig. 2).

The changes in vegetation structure caused by selective grazing, trampling, and manuring are followed by changes in the plant species composition (Van de Laar & Slim, 1979; Bakker et al., 1983; Bakker et al., 1984). Relatively little research has been done on changes in the fauna under extensive grazing. Some information is available for the 'Cranendonck' area, shown in Fig. 1, where rabbits (Oosterveld, 1983), earthworms (Eijsackers, 1983) and mice (Table 2) have been studied. There is strong evidence of increasing species diversity of the mesofauna and macrofauna, directly following the changes in vegetation structure and increasing plant species diversity (Van de Laar & Slim, 1979).

1.3 The reasons for a modelling approach

Research on grazing as a means of creating and maintaining habitat or species diversity in nature reserves with grasslands and other short vegetation types started at the Research Institute for Nature Management (RIN) in the early seventies. Similar studies were soon taken up by a number of university departments, in a cooperation with RIN, in particular the Department of Plant Ecology of Groningen University and the Department of Nature Conservation of the Agricultural University in Wageningen (Thalen, 1984b).

After a decade of study certain generalisations could be made. It then became apparent that a modelling approach to simulating changes in vegetation macrostructure under grazing would be very useful. The main reasons for such a modelling approach are:
a) to gain further insight into the structure and

Table 2. Developments in small-mammal populations as recorded by trapping in the Cranandonck area, and area continuously grazed since 1972 (see text of Fig. 1 and 1.2). From data collected by P.A. Slim.

Species	Years when the trapping was carried out	
	1974, 1976, 1977	1980, 1981, 1982
Apodemus sylvaticus	77	75
Microtus arvalis	23	5
Microtus agrestis	0	12
Sorex araneus	0	6
Sorex minutus	0	1
Clethrionomys glareolus	0	1
Total no. of traps set	2555	2795
Total no. of animals trapped	56	101

Fig. 1. Vegetation development on former arable fields between 1973 (above) and 1982 (next page) in the Cranendonck area, a reserve of some 100 ha with approximately equal areas of formerly arable fields, Pinus plantations and dry heathland with driftsands. The area has been extensively grazed by Icelandic ponies since 1972 (see Oosterveld, 1985).

functioning of the system by recognising the apparently essential processes;

b) to identify gaps in the available knowledge;

c) to develop methods for the systematic and objective prediction of the long-term effects of different grazing regimes;

d) to demonstrate to administrators and reserve managers the probable consequences of various grazing management practices.

1.4 Non-applicability of existing models

A review of already existing and published mathematical models which simulate grazing systems soon showed that most work had been concerned with the effects of management practices on animal production (Breymeyer & Van Dyne, 1980; Christian, 1981). Wherever vegetation changes were involved, the central themes were primary production and biomass rather than vegetation structure (e.g. Noy-Meir, 1978; Parton & Risser, 1979). Most published studies dealt with interactions between large herbivores and grassland. In some cases, microstructural characteristics, related to the morphology of the grass cover, were considered (e.g. Coughenour et al., 1984), but no models were suitable for our needs of treating changes in the macrostructure of vegetation and for a terrain as a whole. The matrix model published by Redetzke & Van Dyne (1976) provided a useful example of a first crude approximation.

The 'Elephants-trees-grass-grazers' model of Van Wijngaarden (1985), developed as part of a study of the Tsavo ecosystem in Kenya, closely approaches the objectives of this study. This work was published after the results of our mod-

Fig. 2. Physiognomy of the vegetation as observed in the Borkener Paradies. The vegetation has been extensively grazed for many years by cattle (see Burrichter et al. 1980).

elling, as presented hereafter, had been completed. Van Wijngaarden's model concentrated on the long-term consequences of increasing elephant density on the structure of the vegetation (total woody cover, tree cover, shrub cover and perennial grass cover) and on other grazers and browsers. His graphical output closely resembles the output of our models. The present paper describes the development and structure of two simple models (the matrix model MAT and the analytical model PROCOS) and results obtained sofar from these models.

2. Some basic concepts and premises

2.1 Vegetation structural classes and succession

In the absence of grazing a number of structural classes can be distinguished in the succession from grassland to forest (see 1.2). Four main classes are recognised here (Fig. 2):

SH - short herbaceous vegetation of grasses and forbs, height <25 cm;
TH - tall herbaceous vegetation of grasses and forbs, height >25 cm;
LW - low woody vegetation dominated by shrubs and small trees, height <2 m;
TW - tall woodland and forest, height >2 m.

It is assumed that, without interference (e.g. by large herbivores), natural succession will follow the sequence SH-TH-LW-TW. A forest area may die off after some time under these conditions and return to one of the herbaceous structural classes, after which a new start of a succession sequence takes place.

The vegetation structure of any terrain can be described quantitatively by the percentage area under each of these structural classes. This can generally be achieved using aerial photographs; for example in The Netherlands stereo-cover is available at a scale 1:20,000 or larger.

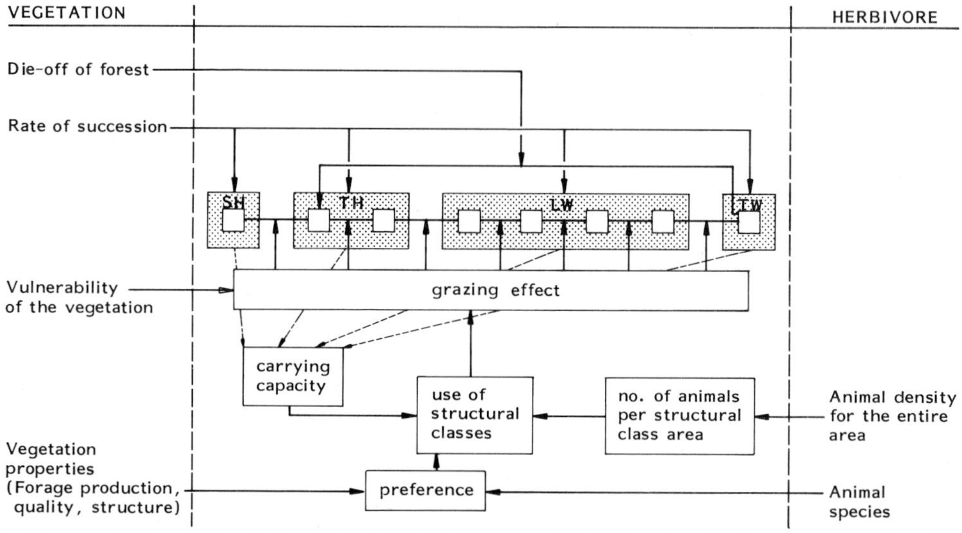

Fig. 3. Schematic presentation of the relations included in the MAT model. (SH = short herbaceous vegetation, TH = tall herbaceous vegetation, LW = low woody vegetation, TW = tall woodland; see 2.1).

2.2 Impact of large herbivores

There is strong evidence (see 1.2) that large grazers, when released into a terrain with different vegetation structural units, soon develop a fixed pattern of use of that terrain. The difference in use of the various structural units leads to differences in succession. The animals may cause a slow-down, a fixation or even a retrogression in the structural succession sequence described under 2.1. The impact of an animal species will, amongst other things, be governed by its preference for a certain structural type and the animal density (number per unit area). The relative preference of the four main livestock species for the four structural classes, as applied in MAT, was estimated from a large number of observations of animal distribution (P. Oosterveld and others, unpublished, see Table 3).

3. The matrix model MAT

3.1 Structure of the model

After an exercise with a linear deterministic matrix model (MAT 1, see Lotz & Poorter, 1983), a non-linear model (MAT 2) was developed. A schematic presentation of the relations that play a role in the simulation is given in Fig. 3. The principle of the matrix model is that in each run (representing a certain time span) the status at time t+1 is reached by multiplying the status at time t with a transition matrix. In the non-linear model, the values used in the transition matrix depend, for each run, on the outcome of the many multiplications performed in the previous run. For further details the reader is referred to relevant textbooks, e.g. Jeffers (1978). Matrix models have certain advantages and disadvantages when applied in ecology; for example Jeffers (1982) concluded that '...'reality' is sacrificed to some extent in order to gain the advantages of the particular mathematical properties of the formulation. The deductive logic of pure mathematics then enables the modeller to examine the consequences of his assumptions without the need for time-consuming 'experimentation' on the model'. This applies to the MAT models, expecially to MAT 1. For MAT 2 the assumptions are given below.

3.2 Assumptions and input

1. The vegetation structure of a terrain (total area 1.0) is characterized by four structural classes: SH, TH, LW and TW. SH and TH do not occur under LW and TW. Every structural class has several developmental phases.

2. In the absence of large herbivores the succession occurs as described under 2.1. Direct transitions SH-LW, SH-TW and TH-TW do not occur. Vegetation development (succession) is simulated in time-runs of at least one year.

3. The various activities performed by animals such as grazing/browsing, trampling, manuring, etc., lead to a change in vegetation structure. The use of the different structural classes is calculated from the preference, which is dependent on animal species and vegetation characteristics (Fig. 3). If a certain vegetation type is not present, it receives the preference value 0. The animals are distributed over the structural types according to their preference (Table 3) and for each structural class the density is calculated.

Table 3. The relative preference of four livestock species for the four macrostructural classes, as used in the simulations. Figures indicate presence in a structural class expressed as percentage of time spent in all classes together. SH = short herbaceous vegetation, TH = tall herbaceous vegetation, LW = low woody vegetation, TW = tall woodland (see 2.1).

Livestock species	Vegetation macrostructural class SH	TH	LW	TW
Cow	80	10	5	5
Sheep	60	20	15	5
Horse	60	15	15	10
Goat	40	15	30	15

4. A use factor is applied to the different structural classes based on the animal density, where the densities lead to intensities of use indicated by the terms 'intensive', 'normal', 'extensive' and 'very extensive' (Table 4). In the calculations cattle and horses were considered to represent one standard livestock unit, and sheep and goats one third.

5. Grazing impact depends on both intensitys of use and structural class (Table 5). Three categories of grazing impact on succession are distinguished: heavy impact - will cause retrogression; moderate impact - inhibits succession; light (or no) impact - allows succession to continue to the next phase.

When the density exceeds the carrying capacity of a structural class the excess number of animals is distributed over the other classes according to the preference (Table 3). The model does not yet simulate spatial effects, so does not take into account the spatial distribution of the structural classes.

Table 4. Use factors (%) of the vegetation macrostructural classes, for five grazing densities, as applied in the simulations with MAT (see text 3.2).

Standard livestock units per hectare of the macrostructural class	Intensity of use Intensive	Normal	Extensive	Very extensive
>1.0	80	15	5	0
0.3–1.0	20	50	25	5
0.1–0.3	5	15	50	30
>0.0–0.1	0	10	25	65
0.0	0	0	0	100

Table 5. Dependence of grazing impact on (1) intensity of use (see Table 4), and (2) vegetation macrostructural class, as presumed in the simulations with MAT; H = heavy impact, M = moderate impact, L = light impact (see text 3.2).

Vegetation macrostructural class		Intensity of use Intensive	Normal	Extensive	Very extensive
Short herbaceous	(SH)	H	H	M	L
Tall herbaceous	(TH)	H	H	M	L
Low woody	(LW)	H	M	M	L
Tall woody	(TW)	M	M	M	M

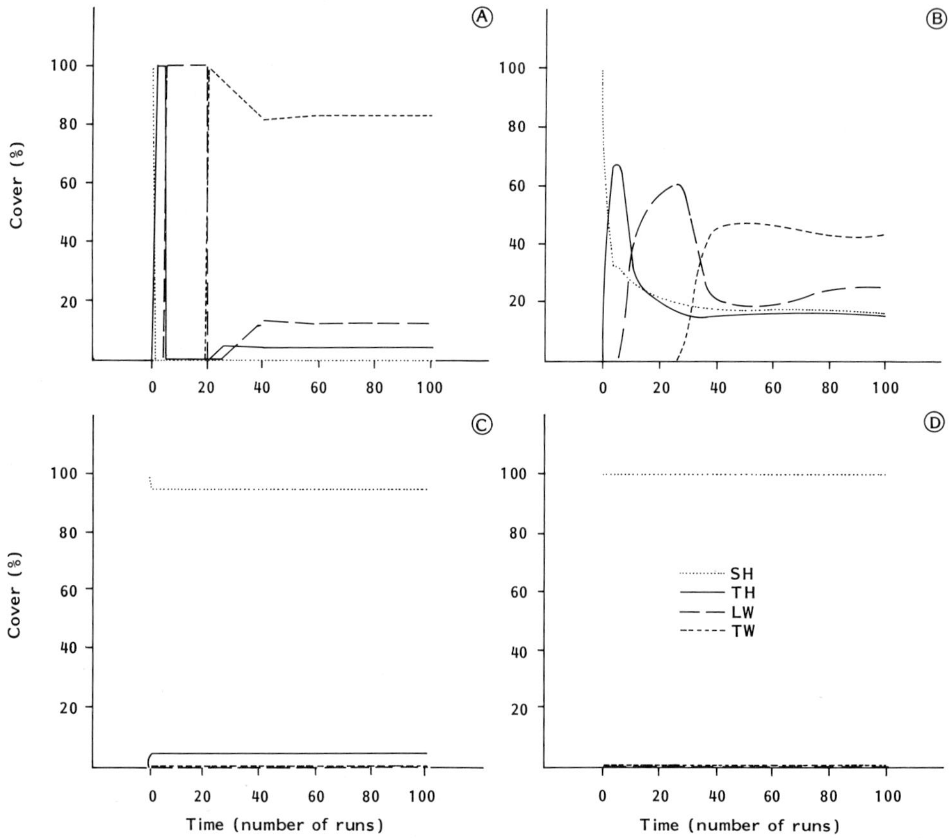

Fig. 4. Results of simulations with the MAT model; vegetation development under different intensities of sheep grazing. Number of animals per hectare: a. 0.0; b. 0.5; c. 2.5; d. 5.0. Start with 100 percent low herbaceous; number of development phases in the structural classes: SH-1, TH-5, LW-15, TW-1. Annually 1 percent of TW changes to TH (die-off of forest).

3.3 Examples of results of simulations

An example may clarify the working of part of the model MAT. A terrain with a vegetation structural cover of 25% each of SH, TH, LW and TW is grazed at a density of one cow per hectare. The preference of cattle for TH is 10% (Table 3). In this structural class there will be 0.1/0.25 = 0.4 cattle per hectare. Table 4 shows the use intensity for this density: intensive 20%, normal 50%, extensive 25% and very extensive 5%. In the structural class TH this leads to the following effects (Table 5: H-70%, M-25% and L-5%): in one run of the programme 70% of TH experiences a retrogression in the direction of LH, 25% remains structurally unchanged and 5% develops in the direction of LW. The results of the simulations can be depicted graphically, e.g. as shown in Fig. 4.

4. The production-consumption model PROCOS

4.1 Structure of the model

The MAT model has the advantage and disadvantage that all factors affecting vegetation structure during succession are condensed in one operator: the transition matrix. Since matrix models, as mentioned earlier, are only descriptive, they do not satisfy one of the prime reasons of modeling, i.e. to gain an insight in the important processes. With this objective a separate analytical model (PROCOS) was developed.

PROCOS stands for PROduction-COnSumption. The model attemps to clarify the effect of large herbivores on vegetation macrostructural changes by simulating the production of aerial phytomass and its consumption by the animals. PROCOS also distinguishes the macrostructural

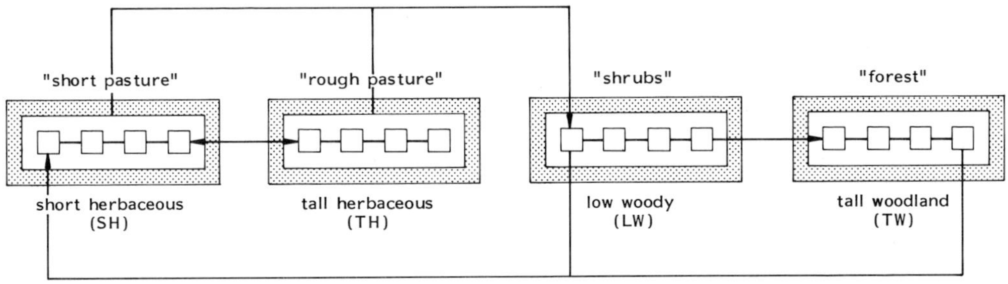

Fig. 5. Schematic presentation of the possibilities of vegetation macrostructural changes as incorporated in the PROCOS model. In the four macrostructural classes development phases (in this case four) are distinguished, each representing a certain standing crop and available forage.

types SH, TH, LW and TW, each divided in a number of development phases (Fig. 5). Each development phase represents a certain above ground biomass (=phytomass) and transitions in development phases take place as a result of changes in biomass. Biomass increment leads to a 'higher' development phase, decrease in biomass through consumption and decomposition leads to a 'lower' phase. The standing crop and its use are central in the reasoning behind the simulation model. Establishment of shrubs and trees may lead to LW and finally TW, forest. Heavy browsing of shrubs may result in a retrogression to herbaceous types, in this model SH (Fig. 5). In PROCOS a correction was also incorporated for the amount of produced biomass lost through animal impact other than grazing in the sense of the defoliation process, e.g. as a result of treading. De Leeuw & Bakker (1986) used the terms 'foraging' and 'occupancy' to make the distinction. Very little quantitative data exist on loss through occupancy. Estimated percentage values were incorporated, based on personal communications and literature data (e.g. Sonnema, 1974).

The relationships between growth, decomposition and consumption as used in PROCOS for herbaceous vegetation are shown diagrammatically in Fig. 6. The only difference for woody vegetation is that no unit 'dead material' was distinguished. As seasonal variation in biomass production and grazing pressure had to be incorporated, the time-step used had to be smaller than for MAT. For the simulations presented hereafter a time-step of one month was used.

4.2 Assumptions and input

The amount of biomass produced by herbaceous vegetation depends on: (1) amount of live material already present, (2) growth rate, (3) time of the year, and (4) amount of dead material. The same applies for woody vegetation except for a distinction between consumable biomass (leaves and twigs) and non-consumable biomass (e.g. trunks and branches). Death and decomposition take place according to time of the year and several other factors. Summarizing, the following material is available for consumption: herbaceous vegetation: young, old and dead material; woody vegetation: young and old material, and fruits, as far as these are within reach of the animals. For each development phase the amount available for each of these categories is calculated.

The food demands of large herbivores depend

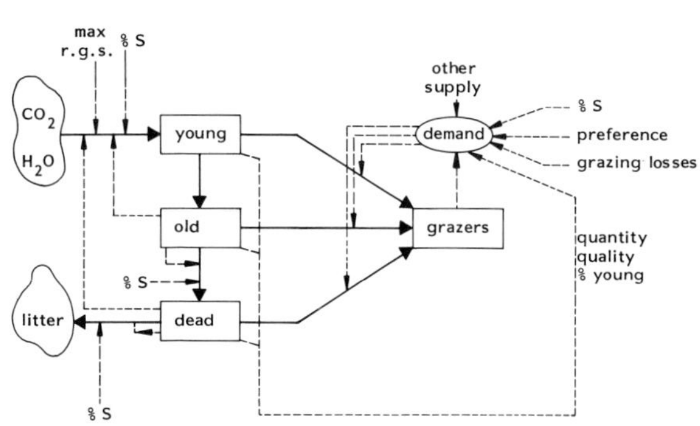

Fig. 6. Schematic presentation of the relations incorporated in the PROCOS model for each of the development phases of the herbaceous classes. %S = seasonal influence, max.r.g.r. = maximum relative growth rate.

on (1) animal species, (2) density, and (3) time of the year. For the demands an integrated biomass weight unit is used. The consumption of each type of forage for each development phase is dependent on the animal's requirement and preference, the biomass of that type of forage available, its quality and the biomass of the other types of forage in that development phase (for details see Lotz & Poorter, 1983). Apart from consumption, biomass disappears as a result of trampling and decomposition (see 4.1).

The establishment of woody species occurs in the herbaceous types, depending on area, standing crop, and time of the year. Woody vegetation disapppears through death, at a certain age, of shrubs and trees. Shrubs younger than one year can be consumed and totally removed; older ones can only be partially eaten. For details and further premises the reader is referred to Lotz & Poorter (1983).

In PROCOS the animal type (grazer, browser or intermediate feeder, see 1.2) is characterized by its annual forage requirement, preference for different types of forage, seasonal pattern of forage requirement, trampling effect, and in a uniformity of grazing in a structural sub-class. Initially the last three factors have been incorporated as constant values or percentage values, because of the lack of reliable data. The values for the annual forage requirements used were taken from the literature (Centraal Veevoederbureau, 1979): horses 3200 kg, cattle 2900 kg, goats 750 kg and sheep 500 kg. The forage preference values used are shown in Table 6.

Parameter values for vegetation production,

Table 6. Forage preference values of four livestock species for different types of forage, as applied in the PROCOS simulations (estimates by P. Oosterveld).

Type of forage	Livestock species			
	Horse	Cow	Goat	Sheep
Herbaceous, young	100	100	100	100
Herbaceous, old	10	10	5	10
Herbaceous, dead	1	1	1	1
Twigs, young	25	10	100	25
Twigs, old	5	5	5	5
Fruits	100	100	100	100

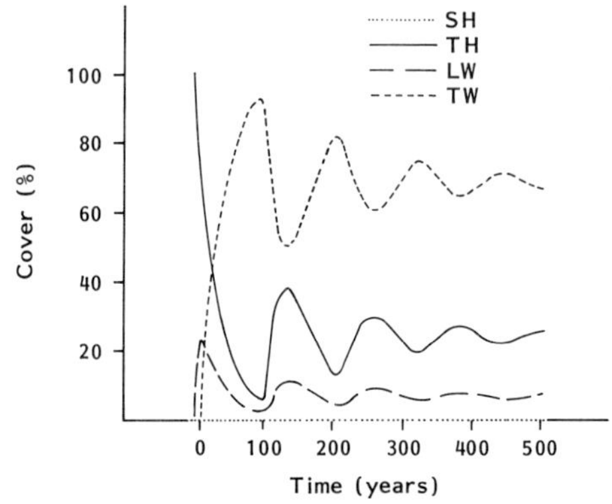

Fig. 7. Simulated changes in the macrostructure of the vegetation over a 500 year period, when ungrazed and starting from a low herbaceous vegetation. The data are the result of a simulation using the PROCOS model.

establishment of woody species and available forage from the classes LW and TW were estimated after a literature search and discussions with experts. Most of these values will differ per terrain type and have yet to be established by measurements in the field (see 6). The assumed values are specified in Lotz & Poorter (1983).

4.3 Examples of output

The graphic output of PROCOS includes changes in percentage cover (relative contribution) and biomass over one year or over many years. The example in Fig. 7 shows for a period of 500 years without grazing the invasion by shrubs (LW), followed by succession to forest (TW). After a century the amount of forest decreases through die-off and is replaced by LW. Thereafter the amplitude of the fluctuations decreases and at some stage an equilibrium would be established between forest die-off and rejuvenation.

The results of a simulation with two grazing intensities, 0.5 sheep and 1.0 sheep per hectare are shown in Fig. 8a,b. The results at low stocking rate (Fig. 8a) are similar to the ungrazed situation with the same cyclic variation. Since forest (TW) had hardly any palatable material, an increase in TW will lead to an increased grazing pressure on the other classes. Establishment of shrubs (LW) decreases and the existing LW is so heavily browsed that no new forest develops. Only with

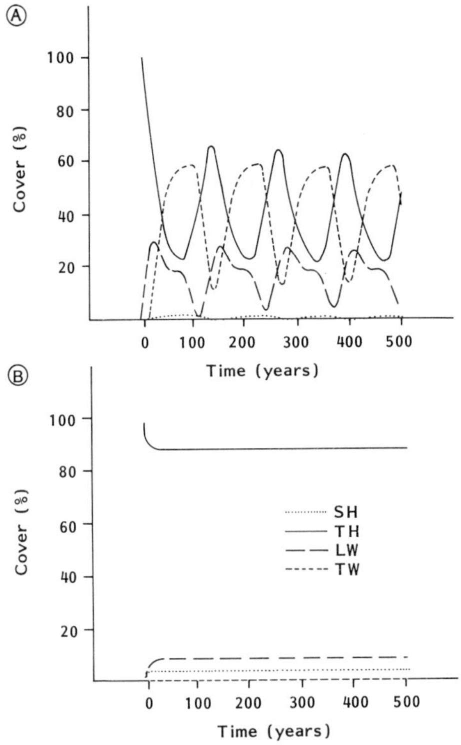

Fig. 8. See Fig. 7. The development under sheep grazing at a density of (a) 0.5 sheep per hectare, and (b) 1.0 sheep per hectare.

the death of trees a new cycle can begin. Under a heavier grazing pressure (Fig. 8b) the shrubs are under such a heavy use pressure that no forest can develop; an equilibrium is reached after 30 years.

The above examples can be used to demonstrate to administrators and reserve managers long/term effects but also serve to demonstrate the temporary nature of some, often unwanted, changes. The common response to such changes is to alter the grazing pressure or introduce other management practices such as mowing of tall forbs or cutting of shrubs. An example of the effect of a change in management is shown in Fig. 9.

An entirely different type of output is the depicting of equilibria finally reached between large herbivores and vegetation, depending on the animal species and density. In case of cyclical changes (e.g. Fig. 8a), average values over a large number of years can be determined, and the equilibria reached, for example under different

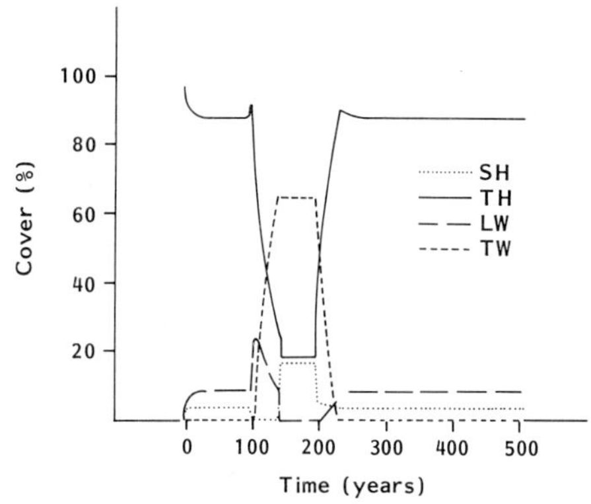

Fig. 9. See Fig. 7. The development under sheep grazing at the following density; yr 0–100: 1.0 sheep per hectare; yr 101–140: no grazing; yr 141–500: 1.0 sheep per hectare.

densities of sheep grazing can be depicted (Fig. 10b). The increased grazing pressure reduces the percentage of woody structural types, but increases the herbaceous types SH and TH. The production of herbaceous vegetation increases under extensive use, but decreases under a heavier pressure (Fig. 10a).

4.4 Sensitivity analysis

The output of a model depends not only on the assumed relations and mathematical equations, but also on the parameter values used. It is of great value to know how changes in parameter values influence the output of the simulation (Swartzman, 1980). Such a sensitivity analysis can indicate how parameters influence other variables or can define critical components of the system and so indicate the need for further experiments or field measurements. For PROCOS two output variables, not mentioned before, were further evaluated: The Mean Structural Value (MSV) and the Total Available Production (TAP). MSV is the weighted mean of the four main structural classes, reached by multiplying the percentage of the total area with a structural value for each class: SH/0, TH-1/3, LW-2/3 and TW-1. The analysis was carried out by comparing the outputs when parameter values were changed by 10 percent. To gain an insight into the interactions, in addition an analysis of variance was done

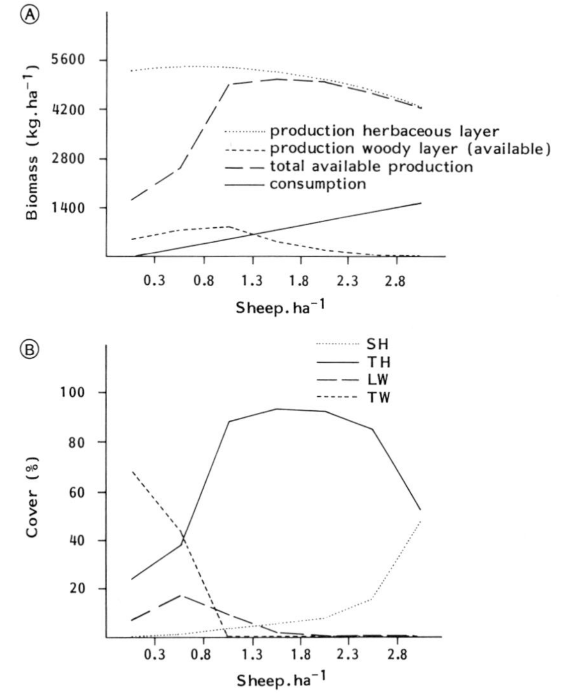

Fig. 10. Results of simulations with PROCOS. The predicted effect of sheep grazing density on the equilibrium values of (a) production of the herbaceous and woody (as far as available to the animal) vegetation, total available forage production and consumption (all in kg per ha per yr); and (b) percentage cover of the four macrostructural classes (see 2.1 and 4.3).

for the perturbations, in a factorial design, of the nine parameters that caused the greatest changes in output. Table 7 gives the main results for these nine tested parameters under two grazing pressures. Under heavy use both, MSV and TAP, are mainly influenced by the relative growth rate of the herbaceous layer and the die-off in this layer. Under extensive use the parameters related to the woody cover are extremely important. The model is relatively stable. The largest deviation from the overall mean can be ascribed to the effect of the herbaceous relative growth rate on TAP and amounts to only 4.2%. All other effects are less. Over 99% of the total variation is explained by the main and first order effects.

5. Application of the models

5.1 General

The most immediate value of the presented grazing models is for the management of conservation areas when grazed with livestock (1.1 and 1.2). The models have considerable demonstration value for administrators and terrain managers, particularly in relation to the time spans that should be considered when applying grazing as a management practice. However, if the models are to be used for predictions of developments in a particular terrain, at least two aspects should be considered:

1. valid and reliable values of input-parameters

Table 7. Main results of the sensitivity analysis for the PROCOS model (see text 4.4). MSV = mean structural value, TAP = total available production, both analysed for two grazing intensities. Percentage of total variation explained by one parameter: +/− = 1–5; ++/−− = 6–20; +++/−−− = 21–100 (+ indicates a positive correlation, − a negative correlation).

Main parameters tested	MSV		TAP	
	Sheep per ha		Sheep per ha	
	0.5	2.0	0.5	2.0
Relative growth rate grasses and forbs		+++	++	+++
Relative growth rate woody species	+++		−−	
Percentage herbaceous dying	−−	−−	++	−−
Establishment of trees	+++		−−	
Percentage wood of twigs/small branches			−−−	
Maximal loss through grazing				
Preference for young herbaceous	+		−	
Preference for old twigs/branches	−−		+	
Consumption by the animals	−−−		+++	−

have yet to be collected for a whole range of terrain types
2. the degree of 'reality' of the models (cf. Jeffers, 1982) has yet to be determined by validating the models under various conditions

It is particularly difficult to validate the models because of the lack of quantitative data on vegetation structural development for areas that have been and preferably still are not too intensively grazed and/or browsed by known numbers of large herbivores. In The Netherlands, records are available for less than two decades, yet observations in the field and preliminary results of the simulations show that data for many decades to some centuries are required, mainly because many of the ecosystems under grazing are poor in nutrients and thus succession is slow. An area for which some useful long-term information exists on the effects of grazing on vegetation structural types is the Borkener Paradies in north-west Germany. A unique opportunity for research and validation may be provided by the grazing management recently imposed on some conservation areas in the newly reclaimed polders with fertile soils. Here successional changes in vegetation structure are fast.

5.2 The Borkener Paradies case: an attempt of validation

The Borkener Paradies is a conservation area of some 30 ha, situated in an old meander of the river Ems, a few kilometers north-west of Meppen in Germany. The area has been under a fairly constant but light summer cattle grazing for many years. The available data were used to run a simulation of vegetation structural development using MAT and PROCOS. The vegetation structural classes, as derived from a vegetation map in Burrichter et al. (1980), were: SH/0.46, TH/0.03, LW/0.08 and TW/0.43. For PROCOS a grazing pressure of 0.3 cattle per hectare from May to October was assumed; for MAT which uses 'annual runs' of continuous grazing, 0.15 cattle per hectare was assumed. The results are shown in Fig. 11a, b. Both model simulations predict an increase in the woody vegetation. It is known that

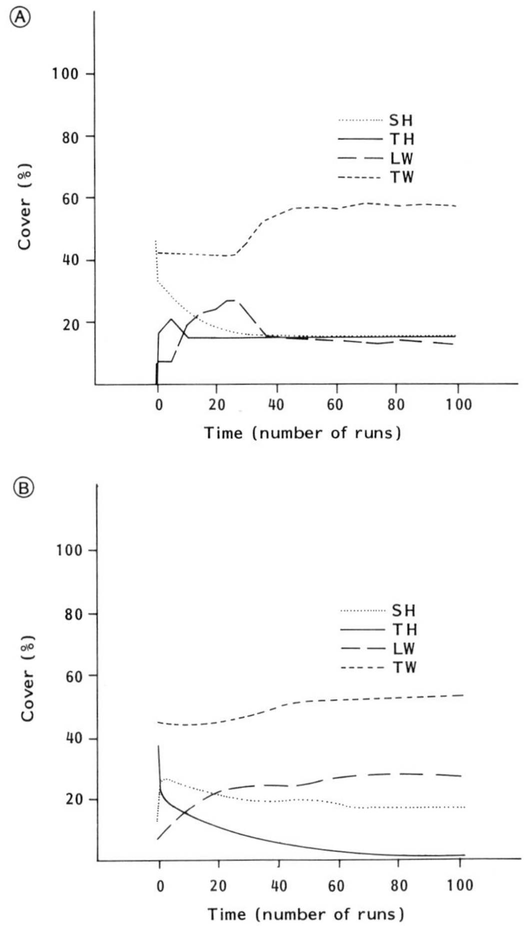

Fig. 11. Simulation of development in the vegetation structure in the Borkener Paradies (see 5.2): (a) results of the MAT model, (b) results of the PROCOS model.

in the past bushes and trees were regularly cut by farmers in order to maintain the open character of the area. A very detailed study on shrub establishment and woodland development was carried out by Jaspers (1983). He estimated an increase in the area under forest of 8 ha over the past 50 years and predicted an increase of another 4 ha in the next 30 years.

The simulation results with MAT and PROCOS are similar to the data collected in the field and found in historic records. This may be as close to a validation as is possible with the available data for any terrain in north-western Europe. However, data from fast-changing areas, such as those found in the new polders, may provide useful validation in the future.

5.3 *The Oostvaardersplassen case: an early use of the prediction possibilities*

The Oostvaardersplassen area is situated in Zuidelijk Flevoland, a new polder in the IJsselmeer. In this polder an area of some 6000 ha has been set aside as a nature reserve; an area of 3600 ha is inundated. The area has great nature conservation values, at present especially as a bird sanctuary. In particular large numbers of the Grey lag Goose (*Anser anser*) use the area. The marsh vegetation, together with the vegetation of the adjacent 2400 ha dyked and drained land, plays an important role in the food supply of geese and other bird species. The future management of the Oostvaardersplassen as a nature conservation area is of particular concern to the management committee. It is especially important to know how different grazing systems (species, densities, seasonal versus continuous grazing) could be used to maintain a more or less predetermined landscape.

PROCOS was used for a first estimate of the consequences of various management practices. The alternatives for which simulations were run included eight possible combinations of the following three alternative possibilities:

a) continuous (=year-round) versus seasonal (May-October) grazing with cattle,
b) with and without geese,
c) with and without part of the area inundated for part of the year.

Specially adapted parameter values were used, that could be partly based on data already collected. The geese grazing was treated the same as other grazing, though with adapted parameter values for preference and food requirements. Fig. 12 shows the results for a selection of four different situations.

The results demonstrate the very large number of cattle that would have to be introduced and managed in this reserve, if seasonal grazing with cattle only was applied, and if the desired large area (70%) of grassland (SH) was to be maintained. Although the use of the model in this context is probably premature it demonstrates the need and scope of use for this type of grazing models.

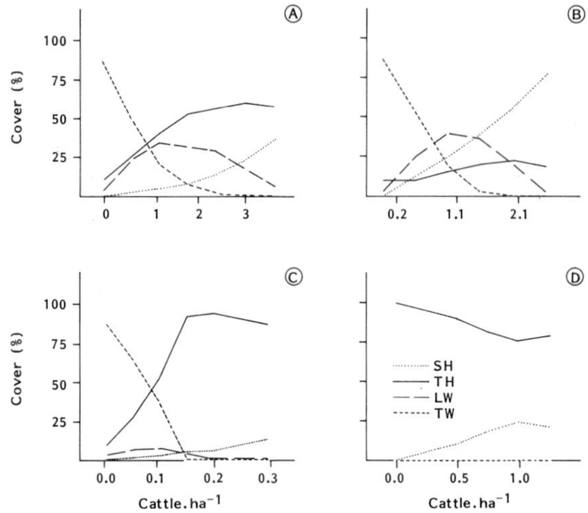

Fig. 12. Results of simulations with the PROCOS model for the Oostvaardersplassen area (see 5.3). The equilibria between the four macrostructural classes are shown, under different systems of cattle grazing: (a) seasonal grazing May-October; (b) seasonal grazing May-October, with 50 percent of the grasslands inundated from December to April; (c) continuous grazing, with inundation as for (b); (d) seasonal grazing May-October, with geese grazing (10 geese per hectare) from November to April. Note the difference in scale of the x-axes. The simulations ranged from no grazing by cattle to the maximum density possible.

6. Further development, improvements

The models MAT and PROCOS, outlined here, are relatively simple and still in an early stage of development. Modelling is an iterative process, whereby the output of simulations should be continuously checked against real values and the model adapted accordingly. There are reasons to continue with both MAT and PROCOS. The first model has the virtue of prediction using a relatively small number of parameters. The second model can provide a further insight in the processes taking place in the various grazing situations, and seems particularly suited to identify key-factors, e.g. with respect to the transitions from one structural class to an other. In PROCOS physiological parameters of the vegetation components and the grazers may have to be incorporated. MAT and PROCOS are, therefore, regarded complementary models. Further development has been considered and is now in progress.

The modelling sofar has two main shortcomings:

1. the important spatial aspects are not yet included,
2. simulation of simultaneous grazing of more than one large herbivore is not yet possible; this aspect, however is of crucial importance, especially in the more natural grazing systems (cf. Drent & Prins, this volume).

The first aspect is at present the major point of attention for MAT. For this model more specifically the following improvements could be made:

a) the density of herbivores should be incorporated as a continuum, rather than discrete classes (cf. Table 4.);
b) a number of vegetation qualities depends on the abiotic conditions of the terrain; such parameters should be accounted for in a simple way, e.g. in the parameter 'carrying capacity';
c) in many reserves grazing is seasonal or supplementary feed is supplied during the winter; the possibility of changing pressures within the year should be incorporated;
d) further a subdivision of the structural classes should be considered.

For PROCOS a number of realistic improvements could also be suggested:

a) a further subdivision could be made in the various structural classes e.g. phenological stages in the classes SH and TH: vegetative, flowering, etc.;
b) important abiotic variables directly influencing productivity should be incorporated;
c) incorporation of a regrowth factor after grazing, probably with a time lag;
d) inclusion of a herbaceous production component in the woody structural types, i.e. simulating grasses and forbs growing under trees;
e) incorporation of forage quality parameters which would greatly improve the degree of reality of the model with respect to the preference of herbivores;
f) incorporation of some key physiological parameters of the animal that determine the intake of food, in order to simulate properly the annual variation of the requirement.

7. Summary and conclusions

For most of Western Europe the climax vegetation is forest, but grazing animals can prevent this succession and create and maintain conditions conducive for other vegetation types, especially short grasslands and shrubland. Experience in The Netherlands has shown that grazing with large herbivores can develop and maintain the structural diversity of vegetation. There is strong evidence that this will, in due time, lead to increased species diversity, of both plants and animals. Grazing can be substituted for, and has advantages over, other practices such as mowing and cutting of shrubs and trees. For economic reasons, but in particular because grazing leads to spatial variation and higher species diversity, grazing with livestock at low densities is now used in several hundred conservation areas in The Netherlands. The changes in vegetation structure caused by grazing animals are slow and methods are needed to predict the consequences of grazing management. Two simple models are described that were developed for this purpose, enabling the simulation of changes in vegetation structure under different grazing regimes. The descriptive matrix model MAT is compact and relatively easy to apply with a low number of input parameters. All the controlling factors are condensed in one operator, the transition matrix. PROCOS is an analytical model in which the impact of large herbivores on the macrostructure of vegetation is approached by simulating the production of aerial phytomass and its consumption or destruction by the grazing and browsing animals.

The MAT model uses four macrostructural classes, in which different development phases can be distinguished, simulating the rate of succession (short herbaceous, tall herbaceous, low woody and tall woody). In the PROCOS model the same four macrostructural types are used for the output, but in this model the relative growth rate is used as the main factor in determining succession. Any terrain can be characterized in terms of the four classes. Grazing may lead to retrogression, steady state or retarded succession in the sequence of the four types as mentioned above.

Validation of the models is difficult, because of lack of historical data on grazing pressures and vegetation development. Both models have been applied to the reserve area Borkener Paradies, for which considerable data exist. This area has been summer grazed by cattle for many years. The results of the MAT and PROCOS simulations are reasonably well in agreement with the information on long-term development in the structure of the vegetation as derived from literature sources.

The models have an important demonstration-value for those involved in policy making, administration, and management practice. The long-term effects can be shown. The models illustrate that it may take many decades or centuries before an equilibrium is reached between the grazing system and the final structural diversity aimed at.

Finally some apparent shortcomings and possibilities for improvement are outlined. In particular the value of incorporating spatial pattern and mixed grazing is considered. Work along these lines is now in progress.

Acknowledgements

We thank J.P. Bakker, H.H.Th. Prins, R.A. Prins, P.A. Slim and S.E. van Wieren for critically reading an early draft of the paper. Their comments have been incorporated in the final version or will be used in the further development of the models. P.A. Slim kindly provided the unpublished data presented in Table 2. The critical remarks and skilful editing of R.W. Snaydon considerably improved the final text.

References

Bakker, J.P., de Bie, S., Dallinga, J.H., Tjaden, P. & de Vries, Y. (1983). Sheep grazing as a management tool for heathland conservation and regeneration in The Netherlands. J. appl. Ecol. 20: 541–560.

Bakker, J.P, de Leeuw, J. & van Wieren, S.E. (1984). Micro-patterns in grassland vegetation created and sustained by sheep-grazing. Vegetatio 55: 153–161.

Breymeyer, A.I. & van Dyne, G.M. (1980). Grassland, systems analysis and man. IBP 19, Cambridge University Press.

Burrichter, E., Pott, R., Raus, T. & Wittig, R. (1980). Die Hudelandschaft 'Borkener Paradies' im Emstal bei Meppen. Abhandlungen aus dem Landesmuseum für Naturkunde zu Münster in Westfalen 42. Jahrgang 1980, Heft 4.

Centraal Veevoederbureau (1979). Verkorte tabel voedernormen voor de landbouwhuisdieren en voederwaarde van veevoeders. Centraal Veevoederbureau in Nederland, Wageningen.

Christian, K.R. (1981). Simulation of Grazing Systems. In: F.H.W. Morley (Ed.) Grazing animals, World Animal Science vol. Bl. pp 361–377. Elsevier Scientific Publishing Company, Amsterdam, Oxford, New York.

Coughenour, M.B., McNaughton, S.J. & Wallace, L.L. (1984). Simulation study of East-African perennial graminoid responses to defoliation. Ecol. Modelling 26: 177–201.

Eijsackers, H.J.P. (1983). Development of earthworm populations in abandoned arable fields under grazing management. In: J.E. Satchell (Ed.) Earthworm Ecology; from Darwin to vermiculture, pp 241–246. Chapman, London.

Hillegers, H.P.M. (1984). Begrazing met Mergelland schapen in Zuid-Limburg. De Levende Natuur 85: 178–184.

Hofmann, R.R. (1973). The Ruminant Stomach · Stomach Structure and Feeding Habits of East African Game Ruminants. East African Literature Bureau Nairobi, Kampala, Dar Es Salaam.

Hofmann, R.R. (1976). Zur adaptieven Differenzierung der Wiederkauer, Untersuchsergebnisse auf der Basis der vergleichenden funktionellen Anatomie des Verdauungstrakts. Der praktische Tierarzt 6: 351–358.

Jaspers, C.J. (1983). Bos- en struweelontwikkeling onder beweiding in het 'Borkener Paradies'. Report no. 724 Department of Nature Conservation, Agricultural University, Wageningen.

Jeffers, J.N.R. (1978). An introduction to systems analysis with ecological applications. Arnold, London.

Jeffers, J.N.R. (1982). Modelling. Outline studies in ecology. Chapman and Hall, London, New York.

Leeuw, J. de & Bakker, J.P. (1986). Sheep-grazing with different foraging efficiencies in a Dutch mixed grassland. J. appl. Ecol. 23: 781–793.

Lotz, B. & Poorter, H. (1983). Natuurtechnische begrazing; een aanzet tot een modelmatige benadering. RIN-rapport 83/2. Research Institute for Nature Management, Leersum, The Netherlands.

Noy-Meir, J. (1978). Grazing and production in seasonal pastures; analysis of a simple model. J. appl. Ecol. 15: 809–835.

Oosterveld, P. (1975). Beheer en ontwikkeling van natuurreservaten door begrazing. Natuur en landschap 29: 161–171.

Oosterveld, P. (1976). Integratie van voormalige landbouwgronden d.m.v. extensief graasbeheer met IJslandse pony's in de Baronie Cranendonck. Contactblad voor Oecologen 12: 99–109.

Oosterveld, P. (1979). Maaien, grazen of stuiven, via natuurbeheer naar meer natuur. Duin 2: 3–8.

Oosterveld, P. (1983). Eight years of monitoring of rabbits and vegetation development on abandoned arable fields grazed by ponies. Acta Zool. Fennica 174: 71–74.

Oosterveld, P. (1985). Grazing as a management tool in nature conservation – grazing management for the integration of formerly arable fields in reserve areas. In: Annual Report 1984 Research Institute for Nature Management, pp 29–32. Leersum, The Netherlands.

Parton, W.J. & Risser, P.G. (1979). Simulated impact of management practices upon tallgrass prairie. In: N. French (Ed.) Perspectives in grassland ecology. Springer Verlag, New York.

Petterson, B. (1965). Gotland and Oland. Two limestone islands compared. Acta Phytogeogr. Suecica 50: 131–140.

Pratt, R.M., Putman, R.J., Ekins, J.R. & Edwards, P.J. (1986). Use of habitat by free-ranging cattle and ponies in the New Forest, Southern England. J. appl. Ecol. 23: 539–557.

Putman, R.J. (1986). Grazing in Temperate Ecosystems: Large Herbivores and the Ecology of the New Forest. Croom Helm, Beckenham.

Redetzke, K.A. & van Dyne, G.M. (1976). A matrix model of a rangeland grazing system. J. Range Manage. 29: 425–430.

Rosen, E. & Sjögren, E. (1973). Sheep grazing and changes of vegetation of the limestone heath of Oland. Zoon. Suppl. 1: 137–150.

Sjögren, E. (1970). Short introduction to Oland especially with remarks on sheep grazing on the limestone heath. Mimeo, 7 pp.

Sonnema, M. (1974). Weidebouw; cultuur en gebruik van grasland. Stam-Robijns, Culemborg, The Netherlands.

Swartzman, G. (1980). Evaluation of ecological simulation models. In: S. Levin (Ed.) Mathematical modelling in biology and ecology, pp 230–267. Springer Verlag, New York.

Thalen, D.C.P. (1984a). Large mammals as tools in the conservation of diverse habitats. Acta Zool. Fennica 172: 159–163.

Thalen, D.C.P. (1984b). Begrazingsbeheer en begrazingsonderzoek: een ontwikkelingsschets. De Levende Natuur 85: 35–40.

Van de Laar, J.A.J. & Slim, P.A. (1979). Veranderingen in flora en vegetatie van de verlaten landbouwgronden in het CRM-reservaat Baronie Cranendonck na vijf jaar begrazing door IJslandse pony's (1972–1979). RIN-report 79/13, Leersum, The Netherlands.

Van de Veen, H.E. (1979). Food selection and habitat use in the Red Deer (Cervus elaphus L.). Ph. D. Thesis, University of Groningen.

Van de Veen, H.E. & van Wieren, S.E. (1980). Van grote grazers, kieskeurige fijnproevers en opportunistische gelegenheidsvreters; over het gebruik van grote herbivoren bij de ontwikkeling en duurzame instandhouding van natuurwaarden. Rapport 80/11 Instituut voor Milieuvraagstukken, VU Amsterdam.

Van Wijngaarden, W. (1985). Elephants - Trees - Grass - Grazers; relationships between climate, soil, vegetation, and large herbivores in a semi-arid savanna ecosystem (Tsavo, Kenya). ITC Publication Number 4. ITC, Enschede, The Netherlands.

Westhoff, V. (1971). The dynamic structure of plant communities in relation to the objectives of conservation. In: Duffey & Watts (Eds.): The scientific Management of Animal and Plant Communities for Conservation, pp 3-14. Blackwell, Oxford, London and Edinburgh.

Westhoff, V. (1977). Botanical aspects of nature conservation in densely populated countries. In: A. Myawaki & R. Tüxen (Eds.): Vegetation Science and Environmental Protection, p 369–374. Mazuren, Tokyo,

CHAPTER 12

Restoration of species-rich grassland after a period of fertilizer application

J.P. Bakker

Detailed contents

1. Introduction

1.1 From degraded grassland to species-rich communities?

Applications of artificial fertilizers, which can be regarded as a disturbance sensu Van Andel & Van den Bergh (Chapter 1), i.c. of a semi-natural biological system, normally convert species-rich grasslands into species-poor communities. The process is well known in the Netherlands. The species composition and species frequency of numerous Dutch grasslands were studied between 1934 and 1958 (Kruijne et al., 1967). These grasslands have been again studied in 1967 and 1968 (Dirven & Wind, 1982). Between these two periods, when N fertilizer use had increased and other agricultural practices had strongly intensified, the abundance of *Poa trivialis, Lolium perenne, Poa annua, Alopecurus geniculatus, Alopecurus pratensis* and *Cardamine pratensis* had significantly increased, whereas the abundance of *Bellis perennis, Bromus hordeaceus, Cerastium fontanum, Achillea millifolium, Leontodon autumnalis, Cynosurus cristatus, Anthoxanthum odoratum, Festuca rubra, Ranunculus acris* and *Trifolium repens* had significantly decreased. Similar results with a sharp decrease in species-diversity were found in experimental plots after long periods of fertilizer application in the Netherlands (Elberse et al., 1983), in England (Williams, 1978) and in Germany (Klapp, 1965).

As the area of species-rich grasslands has declined, interest in these grasslands has increased from a nature conservation point of view. Can these heavily fertilized species-poor grasslands revert to species-rich grassland communities, provided the fertilizer applications ceased and the groundwater table has not been lowered?

1.2 Species-diversity

Nature management is usually aimed at increasing the ecological value of an area, often by increasing species-diversity. This will be termed restoration management. Restoration management usually implies a sudden change from the current management practices, and the introduction of

J. van Andel et al., (eds.), Disturbance in Grasslands. ISBN 90–6193–640–3.

such restoration management can therefore be considered a disturbance, sensu Van Andel & Van den Bergh (Chapter 1); i.c. of a man-made biological system; the subsequent regular management practices will lead to 'new' plant communities dependant on those practices.

Several authors have reviewed factors affecting species-diversity. Grime (1979) related species-diversity to the productivity of the environment, and Huston (1979) extended Grime's model to include the frequency of reduction of above-ground biomass. Naveh & Whittaker (1980) suggested a relationship between species-diversity and grazing intensity and evolutionary time. In each case the authors present equilibrium models, describing species-diversity under constant and predictable conditions (Peet et al., 1983), though continual activity, such as grazing and cutting, may be essential to maintain the equilibrium state, i.e. they are 'system-inherent' factors. The various models do not consider what happens following a sudden change in environment, i.e. 'system-non-inherent' factors or disturbance (Chapter 1). The various models imply that a change in species diversity can be predicted from a simple shift in position of the graphical model. This may be true if a new species equilibrium has been established, but not for short-term changes.

1.3 Soil fertility and management practices

High species-diversity is usually associated with a relatively low standing crop (Grime, 1979); this in turn implies a fairly low soil fertility. Restoration management, therefore, aims at reducing the residual effects of earlier fertilizer applications. This can be achieved by hay-making without fertilizer application, a method of removing nutrients and so impoverishing the soil, which is termed in Dutch 'verschraling', and in German 'Aushagerung'. Grazing at a low stocking density will be less effective because of the return of dung and urine. The success of this process can be measured by changes in soil chemical analysis or by changes in the indicator value of the species with respect to soil fertility. The reverse, i.e. the application or artificial fertilizer, did increase soil fertility measured in soil chemical analyses. Elberse et al. (1983) found a gradual P_2O_5 and K_2O increase in the top soil (0–5 cm) layer over a period of 20 years of fertilizer treatment on heavy clay soil in the Netherlands, whereas the control plots showed low P and K-values.

1.4 Questions to be answered

Does the disturbance under study, i.e. cessation of fertilizer application with later and less frequent hay-making or grazing at a low stocking rate, cause change in the vegetation? How long does it take to return to species-rich grassland? What is the influence on the species composition (number of species and character of the species)? How is the species composition influenced by the substrate (wet or dry soils) and management practices (different hay-making regimes and grazing)? What are the causes of the observed changes in the species composition (e.g. soil fertility and structure of the sward)?

2. Methodology

The effects of restoration management on the vegetation are measured here by changes in several attributes.

The first attribute is the number of species of vascular plants per relevé of 4 m^2. The second attribute is soil nutrient content; simple parameters were adopted: N_{total}, P_{total}, $K_{exchangeable}$, C/N quotient. The availability of macro-nutrients was not measured. Although these analyses are of a very rough character, they do allow comparison with the results of other authors.

An indirect way of quantifying the effects of management could be a species qualification with respect to the nutrient status of the soil. The third attribute is the indicator value of the species present.

Ellenberg's N-figures are, in particular, often applied when describing the synecology of plant communities in combination with stand ordination (Werger et al., 1978; Persson, 1981). Other authors recognize the need to describe environmental changes related to vegetational succession (Loopstra & Van der Maarel, 1984; Melman et al., 1985).

All species occurring in the study area were subdivided into three groups ('poor', 'intermediate' and 'rich' species), indicating the nutrient status of the soils in which they normally occur. Species with nitrogen values of 1 to 3, according to Ellenberg (1979), were grouped as 'poor' species, and values of 7 to 9 as 'rich'; this is similar to grouping of Schiefer (1983). Species with nitrogen values of 1 and 2, according to Klapp (1965), were grouped as 'poor' species, and values of 4 and 5 as 'rich'. The phosphate values according to Kruijne et al. (1967) range form −100 to 100, though only some salt-marsh species show values of over 50. Species with phosphate values of −100 to −50 were therefore grouped as 'poor' species, and values of 1 to 50 as 'rich'. If the authors differed in their assessment of the species by one grouping, majority view was accepted; if they differed by two groupings, the species was categorized as unknown. This differs form the method used by Wind (1980), who took the average of the three authors. In general the authors agreed fairly well (Table 1). The ratio

$$\frac{\text{number of species indicating nutrient poor soils}}{\text{number of species indicating nutrient rich soils}}$$

was used as a qualitative index of soil fertility. A quantification index was calculated by taking into account the cover percentage of the various species.

Although the list of nitrogen values for species, compiled by Ellenberg (1979) referred to Central Europe, it has also been used for S. Sweden (Persson, 1981) and the Netherlands (Sterk et al., 1983; Wind, 1980). Loopstra & Van der Maarel (1984) concluded that Ellenberg's indicator values were suitable for the Netherlands.

Table 1. Frequency of occurence (%) in the groups indicating nutrient poor soil conditions, intermediate- and nutrient rich soil conditions according to literature.

	Kruijne et al. (1967)	Ellenberg (1979)	Klapp (1965)
Nutrient poor soil conditions	33%	39%	39%
Nutrient rich soil conditions	22%	25%	29%
Intermediate soil conditions	45%	36%	32%

Ideally, circular arguments are to be prevented and the species indicator values should be compared with independent parameters. For example, Melman et al. (1985) considered the above-ground standing crop as a good measure of nutrient availability (cf. Grime, 1979), and they found very good correlations (r = 0.90) between above-ground standing crop and the indicator values of both Kruijne et al. (1967) and Ellenberg (1979). In the present study there was also a good correlation between chemical characteristics of the soil and the indicator values of the species present (Table 2). The indicator values of species, therefore, seem to be satisfactory tool of measurement.

3. Species composition

3.1 Wet and dry soils

The restoration management practice used in the valley grasslands in the Nature Reserve 'Stroomdallandschap Drentsche A' is hay-making in July. The results of this practice in wet and dry conditions will first be discussed. The subdivision in wet and dry conditions is based upon the bundles of duration lines obtained in several places (Fig. 1). Duration lines represent the cumulative number of days with a certain groundwater class (Niemann, 1973).

The results of hay-making, without fertilizer application, in several lots have been published by Van Duuren et al. (1981). As an example, vegetational changes (and the species involved) from 1974 onwards are presented on wet peaty soil (Table 3) and on dry sandy soil (Table 4) under hay-making without fertilizer application since 1972. The number of species per relevé increased after the discontinuation of fertilizer application on the wet site and, to a lesser extent, on the dry site (Fig. 2). The number of species

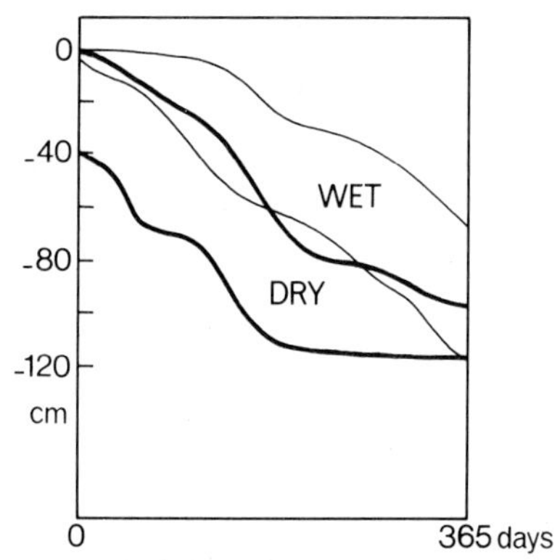

Fig. 1. Lines of duration bundles in wet and dry areas.

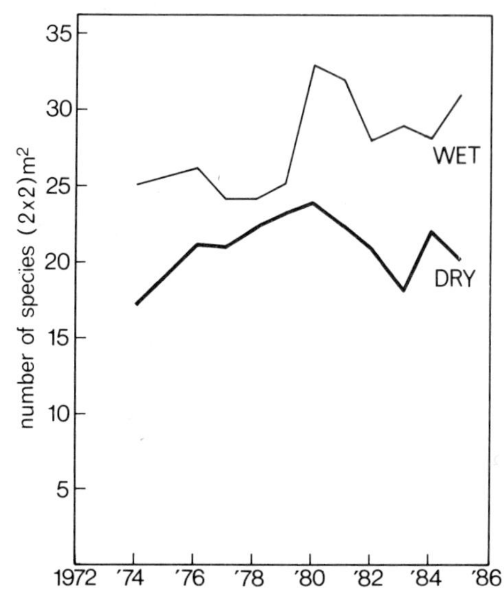

Fig. 2. Changes in the number of species in wet and dry areas cut for hay in July and receiving no fertilizer applications after 1972.

was always greater on the wet site, perhaps because earlier agricultural practices which were more intensive on dry than on wet soils (Proefstation voor Akker- en Weidebouw, 1966).

The changes in the number of rich and poor species are shown in Fig. 3. The poor/rich ratio increased with time on both wet and dry area (Fig. 4), but it was always higher and increased more rapidly in the wet area. The relative importance of the new hay-making regime and the cessation of fertilizer applications will be discussed later in this chapter.

Possibly the new hay-making regime (from several cuts a year after fertilizer application to one late cut without fertilizer dressing) might have changed the sward structure thus giving access to new (mainly 'rich') species over a period of several years. This mixture of supposed nutrient removal and changes in canopy structure will be discussed later in this chapter.

The quantitative poor/rich ratio, based upon the coverage of the occurring species, shows larger fluctuations than the qualitative quotients; Schmidt (1985) obtained similar results, possibly caused by large year to year fluctuations in the coverage of some species, though Durwen (1982), Böcker et al. (1983) and Melman et al. (1985) found little differences between the two methods. Whittaker (1954) and Persson (1981) stressed the importance of taking into account the coverage of species in such situations, but did not consider variation in time.

In this study the qualitative poor/rich ratio will be used, because of its greater year to year consistency.

Table 2. The nutrient content in autumn of the surface soil (0–5 cm) beneath some plant communities in the Netherlands. The number of species indicating nutrient-poor and nutrient-rich soil conditions in each relevé (2×2)m² is also shown. The plant communities chosen are supposed to succeed each other in the given sequence during the process of hay-making wihout fertilizer application (Everts et al., 1984).

Community	mg.100 g dry soil^{-1}			C/N	Number of species		Poor/Rich
	N_{total}	P_{total}	$K_{exchangeable}$		Poor	Rich	species ratio
Poo-Lolietum	1880	510	71	14	1	7	0.1
Poo-Lolietum/ Lolio Cynosuretum	2320	725	65	15	2	6	0.3
Lolio-Cynosuretum	1930	575	52	14	6	5	1.2
Senecioni-Brometum	1070	189	47	16	8	5	1.6
Juncetum acutiflori	390	80	41	15	12	4	3.0

3.2 Comparison of hay-making regimes

Different cutting regimes were started in 1972 to study the effects of the timing and frequency of hay-making in the absence of fertilizer application. The study area had a homogeneous substrate and fertilizer application had stopped in 1971. The grassland contained *Agrostis stoloni-*

Table 3. Changes in the abundancy (cover %) of species in a permanent plot (2×2)m² on wet peaty soil under hay-making and receiving no fertilizer after 1972. R: species indicating nutrient-rich soils; P: poor soils; I: intermediate soils; X: unknown.

	Year	1974	76	77	78	79	80	81	82	83	84	85
R	*Poa annua*	2	·	·	·	·	·	·	·	·	·	·
I	*Cerastium fontanum*	4	1	1	1	1	1	1	1	1	1	1
R	*Trifolium repens*	4	·	·	·	1	2	1	1	2	1	4
I	*Sagina procumbens*	4	4	·	·	·	1	1	1	1	·	1
R	*Alopecurus geniculatus*	10	1	1	8	1	1	1	1	1	1	·
R	*Poa trivialis*	·	1	4	1	2	8	2	2	4	4	1
R	*Festuca pratensis*	·	2	1	4	8	1	1	1	·	1	1
P	*Juncus effusus*	·	1	1	1	8	4	1	1	2	4	1
P	*Juncus articulatus*	·	1	·	1	·	1	1	1	1	2	1
R	*Agrostis stolonifera*	·	·	1	10	1	1	2	2	1	1	·
R	*Ranunculus repens*	4	4	12	30	20	20	12	12	30	12	4
P	*Rhinanthus angustifolius*	·	·	·	4	40	12	20	30	8	40	12
I	*Galium palustre*	·	·	·	1	1	1	1	1	1	1	1
X	*Caltha palustris*	·	·	·	·	1	2	1	1	2	4	4
P	*Ranunculus flammula*	·	1	·	·	·	1	1	1	1	1	2
I	*Holcus lanatus*	4	20	40	2	·	2	8	4	20	4	1
R	*Poa pratensis*	2	·	·	·	·	·	1	·	·	·	1
R	*Lolium perenne*	4	·	·	·	1	1	1	·	·	·	·
I	*Agrostis capillaris*	2	·	·	·	·	1	·	·	·	·	·
I	*Cynosurus cristatus*	4	2	2	·	2	4	1	2	4	2	1
I	*Anthoxanthum odoratum*	10	1	1	2	8	12	4	4	8	8	10
I	*Glyceria fluitans*	4	2	1	1	2	8	1	1	1	1	·
I	*Festuca rubra*	4	2	1	·	·	1	1	·	·	·	1
I	*Rumex acetosa*	2	1	2	1	2	4	2	1	2	1	2
I	*Cardamine pratensis*	2	1	1	1	1	2	1	1	1	1	1
I	*Ranunculus acris*	1	·	1	·	1	1	1	1	1	1	1
I	*Bellis perennis*	2	1	1	1	1	2	1	1	·	1	·
X	*Taraxacum sp.*	2	1	1	1	1	1	1	1	1	1	2
P	*Equisetum palustre*	2	1	1	4	2	1	1	2	2	8	50
I	*Stellaria glutinosa*	2	1	1	1	·	1	1	1	1	1	1
I	*Lychnis flos-cuculi*	1	·	1	1	1	2	1	1	1	1	1
I	*Mentha aquatica*	·	1	1	·	1	1	1	·	1	·	·
I	*Myosotis palustris*	·	·	·	1	·	1	·	1	·	·	·
I	*Veronica serpyllum*	1	·	·	·	·	·	·	·	·	·	·
R	*Plantago major*	·	1	·	·	·	·	·	·	·	·	·
X	*Epilobium obscurum*	·	1	·	·	·	·	·	·	·	·	·
I	*Filipendula ulmaria*	·	1	·	·	·	·	·	·	·	·	1
X	*Lythrum salicaria*	·	·	1	·	·	·	·	·	·	·	·
P	*Epilobium palustre*	·	·	·	1	·	·	·	·	1	·	·
I	*Polygonum hydropiper*	·	·	·	·	·	·	·	·	1	·	·
I	*Equisetum fluviatile*	·	·	·	·	·	·	·	·	·	·	1
I	*Plantago lanceolata*	·	·	·	·	·	·	·	·	·	·	1
I	*Juncus bufonius*	·	·	·	·	·	·	·	·	·	·	1
P	*Juncus conglomeratus*	·	·	·	·	·	·	·	·	·	·	1

fera, *Poa pratensis, Holcus lanatus, Poa trivialis, Festuca pratensis, Ranunculus repens, Festuca rubra* and *Lolium perenne,* in order of decreasing abundancy. The treatments were: no cutting, hay-making in July, in September and in both July and September.

Changes in species composition, resulting from the different treatments have been described by Bakker et al. (1980) and Bakker & De Vries (1985A, 1985B). The number of species initially declined in the abandoned plots but then increased slightly (Fig 5). There was also an initial decline in the number of species on the various cut plots, but then a sharp increase after about five years. In general the number of species was greatest in plots cut in September.

No 'poor' species appeared in the abandoned site (Fig. 6). The poor/rich ratio was highest in the area cut in both July and September.

The above-ground standing crop steadily declined after fertilizer applications ceased in 1971 with no consistent differences between cutting regimes (Fig. 7). However, there were also year to year variations, presumably caused by weather

Table 4. Changes in the abundancy (cover %) of species in a permanent plot (2×2)m² on dry sandy soil under hay-making and receiving no fertilizer after 1972. R: species indicating nutrient-rich soils; P: poor soils; I: intermediate soils; X: unknown.

	Year	1974	76	77	78	79	80	81	82	83	84	85
R	*Lolium perenne*	40	4	4	2	1	1	1	·	·	·	1
R	*Alopecurus geniculatus*	10	2	2	2	2	2	1	2	1	1	1
I	*Cerastium fontanum*	4	1	1	1	1	1	1	1	1	1	1
R	*Bromus hordeaceus*	10	2	20	4	1	1	1	1	·	·	·
R	*Trifolium repens*	·	2	2	4	2	2	1	4	·	1	1
R	*Poa pratensis*	·	1	1	1	1	2	1	2	·	1	1
R	*Festuca pratensis*	·	1	1	1	1	1	·	1	2	1	1
R	*Agrostis stolonifera*	·	·	1	8	20	12	8	8	1	8	·
R	*Poa trivialis*	·	1	8	20	12	20	4	4	4	12	4
R	*Ranunculus repens*	4	2	4	8	20	30	20	12	30	20	20
P	*Rhinanthus angustifolius*	·	·	·	·	·	1	1	8	8	12	10
I	*Anthoxanthum odoratum*	·	1	1	1	2	2	4	8	8	12	20
I	*Cynosurus cristatus*	2	1	1	1	1	1	1	2	4	2	2
I	*Glyceria fluitans*	·	·	·	·	·	·	·	·	1	1	·
I	*Holcus lanatus*	4	10	30	30	2	4	8	30	30	20	10
I	*Rumex acetosa*	4	2	8	1	20	2	2	1	2	2	2
R	*Phleum pratense*	1	·	·	·	·	2	1	·	·	1	·
I	*Agrostis capillaris*	2	2	·	·	·	2	1	·	·	1	1
I	*Cardamine pratensis*	1	1	1	1	1	1	1	2	1	1	1
I	*Bellis perennis*	2	1	1	1	2	1	1	1	·	1	·
X	*Taraxacum sp.*	2	1	1	1	1	·	1	1	·	1	1
R	*Rumex obtusifolius*	·	1	4	2	2	·	1	·	·	·	·
R	*Elymus repens*	·	2	·	1	1	1	·	·	·	·	·
R	*Stellaria media*	·	·	1	·	1	·	·	·	·	·	·
R	*Plantago major*	·	·	·	1	1	1	·	1	·	·	·
I	*Trifolium dubium*	2	·	·	·	·	·	·	·	·	·	·
I	*Veronica arvensis*	4	·	·	·	·	·	·	·	·	·	·
I	*Myosotis palustris*	·	·	·	·	·	1	·	·	·	1	·
I	*Rhinanthus minor*	·	·	·	·	·	·	·	1	·	·	·
X	*Caltha palustris*	·	·	·	·	·	·	·	·	2	·	·
I	*Lychnis flos-cuculi*	·	·	·	·	·	·	·	·	1	·	·
I	*Mentha aquatica*	·	·	·	·	·	·	·	·	1	·	·
P	*Juncus effusus*	·	·	·	·	·	·	·	·	·	·	1
I	*Plantago lanceolata*	·	·	·	·	·	·	·	·	·	·	1

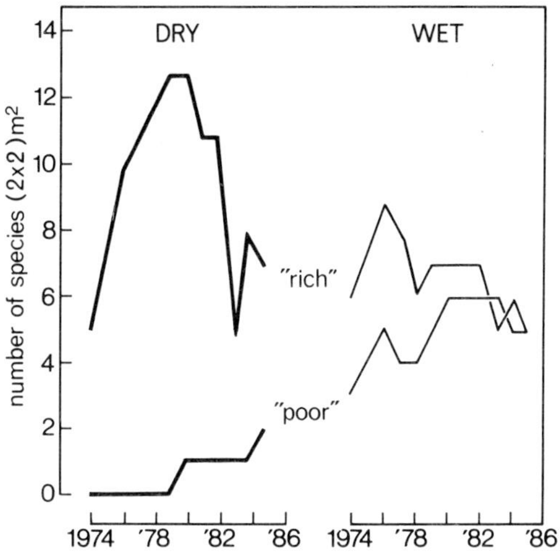

Fig. 3. The number of 'poor' and 'rich' species in wet and dry areas cut for hay in July and receiving no fertilizer applications after 1972.

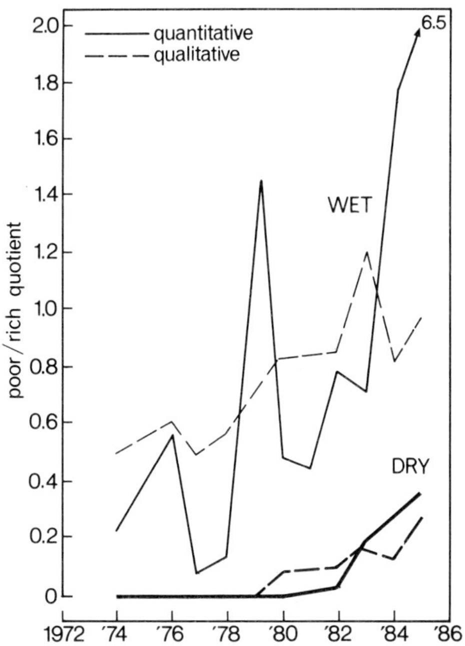

Fig. 4. Changes in the ratios $\frac{\text{abundance of poor species}}{\text{abundance of rich species}}$, based either on the number of species (qualitative) or the cover % of species (quantitative) in wet and dry areas cut for hay in July and receiving no fertilizer applications after 1972.

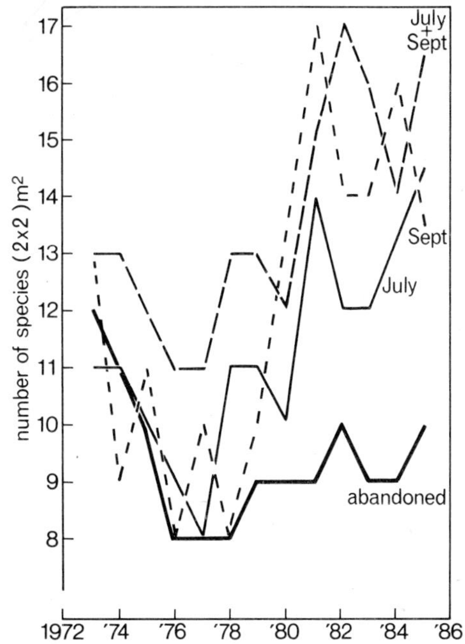

Fig. 5. The number of species in areas with different hay-making regimes and receiving no fertilizer applications after 1973.

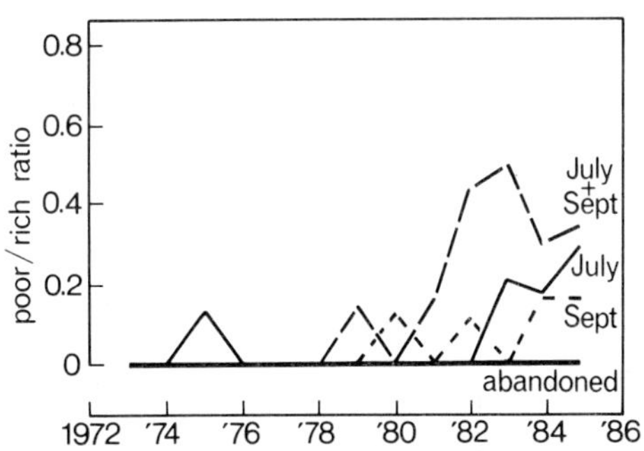

Fig. 6. The ratios of 'poor' to 'rich' species on plots with different hay-making regimes and receiving no fertilizer applications after 1973.

conditions. Schreiber & Schiefer (1985) and Elberse et al. (1983) also found a general decrease in yield after cessation of fertilizer application, though they and Schiefer (1984) reported large fluctuations in yield from year to year. Oomes (1977) and Oomes & Mooi (1981) found a sharp decline from 800 g dw.m^{-2} to 300–400 g dw.m^{-2} when different hay-making regimes were imposed on grassland on clay in the Netherlands followed by fluctuations around that level. Wind (1980) found examples of both gradual decrease and sudden decrease, the former on clay and the latter on sandy soil.

Despite the removal of considerable amounts of biomass, the content of N, P and K in the soil showed no general decline over the years (Fig. 8).

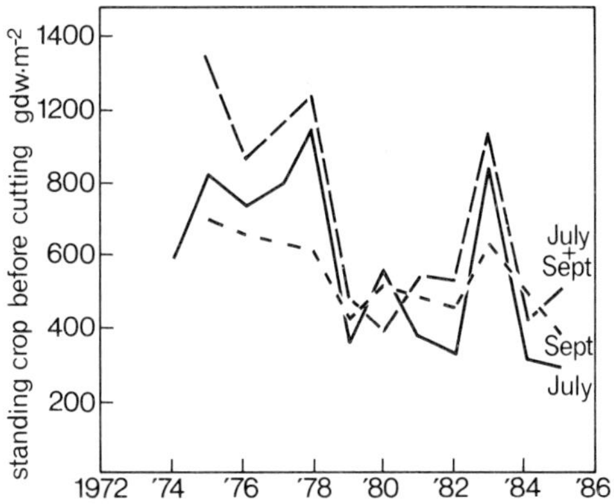

Fig. 7. Above-ground standing crop just before cutting in plots with different hay-making regimes and receiving no fertilizer applications after 1973.

3.3 Comparison of grazing and hay-making

July hay-making and sheep-grazing (including the return of dung) were compared from 1972 onwards, fertilizer applications having ceased in 1971. The area had been reclaimed from heathland during the 1950's and sown to grass. The initial vegetation was characterized by *Holcus lanatus, Lolium perenne, Poa pratensis, Phleum pratense, Agrostis capillaris, Cerastium fontanum* and *Trifolium repens* in order of decreasing abundancy.

Some results have been described by Bakker et al. (1983; 1984). Grazing at a low stocking rate (3 sheep. ha^{-1}) resulted in locally different efficiencies of foraging (the efficiency of foraging expresses the herbage removed as a proportion of the herbage accumulated) (Hodgson, 1979). The closely grazed areas with *Leontodon autumnalis* had a foraging efficiency of 0.90, the lightly grazed areas, dominated by *Agrostis capillaris* had a foraging efficiency of 0.10, and the intermediately grazed areas, dominated by *Holcus lanatus* had a foraging efficiency of 0.30 (De Leeuw & Bakker, 1986).

The number of species in the lightly grazed area declined sharply in the first three years (Fig. 9). The number of species initially increased in the heavily grazed area, but decreased again later to the original level. The number of species in the hay-making treatment remained fairly constant.

The poor/rich species ratio was slightly greater

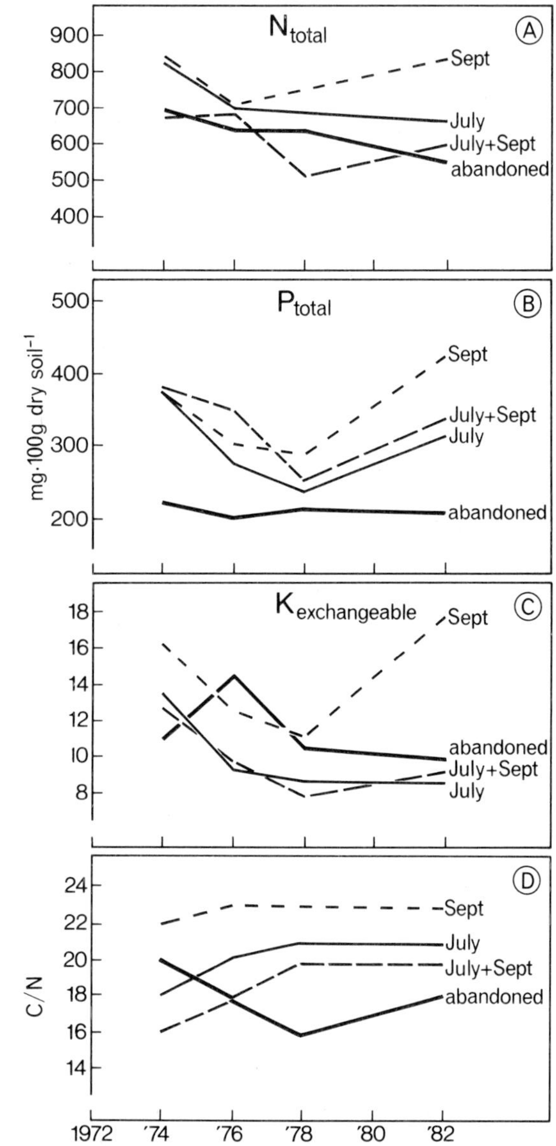

Fig. 8. The contents of N_{total}, P_{total}, $K_{exchangeable}$ and the C/N quotient in October in the topsoil (0–5 cm) in plots with different hay-making regimes and receiving no fertilizer applications after 1973.

in the heavily grazed area than under the hay-making regime (Fig. 10), despite the return of dung in the grazed area, but the ratio was much greater in the lightly grazed area. The large fluctuations in the lightly grazed area are due to the very low total number of species present.

There was no difference in soil chemical composition (Fig. 11) between the heavily grazed area and the hay treatment. The 'rich' species

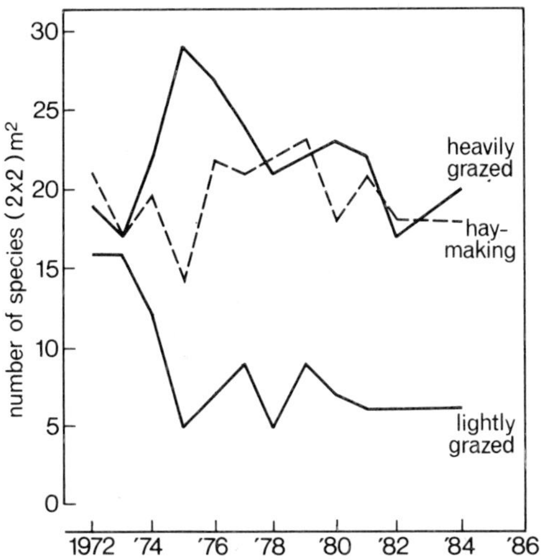

Fig. 9. The number of species on plots cut for hay in July and different grazing intensities and receiving no fertilizer applications after 1971.

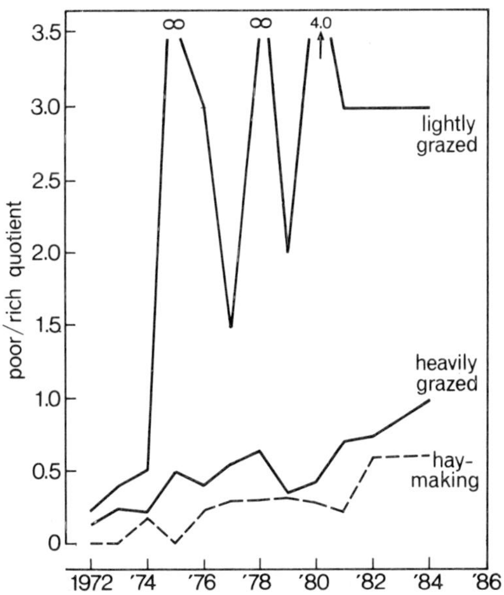

Fig. 10. The ratios of 'poor' and 'rich' species in plots cut for hay in July and different grazing intensities and receiving no fertilizer applications after 1971.

Lolium perenne persisted better in the grazed than in the cut area (Fig. 12A) apart from an initial difference in coverage. The 'intermediate' species *Anthoxanthum odoratum* (Fig. 12B) increased sharply in the cut area, but increased only slightly in the heavily grazed area. These findings seem to be in accordance with a more fertile soil in the grazed area, because grazing

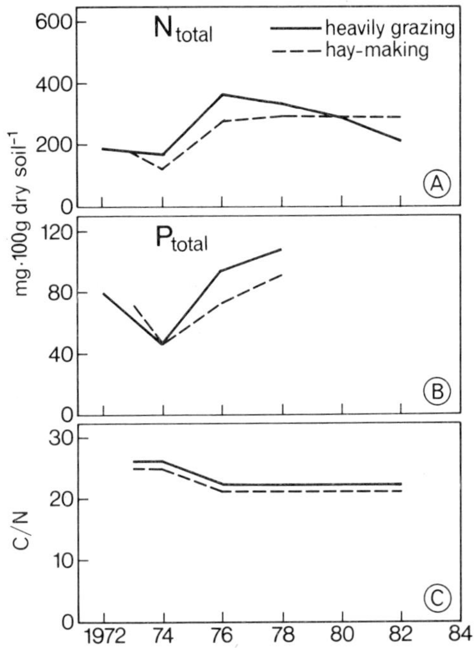

Fig. 11. The contents of N_{total}, P_{total} and the C/N quotient in October in the top soil (0–5 cm) on plots cut for hay and heavily grazed, and receiving no fertilizer applications after 1971.

stimulates the decomposition of organic matter (Perkins et al., 1978) and hence mineral cycling of N (Floate, 1970), P (Harrison, 1978), Ca, Mg, Mn and P (Bülow-Olsen, 1980). The coverage, however, of the 'poor' species *Hypochaeris radicata* initially increased sharply (Fig. 12C), especially on the heavily grazed area, but later declined again. Reasons for these changes will be discussed in the following sections.

4. Causes of changes in species composition

4.1 Mineral removal

The amount of nitrogen removed was similar in 1975 and 1983, though considerably less phosphate was removed in 1983 than in 1975 (Table 5). The amount of potassium removed was generally similar in 1975 and 1983, except for the values in the September cutting regime in 1975. More minerals were removed in the twice-cut treatment because of the greater amount of biomass removed in 1975, but of the higher mineral

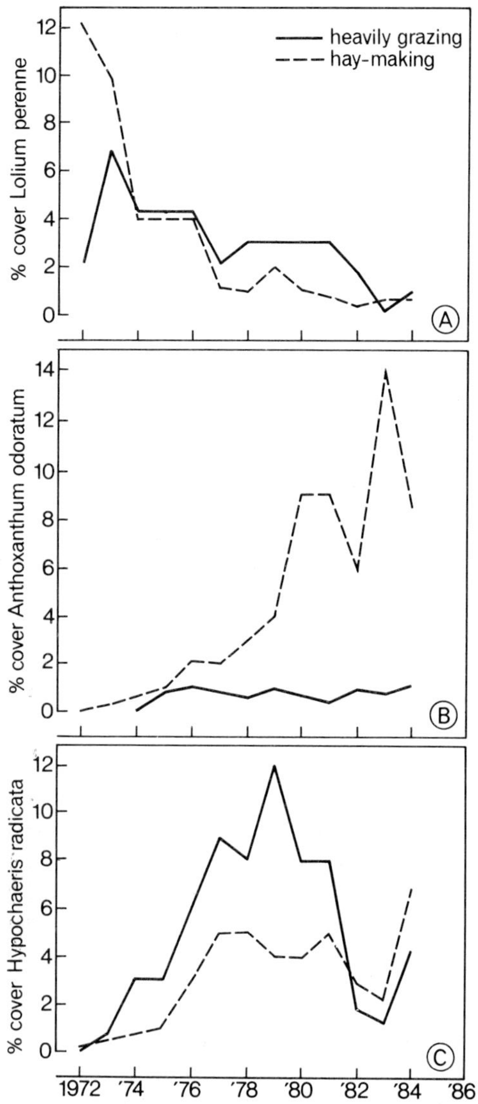

Fig. 12A-C. Changes in the cover % of *Lolium perenne* (A), *Anthoxanthum odoratum* (B) and *Hypochaeris radicata* (C) in plots cur for hay in July or heavily grazing. The plots had received no fertilizer applications since 1971.

Table 5. The amount of nutrients (kg.ha^{-1}) removed under different hay-making treatments in 1975 and in 1983. No fertilizer was applicated after 1973.

	N		P		K	
	1975	1983	1975	1983	1975	1983
July	107	122	41	19	55	40
September	104	108	44	21	73	39
July+September	156	176	66	30	91	93

content in 1983. In general the amount of minerals removed was not closely correlated with the above-ground standing crop. Schmidt (1985) reached a similar conclusion in comparing two, four and eight cuts per year with yields ranging from only 350 to 400 g dw.m^{-2}, but nitrogen removal of 60, 10 and 105 kg N.ha^{-1}, respectively. Schiefer (1984) and Schmidt (1985) also found large variations in the contents of hay with different cutting frequencies. Only between 1 and 3% in the nutrient pool (0–20 cm soil depth) was removed each year (Table 6). The highest percentage (2.5%) was for N, but this is an overestimate, since about 40 kg N.ha^{-1}.yr^{-1} is returned via 'acid rain' (Van Dam, 1983). Rather more minerals were removed by July hay-making than by heavy grazing in the period from April to August 1975 (Table 7), but the differences in removal were still greater since the mineral return from urine was not measured. Perkins et al. (1978) found that the N, P and K content of plants and grazing sheep amounted to only 0.5 to 1.5% of that in the soil (0–30 cm) beneath an *Agrostis capillaris* stand. Bülow-Olsen (1980) estimated values of between 0.3 and 2% of P in plants in a

Table 6. The mineral content (kg.ha^{-1}) of soil (0–20 cm) in 1972 and vegetation (average 1975–1983), and the percentage removed from the soil pool by hay-making.

	Soil 0–20 cm	Vegetation	% Removal by hay-making
N_{total}	4750	110	2.5
P_{total}	2000	20	1.0
$K_{exchangeable}$	5300	60	1.1

Table 7. The mineral pool in the soil (0–5 cm) and mineral removal (kg.ha^{-1}) under July hay-making and heavy grazing regimes in the period of April to August 1975. Dung return was subtracted from the mineral removal, urine return was not measured in the grazing regime. The treatments were started in 1972.

	Mineral pool		Mineral removal	
	N	P	N	P
July hay-making	1280	370	35	18
Heavily grazing	1600	420	21	6

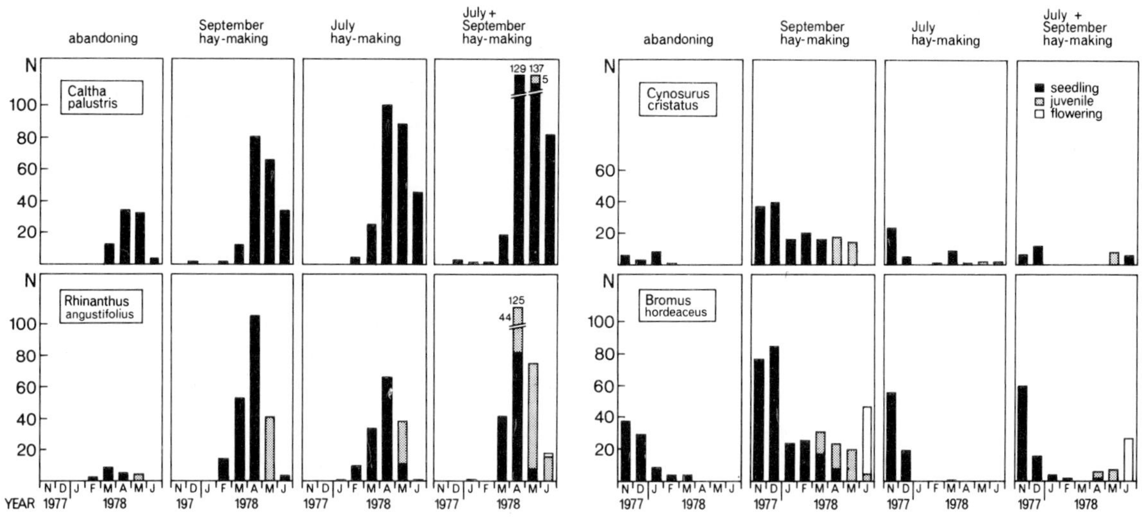

Fig. 13. The effect of cutting regimes on the total number of seedlings in 1978, when sown in November 1977. The hay-making was first applied in 1973 and the plots had received no fertilizer applications since then.

Deschampsia flexuosa stand compared to the soil (0–10 cm) pool, K in the plants amounted to between 30 and 75% of its soil pool, probably due to the very infertile soil.

Other studies have investigated nutrient removals from other grassland ecosystems. Wells (1980) measured an average annual nutrient removal of 30 kg $N.ha^{-1}$, 2 kg $P.ha^{-1}$ and 28 kg $K.ha^{-1}$ from chalk grassland in Great Britian that had a mean annual yield of 170 g $dw.m^{-2}$. The N-losses would have been replaced by nutrients in rainfall, but removal of P and K caused a significant decrease in extractable contents over 10 years. Bülow-Olsen (1980) found that sheep-grazing did not cause any significant decline in the amounts of available Ca, Mg, K, Mn, Fe and P in the soil beneath a *Deschampsia flexuosa* stand in Denmark. Apart from P, the deposition of nutrients from precipitation was greater than the amounts removed from the area in the carcasses of sheeps.

Nutrient depletion is apparently a very slow process, and is unlikely to account for accordance of the appearance of 'poor' species within a few years after the start of hay-making or grazing without fertilizer application in this study. Other factors are likely to be important.

The whole process of impoverishing the soil can be seriously countered by drainage. Grootjans et al. (1985) mentioned that a lowering of the groundwater table in a poorly fertilized hay-field stimulated the net N-mineralization by 5 to 10 fold.

The relationship between nutrient pool, nutrient availability, above-ground standing crop and nutrient removal will be discussed in the concluding section of this chapter.

4.2 Structure of the sward

The structure of the sward is expected to affect the ability of species to invade. For example artificially sown seeds of several species, sown in November 1977, were recorded in various treatments without fertilizer application. Although seedlings emerged in all treatments, the emergence and survival to juvenile and flowering individuals were highest in both September and twice cut plots and lowest in the uncut plots (Fig. 13). This could have been due to the low coverage of litter in winter and early spring on plots cut both in July and September (0–40%) as opposed to September cutting (20–50%), July cutting (30–80%), and in uncut plots (70–95%) (Bakker et al., 1980). There was also less penetration of light into the canopy of July cut and uncut plots (Bakker & De Vries, 1985A). Since both a 'poor' species (*Rhinanthus angustifolius*) and a 'rich' species (*Bromus hordeaceus*) reacted in a similar way to cutting regimes it is likely that the structure of the sward plays a more important role than the nutrient status of the soil in this experiment.

Grazing intensity also influences the emergence and the fate of the seedlings. When seed was sown in January 1983, more seedlings emerged and survived until October in the short turf of the heavily grazed area than in the lightly grazed patches (Table 8). This was again correlated with the amount of bare soil (5% in the heavily grazed area, and 1% in the lightly grazed patches), and litter coverage (10% in the heavily grazed area, and 90% in the lightly grazed patches). The soil in the lightly grazed patches received 20% full day light, whereas the soil in the heavily grazed area received 100%.

Litter accumulation can prevent seedling establishment (Silvertown, 1980; Willems, 1983), but can also increase microbial attack of emerged seedlings (Grime, 1979) and cause desiccation before their roots can reach the soil (Grubb, 1977). Rosette plant species such as *Hypochaeris radicata* (Fig. 12C), and also *Taraxacum sp., Leontodon autumnalis, Bellis perennis* and *Plantago major* in particular established better in heavily grazed areas. These species have a higher frequency percentage in a grazed area than in a hayfield (Elberse et al., 1983; Willems, 1985). It can be assumed that the short turf in the closely grazed sward provided more safe sites for these rosette plants than the hay-making regime with its taller canopy. Indeed Blom (1977), Harper (1977), Mølgaard (1977), Fenner (1978) and Schenkeveld & Verkaar (1984) demonstrated a better development of *Plantago major, Bellis perennis, Taraxacum sp., Hypochaeris radicata* and chalk grassland species, respectively, in a short turf than in a dense sward. The experimental creation of open patches in a dense sward or litter layer also favoured seedling establishment of *Rhinanthus angustifolius* (De Hullu, 1985), of *Solidago spp.* (Goldberg & Werner, 1983) and several biennial species (Gross & Werner, 1982).

The emergence and survival of seedlings is positively correlated to the percentages of full sunlight reaching the soil surface (Mølgaard, 1977; Goldberg & Werner, 1983; Schenkeveld & Verkaar, 1984). The leaf canopy not only changes the quantity of light but also its quality at soil level. The lower red: far red ratio under a plant canopy can reduce the emergence and seedling establishment of various species (Silvertown, 1980).

The highest species-diversity (20–30 species.0.25 m^{-2}) occurs in grasslands with aboveground standing crop (including litter) of between 500 and 600 g dw.m^{-2} in the United Kingdom (Grime, 1979) or at standing crop (excluding litter) of 400 tot 500 g dw.m^{-2} in the Netherlands (Vermeer & Berendse, 1983). The plant communities studied in both reports had been under a stable management for a long period of time.

The highest species-diversity, however, in experimental plots where management has recently changed (or sometimes c. 20 years ago) is related to less than 500 g dw.m^{-2}. Willems (1980) found that the highest species-diversity occurred in chalk grassland vegetation with standing crop between 150 and 350 g dw.m^{-2}, while Schiefer

Table 8. The seedlings emerged and survival until October 1983 after sowing 150 seeds of each species in January of the same year in heavily grazed areas and lightly grazed patches.

	Heavily grazed Emergence	Eearly establishment	Lightly grazed Emergence	Early establishment
Nardus stricta	17	13	4	0
Narthecium ossifragum	10	0	2	0
Danthonia decumbens	16	12	0	0
Gentiana pneumonanthe	14	0	0	0
Potentilla erecta	0	0	1	1
Succisa pratensis	29	24	34	12

(1984) found that new species only occurred in German grasslands when standing biomass was below 350 g $dw.m^{-2}$.

4.3 Seed dispersal

Oomes & Mooi (1981; 1985) mentioned a sharp decline of standing biomass from 1000 to 400 g $dw.m^{-2}$ after two years and later on stable yields between 300 and 500 g $dw.m^{-2}$ when fertilizer applications ceased. This reduction in yield was associated with only a limited increase of the species-diversity. Some newly established species emerged from the seed bank, but others had to invade from elsewhere. The absence of species in the seed bank can explain the time lag between the decreasing yield and the increase of species-diversity. Schenkeveld & Verkaar (1984) also mentioned the absence of numerous short living chalk grassland species in the seed bank of a former chalk grassland that had been fertilized. Numerous species typically of hay-field were not present in the persistent seed bank in our study area (Bakker, 1983). Some of these species were able to establish under the different cutting regimes (Fig. 13), and grazing regimes (Table 8), if seed was present. Their appearance therefore seems to depend more on dispersion than on factors determining their germination.

5. Conclusions

The application of fertilizers to a semi-natural species-rich grassland is a disturbance of the biological system but, conversely, the cessation of artificial fertilizer application is also a disturbance of the man-made biological system. The nutrient status of the soil is likely to decline after application of fertilizer ceases. The process is likely to be speeded up when nutrients are removed with the hay crop. This will lead to lower above-ground standing crop and hence a more open structure of the sward, facilitating the establishment of species adapted to nutrient-poor soil condition. In this, and several other published studies, the above-ground standing crop usually did decrease, and the yields were strongly influenced by weather conditions. There was little relationship between current mineral yield of standing crop. In general, the removal of minerals was very low, compared to the mineral pool in the soil, and mineral inputs by precipitation were usually comparable to mineral removal.

The above-ground standing crop may be used as a measure of the availability of nutrients (Melman et al., 1985; Vermeer & Berendse, 1983). The mineralization rate is likely to be a good measure of nutrient availability (Vermeer & Berendse, 1983), but it is difficult to measure and likely to be affected by other factors such as drainage of peaty soils (Grootjans, 1985). Since a poor relationship was found between the N, P and K content in the soil and the biomass yield, impoverishment of the soil could be measured by mineral removal (Oomes & Mooi, 1985). Moreover, the strong fluctuations in an above-ground standing crop suggest the influences of weather conditions upon mineralization rates of dying plant material and litter. In search for finding the right parameter in the process of impoverishing the soil it seems important to answer two more questions: (i) how do the mineralization rates change in our study areas under different management practices after the cessation of fertilizer application, and (ii) what will different plant species, used as bio-assays, indicate on the nutrient status of the soil under different management regimes (Chapter 4).

Mineral removal by grazing or hay-making without fertilizer applications seems to be a very slow process, as Wells (1980) suggested in chalk grassland experiments. Nevertheless in this and several other studies the species-diversity increased a little and the proportion of species indicating nutrient-poor soil conditions increased relatively quickly. The performance of these nutrient-poor species depended on safe sites created in the sward, in which the timing and frequency of removal of the above-ground standing crop was very important, provided the seeds of species to establish were present.

Finally, it can be concluded that the disturbance caused by starting hay-making or grazing

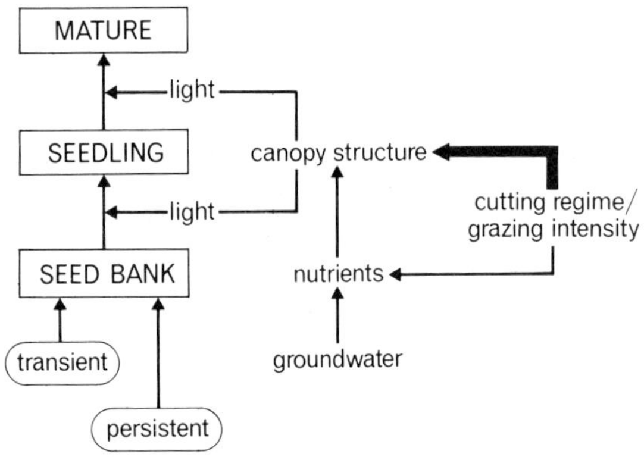

Fig. 14. Diagrammatic representation of the environmental factors influencing the establisment of new species. The thickness of the arrow indicates the importance of the effect.

without fertilizer application can be measured by a slow overall increase of species-diversity and a quicker increase of species indicating nutrient-poor soil conditions. These responses are mainly due to changes in the structure of the canopy as a result of timing and frequency of above-ground standing crop removal and hardly to impoverishment of the soil by nutrient removal which seems to be a long-term process (Fig. 14). The same conclusion was reached by Oomes & Mooi (1985). The effect of a disturbance of this kind takes at least 15 years as reflected by the vegetational succession that is still going on and indicates that an equilibrium had not as yet been established with the new management practices.

References

Bakker, J.P. (1983). Seed bank and species-diversity in moist grassland communities. Acta Bot. Neerl. 32: 246.

Bakker, J.P., Dekker, M. & Vries, Y. de (1980). The effect of different management practices on a grassland community and the resulting fate of seedlings. Acta Bot. Neerl. 29: 469–482.

Bakker, J.P., Bie, S. de, Dallinga, J.H., Tjaden, P. & Vries, Y. de (1983). Sheep-grazing as a management tool for heathland conservation and regeneration in the Netherlands. J. Appl. Ecol. 20: 541–560.

Bakker, J.P., Leeuw, J. de & Wieren, S.E. van (1984). Micro-patterns in grassland vegetation created and sustained by sheep-grazing. Vegetatio 55: 153–161.

Bakker, J.P. & Vries, Y. de (1985A). Ueber die Wiederherstellung artenreicher Wiesengesellschaften unter verschiedenen Mahdsystemen in den Niederlanden. Natur und Landschaft 60: 292–296.

Bakker, J.P. & Vries, Y. de (1985B). The results of different cutting regimes in grassland taken out of the agricultural system. In: Schreiber, K.F. (Ed.). Sukzession auf Grünlandbrachen, pp 51–57. Münstersche Geographische Arbeiten 20. Schöningh, Paderborn. 230 pp.

Blom, C.W.P.M. (1977). Effects of trampling and soil sompaction on the occurrence of some *Plantago* species in coastal sand dunes. II. Trampling and seedling establishment. Oecol. Plant. 12: 363–381.

Böcker, R., Kowarik, I. & Bornkamm, R. (1983). Untersuchungen zur Anwendung der Zeigerwerte nach Ellenberg. Verh. Gesellsch. f. Ökologie XI: 35–56.

Bülow-Olsen, A. (1980). Nutrient cycling in grassland dominated by *Deschampsia flexuosa* and grazed by nursing cows. Agro-Ecosystems 6: 209–220.

Dam, D. van (1983). Invloed van luchtverontreiniging op de Nederlandse flora en op verzuring van de bodem. Report, Research Institute for Nature Management, Leersum, (mimeo).

Dirven, J.G. & Wind, K. (1982). Intensivering van de grasland exploitatie en de plantkundige samenstelling van de grasmat. Meded. 65 Vakgroep Landbouwplantenteelt en Graslandkunde, L.U. Wageningen (mimeo).

Durwen, K.J. (1982). Zur Nutzung von Zeigerwerten und artspezifische Merkmalen der Gefässpflanzen Mitteleuropas für Zwecke der Landschaftsökologie und -Planung mit Hilfe der EDV. Arb.ber. Lehrst. Landschaftsökol. Münster 5: 1–138.

Duuren, L. van, Bakker, J.P. & Fresco, L.F.M. (1981). From intensively agricultural practices to hay-making without fertilization. Vegetatio 47: 241–258.

Elberse, W.Th., Bergh, J.P. van den & Dirven, J.G.P. (1983). Effects of use and mineral supply on the botanical composition and yield of old grassland on heavy clay soil. Neth. J. Agric. Science 31: 62–88.

Ellenberg, H. (1979). Indicator values of vascular plants in Central Europe. Scripta Geobot. 9: Goltze, Göttingen.

Everts, F.H., Grootjans, A.P. & Vries, N.P.J. de (1984). Vegetatiekartering van de Drentsche Aa. Internal Report Department of Plant Ecology University Groningen, State Forestry Commission, Assen (mimeo).

Floate, M.J.S. (1970). Mineralization of nitrogen and phosphorus from organic materials of plant and animal origin and its significance in the nutrient cycle in grazed uplands and hill soils. J. Br. Grassl. Soc. 25: 295–302.

Fenner, M. (1978). A comparison of the abilities of colonizers and close-turf species to establish from seed in artificial swards. J. Ecol. 66: 953–963.

Goldberg, D.E. & Werner, P.A. (1983). The effects of size opening in vegetation litter and litter cover on seedling establishment of golden rods (*Solidago* spp.). Oecologia (Berl.) 60: 149–155.

Grime, J.P. (1979). Plant strategies and vegetation processes. Wiley, Chichester.

Grootjans, A.P., Schipper, P.C. & Windt, H.J. van der (1985). Influence of drainage on N-mineralization and vegetation response in wet meadows I. *Calthion palustris* stands. Acta Oecol./Oecol. Plant. 6: 403–417.

Gross, K.L. & Werner, P.A. (1982). Colonizing abilities of 'biennial' plant species in relation to ground cover: implications for the distribution in a succession field. Ecology 63: 921–931.

Grubb, P.J. (1977). The maintenance of species-richness in plant communities: the importance of the regeneration niche. Biol. Rev. 52: 107–145.

Harper, J. (1977). Population biology of plants. Academic Press, London.

Harrison, A.F. (1978). Phosphorus cycles of forest and upland grassland ecosystem and some effects of land management practices. In: Potter, R. & Fitzsimmon, D.W. (Eds.). Phosphorus and the environment: its chemistry and biochemistry, pp 175–199. Elsevier, Amsterdam.

Hodgson, J. (1979). Nomenclature and definitions in grazing studies. Grass and Forage Science 14: 11–18.

Hullu, P.C. de (1985). The influences of sward density on the population dynamics of *Rhinanthus angustifolius* in a grassland succession. Acta Bot. Neerl. 34: 23–32.

Huston, M. (1979). A general hypothesis of species diversity. American Naturalist 113: 81–101.

Klapp, E. (1965). Grünlandvegetation und Standort. Parey, Berlin. 284 pp.

Kruijne, A.A., Vries, D.M. de & Mooi, H. (1967). Bijdrage tot de oecologie van de Nederlandse graslandplanten. Versl. van Landb. Onderz. 696. Pudoc, Wageningen.

Leeuw, J. de & Bakker, J.P. (1986). Sheep-grazing with different foraging efficiencies in a Dutch mixed grassland. J. Appl. Ecol. 23: 781–793.

Loopstra, I.L. & Maarel, E. van der (1984). Toetsing van de ecologische soortengroepen in de Nederlandse flora aan het systeem van indicatiewaarden volgens Ellenberg. Rapport 381. Rijksinstituut voor onderzoek in de bos- en landschapsbouw 'De Dorschkamp', Wageningen (mimeo).

Melman, Th.C.P., Clausman, P.H.M.A. & Haas, H.A. Udo de (1985). Voedselrijkdom indicatie van graslanden. Mededeling 19. Centrum voor Milieukunde Leiden, Leiden (mimeo).

Mølgaard, P. (1977). Competitive effects on establishment and performance of *Taraxacum officinale*. Oikos 29: 370–382.

Naveh, Z. & Whittaker, R.H. (1980). Structural and floristic diversity of shrublands and woodlands in northern Israel and other Mediterranean areas. Vegetatio 41: 171–190.

Niemann, E. (1973). Grundwasser und Vegetationsgefüge. Nova Acta Leopold 6: 1–147.

Oomes, M.J. (1977). Cutting regime experiments on extensively used grasslands. Acta Bot. Neerl. 26: 265–266.

Oomes, M.J. & Mooi, H. (1981). The effect of cutting and fertilizing on the botanical composition and production of an *Arrhenatherion elatioris* community. Vegetatio 47: 233–239.

Oomes, M.J.M. & Mooi, H. (1985). The effect of management on succession and production of formerly agricultural grassland after stopping fertilization. In: Schreiber, K.F. (Ed.). Sukzession auf Grünlandbrachen. Münstersche Geographische Arbeiten 20, pp 59–67. Schöningh, Paderborn.

Peet, R.K., Glenn-Lewin, R.C. & Walker Wolf, J. (1983). Prediction of man's impact on plant species diversity. In: Holzner, W., Werger, M.J.A. & Ikusima, I. (Eds.). Man's impact on vegetation, pp 41–54. Junk, The Hague.

Perkins, D.F. (1978). The distribution and transfer of energy and nutrients in the *Agrostis-Festuca* system. In: Heal, O.W. & Perkins, D.F. (Eds.). Production ecology of British moors and montane grasslands, pp 375–395.

Persson, S. (1981). Ecological indicator values as an aid in the interpretation of ordination diagramms. J. Ecol. 69: 71–84.

Proefstation voor de Akker- en Weidebouw (1966). Een graslandvegetatiekartering van het ruilverkavelingsgebied 'Anlo'. Report 211. Wageningen (mimeo).

Schenkeveld, A.J.M. & Verkaar, H.J.P.A. (1984). On the ecology of short-lived forbs in chalk grasslands. Ph.D. Thesis, Utrecht.

Schiefer, J. (1983). Ergebnisse der landschaftspflegeversuch in Baden-Württenberg: Wirkungen des Mulchens auf Pflanzenbestand und Streuersetzung. Natur und Landschaft 58: 295–300.

Schiefer, J. (1984). Möglichkeiten der Aushagerung von nährstoffreichen Grünlandflächen. Veröff. Naturschutz Landschaftspfl. Bad.-Wurtt. 57/58: 33–62.

Schreiber, K.F. & Schiefer, J. (1985). Vegetations- und Stoffdynamik in Grünlandbrachen, 10 Jahre Bracheversuche in Baden-Württenberg. In: Schreiber, K.F. (Ed.). Sukzession auf Grünlandbrachen, pp 111–153. Münstersche Geographische Arbeiten 20. Schöningh, Paderborn.

Schmidt, W. (1985). Mahd ohne Düngung – Vegetationskundliche und ökologische Ergebnisse aus Dauerflächenuntersuchungen zur Pflege von Brachflächen. In: Schreiber, K.F. (Ed.). Sukzession auf Grünlandbrachen, pp 81–99. Münstersche Geographische Arbeiten 20. Schöningh, Paderborn.

Silvertown, J.W. (1980). Leaf canopy-induced seed dormancy in a grassland flora. New Phytologist 85: 109–118.

Sterk, A.A., Groenhart, M.C. & Mooren, J.F.A. (1983). Aspects of the ecology of some microspecies of *Taraxacum* in the Netherlands. Acta Bot. Neerl. 32: 385–415.

Vermeer, J.G. & Berendse, F. (1983). The relationship between nutrient availability, shoot biomass and species richness in grassland and wetland communities. Vegetatio 53: 121–126.

Wells, T.C.E. (1980). Management options for lowland grassland. In: Rorison, I.H. & Hunt, R. (Eds.). Amenity grassland, pp 175–195. Wiley, Chichester.

Werger, M.A., Smeets, P.J.A.M., Helsper, H.P.G. & Westhoff, V. (1978). Ökologie der subalpinen Vegetation des Lauschbachtales. Tirol. Verh. Zool.-Bot. Ges. Wien 116/117: 111–125.

Whittaker, R.H. (1954). Plant populations and the basis of

plant indication. Angewandte Pflanzensociologie 1: 183–206.

Willems, J.H. (1980). Observations on north-west European limestone grassland communities. An experimental approach to the study of species diversity and above-ground biomass in chalk grassland. Proceedings Kon. Ned. Akad. Wetensch. Series C. 83: 279–306.

Willems, J.H. (1983). Species composition and above ground phytomass in chalk grassland with different management. Vegetatio 52: 171–180.

Willems, J.H. (1985). Growth form spectra and species diversity in permanent grassland plots with different management. In: K.F. Schreiber (Ed.), Sukzession auf Grünlandbrachen, pp 35–43 Münster Geogr. Arb. 20.

Williams, E.D. (1978). The Botanical Composition of the Park Grass Plots at Rothamsted 1856–1976. Rothamsted Experimental Station, Harpenden.

Wind, K. (1980). Botanische samenstelling van grasland bij extensivering van het gebruik. Med. 52 Vakgroep Landbouwplantenteelt en Graslandkunde L.U. Wageningen (mimeo).

CHAPTER 13

The function of plant pathogenic fungi in natural communities

J.C. Zadoks

Detailed contents

1. Introduction

There seems to be some discrepancy in opinion between vegetation scientists (e.g. Harper, 1977), talking about the effects of plant disease on vegetation, and plant pathologists (e.g. Burdon & Shattock, 1980; Dinoor & Eshed, 1984; Thresh, 1981), talking about the effects of vegetation on diseases. The very notion of disease seems to differ between the two disciplines. One vegetation scientist, his identity does not matter here, summarized his view plainly: 'A plant becomes diseased when it is weakened by other causes than pathogens'. This does indeed happen (e.g. Bachi & Peterson, 1985). For example, the fungus *Sphaeropsis sapinea* (Fr.) Dyko & Sutton has become a common disease in Dutch pine plantations since 1982, especially when trees are weakened (Boerema et al., 1985). Nevertheless, professional plant pathologists have a rather more complex view to offer.

Disease is a condition of an organism, in which the organism does not function normally. The definition implies, that the borderline between diseased and healthy is indistinct and subjective. This paper will be limited to diseases caused by fungal pathogens. The plants affected by a pathogen are called host plants. Fungal diseases of wild plants are well documented (e.g. Brandenburger, 1985).

Any visible deviation of a host plant is called an injury. Injury may not necessarily cause damage, which is a deviation from the normal production of the plant. Damage in turn may not necessarily have an economic impact, called loss (Zadoks, 1985). For natural vegetation, the notion of loss, but not that of damage, may be ignored. In crop plants, damage can be recognized as a reduction in yield and/or quality of the desired product. It is more difficult to define damage in natural vegetation, but two types of damage, both referring to reproduction, are relevant. Primary damage is a decrease of reproductive capacity in the year of injury, while secondary damage is a decrease of reproductive capacity in the following years (Zadoks & Schein, 1979).

One more concept from phytopathology must be introduced here, the concept pathosystem (Robinson, 1976). A population of a pathogenic fungus species interacts with a population of a host plant species in such a way that they exert mutual selection pressures on each other. As a re-

J. van Andel et al., (eds.), Disturbance in Grasslands. ISBN 90-6193-640-3.

sult, the two species may develop a subtle genetic interplay in time and space, which is influenced by the environment. The rules of the game are such that both species survive under a given set of ecological conditions, but that the equilibrium between the two species can be upset if the conditions are changed. Either species may be eliminated, but usually the pathogen wins, at least temporarily. Agriculture is troubled by pathogens because man has changed the ecological conditions prevailing in nature and so has disturbed the previously existant equilibrium (a.o. Zadoks & Schein, 1979).

In biblical times, pests and diseases of crop plants were considered to be punishments of God. This notion continued well into the twentieth century. One plant pathologist called pathogens 'organisms out of place' (Yarwood, 1967). Plant pathogens were associated with evil. They were to be 'banished' (Stakman & Harrar, 1957, page 541). Some plant pathologists took a different stand, stating that pathogens were part of nature. Consequently, plant pathogens should be managed where necessary, not eradicated (Zadoks & Schein, 1979). Since cultivated plants have evolved from wild plants, their pathogens have evolved from natural pathogens. New pathogens, that is pathogens appearing de novo, are extremely rare (Zadoks, 1967).

2. Ecology of plant inhabiting microorganisms

Many microorganisms inhabit plants, but relatively few are pathogens. In this review, all epiphytic microorganisms are ignored, even though some may occasionally do damage. The plant-invading microorganisms, which as a rule are beneficial, such as nitrogen fixing bacteria and mycorrhizal fungi, are briefly considered. The fungi, that invade weakened and dead plants and primarily function as decomposers, are not discussed. The review focuses on those microorganisms that are real consumers.

Although the borderlines between ecologically different groups of microorganisms, including various types of pathogens, are not clear-cut, it is often important to distinguish between these groups. There are differences indeed between root-invading pathogens and shoot-invading pathogens. It is relevant to consider whether the pathogenic fungus first kills the host cells and then consumes them, the perthotrophic fungus, or first consumes them and in doing so kills the host cells, the biotrophic fungus. The biotrophs can only survive on living hosts and florish best when the host is growing in optimal condition. The perthotrophs, which often have a saprobic phase, may profit when the host is in a relatively poor condition.

Phytopathology cannot offer the vegetation scientists one simple and straightforward view on plant disease. Fortunately, phytopathologists have recently become more and more interested in the function of pathogens in natural vegetation, for two reasons. One is the feeling that some of the mechanisms to establish equilibrium between host and pathogen, which operate in nature, have been lost in agriculture (Van Leur, 1981). The other is the need to replace chemical control of weeds by biological control using pathogens when and where that is feasible (e.g. Scheepens & Van Zon, 1982).

3. Pathogens in plant communities

Pathogens can affect vegetation in different ways. Some pathogens are specialized on seedlings. They are typical opportunists, following the r-strategy, attacking rapidly when host and circumstances are favourable. They require moist spring conditions at soil level, under a dense canopy. Their role in natural vegetation is not yet well established, but there is evidence that selection for resistance in the host rarely occurs (De Nooij & Merkx, 1984). Probably, the regulatory function of these opportunists is not too important to plant communities as most of the seedlings would have died anyhow. The situation changes when many plants, even genetically different plants, of one species are placed together for agricultural purposes. For example, dense experimental stands of *Aster tripolium* L., grown as a delicacy for salad making, were rapidly eliminated by damping-off fungi (Knip, 1984).

Intraspecific effects of pathogens on population density and genetic diversity should be separated from interspecific effects of pathogens on species composition of plant communities. The statement 'Diversity is good' used to be a dogma in ecology. The plant pathologist can only make an inverse statement: 'Lack of diversity is a risk' (Zadoks & Schein, 1979). This lack of diversity is likely to occur when the coevolution of host and pathogen has been interrupted by temporal, spatial, or genetic isolation (Buddenhagen, 1977). For example, the pine forests of the Rocky Mountains, the birch taiga of the USSR, the *Limonium vulgare* fields of the Dutch island of Terschelling, the *Calluna vulgaris* heaths of north-west Europe have persisted for long periods. They represent healthy populations of a single dominant species, in spite of the presence of diseased individuals. In an agricultural context, the distinction between interspecific and intraspecific diversity is relevant too. In agriculture, interspecific diversity is often useful, and sometimes dangerous. Intraspecific diversity is usually beneficial, at least with respect to diseases affecting the shoot (Wolfe, 1985; Zadoks & Schein, 1979).

The role of pathogenic fungi in natural vegetation can be considered at two levels, the population and the vegetation. Each level can be considered at two time scales, one with a time horizon of a few years only, and the other with a horizon between a decade and thousands of years (Table 1).

Table 1. Possible effects of plant pathogenic fungi in natural vegetations according to complexity level and time horizion.

Time horizon	Complexity level	
	Population	Vegetation
Short	Population control	Patchiness, microsuccession
Long	Evolutionary segregation	Acceleration of succession

3.1 Population control

It is still uncertain whether pathogens are able to control population density of wild plant species under near-equilibrium conditions. The issue is a central question in the agricultural domain, where biological control of weeds by means of parasitic fungi is studied. Pathogens rarely kill wild plants immediately (e.g. Kranz, 1968). One reason may be the lack of inoculum hitting the target plant (Vanderplank, 1982). The other reason is the presence of various defense mechanisms in the infected plant, at least in the older plant. Seedlings are relatively vulnerable. Damping-off fungi have already been mentioned. These are fungi from different taxonomic groups which cause a rotting of seedlings at soil level. For example, Ernst (1983) stated that 20 per cent of the seedlings of *Anemone nemorosa* were killed by the fungus *Stachybotrys chartarum* (Ehrenb. ex Link) Hughes 1 before emergence.

The foregoing referred to elimination of individuals within the growing season. Greater damage can occur over successive seasons, when reproduction or reproductive capacity is diminished. Ernst (1983) provides examples of damage to the perennial *A. nemorosa* by different pathogens. The ascomycete *Sclerotinia tuberosa* (Fries) Fuckel produces a sclerotium, a firm mass of hyphae, that contains some 30 per cent of the host's rhizome biomass. Flower formation is suppressed and within a few years the host plant dies. The rust fungus *Tranzschelia anemones* (Pers.) Nannf. inhabits the leaf lamina. The infected plants fail to fruit and the annual weight increase of the rhizome is reduced. The rust fungus *Ochropsora ariae* (Fuch.) Ramsb. halves fruiting and severely reduces the annual weight increase of the rhizome. All three pathogens reduce host multiplication and endanger host survival.

Jennersten et al. (1983) demonstrate a regulatory effect of the smut fungus *Ustilago violacea* (Pers.) Roussel on the perennial *Viscaria vulgaris* (Caryophyllaceae). The fungus effectively sterilizes the plant. The host plants grow in isolated patches, and the larger the patches and the shorter the distances between them, the higher the proportion of sterilized plants. It seems as if the fungus slows down the population growth of a patch, once the patch reaches a certain size.

Notwithstanding these selected examples, the effect of pathogenic fungi on population density or size is often limited. Downy mildew (*Peronospora farinosa* (Fries, 1832) Fr.) can seriously attack the weed *Chenopodium album* causing premature death of the infected leaves, emergency ripening of the host plant, and reduced seed set. However, the next generation of *C. album* is usually as dense as the foregoing (Frinking and Linders, 1986).

3.2 Patchiness and microsuccession

The rust fungus *Gymnosporangium juniperinum* (L.) Fr. is heteroecious. The asexual phase of the rust multiplies on *Sorbus aucuparia*, whereas the sexual phase inhabits *Juniperus communis*. The two hosts, which can grow side by side, are linked together with and by the fungus in a tripartite pathosystem. *Sorbus* can grow taller than *Juniperus* and shade it to the degree that *Juniperus* is killed. However, if the two species come too near to each other the rust may interfere. The rust disperses annually from *Juniperus* as a source to the nearby *Sorbus* as a target, and vice versa. *Sorbus* may be attacked to the extent that leaves, infected by the rust, are dropped prematurely (J.J. Barkman, personal communication). The juniper is not much damaged by the rust. Without rust the *Sorbus* may kill the *Juniperus* by shading. With rust the *Juniperus* may be instrumental in the death of the *Sorbus,* since repeated premature defoliation weakens trees and renders them susceptible to adverse environmental effects, such as drought and frost, and to fungi attacking weakened plants (e.g. Ostry & McNabb, 1985; Zadoks & Schein, 1979). So, the heteroecious rust can influence the distribution of its two hosts.

Natural vegetation is often patchy. For example, a poor meadow may have patches of *Trifolium pratense*, *Holcus lanatus*, and other species. The patches are not necessarily stable. W.H. van Dobben (personal communication) suggested a possible role of pathogens. Red clover is susceptible to various species of nematodes and fungi and highly susceptible to several virus diseases. Nematode-induced death comes slowly, but virus kills rapidly. The decline of the clover would allow other species to attain dominance. Similarly, the centre of patches of *Limonium vulgare* is sometimes overgrown by *Festuca rubra.* A pathogen may be involved but the evidence is not conclusive. The sequence may prelude the transition from Limonietum to Festucetum (J.G.A. van Leur, unpublished).

Plant pathogens may play a part in these processes of microsuccession, where species succeed each other without changing the aspect of the vegetation as a whole. One can only hypothesize because phytopathological research in microsuccession has seldom if ever been published.

3.3 Acceleration of succession

The role of plant pathogens in succession is sometimes very apparent. For example, the willows *Salix pulchra* and *S. alaxensis* are pioneer species forming nearly pure stands on gravel banks of the river Yukon in Alaska, once the ice has receeded. The willow rust *Melampsora bigelowii* Thuem appears soon and kills manu seedlings. On higher grounds, willows persist as pure stands. Some fifty years after the establishment of willow, birch and spruce appear, and after about 175 years, the climax forest consists of birch and spruce, with a few relict willows. During the succession the amount of rust declines but various canker and heart rot fungi become abundant and kill the willows (Baxter & Wadsworth, 1939). The Alaskan season is short, and the process is so slow that the population dynamics of willow and rust can be seen, as if in slow motion, over a period of nearly two centuries. The pathogen apparently accelerated the decline in dominance of willow permitting other species to increase, so accelerating the succession. Note that the foregoing is an interpretative description only; there has been no rust-free control experiment.

The possible acceleration of succession by pathogens was studied in a vegetation of *Limonium vulgare* on the Dutch island of Terschelling. The rust *Uromyces limonii* (D.C.) Lev. may kill an incidental plant but it does not cause an acceleration of succession. Powdery mildew (*Erysiphe communis* Wallr.) Link.) is locally so severe, that succession seems to be accelerated indeed. As Limonium is a slow growing perennial, the process goes in slow motion again.

Pathogenic catastrophies are rare in nature. Coevolution of host and pathogen equilibrate the pathosystem. Only when host and pathogen meet again after long separation, does a 'new encounter disease' (Buddenhagen, 1977) occur and an explosive epidemic may start. Several examples of such explosive epidemics are known. In North America the fungus *Endothia virginiana* P.J. and H.W. Anderson caused an unimportant disease on American chestnuts (*Castanea dentata* (Marsh.) Borkh.), while in China, the fungus *Endothia parasitica* (Murr.) P.J. and H.W. Anderson caused an insignificant disease on East Asian chestnut (*C. mollissima* Bl., *C. crenata* Sieb. et Zucc.). Around 1900, *E. parasitica* was inadvertently introduced in North America. Since *C. dentata* had no resistance against this fungus the new encounter caused a tragedy. By 1930, the chestnut forests of the Apalachian Mountains, over a million hectares, did no longer exist. Only secondary oak forest survived (Zadoks & Schein, 1979).

Another devastating epidemic occurred on *Zostera marina*. Around 1935 some 15000 hectares of evergreen submarine pastures were cleared away in the Dutch Shallows. The epidemic began in Portugal and reached the United Kingdom and Scandinavia within a few years. The pathogen, supposedly *Labyrinthula macrocystis* Cienk., already occurred on *Zostera* at the east coast of the USA (Zadoks, 1967). The pathogen may have crossed the Atlantic Ocean aboard ship. (Note that some authors disagree with the disease theory, e.g. Rasmussen, 1973).

Probably both these pathogenic catastrophies were anthropogenic, i.e. the new encounter was man-made. New encounter diseases in crops can have enormous economic consequences. A modest but recent example of a new encounter is an outbreak of downy mildew on an ornamental crop called Texas-bluebell, *Lisianthus russellianus* (Gentianaceae), originating from the USA and grown in the greenhouse district south of The Hague, in the Netherlands. The causal fungus, *Peronospora chlorae* de Bary, was traced to *Centaurium pulchellum* (Gentianaceae), which grows naturally in the dunes separating the greenhouse district from the North Sea (Boerema et al., 1984).

3.4 Evolutionary segregation

Longer term evolutionary processes can be reconstructed from shorter term processes (Harlan, 1976). Grasses of the genus *Hordeum* are parasitized by a rust of the genus *Uromyces*. The rust is heteroecious; its dikaryotic phase lives on grasses and its haploid phase lives on species of the Liliaceae. Both the haploid and the diploid phases are able to specialize on one or more host species. A highly complex pattern of mutual infections between barley species and liliaceous species has evolved. The complexity is greatest in the gene centre of *Hordeum*, the Near East, where it has been studied in detail (Anikster & Wahl, 1979). The environmental effect on speciation has been considered too. *Leopoldia eburnea*, a bulbous plant of the Negev desert, has received special attention. It flowers only if seasonal rainfall is sufficient, and in very dry seasons the plants do not emerge at all. In the Negev, *Leopoldia* is never rusted, but in experiments *Leopoldia* was found to be susceptible to all *Uromyces* species from *Hordeum* spp. (Anikster et al., 1980; Anikster & Wahl, 1984). Consequently, *Leopoldia* is supposed to be the primordial host of the haploid *Uromyces*. It seems that *L. eburnea* has been eliminated from *Hordeum's* gene centre and now occurs only in a peripheral area, where ecological conditions are such that *Hordeum* can barely survive and the rust not at all. It appears that, in its struggle for life, *L. eburnea* has lost all its resistance genes arriving at a state in which it is very susceptible to and, at the same time, free from infection. The pathogen in the course of its evolution has apparently forced its host into an ecologically marginal environment, where the pathogen itself cannot survive. This process I call 'evolutionary segregation', the converse of coevolution. It is an audaceous thought, for which no direct proof can yet be given, but which may stimulate further study.

4. Concluding remarks

From an agricultural point of view, the most interesting aspects of pathogens in natural vegetation are found in the area of pathosystem genetics. These can only be indicated superficially. Extensive studies have been made of wild grass stands in Israel (Browning, 1974). The effect of pathogenic fungi on the genetic composition of plant populations, i.e. on within-species diversity, is manifest, though it can be moderated by environmental conditions.

The effect of pathogenic fungi on species composition of a vegetation, i.e. on between-species diversity, is difficult to prove. Acceleration of succession seems to be the most evident effect, but its frequency is unknown. The causal relation between pathogens and patchiness can be studied experimentally if a period of five to ten years is available. The possibility of pathogen-induced 'evolutionary segregation' is proposed, but the collection of data and the performance of supporting experiments requires decades rather than years.

New encounters have been organized to control weeds, rather successfully in Australia, where the imported composite *Chondrilla juncea* is effectively controlled by the imported rust *Puccinia chondrillina* Bubak & Syd., but without much success for the same pathosystem in the Pacific northwest of the USA (Adams & Line, 1984). The difference is probably in the environmental conditions. For a new encounter a single introduction should be sufficient. Where host and pathogen have coevolved, another approach is needed. The 'inundation technique' can be successful, as with the leaf flecking fungus *Colletotrichum gloeosporioides* (Penz.) Sacc. used against *Aeschynomene virginica*, a weed of rice in the USA (Scheepens and Van Zon, 1982). The fungus is massively applied as a microbial herbicide (Templeton et. al., 1984) under the trade name Collego.

Studies of the biological control of weeds have led to a deepening of our insight into pathosystems of natural biocoenoses. The present state of knowledge allows us to reach only preliminary conclusions. Pathogenic fungi are an essential part of every biocoenosis, but their impact on vegetation is of incidental importance only. Such a conclusion is not amazing in view of the often mentioned coevolution of host and pathogen. Coevolution has proceeded to the degree that the pathogen produces altruistic genes, genes which limit the diseasing capacity of the pathogen with respect to the number of infectible host individuals. Genes for virulence can be hypostatically hidden, ready for use if the need arises. The pathogen pays a price to survive for ever.

Acknowledgements

Thanks are due to Professor J.J. Barkman for suggestions and constructive criticism, to Dr. W.G. Beeftink for comments, and to Dr. J. Anikster for information on *Leopoldia*.

References

Adams, E.B. & Line, R.F. (1984). Epidemiology and host morphology in the parasitism of rush skeleton weed by *Puccinia chondrillina*. Phytopathology 74: 745–748.

Anikster, Y., Moseman, J.G. & Wahl, I. (1980). Development of basidia and basidiospores in *Uromyces* species on wild barley and Liliaceae in Israel. Trans. Br. mycol. Soc. 75: 377–382.

Anikster, Y. & Wahl, I. (1979). Coevolution of the rust fungi on Gramineae and Liliaceae and their hosts. Annu. Rev. Phytopathology 17: 367–403.

Anikster, Y. & Wahl, I. (1984). *Puccinia hordei* and *Uromyces* species on *Hordeum* and their relationship. Rept. Tottori Mycol. Inst. (Japan) 22: 97–101.

Bachi, P.R. & Peterson, J.L. (1985) Enhancement of *Sphaeropsis sapinea* stem invasion of pines by water deficits. Plant Disease 69: 798–799.

Baxter, D.V. & Wadsworth, F.H. (1939). Forest and fungus succession in the lower Yukon Valley. Bull. Univ. Michigan School of Forestry and Conservation, Ann Arbor, No. 9, 52 pp.

Boerema, G.H., Loerakker, W.M. & Kesteren, H.A. van (1984). Valse meeldauw bij *Lisianthus russellianus*. Versl. Meded. Plziektenk. Dienst 162 (Jaarboek 1983): 16–18.

Boerema, G.H., Kesteren, H.A. van & Loerakker, W.M. (1985). Vermeldenswaardige schimmelaantastingen in de periode 1980–1984. Gewasbescherming 16: 103–119.

Brandenburger, W. (1985). Parasitische Pilze an Gefaess-pflanzen in Europa. Stuttgart: Fischer. 1248 pp.

Browning, J.A. (1974). Relevance of knowledge of about natural ecosystems to development of pest management programs for agro-ecosystems. Proc. Amer. Phytopathol. Soc. 1: 191–199.

Buddenhagen, I.W. (1977). Resistance and vulnerability of tropical crops in relation to their evolution and breeding. Ann. New York Acad. Sciences. 287: 309–326.

Burdon, J.J. & Shattock, R.G. (1980). Disease in plant communities. Applied Biology 5: 145–219.

Dinoor, A. & Eshed, N. (1984). The role and importance of pathogens in natural plant communities. Annu. Rev. Phytopathology 22: 443–466.

Ernst, W.H.O. (1983). Population biology and mineral nutrition of *Anemona nemorosa* with emphasis on its parasitic fungi. Flora (DDR) 173: 335–348.

Frinking, H.D. & Linders, E.G.A. (1986). A comparison of two pathosystems: downy mildew on *Spinacia oleracea* and on *Chenopodium album*. Netherlands J. Plant Pathology 92: 97–106.

Harlan, J.R. (1976). Diseases as a factor in plant evolution. Annu. Rev. Phytopathology 14: 31–51.

Harper, J.C. (1977). Population-biology of plants. Academic Press, London. 892 pp.

Jennersten, O., Nilsson, S.G. & Wastljung, U. (1983). Local plant populations as ecological islands: the infection of *Viscaria vulgaris* by the fungus *Ustilago violacea*. Oikos (Acta Oecol. Scand.) 41: 391–395.

Knip, K. (1984). Lamsoor en breeksla uit eigen tuin. NRC-Handelsblad 14 juli 1984.

Kranz, J. (1968). Eine Analyse von annuellen Epidimieen pilzlicher Parasiten. II. Qualitative und quantitative Merkmale von Befallskurven. Phytopathol. Z. 61: 171–190.

Leur, J.G.A. van (1981). Onderzoek naar het evenwicht tussen waard en pathogenen in een natuurlijke situatie. Vakblad Biologen 61: 466–471.

Nooij, M.P. de & Merkx, T. (1984). The influence of parasitic soil micro-organisms on seedling establishment in *Plantago lanceolata*. Progress Rep. Inst. Ecol. Res. 1983: 53–55.

Ostry, M.E. & McNabb Jr, H.S. (1985). Susceptibility of *Populus* species and hybrids to disease in the North Central United States. Plant Disease 69: 755–757.

Rasmussen, E. (1973). Systematics and ecology of the Isefjord marine fauna (Denmark). With a survey of the Eelgrass (*Zostera*) vegetation and its communities. Ophelia 11: 1–507.

Robinson, R.A. (1976). Pathosystems. Berlin: Springer. 184 pp.

Scheepens, P.C. & Zon, H.C.J. van (1982). Microbial herbicides. In: Kurstak, E. (Ed.), Microbial and viral pesticides, pp 623–641. Marcel Dekker, New York. 720 pp.

Stakman, E.C. & Harrar, J.G. (1957). Principles of plant pathology. Ronald Press, Cy. New York. 581 pp.

Templeton, G.E., TeBeest, D.O. & Smith jr., R.J. (1984). Biological weed control in rice with a strain of *Colletotrichum gloeosporioides* (Penz.) Sacc. used as a mycoherbicide. Crop Protection 3: 411–424.

Thresh, J.M. (Ed.) 1981). Pests, pathogens and vegetation. The role of weeds and wild plants in the ecology of crop pests and diseases. Pitman, London. 517 pp.

Vanderplank, J.E. (1982). Host-pathogen interactions in plant disease. Academic Press, New York. 207 pp.

Wolfe, M.S. (1985). The current status and prospects of multiline cultivars and variety mixtures for disease resistance. Annu. Rev. Phytopathology 23: 251–273.

Yarwood, C.E. (1967). Pathogens as organisms out of place. Phytopathol. Z. 58: 305–314.

Zadoks, J.C. (1967). International dispersal of fungi. Neth. J. Pl. Path. 73: Suppl. 1: 61–80.

Zadoks, J.C. (1985). On the conceptual basis of crop loss assessment: the threshold theory. Annu. Rev. Phytopathology. 23: 455–473.

Zadoks, J.C. & Schein, R.D. (1979). Epidemiology and plant disease management. Oxford University Press, New York. 427 pp.

PART 4

Population and species responses

Snaydon (Chapter 2) has pointed out that the response of plant populations to environmental variation, whether inherent or caused by external factors, depends on the pattern and magnitude of environmental variation and on biological attributes of the plant species. The most common responses to rapid environmental changes in time (much smaller than the lifespan of the plant species) and rapid changes in space (smaller than plant size), are phenotypic plasticity and physiological acclimation. Temporal changes occurring with a frequency similar to the duration of lifespan, and spatial variation about plant size are usually associated with polymorphism within populations. Temporal changes with a frequency substantially longer than the lifespan, and spatial changes over distances substantially greater than the size of an individual plant, give rise to population differentiation. In general, responses at the higher levels of organisation (i.e. population differentiation and speciation) have received the greatest attention and have been studied for a longer time, however, including the gaps in our knowledge to be mentioned. More recently, polymorphism, phenotypic plasticity and physiological acclimation have received some attention, but information on the latter response is still very inadequate, indicating a gap in our knowledge.

The various disturbances discussed in Part 4 are probably best categorized by the various mechanisms of response. The mechanisms of response to disturbance can be broadly classified into: population differentiation (Ernst, Chapter 14), physiological response (Lambers & Dijkstra, Chapter 16), mechanisms underlying speciation (Urbanska, Chapter 20).

Ernst (Chapter 14) deals with the selective forces resulting in population differentiation. Population differentiation is essentially change in genotypic composition; any selective force which favours certain genotypes over others will change the structure of the population. Various types of disturbance can change genotypic composition and hence cause population differentiation. The most common changes imposed on grassland species are fertilizers, herbicides and other modern chemicals, though the frequency, duration and intensity of grazing and mowing have also changed. The dose, exposure time, and interval between applications will determine the selection pressure, and thus the extent of population differentiation. Genetic variability within a species is a prerequisite for population differentiation to occur, and is usually present within most grassland species, because of their outbreeding habit. Population differentiation can occur by directional, stabilizing or disruptive selection. Directional selection favours genotypes of one extreme, while stabilizing selection favours the average type at the expense of the extremes. Disruptive selection gives advantages to both extremes, so increasing the variance of the populations. Examples of response to changing soil acidity, water supply, availability of mineral elements, environmental pollution, vegetation structure, plant-micro-organism interactions and plant-herbivore interactions are given.

Lambers & Dijkstra (Chapter 16) consider whether there may be an advantage in low relative growth rate (RGR) for genotypes in less productive habitats. They point out that some genotypes from infertile habitats have a high RGR, and that competitive ability is not necessarily correlated with a high RGR. For example genotypes from infertile or dry habitats tend to invest more biomass in roots giving a low leaf weight ratio (LWR) and hence a low RGR. However, the low RGR may be compensated for by a high net assimilation rate (NAR), so that RGR may still be high. Selective pressure on NAR or LWR may affect RGR of a species; slow growth is thus considered an indirect effect of selective forces.

In addition to the more general chapters, there are several chapters dealing with specific disturbances. Kik (Chapter 15) considers the survival and vegetative/generative reproduction of *Agrostis stolonifera* populations in inland meadow and coastal sand dune habitats. The results show that the meadow population reproduced almost exclusively by vegetative means, so that its response to disturbance is expected to be primarily restricted to phenotypic plasticity. By contrast, the sand dune population has a strong component of sexual reproduction and therefore can respond to a disturbance by both genetic and plastic responses. Biere (Chapter 17) focuses on size variation within plant populations. He shows considerable size inequality in populations of twelve naturally co-occurring meadow species, grown at low and high nutrient levels in the glasshouse, though there was less inequality under optimal nutrient conditions. He also attempts to evaluate the ecological significance of size distributions in response to disturbance by manipulating the initial size distributions of populations under optimal and low nutrient levels. De Hullu (Chapter 18) describes the rise and fall of *Rhinanthus angustifolius* populations under a hay-making regime without fertilizer applications in formerly fertilized meadows. The vegetational response resulting from this sudden change in management practices has already been considered by Bakker (Chapter 12). The response of *Rhinanthus* populations largely reflected the vegetational response and the expansion of *Rhinanthus* populations may often be indicative of the disturbance of vegetation. Kwak (Chapter 19) considers the disturbance in gene flow within *Rhinanthus angustifolius* populations resulting from change in bumblebee pollination. Particularly dense populations of *Rhinanthus angustifolius* attract honeybees which compete with bumblebees for pollen and nectar. Considered from the point of view of *Rhinanthus angustifolius*, pollen-collecting honeybees increase outbreeding by increasing pollen transfer between flowers of different plants. However, foraging honeybees cover a smaller area than bumblebees at high flower density and may therefore increase inbreeding.

Urbanska (Chapter 20) discusses hybridization and speciation resulting from disturbance. Hybrid plants can become established only if they are adaptively superior to their parents in ancestral habitats, or if they find a convenient ecological niche not inhabited by the parent taxa. In a closed and settled community, no such niches are available to hybrids, and selection operates in favour of established parent genotypes. If, however, the ancestral habitats are disturbed, selective pressures change and new niches may emerge; in this case, hybrid establishment may become possible. In addition, geographical and ecological isolation barriers may be broken down by human activities. Man has often introduced, sometimes inadvertently, plants into new areas resulting in hybridization between native and alien taxa, while man-made disturbance frequently results in the formation of intermediate or open habitats, or even entirely new ecological situations that are free to be colonized. These changes commonly result in the occurrence of hybrids and hybrid swarms.

We need also to indicate the various gaps in our knowledge. Firstly, most studies consider response to a single kind of disturbance, yet misfortunes rarely come singly, since various kinds of disturbance commonly coincide or follow each other in a short sequence of events. This situation is well illustrated by *Tragopogon* and *Carduus* in North America where the introduction of the Old World species was accompanied by other kinds of human interference promoting hybrid establishment. Similarly, the spectacular hybrid speciation in *Cardamine* at Urnerboden, Switzerland, is associated with the combined effects of drainage, fencing, cutting and fertilizer applications (Urbanska, Chapter 20). At a lower level of organization, the metabolic consequences of various associated disturbances may result in synergistic, additive or antagonistic effects, which can affect the competitive ability of genotypes (Ernst, Chapter 14). Antagonistic effects tend to neutralize the metabolic costs and so will not lead to population differentiation, while synergism magnifies the effects, and so hastens population differentiation or somehow adaptive response.

CHAPTER 14

Population differentiation in grassland vegetation

W.H.O. Ernst

Detailed contents

1. Introduction

Most temperate grasslands are the result of human impact on, and management of, former forests and woodlands. They came to existence, and remain in their present state, because of human disturbance; few natural grasslands occur in the temperate zone of Europe. I will use the term disturbance here as 'the change in conditions interfering with a normal functioning of a given biological system' (cf. Chapter 1). 'Disturbance', as other biological terms, may have a different meaning at the level of the cell, the individual, the population, the plant community of the ecosystem, but here I shall consider mainly populations. Cyclic events, such as seasonal changes in temperature or precipitation, will not be considered per se as disturbance. In agreement with Hedrick (1984) a population will be defined as 'a group of interbreeding individuals of the same species that coexist together in time and space'.

Population differentiation is based on change in genotypic composition; any selective force which favours certain genotypes over others will change the structure of the population. This implies that population differentiation in most temperate grassland is caused by disturbance. A certain environmental condition may maximize the fitness of certain genotypes, but will minimize the fitness of other genotypes, as has been demonstrated for zinc tolerant genotypes of *Agrostis capillaris* and *A. canina* in a grassland before and after air and soil pollution by a zinc smelter (Ernst, 1972, 1982b). Such an environmental pollution has changed the physiology of a zinc tolerant plant so thoroughly that in this case a high dosis of zinc is necessary for normal functioning of zinc tolerant genotypes. The same zinc dosis is an

J. van Andel et al., (eds.), Disturbance in Grasslands. ISBN 90–6193–640–3.

agent of disturbance for all non-zinc-tolerant genotypes in the population.

In this contribution I will discuss some of the necessary conditions for population differentiation, and give some examples of its occurrences.

2. Some basic principles of population differentiation

Since the studies of Turesson (1922) the interpopulation differentiation within plant species has received more and more attention. A variety of morphological, physiological and biochemical characteristics have been shown to differ between populations. In view of this abundance I will focus on some prerequisites for the basic concepts of population differentiation without reviewing the complete literature.

2.1 Genetic variability at the species level

The first of all prerequisites for population differentiation is a substantial amount of genetic variability within a species. This genetic variability is present in many plant species (Hamrick, 1979 and Ennos, 1983); natural and man made grasslands are no exceptions. The range of variability within species may extend from a conspicious polymorphism in flower colour down to the invisible molecular variation of proteins. However, not all of this variability is involved in population differentiation, it may simply occur within populations.

The most obvious form of polymorphism is that of flower colour within populations. Multicoloured populations of *Crocus albiflorus, Cirsium palustre, Dactylorchis maculata, Polygala serpyllifolia, Viola curtesii*, and other plant species from grassland, are well known for their white, yellow, red, pink, blue and blue-violet flowers. A very precise analysis of such multicoloured population of *Cirsium palustre* has detected sharp local changes of the flower colour frequency within distances of only 200 m (Mogford, 1978), though ecological factors, affecting such variation are not known.

A considerable amount of genetically based physiological variation is maintained at the intrapopulational level; in part, this may prevent intense competition between individuals of a population, since they will not exploit precisely environmental resources. Because of the key role of the mineral nutrients in many metabolic pathways, slightly different requirements for various mineral nutrients may be an excellent target for selection, and thus for population differentiation (Table 1).

At the molecular level, extensive variation of the molecular form of enzymes (allozymes) have been detected in plant species (Hamrick et al., 1979; Verkleij, 1980), though there is a great lack of information on the interpopulation level (Allard et al., 1978; Jain, 1976; Johnson, 1979). There is still much to learn about the biochemical machinery and the ecological relevance of this enzyme polymorphism (Fig. 1), but allozyme variation between populations may prove to be a rewarding technique for studies in population dynamics and population genetics.

2.2 Genetic variability within a population

Many environmental factors, especially climatic factors, in temperate regions fluctuate, so that the physiological processes of various genotypes will be differently affected over different time

Table 1. The amount of mineral nutrients in even-aged individuals of a population of *Senecio sylvaticus*, grown in potting compost in the greenhouse as far as the rosette stage. Each individual has been collected as fruits from another mother plant.

Plant dry	Amount of nutrients (μ mol per plant)							
weight mg	P	K	Ca	Mg	Na	Fe	Mn	Zn
260	14	136	52	28	134	1,9	1,1	0,6
270	25	264	72	29	201	1,6	1,3	0,8
810	85	547	249	111	562	5,7	3,7	1,3
817	27	432	236	98	869	6,8	4,6	2,0
1576	133	963	446	194	1407	6,3	5,6	3,8
1578	134	940	528	214	924	7,8	6,8	4,2
2511	210	1254	758	437	1910	9,8	11,1	7,1
2521	274	1679	678	245	1508	14,2	8,8	5,5
4106	325	2977	1016	481	3299	16,4	11,9	8,8
8489	654	6133	3256	1230	6323	22,4	36,2	20,5

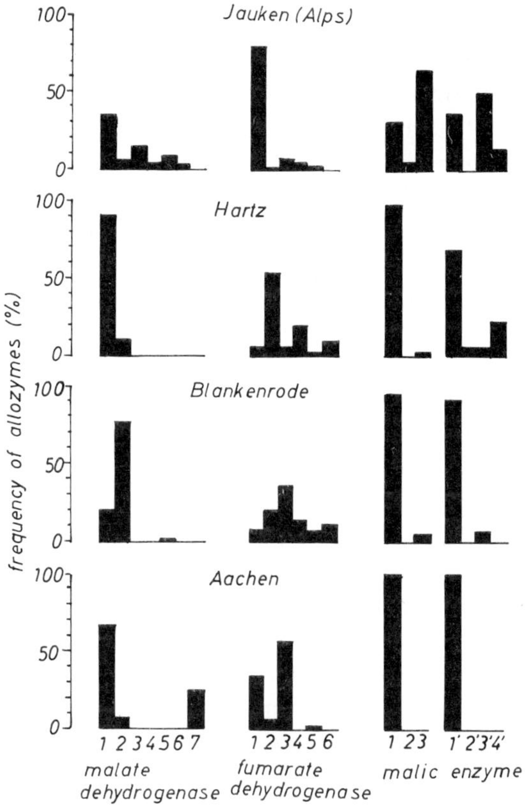

Fig. 1. The variation of allozyme frequency of malate dehydrogenase, fumarate dehydrogenase and malic enzyme in populations of *Silene cucubalus*, growing on zinc enriched soils from the hilly (Aachen, Blankenrode), mountain (Hartz) and alpine region (Jauken). The above mentioned enzymes may play a role in the production of malate, being part of the zinc tolerance mechanism. Nevertheless, each population can be distinguished by its frequency of allozymes. Data derived from Verkleij (1980).

scales. A stable genetic composition of a population will only be achieved if the various genotypes have sufficient phenotypic plasticity to counterbalance the effects of fluctuating factors. Experiments with genetically identical individuals, such as those of *Anthoxanthum odoratum* by Antonovics (1984), give the impression that phenotypic plasticity in life-span and reproductive output is insufficient to prevent at least short-term changes in the genotypes frequency within a population. As a result, randomly fluctuating factors will ensure a stable genetic composition of a population, as long as one of the theorems of evolutionary biology, i.e. maximizing fitness by maximizing final

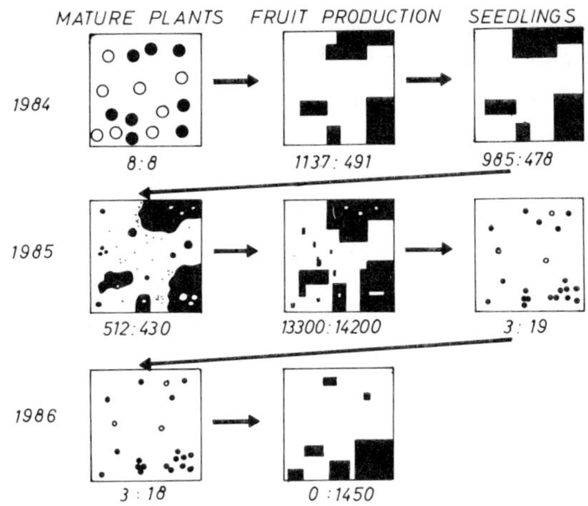

Fig. 2. Changes of two morphologically distinct genotypes of *Phleum arenarium* during two successive years. Filled area: less reproductive morph; open area: highly reproductive morph. Due to the dry autumn 1985 and an early frost period in November 1985 the population was drastically reduced. Heavy predation of the ear of the highly reproductive morph by mice gave rise to a complete loss of this morph in the third generation (Ernst, unpublished).

yield (Cohen, 1971, 1976; Denholm, 1975; Caswell, 1982) will not operate at the genotype level.

The effects of randomly fluctuating environmental conditions on genetic variation have been studied experimentally using two morphs of the cleistogamous dune grass *Phleum arenarium* differing in reproduction rate (Ernst, 1985a). Eight seedlings of each morph, being derived from one parent plant, were planted 2,5 cm apart in open patches of a Tortulo-Phleetum arenarii (Fig. 2). At the end of the first growing season the eight individuals of the high productive morph produced 2.3 times more seed than the low reproductive one, but after germination and seedling growth in the following season, there were only 1.2 times as many of these which gave less seed than the less productive morph, resulting in a progeny ratio 0.94. This diminished contribution of the highly reproductive morph was due to an increased mortality between flower emergence and grain ripening. This ongoing experiment shows that differences in fertility may be counterbalanced by differences in sensitivity to drought, frost, wind, or crowding.

Most models of the population dynamics of annual plants concentrate on quantitative processes i.e. changes in numbers (De Angeles et al., 1979; Watkinson, 1980; Pacala & Silander, 1985), rather than on qualitative processes, i.e. they underestimate the genetic variability within populations. Holsinger & Roughgarden (1985) ascribe the variation of relative growth rate (RGR) within populations only to spatial heterogeneity in the environment. However, as shown for *Erophila verna* (Rozijn, 1984; Van Andel et al., 1986) there is also considerable genetic variation in RGR which is also dependent on development; the greatest differences between morphs and families occur during bloom (Tab. 2). It seems therefore, that models of population dynamics are based on ecologically oversimplified assumptions, and the statement made by Holsinger & Roughgarden (1985, p. 309): 'In conjunction with models of neighbourhood interactions, the models of population dynamics may form a foundation on which a theory of plant population dynamics can be built' seems premature.

2.3 Selective effects of environmental exposure

In addition to the presence of a genetic variability within a species, a further prerequisite of population differentiation is that the species is exposed to a variety of combinations of ecologically important factors. Population differentiation can therefore be expected to occur more frequently if the species has a broad ecological range. There is no doubt that different species, and different populations within species, differ in their ability to cope with a range of environmental conditions. Some species may not have the appropriate variability to cope with extreme environmental conditions, this has recently been described as 'genostasis' (Bradshaw, 1984). Until now, the ecophysiological knowledge of plant species or even that of one plant population has been too poor to predict the adaptive potential of the species or population. I will demonstrate this problem with two grassland species with a broad ecological range.

Lolium perenne is a perennial grass species with considerable ecological tolerances: it can withstand irradiation conditions from high intensities of a Lolio-Plantaginetum majoris to the shaded conditions of a Solano-Rubetum ulmifolii; it can also resist a range of salinity from the brackish Artemisietum maritimae to the non-saline Bromo-Hordeetum murini, and a wide range

Table 2. Variability of relative growth rate due to genetic variation within five families of two morphs (L = low reproduction rate, H = high reproduction rate) of the dune annual *Erophila verna* (based on partially published data, Van Andel et al., 1986). CV = coefficient of variation.

Family number	(Weeks after germination) 3	6	9	12	15	18
L1	0.1143	0.1444	0.0785	0.0523	0.0213	0.0069
L2	0.1025	0.1738	0.0938	0.0204	0.0214	0.0105
L3	0.1370	0.1280	0.0765	0.0479	0.0185	0.0067
L4	0.1318	0.1368	0.0908	0.0255	0.0326	0.0094
L5	0.1191	0.1561	0.0969	0.0209	0.0151	0.0096
Mean L-families	0.1209	0.1478	0.0873	0.0334	0.0218	0.0086
	±0.0138	±0.0178	±0.0092	±0.0154	±0.0066	±0.0017
	CV=11.42	CV=12.05	CV=10.57	CV=46.25	CV=30.11	CV=19.77
H1	0.0986	0.1559	0.0852	0.0341	0.0287	0.0015
H2	0.0984	0.1438	0.0845	0.0251	0.0320	0.0096
H3	0.1070	0.1254	0.0803	0.0352	0.0295	0.0125
H4	0.1116	0.1119	0.0771	0.0538	0.0047	0.0108
H5	0.1216	0.1288	0.0688	0.0486	0.0250	0.0024
Mean H-families	0.1074	0.1332	0.0792	0.0394	0.0240	0.0074
	±0.0097	±0.0170	±0.0067	±0.0116	±0.0111	±0.0050
	CV=9.05	CV=12.79	CV=8.44	CV=29.53	CV=46.10	CV=67.57

of nitrogen supply from the poorly fertilized Medicagini-Avenetum pubescentis to the heavily fertilized Poo-Lolietum (Westhoff & Den Held, 1969). However, it cannot endure severe frost and is therefore absent in alpine meadows (Oberdorfer, 1970). After long exposure to high SO_2-levels *Lolium* populations in Central England have evolved SO_2-resistance (Bell & Mudd, 1976), at the same time reducing their medium frost resistance (Davison & Bailey, 1982). In contrast to its capacity for the evolution of SO_2-resistance, *L. perenne* can not withstand high concentrations of heavy metals (Gartside & MacNeilly, 1974; Ernst, 1974; Lolkema, 1985).

The perennial grassland herb *Silene vulgaris* seldom co-occurs with *Lolium perenne*, even though it can withstand a wide range of irradiance from the open chalk debris of the high alpine Thlaspietea rotundifolii to the shade of a Cephalanthero-Fagetum, it can also withstand various levels of salinity from the saline coastal shore to the non-saline chalkgrassland, a wide range of nitrogen supply from the nitrogen poor Gentiano-Koelerietum to the highly fertilized alpine meadows of the Geranio-Trisetetum. In contrast to *L. perenne* it has developed a high frost resistance, allowing to occur in alpine vegetation even above 2300 m sea level (Oberdorfer, 1970). Certain populations demonstrate a high SO_2-resistance (Ernst et al., 1985). In contrast to *L. perenne, Silene vulgaris* can tolerate the heavy metals zinc, copper, cadmium, lead and nickel (Ernst, 1974, 1982b; Sasse, 1976; Lolkema, 1985), though certain populations, especially those from the chalkgrassland in the Teutoburg Forest, show a genostasis for the evolution of zinc and copper tolerance.

Population differentiation should be characterized by as many attributes as possible. For example 'heavy metal' tolerance involves tolerances of many different elements. Experimental screening work (Gregory & Bradshaw, 1965; Ernst, 1974), and physiological studies of the resistance mechanism for zinc (Ernst, 1975; Mathys, 1977) and copper (Lolkema et al., 1984; Lolkema, 1985), have shown that tolerance is highly specific to the particular element concerned. Similarly populations of *Anthoxanthum odoratum* (Davies & Snaydon, 1973, 1974, 1975) are specifically adapted to the macronutrients K, Ca, and P, but not to Mg and to the metabolically non-essential aluminium; this population differentiation has occurred over relatively short time periodes (<50 years) and over short distances (<30 m). Although such consideration of more than one ecological factor will demand a very intensive research, it will help to elucidate compatible and incompatible pathways in the adaptation possibilities of populations and species, and to predict the selective forces of certain environmental factors.

2.4 *Population isolation*

Population differentiation between adjacent populations will be speeding up if gene flow, by pollen of diaspores, is prevented either by biological or environmental events. This process of isolation at the population level can be compared with the process of species isolation, as defined in the theories of island biogeography (MacArthur & Wilson, 1967). Except in cleistogamous, parthenogenetic or predominantly vegetatively reproducing plants, gene flow by pollen is the most effective manner of gene flow (Levin & Kerster, 1974). The efficacy of gene flow is dependent on the activity of the biological pollinator of zoogamous species, or on the velocity, frequency and direction of the wind in anemogamous species, as demonstrated for *Festuca rubra* in a gradient from coastal dunes to a salt marsh (Rhebergen, 1985). Differences in flowering times between adjacent populations as observed in *Anthoxanthum odoratum* (McNeilly & Antonovics, 1968) and incompatability systems operating at the same flowering time, can also enhance the process of genetic differentiation.

Dispersal of seeds or fruits can also bring about gene flow. Small seeds are usually dispersed further (Harper, 1977), though dispersal structures such as plumes, wings and seed shape, and the timing of seed dispersal in relation to climatic conditions or surrounding vegetation will modify the effects of seed size, as has been found in *Phleum arenarium* (Ernst, 1981), *Schoenus nigri-*

cans, *Ammophila arenaria*, *Aira praecox* and other plant species in natural grassland. A thorough study of these aspects in relation to the isolation of plant population will help to come to realistic models of neighbourhood size and population dynamics.

2.5 The ecological experiences of grassland species in the past

For a proper evaluation of the possibilities of population differentiation in grassland, the ecological history of a species in the past, i.e. since the postglacial period, is of great interest, especially for the species of the semi-natural and man-made grassland. Some grassland species such as *Dactylis glomerata* and *Bromus erectus* originally occurred and still grow in the half-shade of the forest canopy of the Carpinion, the Cephalanthero-Fagion and the Quercion pubescentis, alliances on dry basic soils, though at certain times of the day they are exposed to full or nearly full sunlight, if it penetrates through the canopy gap (Schulze, 1972). A change from the forest to the open grassland will not necessarily demand an adaptative change of these species to light. It is therefore not surprising that Roy (1985) did not detect ecotypic differentiation in populations of *Dactylis glomerata* and *Bromus erectus* from an ecological gradient from *Quercus pubescens* woodland to abandoned vineyard, despite the well-known genetic variability of *Dactylis glomerata* for other ecological factors (Ashenden et al., 1975). On the other hand, species which have entered man-made grassland from natural grassland, such as *Festuca ovina* or *Linum catharticum* will have to compete for light with the taller grasses and herbs. The genostasis of *Festuca ovina*, with regard to irradiance, may have prevented its spread to shady habitats, when other species, such as *Plantago major* (Kuiper & Smid, 1985) and *Solidago virgaurea* (Björkman & Holmgren, 1963), have been very successful. Because of the various origins of grassland species, with regard to their ecological background (Fig. 3), adaptation to other ecological constraints of man-made grassland may already be present in

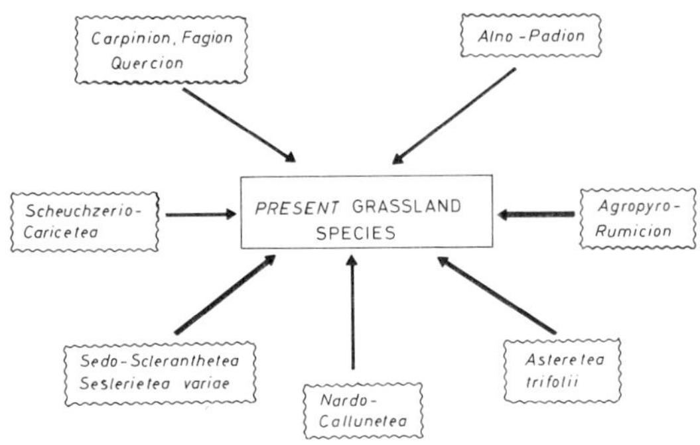

Fig. 3. Origin of grassland species in man-made grassland in temperate Europe.

the genetic or physiological reservoir of each population and will influence population differentiation.

Grassland species derived from forests and woodlands will have experienced extensive grazing by cattle since the neolithic period (Ellenberg, 1963). Adaptation to defoliation, treading and very local fertilization by urine and dung has been underway for more than 6000 years, i.e. some hundreds or thousands of generations. Even intense grazing pressure, as that on *Festuca rubra* by barnacle geese in salt marshes (Ydenburg & Prins, 1981) may have been a common experience of the current species of agricultural grassland in their original environment. In contrast to grazing, adaptation to mowing has a shorter history. Extensive mowing with removal of biomass (oligotrofication) has occurred only in the last thousand years in Central Europe (Sukopp, 1969), while intensive mowing with the applications of organic or inorganic fertilizer has occurred only in the last 100 to 150 years, so that the selective agents of grazing and mowing have operated for different time periods. The main new constraints imposed on grassland species are fertilizers, herbicides and other modern chemicals, however, the frequency, duration and intensity of other factors have also changed. This means that the exposure time and dose, the interval between applications, and their simultaneity will determine the selection pressure, and thus the degree of population differentiation.

3. Population differentiation in grassland

Changes in environmental factors in grassland may be caused by pedogenesis, vegetation succession, herbivores and grassland management. In natural populations, population differentiation can occur by directional and disruptive selection. Directional selection favours genotypes of one extreme and changes the mean of the ecological response in comparison to stabilizing selection. Disruptive selection gives advantages to both extremes, increasing the variance of the population and slowing down the speed of differentiation (Hedrick, 1984).

Soil formation and vegetation succession occur so slowly that population differentiation in response to these factors cannot be observed directly. However, it is vere clear, that population differentiation can occur in response to non sudden man-made changes in the environment (MacNair, 1981; Radosevich & Holt, 1984). It also seems evident that differentiation of populations in space (Claussen et al., 1948), reflects population differentiation in time, in much the same way that spatial variation in species composition has been used to infer successional changes in time (Ellenberg, 1963; Ernst, 1966, 1974).

3.1 Pedogenesis and population differentiation

3.1.1 Soil acidification. In temperate Europe the surplus of precipitation over evaporation governs soil development, which is characterized by progressive leaching and consequently by acidification of drained soils. This climatically driven soil process can be modified by plant populations and by vegetation types (for a review: Miles, 1985). One of the most obvious changes in soil chemistry occurs during vegetation succession on orginally calcareous coastal dunes. Calcium carbonate is progressively leached from the soil. A moss cover of *Tortula ruraliformis* can greatly accelerate this decalcification progress (Rozema et al., 1985a). If this decalcification is followed by acidification, the availability of iron, manganese and aluminium to plants will increase; this may give rise to the selection of genotypes that are tolerant of alu-

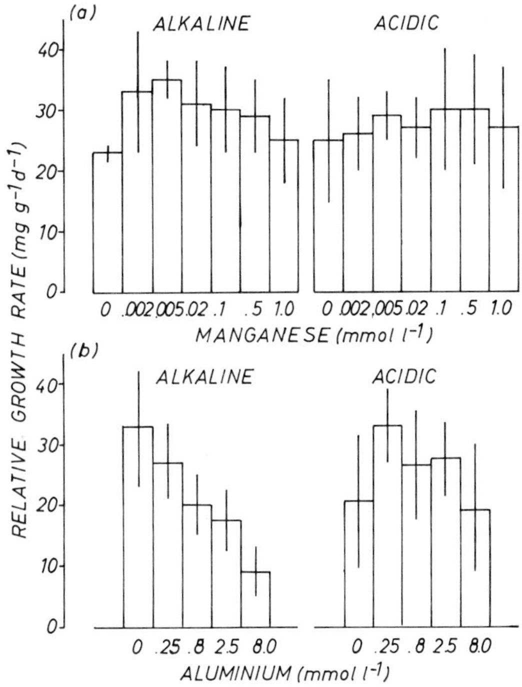

Fig. 4. Response of adjacent populations of *Festuca ovina*, collected from acidic and alkaline coastal dune sand, to increasing concentrations of (a) manganese and (b) aluminium in the nutrient solution (after Laan in Rozema et al., 1985a).

minium, iron and/or manganese. Such differences have been found within a few meters for genotypes of *Festuca ovina* in coastal dunes (Rozema et al., 1985); the selection for aluminium tolerance was apparently stronger than that for manganese tolerance (Fig. 4). Such physiologically distinct populations adapted to small scale differences in soil chemistry, whether naturally occurring or man-made, occur in many grassland species and grassland vegetation, e.g. in *Melilotus alba* (Ramakrishnan, 1968), *Teucrium scorodonia* (Hutchinson, 1968) and *Anthoxantum odoratum* (Davies & Snaydon, 1973, 1974; Davies, 1975).

3.1.2 Increasing water capacity. As a consequence of pedogenesis, organic material can accumulate in the top soil layer and so increase the water-holding capacity. This process has been termed as facilitation (Connell & Slatyer, 1977). If genotypes differ in the insensitivity to soil moisture, the frequency of genotypes within a population may change; this has been shown by Ashenden et

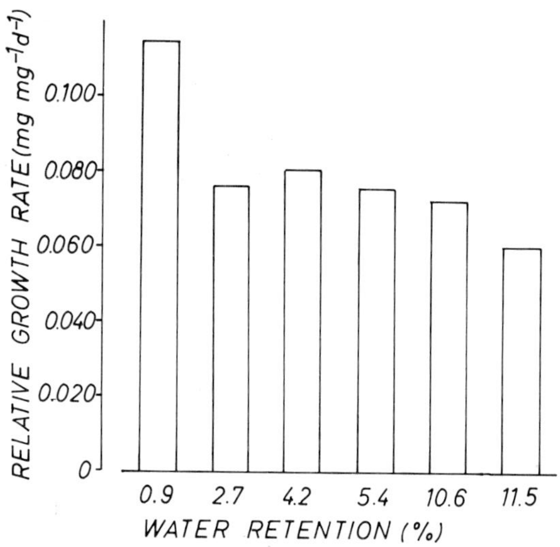

Fig. 5. Relative growth rates of seedling populations of *Dactylis glomerata* collected from dune soils with different water retention when grown in a low watering regime (after Ashenden et al., 1975).

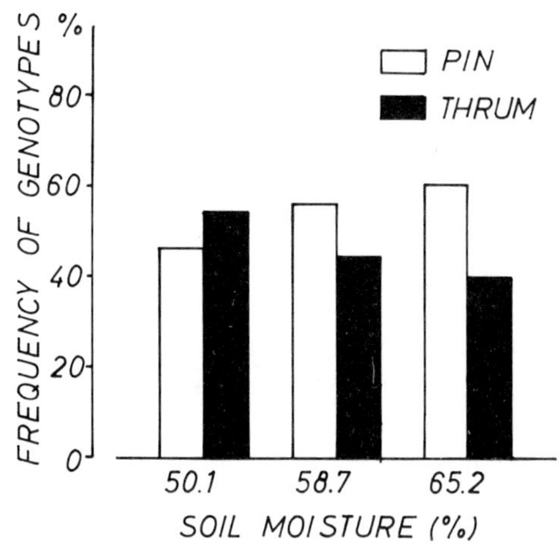

Fig. 6. The percentage frequency of pin and thrum genotypes in *Primula elatior* in adjacent sites with different in soil moisture during spring and early summer (Ernst, unpubl.).

al. (1975) for population of *Dactylis glomerata* on sand dunes (Fig. 5). Sometimes differences in genotype frequency will be only recognized if the genotypes also differ morphologically. For example, in the grassland of the Teutoburg Forest, areas with a different water capacity were characterized by patches in the frequency of pin and thrum genotypes of *Primula elatior*. With increasing soil moisture pin genotypes became more frequent than the thrum genotypes (Fig. 6).

3.1.3 Availability of mineral elements. The increase of soil organic matter during pedogenesis influences not only the water capacity of the soil, but also the availability of nutrients. Soil organic matter does not only change the cation exchange capacity of the soil, but it can alter the chemical form of a mineral element. This may be especially important on heavy metal soils (Ernst, 1974, 1976, 1982b). High concentrations of heavy metals slow the process of pedogenesis, and cause strong gradients in the metal concentration between an ore outcrop and its surroundigs (Ernst, 1966, 1974, 1982b). The most interesting aspect is the chemical speciation of the water-soluble heavy metals in the soil and its selective effects on the plant populations. The only genetic analysis of plant populations on an undisturbed heavy metal soil gradient is that of *Silene cucubalus* in a heavy metal vegetation, belonging to the Violetum calaminariae, and in an adjacent chalkgrassland, belonging to the Gentiano-Koelerietum (Ernst, 1976). *Silene cucubalus* from the heavy metal vegetation has a high zinc tolerance (Tab. 3), whereas the plants on the chalkgrassland, only 6 m away, have a low zinc tolerance. Plants of the boundary have a medium to low tolerance, despite the fact that the total concentration of water soluble zinc does not differ from that of the Violetum calaminariae. However, at the boundary the zinc is completely bound to organic material, i.e. phenolic, fulvic and humic acids (Van der Werff, 1981). Since organically bound zinc has a low toxicity for higher plants (Ernst, 1968; Marquenie-van der Werff & Out, 1981) the population of *Silene cucubalus* at the borderline is exposed to a low selection pressure for zinc tolerance.

Similar differences in selection pressure, being caused by the chemical status of an element, can be found in populations of other plant species, growing in gradients from juvenile to mature soils; however, only a sophisticated approach to the qualitative analysis of soil solution will help to elucidate such processes of populations differentiation.

3.2 Biotic factors

3.2.1 Vegetation structure. Dense and tall vegetation affects the environmental conditions for small plants in the ground layer by reducing light receipt, increasing air humidity and diminishing exposure to wind velocity (cf. Grace, 1977). Low-growing plant species may therefore be exposed to different selective forces in a tall grassland and in short or open grassland. In the Dutch coastal dunes the cleistogamous grass species *Phleum arenarium* occurs in different growth forms (Fig. 7) with tall genotypes in the tall vegetation of an Elymo-Ammophiletum festucetosum (Ernst, 1985a, Rhebergen, 1985) and with a small 'normal' genotype in the open vegetation of a Tortulo-Phleetum (Ernst, 1981). The character of stem height is under genetic control and individual fecundity is closely related with plant weight and height, tall genotypes produce more than twice as many caryopses per spike. Since fitness in annual plants is said to be dependent on final seed yield (Cohen, 1971, 1976; Denholm, 1975; King & Roughgarden, 1982a, b) tall genotypes should have a competitive advantage over small plants. However, tall genotypes are restricted to the tall grassland habitat. A transplant experiment has shown the possible factor limiting the tall genotypes in the Tortulo-Phleetum. During windy summer periods, the panicles of the tall genotypes were blown to the ground, where caryopses germinated within the spikelet in late autumn, but soon died, due to shortage of nutrients. By contrast, most of the caryopses of the normal

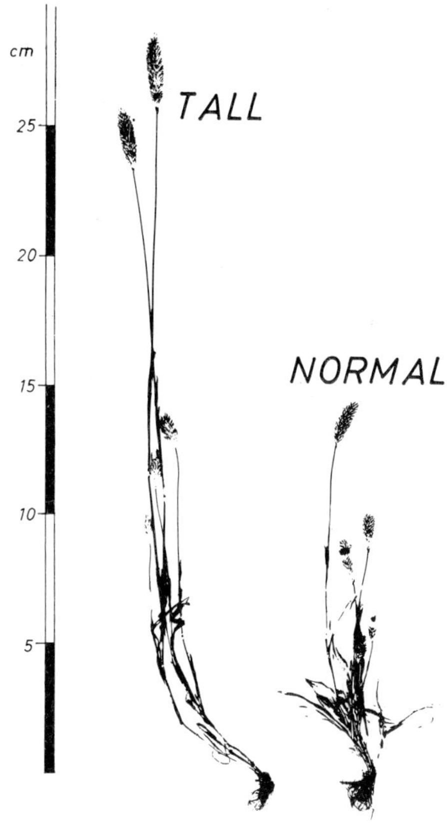

Fig. 7. Morphology of the tall and normal genotypes of *Phleum arenarium*, collected from an Elymo-Ammophiletum and a Tortulo-Phleetum, respectively on Wadden island Schiermonnikoog.

genotypes were scattered around the mother plant, where they germinated and survived the winter period (Ernst, 1981). In the Elymo-Ammophiletum there was less wind and panicles

Table 3. The zinc tolerance of *Silence cucubalus* populations in relation to the chemical form of zinc in the soil solution in a gradient from an ore outcrop with a typical heavy metal vegetation (Violetum calaminariae) and a surrounding chalkgrassland (Gentiano-Koelerietum, Data from Ernst, 1976, unpubl., and van der Werff, 1981).

	Violetum	Boundary	Gentiano-Koelerietum
Zinc tolerance of *Silene cucubalus*			
(mmol Zn l^{-1})	10–100	0.1–1.0	0.01–0.10
Water-soluble zinc (mmol kg^{-1} dry soil)			
total fraction	9.8±1.7	10.2 ±2.1	0.14 ±0.15
Ionic zinc	9.7±1.7	0.01±0.01	0.00 ±0.00
Phenolic zinc	0.1±0.0	0.1 ±0.0	0.015±0.010
Humic+ fulvic zinc	0.0±0.0	10.1 ±2.2	0.15 ±0.15

were supported by surrounding vegetation; they therefore remained upright and released their caryopses normally, leading to normal germination and establishment. It would seem therefore that wind is a selective factor differentiating between Elymo-Ammophiletum and Tortulo-Phleetum. Light may also be a selection factor in the Elymo-Ammophiletum.

3.2.2 Plant-micro-organism interactions. Many grassland species have the ability to form a symbiosis with vesicular-arbuscular mycorrhizal fungi (VAM) and/or with nitrogen-fixing bacteria, so that co-adaptive population differentiation may occur in pairs of species. Wilson (1980) speculated that only weak selection pressure should occur for mutualistic associations, but there is now considerable specific adaptation between populations of *Rhizobium* and of legumes, as measured by effectiveness of the symbiotic association (Tab. 4), as shown by Wood et al. (1984) and Jones & Hardarson (1979).

Differences in susceptibility to the VAM fungus *Glomus fasciculatus* has been found between populations of the perennial grass species *Calamagrostis epigeios* from an outerdune area compared with an innerdune area (Ernst et al., 1984). These differences in infection give rise to differences in nutrient uptake and translocation. Obviously in addition to parasite-host interactions (see Zadoks, Chapter 13) there are other important areas where correlated population differentiation in plants and their associated microorganisms await study (Dueck et al., 1986 b.).

Table 4. The percentage nodulation of the varieties of *Trifolium repens* infected with three strains of *Rhizobium trifolii* (after Wood et al., 1984).

Rhizobium strain	Percentage nodulation of the *Trifolium repens* varieties				
	Donna	S184	Kent Wilt White	Huia	S100
HIDL	100	100	67	50	17
HR 1	100	67	100	83	83
HD 3	83	83	83	83	17

3.2.3 Plant-herbivore interactions. Plants always experience some defoliation, because herbivores are a component of ecosystems. Herbivory can sometimes extend to complete defoliation, as sometimes in oak by the caterpillar of the winter moth *Tortrix viridana* (Feeny, 1970) or in the ragwort *Senecio jacobaea* by the larvae of the cinnabar moth *Tyria jacobaeae* (Van der Meijden, 1979). Such severe defoliation of a plant species can regulate the herbivore population in the following growing season, as demonstrated for the larvae of the stemborer *Archanara geminipuncta* in common reed *Phragmites australis* (Van der Toorn & Mook, 1982). Despite these severe impacts, and an arsenal of possible chemical defense products such as alkaloids, phenolics and cyanogenic glycosides (Rosenthal & Janzen, 1979; Crawley, 1983) no obvious differences involving secondary plant metabolites have been reported between heavily and slightly defoliated plant populations. However, the frequency and intensity of defoliation can cause population differentiation for several aspects of life history and plant allocation strategies (Tab. 5). Defoliation normally reduces the photosynthetic capacity of plants and adaptation to defoliation may involve the development of photosynthetic active system nearer the soil surface, when the leaf area and meristematic region are now protected from grazing. Population differentiation may therefore be re-

Table 5. Consequences of heavy defoliation and possibilities for selection in grassland species.

Loss of	Selection for
photosynthetic capacity	photosynthetic activities near soil surface subsurface storage systems defoliation resistance
nutrients for retranslocation	nutrient efficiency
water	drought resistance heat resistance
reproductive capacity	earliness or lateness of flowering perennality with emphasis on vegetative reproduction sensitivity of seed predation after changes of the herbivorous insects

lated to the grazing characteristics of the herbivores. For example, Chapman (1983) found population differentiation of *Trifolium repens* as response to cattle and sheep grazing. Another important effect of grazing is that on nutrient cycling. Usually, more than 90 per cent of the nutrients consumed by the animal is excreted in the dung and urine (Wilkinson & Lowry, 1973). Nutrients in urine are immediately available to plants; nutrients in dung are released more slowly, giving rise to a special coprophilic flora (Dickinson & Underhay, 1977) and fauna (Brussaard, 1985) during decomposition. In addition to these changes of nutrient turnover in grassland, grazing animals can diminish the nutrient reservoir of some areas of a grassland in a similar manner to that of cutting for hay (Tab. 6). This occurs because animals graze herbage from some parts of the grassland and deposit excreta in other parts so that some parts are depleted and other parts are enriched (Wilkinson & Lowry, 1971). This 'oligotrophication' by harvesting is compensated or even overcompensated in modern times by fertilisation. Additional supplies of fertiliser can cause rapid population differentiation, as shown for populations of *Anthoxanthum odoratum* (Davies & Snaydon, 1973, 1974; Davies, 1975). In addition to the effect on the photosynthetic system, defoliation interrupts the uptake or/and retranslocation of mineral nutrients; this is especially important for nutrients that are abundant in young growing tissues, i.e. N, P, K, Zn and Cu.

Table 6. The amount of biomass and nutrients (kg ha^{-1}) removed by hay and by grazing ([1] Dawson, 1956; [2] Leon, 1968; [3] Loach, 1968; [4] Russel, 1973; [5] Elzebroek, 1982).

Pasture-type	Yield	N	P	K
Hayed				
Festuca rubra[1]	2063	61	7	35
Molina coerulea[2,3]	2500	33	15	28
Trifolium pratense[4]	4215	110	12	78
Mixed meadow[4]	160	55	6	48
Mixed meadow[5]	4497	79	12	69
Mixed meadow[5]	6370	118	22	77
Grazed				
Mixed pasture[5]	8080	172	29	101

A loss of these mineral nutrients by defoliation might be compensated for by accelerated uptake, but defoliation also reduces nutrient uptake (Wilkinson & Lowry, 1973).

Defoliation can also affect the reproductive performance of plants, especially when grazing occurs during the period between flower initiation and seed ripeness. As a result, there may be selection of genotypes which flower earlier or later. Using models, Cohen (1971, 1976) has shown that a fairly sharp transition between the vegetative and reproductive growth phases will maximize total reproductive biomass if the environment has a fixed length. Such a reaction has been found in some annual plant species in an ungrazed situation (King & Roughgarden, 1983; Ernst, 1985a, b), but may also be very relevant to perennial plants at various frequencies of defoliation. If vegetative growth and persistence is as important as the seed production (Cohen, 1976), then a graded biomass allocation should occur (King & Roughgarden, 1982b) or multiple switches occur between the two phases (King & Roughgarden, 1982a). Breese & Charles (1962) have shown that populations of *Lolium perenne* from habitats with a high frequency of cutting for hay or silage were earlier heading than those for continuously grazed habitats. Similarly, Humphreys (1980) found that populations of *Lolium perenne* grazed by rabbits were earliest and grazed by sheep latest (Tab. 7). However, care should be taken in the interpretation of these results, flowering time may be selected for in response to various abiotic factors (Rhebergen, 1985); only a comparison of populations in the same environment but under different grazing regimes will be conclusive.

Selection for earliness or lateness of flowering

Table 7. The heading of *Lolium perenne* in relation to the grazing species (from Humphreys, 1980).

Grazer	Heading time
sheep	late
horse	medium - late
cattle	early - medium
rabbit	early

may be expected to be more rapid in wind-pollinated on account of high pollen dispersal and in autogamous plant species. In entomogamous species, stabilizing selection by both pollinators and seed predators should maintain a low variance in timing of flowering (Augspurger, 1981) and thus hamper a rapid selection for other flowering periods. Changes in flowering date may occur if other pollinators can substitute for the conventional ones, and if seed predation is reduced by hyperparasites of the seed predators, as found in the *Orobanche ramosa* - *Phytomyza orobanchia* - *Crataepiella carlinarum* system (Mihajlovic, 1986).

Morphological adaptation to predation has been found in adjacent populations of *Silene cucubalus* in chalk grassland adjacent to a heavy metal vegetation on copper rich soils (Ernst, 1985c). Seed capsules of the chalk grassland population were frequently devastated by the caterpillar of the noctuid *Hadena cucubali*, whereas the seed capsules of the copper tolerant population were not attacked, even though the heavy metal concentration of the seeds and the seed capsules of the two populations did not differ (Ernst, 1974, 1985c). Transplant experiments, using the caterpillar of *Hadena cucubali*, showed that the seed capsule of the copper-tolerant plants were too small for the full development and protection of the caterpillar up to the puppae stage (Fig. 8). As a result, after the third instar, the caterpillar on the copper-tolerant population was forced to leave the capsule and to graze on the copper-rich leaves, which are toxic for this noctuid. In this case population differentiation seems to involve both, a small capsule with poor seed production (Dueck et al., 1986a) and a toxic concentration of copper in the leaves.

4. Environmental pollution

Acids and heavy metals may be deposited far from their original site of emission. At the deposition site they may cause population differentiation, such as that of SO_{-2} tolerance in *Lolium perenne* (Bell & Mudd, 1976), *Silene cucubalus*

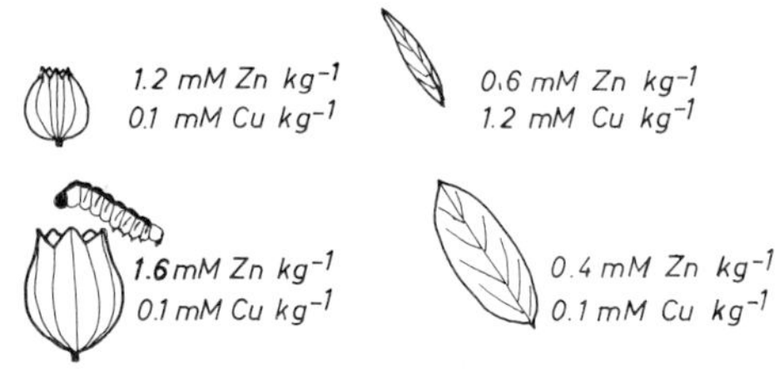

Fig. 8. The shape and heavy metal concentration)mM kg^{-1} dry weight) of capsules and leaves of a copper tolerant population or normal population of *Silene cucubalus*. Caterpillars of the noctuid *Hadena cucubali* attack only the normal capsules.

(Ernst et al., 1985; Dueck et al., 1986a), or *Trifolium repens* (Dueck, 1986). Similarly copper fallout from high-tension lines, and deposition of metal fallout by smelters, refineries, and cement kiln (Ernst, 1972, 1982a, 1983; MAGS, 1975; Wu et al., 1975; Dueck et al., 1984) and possibly application of copper-enriched pig slurry to grassland (Van Rhee, 1975; Dueck, pers. communication) can cause population differentiation in grassland.

Up to now, nearly no research has been carried out which can give indications on the selectivity of low and medium concentrations of these pollutants if they enter grassland vegetation. Thorough investigations at the interpopulation and interspecies level will unveil if 'weak' disturbances will alternate stable states of if they will be resilient.

5. Concluding remarks

It is clear from the examples outlined above that genetic responses occur in plant populations as a consequence of autogenic processes, such as pedogenesis, and allogenic events, such as grassland management and environmental pollution. Our present framework of reference is response to one kind of disturbance, which is normally allogenic, but various kinds of disturbance may occur, sometimes simultaneously, sometimes sequentially. The metabolic consequences of these disturbances may result in synergistic, additive, or antagonistic effects, thus influencing the competitive ability of genotypes within and between populations.

Antagonistic effects tend to neutralise the metabolic costs and will rarily contribute to population differentiation. There are few examples of such antagonism, but evidence is accumulating that nitrogen fertilisation can ameliorate the toxic effects of sulphur dioxide pollution (Cotrufo & Berry, 1970; Milchunas et al., 1981), even if nitrogen is supplied as nitrogen oxide (Dueck et al. 1986a). In the latter case, the balance between negative and positive effects on plant growth and reproduction depends on the ecologic background of the population.

The combined effects of disturbances may also be additive. If two disturbing factors affect different pathways in the cell, it seems very likely that their effect will be additive. For example, zinc tolerance and copper tolerance in plants is regulated by two quite different forms of compartmentation, whereas zinc is predominantly accumulated in the vacuole of zinc tolerant plant cells (Ernst, 1974), copper is retained in the cytoplasm and metabolically inactivated by copper binding organic compounds with metallothione-like properties (Lolkema et al., 1984; Rauser, 1984; Robinson & Thurman, 1986). Plants possessing both tolerances, i.e. multiple tolerances, are less productive than plants with only one tolerance, due to the metabolic costs of the tolerance mechanisms (Ernst, 1983, 1986). As a result, the only genotypes which survive in the population are those with the right balance between disturbance exposure and metabolic responses, and hypertolerant genotypes are rare in populations (Wu et al., 1975; Dueck et al., 1984).

Synergism, i.e. when the combined effect of various disturbance exceeds the sum of the individual effects, may be expected to occur if multiple disturbance affect the same metabolic pathway. Davison and Bailey (1982) demonstrated that exposure to sulphur dioxide reduced the resistance of *Lolium perenne* to sub-zero temperatures. It may be speculated that both, frost and SO_2 affect biomembrane integrity and thus may be very effective selective agents. However, more work is needed to identify possible interactions among multiple disturbances.

Changes in the genetic constitution of populations may have important consequences at the community level. Local disturbances, *e.g.* mole hills, defecation plots, hoof marks, below or aboveground grazing, change the abiotic environment, *e.g.* microclimate, soil moisture, nutrient availability, aeration. Dependent on the metabolic impact of local disturbances and the metabolic potential of the genotypes within a population, they may imply selection and population differentiation. Change in genetic constitution of a plant population will affect the responses to herbicides, parasites, symbionts, and other associated populations, thus stimulating niche differentiation in the community.

References

Allard, R.W., Miller, R.D. & Kahler, A.L. (1978). The relationship between degree of environmental heterogeneity and genetic polymorphism. In: Freijsen, A.H.J. & Woldendorp, J.W. (Eds.), Structure and Functioning of Plant Populations, pp 49–73. North-Holland Publ. Comp., Amsterdam.

Antonovics, J. (1984). Genetic variation within populations. In: Dirzo, R. & Sarukhán, J. (Eds.), Perspectives on Plant Population Ecology, pp 229–241. Sinauer Ass. Publ., Sunderland.

Ashenden, T.W., Stewart, W.S. & Williams, W. (1975). Growth responses of sand dune populations of *Dactylis glomerata* L. to different levels of water stress. J. Ecol. 63: 97–107.

Augspurger, C.K. (1981). Reproductive synchrony of a tropical shrub: Experimental studies on effects of pollinators and seed predators on *Hybanthus prunifolius* (Violaceae). Ecology 62: 775–788.

Bell, J.N.B. & Mudd, C.H. (1976). Sulphur dioxide resistance in plants: a case study of *Lolium perenne*. In: Mansfield, T.W. (Ed.), Effects of Air Pollutants on Plants, pp 87–103. Cambridge Univ. Press., Cambridge.

Björkman, O. & Holmgren, P. (1963). Adaptability of the photosynthetic apparatus to light intensity in ecotypes from exposed and shaded habitats. Physiol. Plant. 15: 883–914.

Bradshaw, A.D. (1984). Ecological significance of genetic variation between populations. In: Dirzo, R. & Sarukhán, J. (Eds.), Perspective on Plant Population Ecology, pp 213–228. Sinauer Ass. Publ., Sunderland.

Breese, E.L. & Charles, A.H. (1962). Population studies in ryegrass. Rep. Welsh Plant Breed. Stat. 1961: 31–34.

Brussaard, L. (1985). A pedobiological study of the dung

beetle *Typhaena typhoeus* (Coleoptera, Geotrupidae). Ph.D. Thesis, Agricultural University Wageningen.

Caswell, H. (1982). Optimal life histories and the maximization of reproductive value: a general theorem for complex life strategies. Ecology 63: 1218–1222.

Chapman, D. (1983). Growth and demography of *Trifolium repens* stolons in grazed hill pastures. J. Appl. Ecol. 20: 597–608.

Claussen, J., Keck, D.D. & Hiesey, W.M. (1948). Experimental studies on the nature of species. III. Environmental response of climatic races of *Achillea*. Carnegie Inst. Washington Publ. 581.

Cohen, D. (1971). Maximizing final yield when growth is limited by time or by limiting resources. J. Theor. Biol. 33: 299–307.

Cohen, D. (1976). The optimal timing of reproduction. Am. Nat. 110: 801–807.

Connell, J.H. & Slatyer, R.O. (1977). Mechanisms of succession in natural communities and their role in community stability and organization. Am. Nat. 111: 1119–1144.

Cotrufo, C. & Berry, C.R. (1980). Some effects of a soluble NPK fertilizer on sensitivity of Eastern white pino to injury from SO_2 air pollution. Forest Sci. 16: 72–73.

Crawley, M.J. (1983). Herbivory. The Dynamics of Animal-Plant Interactions. Oxford: Blackwell Scientific.

Davies, M.S. (1975). Physiological differences among populations of *Anthoxanthum odoratum* L. collected from the park grass experiment, Rothamsted. IV. Response to potassium and magnesium. J. Appl. Ecol. 12: 953–974.

Davies, M.S. & Snaydon, R.W. (1973a). Physiological differences among populations of *Anthoxanthum odoratum* L. collected from the park grass experiment, Rothamsted. I. Response to calcium. J. Appl. Ecol. 10: 33–45.

Davies, M.S. & Snaydon, R.W. (1973b). Physiological differences among populations of *Anthoxanthum odoratum* L. collected from the park grass experiment, Rothamsted. II. Response to aluminium. J. Appl. Ecol. 10: 47–55.

Davies, M.S. & Snaydon, R.W. (1974). Physiological differences among populations of *Anthoxanthum odoratum* L. collected from the park grass experiment, Rothamsted. III. Response to phosphate. J. Appl. Ecol. 11: 699–708.

Davison, A.W. & Bailey, J.F. (1982). SO_2-pollution reduces the freezing resistance of rye grass. Nature (Lond.) 297: 400–402.

Dawson, R.B. (1956). Practical Lawn Craft. Crosby Lockwood, London.

De Angeles, D.L., Travis, C.C. & Post, W.M. (1979). Persistence and stability to seed-dispersed species in a patchy environment. Theor. Pop. Biol. 16: 107–125.

Denholm, J.V. (1975). Necessary condition for maximum yield in a senescing two-phase plant. J. Theor. Biol. 52: 251–254.

Dickinson, C.H. & Underhay, V.H.S. (1977). Growth of fungi in cattle dung. Trans. Brit. Mycol. Soc. 69: 473–477.

Dueck, Th.A., Ernst, W.H.O., Faber, J. & Pasman, F. (1984). Heavy metal imission and genetic constitution of plant populations in the vicinity of two metal emission sources. Angew. Bot. 58: 47–59.

Dueck, Th.A., Ernst, W.H.O., Mooi, J. & Pasman, F.J.M. (1986a). Effects of SO_2, NO_x and O_3 in combination on the yield and reproduction of *Silene cucubalus* populations. J. Plant. Physiol. 122: 97–106.

Dueck, Th. A, Visser, P., Ernst, W.H.O. & Schat, H. (1986b). Vesicular-arbuscular mycorrhizae decrease zinc-toxicity to grasses growing in zinc-polluted soil. Soil Biol. Biochem. 18: 331–333.

Dueck, Th.A. (1986). Impact of Heavy Metals and Air Pollutants on Plants. Ph.D. Thesis, Amsterdam.

Ellenberg, H. (1963). Vegetation Mitteleuropas mit den Alpen. E. Ulmer, Stuttgart.

Elzebroek, A.T.G. (1982). Evolutie en gebruiksmogelijkheden van grasvegetaties bij extensivering van de exploitatie. Meded. 69. Wageningen: Agricultural University, Vakgroep Landbouwplantenteelt en Graskunde.

Ennos, R.A. (1983). Maintenance of genetic variation in plant populations. In: Hecht, M.K., Wallace, B. & Peerce, G.T. (Eds.), Evolutionary Biology 16, 129–155. Plenum Press., New York.

Ernst, W.H.O. (1966) Ökologisch-soziologische Untersuchungen an Schwermetallpflanzengesellschaften Südfrankreichs und des östlichen Harzvorlandes. Flora B 156: 301–318.

Ernst, W. (1968). Der Einfluß der Phosphatversorgung sowie der Wirkung von ionogenem und chelatisierten Zink auf die Zink- und Phosphataufnahme einiger Schwermetallpflanzen. Physiol. Plant. 21: 323–333.

Ernst, W.H.O. (1972). Zink- und Cadmium-Immissionen auf Böden und Pflanzen in der Umgebung einer Zinkhütte. Ber. Deutsch. Bot. Ges. 85: 295–300.

Ernst, W.H.O. (1974). Schwermetallvegetation der Erde. G. Fischer, Stuttgart.

Ernst, W.H.O. (1975). Physiology and heavy metal resistance in plant. Int. Conf. Heavy Metals in the Environment, Toronto 1975, Vol. 2: 121–136.

Ernst, W.H.O. (1976). Ökologische Grenze zwischen Violetum calaminariae und Gentiano-Koelerietum. Ber. Deutsch. Bot. Ges. 89: 381–390.

Ernst, W.H.O. (1981). Ecological implication of fruit variability in *Phleum arenarium* L., an annual dune grass. Flora 171: 387–398.

Ernst, W.H.O. (1982a). Monitoring of particulate pollutants. In: Steubing, L. & Jäger, H. (Eds.), Monitoring of Air Pollutants by Plants. pp 121–128. Junk, The Hague.

Ernst, W.H.O. (1982b) Schwermetallpflanzen. In: Kinzel, H. (Ed.), Pflanzenökologie und Mineralstoffwechsel, pp 472–506. E. Ulmer, Stuttgart.

Ernst, W.H.O. (1983). Ökologische Anpassungsstrategien an Bodenfaktoren. Ber. Deutsch. Bot. Ges. 96: 49–71.

Ernst, W.H.O. (1985a). Some considerations of and perspectives in coestal ecology. Vegetatio 62: 533–545.

Ernst, W.H.O. (1985b). The effects of forest management on the genetic variability of plant species in the herb layer. In:

Gregorius, H.R. (Ed.), Population Genetics in Forestry. Lecture Notes in Biomathematics 60, pp 200–211. Springer, Berlin.

Ernst, W.H.O. (1985c). Schwermetallimmissionen - ökophysiologische und populations-genetische Aspekte. Düsseldorf. Geobot. Koll. 2: 43–57.

Ernst, W.H.O., Verkleij, J.A.C. & Vooijs, R. (1983). Bioindication of a surplus of heavy metals in terrestrial ecosystems. Environ. Monit. Assessment 3: 297–305.

Ernst, W.H.O., Van Duin, W.E. & Oolbekking, G.T. (1984). Vesicular-arbuscular mycorrhiza in dune vegetation. Acta Bot. Neerl. 32: 151–160.

Ernst, W.H.O., Tonneijck, A.E.C. & Pasman, F.J.M. (1985). Ecotypic response of *Silene cucubalus* to air pollutants (SO_2, O_3). J. Plant Physiol. 118: 439–450.

Ernst, W.H.O. (1986). Longterm pollution and selection. 2nd Int. Conf. Environmental Pollution, Amsterdam 1986. CEP Consultants Edinburgh, pp 10–15.

Feeny, P. (1970). Seasonal changes in the oak leaf tannins and nutrients as a cause of spring feeding by winter moth caterpillars. Ecology 51: 565–581.

Gartside, D.W. & McNeilly, T. (1974). The potential of heavy metal tolerance in plants. Heredity 32: 335–348.

Grace, J. (1977). Plant Response to Wind. Academic Press., London.

Gregory, R.P.G. & Bradshaw, A.D. (1965). Heavy metal tolerance in populations of *Agrostis tenuis* Sibth. and other grasses. New Phytol. 64: 131–143.

Hamrick. J.L. (1979). Genetic variation and longevity. In: Solbrig, O.T., Jain, S., Johnson, G.B. & Raven, P.H. (Eds.), Topics in Plant Population Biology, pp 84–113. Columbia Univ. Press., New York.

Hamrick, J.L., Linhart, Y.B. & Mitton, J.B. (1979). Relationship between life history characteristics and electrophoretically detectable genetic variation in plants. Ann. Rev. Ecol. Syst. 10: 173–200.

Harper, J.L. (1977). Population Biology of Plants. Academic Press., London.

Hedrick, P.W. (1984). Population Biology. The Evolution and Ecology of Populations. Jones & Bartlett, Boston.

Holsinger, K.E. & Roughgarden, J. (1985). A model for the dynamics of an annual plant population. Theor. Pop. Biol. 28: 288–313.

Humphreys, M.O. (1980). Grass breeding: objectives, principles, and potentials. In: Rorison, I.H. & Hunt, R. (Eds.), Amenity Grassland. An Ecological Perspective, pp 57–67. J. Wiley & Sons, Chichester.

Hutchinson, T.C. (1968). A physiological study of *Teucrium scorodonia* ecotypes which differ in their susceptibility to lime-induced chlorosis and iron-defficiency chlorosis. Plant Soil 28: 81–105.

Jain, S.K. (1976). Patterns of survival and microevolution in plant populations. In: Karlin, S. & Nevo, E. (Eds.), Population Genetics and Ecology, pp 49–89. Academic Press., New York.

Johnson, G.B. (1979). Enzyme polymorphism: Genetic variation in the physiological phenotype. In: Solbrig, O.T., Jain, S., Johnson, G.B. & Raven, P.H. (Eds.), Topics in Plant Population Biology, pp 62–83. Columbia Univ. Press., New York.

Jones, D.G. & Hardarson, G. (1979). Variation within and between white clover varieties in their preference for strains of *Rhizobium trifolii*. Ann. Appl. Biol. 92: 221–228.

King, D. & Roughgarden, J. (1982a). Multiple switches between vegetative and reproductive growth in annual plants. Theor. Pop. Biol. 21: 194–204.

King, D. & Roughgarden, J. (1982b). Graded allocation between vegetative and reproductive growth for annual plants in growing season of random length. Theor. Pop. Biol. 22: 1–16.

King, D. & Roughgarden, J. (1983). Energy allocation patterns of the California grassland annuals *Plantago erecta* and *Clarkia rubicunda*. Ecology: 16–24.

Kuiper, D. & Smid, A. (1985). Genetic differentiation and phenotypic plasticity in *Plantago major* ssp. *major*. 1. The effect of differences in level of irradiance on growth, photosynthesis, respiration and chlorophyll content. Physiol. Plant. 65: 520–528.

Léon, R. (1968). Balance d'eau et d'azote dans les prairies à litière des alentours de Zurich. Veröff, Geobot. Inst. ETH Rübel 41: 2–68.

Levin, D.S. & Kerster, H.W. (1974). Gene flow in seed plants. In: Dobzhansky, T., Hecht, M.K. & Steele, W.S. (Eds.), Evolutionary Biology 7, pp 139–220. Plenum Press., New York.

Loach, K. (1968). Seasonal growth and nutrient uptake in a Molinietum. J. Ecol. 56: 433–444.

Lolkema, P.C. (1985). Copper resistance in higher plants. Ph.D. Thesis, Free University Amsterdam.

Lolkema, P.C., Donker, M.H., Schouten, A.J. & Ernst, W.H.O. (1984). The possible role of metallothioneine in copper tolerance of *Silene cucubalus*. Planta 162: 174–179.

MacArthur, R.H. & Wilson, E.O. (1967). The Theory of Island Biogeography. Princeton University Press., Princeton.

MacNair, M.R. (1981). Tolerance of higher plants to toxic materials. In: Bishop, J.A. & Cook, L.M. (Eds.), Genetic Consequences of Man made Change, pp 177–207. Academic Press., London.

MAGS - Ministerium Arbeit, Gesundheid, Soziales Land NRW (1977). Unweltbelastung im Raum Datteln. Analyse und Problem lösung. Landesregierung, Düsseldorf.

Marquenie-van der Werff, M. & Out, T. (1981). The effect of humic acid as Zn complexing agent on water culture of *Holcus lanatus*. Biochem. Physiol. Pflanz. 176: 274–282.

Mathys, W. (1977). The role of malate, oxalate, and mustard oils glucoside in the evolution of zinc-resistance in herbage plants. Physiol. Plant. 40: 130–136.

McNeilly, T. & Antonovics, J. (1968). Evolution in closely adjacent plant populations. IV. Barriers to gene flow. Heredity 23, 205–218.

Mihajlovic, Lj. (1986). Result of investigation on *Orobranche* spp. entomofauna in Yugoslavia and the possibility of using insects for biological control. In: Ter Borg, S. (Ed.), Biology and Control Orobanche LH/VPO Wageningen, pp 118–126.

Milchunas, D.G., Lauenroth, W.K., Dodd, J.L. & MacNary, F.J. (1981). Effects of SO_2 exposure with nitrogen and sulphur fertilization on the growth of *Agropyron smithii*. J. Appl. Ecol. 18: 291–302.

Miles, J. (1985). The pedogenic effects of different species and vegetation types and the implications of succession. J. Soil. Sci. 36: 579–584.

Mogford, D.J. (1978). Pollination and flower colour polymorphism, with special reference to *Circium palustre*. In: Richards, W.J. (Ed.), The Pollination of Flowers by Insects, pp 191–200. Academic Press., London.

Oberdorfer, E. (1970). Pflanzensoziologische Exkursionsflora für Süddeutschland. Ulmer, Stuttgart.

Pacala, S.W. & Silander, J.A. (1985). Neighbourhood models of plant population dynamics. I. Single-species models for annuals. Am. Nat. 125: 385–411.

Radosevich, S.R. & Holt, J.S. (1984). Seed Ecology. Implications for Vegetation Management, Wiley & Sons, New York.

Ramakrishnan, P.S. (1968). Nutritional requirements of the edaphic ecotypes of *Melilotus albus* Medic. New Phytol. 67: 145–157, 301–308.

Rauser, W.E. (1984). Copper binding protein and copper tolerance in *Agrostis gigantea*. Plant Sci. Lett. 33: 239–247.

Rhebergen, H. (1985). Microevolutionary processes in *Festuca rubra*. Ph.D. Thesis, Free University Amsterdam.

Robinson, N.J. & Thurman, D.A. (1986). Involvement of a metallothionein-like copper complex in the mechanism of copper-tolerance in *Mimulus guttatus*. Proc. R. Soc. Lond. B 227: 493–501.

Rosenthal, G.A. & Janzen, D.H. (Eds.) (1979). Herbivores. Their Interaction with Secondary Plant Metabolites. Academic Press., New York.

Roy, J. (1985). Comparison of *Dactylis glomerata* and *Bromus erectus* populations from contrasted successional stages. In: Haeck, J. & Woldendorp, J.W. (Eds.), Structure and Functioning of Plant Populations. 2. Phenotypic and Genotypic Variation in Plant Populations, pp 51—64. North-Holland Publishing Comp., Amsterdam.

Rozema, J., Laan, P., Broekman, R., Ernst, W.H.O. & Appelo, C.A.J. (1985a). On the lime transition and decalcification in the coastal dunes of the province of North Holland and the island Schiermonnikoof. Acta Bot. Neerl. 34: 393–411.

Rozema, J., Luppes, E. & Broekman, R. (1985b). Differential response of salt-marsh species to variation of iron and manganese. Vegetatio 62: 93–301.

Rozijn, N.A.M.G. (1984). Adaptive strategies of some dune annuals. Ph.D. Thesis, Free University Amsterdam.

Russel, E.W. (1973). Soil Conditions and Plant Growth. Longman, London.

Sasse, F. (1976). Ökologische Untersuchungen der Serpentinvegetation in Frankreich, Italien, Österreich und Deutschland. Ph.D. Diss., Universität Münster.

Schulze, E.D. (1972). Die Wirkung von Licht und Temperatur auf den CO_2-Gaswechsel verschiedener Lebensformen aus der Krautschicht eines montanen Buchenwaldes. Oecologia 9: 225–258.

Sukopp, H. (1969). Der Einfluß des Menschen auf die Vegetation. Vegetatio 17: 360–371.

Turesson, G. (1922). The genotypical response of the plant species to the habitat. Hereditas 3: 211–350.

Van Andel, J., Rozijn, N.A.M.G., Ernst, W.H.O. & Nelissen, H.J.M. (1986). Variability in growth and reproduction in F_1-families of an *Erophila verna* population. Oecologia 69: 79–85.

Van der Meijden, E. (1979). Herbivore exploitation of a fugitive plant species: local survival and extinction of cinnabar moth and ragwort in a heterogenous environment. Oecologia 42: 307–323.

Van der Toorn, J. & Mook, J.H. (1982). The influence of environmental factors and management of stands of *Phragmatis australis*. J. Appl. Ecol. 19: 477–499.

Van der Werff, M. (1981). Ecotoxicity of heavy metals in aquatic and terrestrial higher plants. Ph. D. Thesis, Free University Amsterdam.

Van Rhee, J.S. (1975). Copper contamination effects on earthworms by disposal of pig waste in pastures. In: Vanek, J. (Ed.), Progress in Soil Zoology, pp 451–457. Academia, Prague.

Verkleij, J.A.C. (1980). Isoenzyme variation in pioneer species from ecologically different habitats. Ph.D. Thesis, Free University Amsterdam.

Watkinson, A.R. (1980). Density-dependence in single-species populations of plants. J. Theor. Biol. 183: 345–357.

Westhoff, V. & den Held, A.J. (1969). Plantengemeenschappen in Nederland. Thieme & Cie, Zutphen.

White, D.J.B. (1961). Some observations on the vegetation of Blakeny Point, Norfolk, following the disappearance of rabbits in 1954. J. Ecol. 49: 113–118.

Wilkinson, S.R. & Lowry, R.W. (1973). Cycling of mineral nutrients in pasture ecosystem. In: Butler, G.W., & Bailey, R.W. (eds), Chemistry and Biochemistry of Herbage, pp 247–315. Academic Press., London.

Wilson, D.S. (1980). The Natural Selection of Populations and Communities. Cummings, Meno Park.

Wood, M., Cooper, J.E. & Holding, A.J. (1984). Soil acidity factors and nodulation of *Trifolium repens*. Plant Soil 78: 381–391.

Wu, L., Bradshaw, A.D. & Thurman, D.A. (1975). The potential for evolution of heavy metal tolerance in plants. Heredity 34: 165–187.

Ydenburg, R.C. & Prins, H.H.Th. (1981). Spring grazing and the manipulation of food quality by barnacle geese. J. Appl. Ecol. 18: 443–453.

CHAPTER 15

Population responses in *Agrostis stolonifera* to selective forces in inland and coastal habitats

C. Kik

Detailed contents

1. Introduction

Coastal habitats, such as salt marshes and sand dunes impose different selection forces on plant species than do inland habitats. Coastal populations of grassland species are affected by inundation of seawater or by airborne salinity, by drought, by low nutrient availability and by sand. Inland populations of grassland species are primarily influenced by soil chemical factors (Pegtel, 1983) and by groundwater regimes (Grootjans, 1985), which in turn are dominated by a grazing and/or mowing regime (Bakker et al., 1980, 1983). These differences in selection regime can lead to the formation of genetically distinct and locally adapted populations (Aston & Bradshaw, 1966; Silander & Antonovics, 1979) and/or to the dominance of genotypes which possess a high degree of phenotypic plasticity (Bradshaw, 1965; Waite & Hutchings, 1982). The underlying genetical and evolutionary processes of population differentiation are, however, still poorly understood (Antonovics, 1978; Dingle & Hegmann, 1982).

The work described in this paper aims to analyse:

1. The extent of genetical and environmental determinants of survival and reproduction (i.e. parameters of fitness) in populations of *Agrostis stolonifera* L.*
2. The influence of genetic variation in an important ecological trait (salt tolerance) upon the fitness (survival) of an individual transplant under natural conditions (salt marsh). Salt tolerance was chosen, because populations of *Agrostis stolonifera* are known to differ markedly for this attribute (Ahmad & Wainwright, 1977; Rozema & Blom, 1977; Wu, 1981).

2. Study area

Plants were collected from three contrasting sites (Fig. 1).

The meadow population was collected from the brook valley system of the Drentsche Aa, where

J. van Andel et al., (eds.), Disturbance in Grasslands. ISBN 90–6193–640–3.

* Nomenclature follows Heukels & Van Ooststroom (1977)

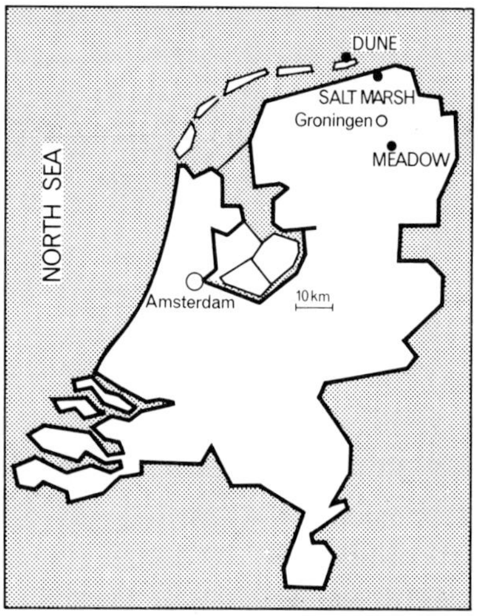

Fig. 1. Location of the sampled *Agrostis stolonifera* populations in the Netherlands.

a nutrient rich topsoil (Table 1) overlies sand. The soil is moist throughout the year and waterlogged during winter. The meadow is cut for hay in August/September.

The salt marsh population was collected on the zone just above mean high water level, from a calcareous clay soil, relatively saline throughout the year and with a nitrogen content intermediate between the inland meadow and sanddune soil (Table 1). The site is infrequently flooded by the sea throughout the whole year. In summer it is grazed by cattle.

The dune population was collected from an embryo dune; the sandy soil was deficient in macronutrients and had a low level of soil moisture and salt (Table 1). The accretion and abrasion of sand in this habitat are important.

3. Experimental design

3.1 Sampling procedure

Fifty tillers were randomly collected in each site in April 1983 and grown in a heated glasshouse in potting compost before being used in experiments.

3.2 Reciprocal transplant

The three populations were transplanted into each of the three collection sites in June 1983. Each population was represented by approximately forty individuals, each planted as a single tiller at each transplantation site; individuals were distributed at random among four replications in a completely randomized block design (Cochran & Cox, 1957). Each transplant was placed in a four centimeter deep hole, prepared with a knife, and marked with yellow adhesive tape. They could easily be traced using a plotter (Cullen et al., 1978). An acclimatization period of two months was included, to allow establishment of transplants. Survival of every transplant was recorded from August 1983 onwards, tiller and inflorescence production from May 1984.

3.3 Salt tolerance experiment

A modification of the rooting technique by Hannon & Bradshaw (1968) was used. The same individuals of the three populations were used as in the reciprocal transplant experiment, so that salt tolerance could be related to survival in the salt marsh site.

Each individual transplant was seperated into

Table 1. Soil characteristics (0–5 cm) of the three sites. The range of values presented are the minimum and maximum values found when soils were sampled every two months during one year. All soil analysis were performed according to Hofstee & Fien (1971).

Population	N-mineral mg N kg^{-1} dry soil	P mg P_2O_5 kg^{-1} dry soil	K^+ mg K^+ kg^{-1} dry soil	Soil moisture g H_2O kg^{-1} wet soil	NaCl g NaCl l^{-1} soil moisture
Meadow	32–67	350–530	110–280	32–32	0.06– 0.20
Salt marsh	13–23	330–570	520–770	20–40	2.00–29.00
Sand dune	0– 4	30– 40	20– 30	0–15	0.07– 6.00

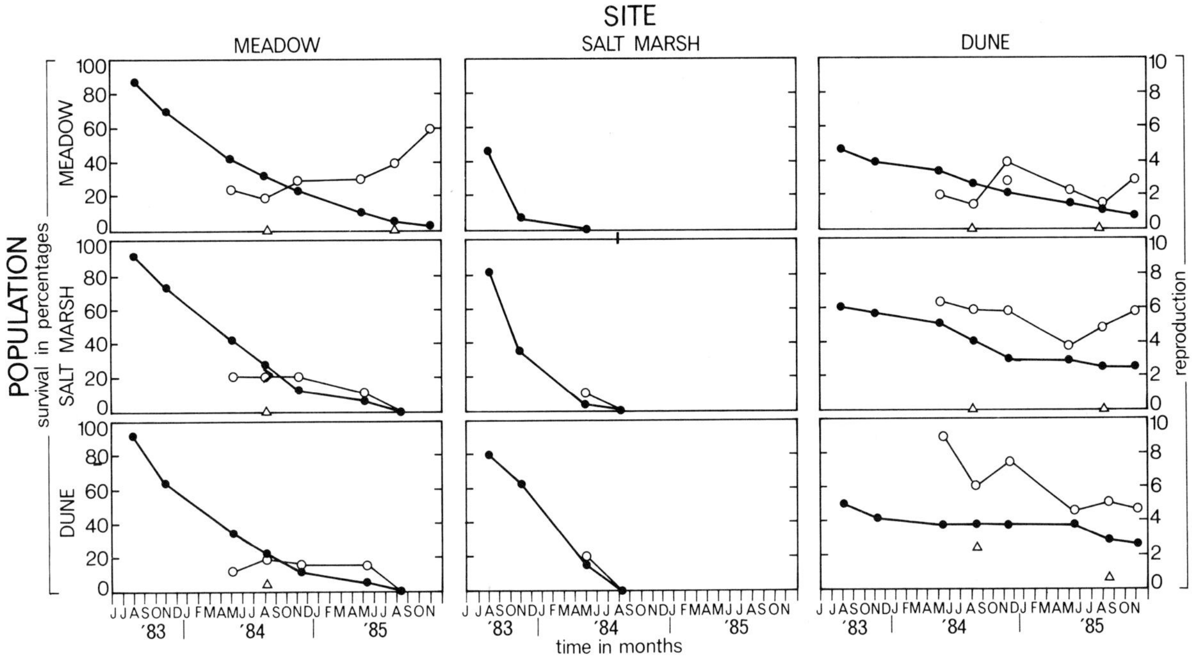

Fig. 2. The survival (●), and reproduction (number of tillers per survivor (0) and number of inflorescences per survivor (△) of three populations of *Agrostis stolonifera* transplanted into each of the three sites of origin.

ten similar tillers, which were grown in a Steiner solution (Steiner, 1961) for one week; the length of the longest root of each tiller was then measured. The salt concentration of the culture solution of five tillers per individual was then raised stepwise to 200 mM NaC1 in four days, while the other five tillers remained in the control solution. Twelve days after the first measurement, the final length of the longest root per tiller was measured.

The salt tolerance per individual was calculated as the ratio of the mean extention of the longest root in the salt solution and the mean extension of the longest root in the control solution. On basis of the ratio's per individual of the growth of the longest root in the salt solution and the mean growth of the longest root in the control solution the repeatability (Falconer, 1981) per population was calculated.

4. Results

4.1 Reciprocal transplant

4.1.1 Survival

Site effects on survival were statistically significant (Table 2). Survival was least in the salt marsh site, where no plants survived after one year (Fig. 2). Survival in the sand dune proved to be highest, while the meadow site was intermediate between both other sites (Fig. 2). All comparisons between survival in the different sites were statistically significant.

Seasonal effects on survival were also appar-

Table 2. Survivorship analysis of three populations of *Agrostis stolonifera* transplanted into each site of origin.

Source of variation	df	MS	P
Population	2	22.103	0.620
Site	2	3677.506	0.001
Population × site	4	105.398	0.061
Total	270	73.638	

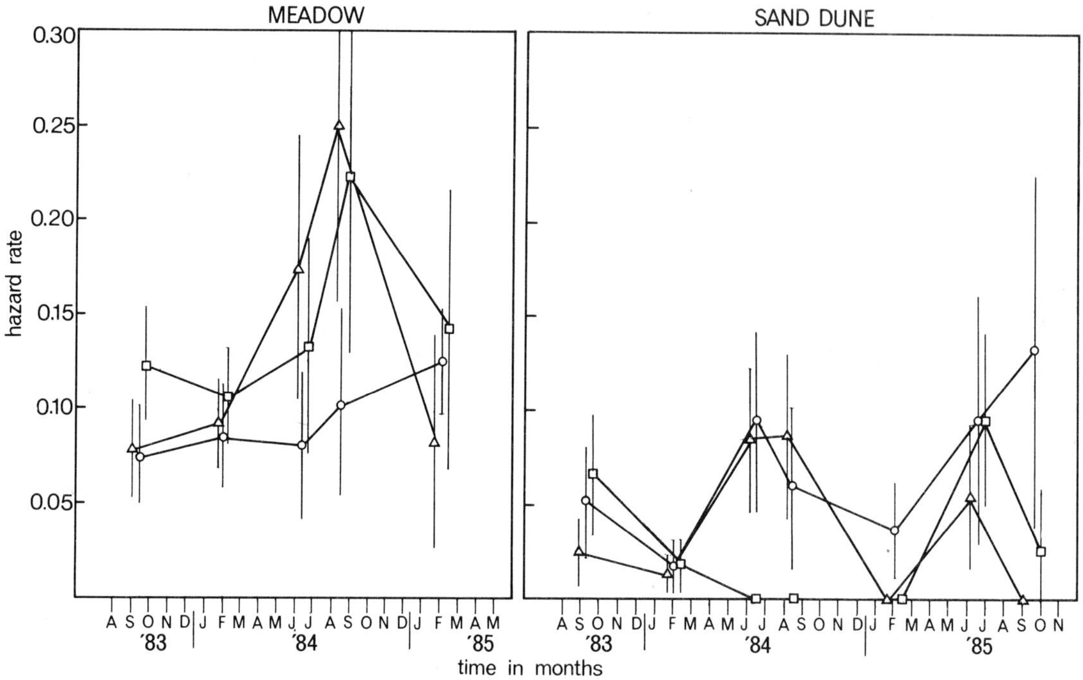

Fig. 3. The hazard rates (means and their standard errors) per time period of three populations (0: Meadow, △: Salt Marsh, □: Sand dune) of *Agrostis stolonifera* transplanted into two sites of origin (Meadow and Sand dune).

ent, as was revealed by the hazard rate (Fig. 3). The hazard rate, as defined by Gehan (1965), is the chance that an individual will die in a certain time interval, given it has survived up until the beginning of the interval. In the sand dune site the alien populations showed an alternating pattern of high hazard rates in summer (May-November) and low hazard rates in winter (November-May) (Fig. 3). The home population did not show such a pattern (Fig. 3). Especially in the summer of 1984 the differences in hazard rate between both aliens and home population were statistically significant. The same trends could be seen in the meadow site, although to a lesser degree (Fig. 3). In the period between November 1983 and November 1984 the hazard rates of the aliens were in summer higher than in winter. The home population did not show such a response: it is increased slightly (Fig. 3). Once again in the summer period of 1984 the differences between alien populations and the home population were statistically significant.

Population effects on survival were not statistically significant (Table 2). Population x site interaction on survival was almost statistically significant (Table 2), indicating local adaptation.

4.1.2 Fecundity

Vegetative propagation. In May, August and No-

Table 3. Analysis of the number of tillers on four dates of three populations of *Agrostis stolonifera* transplanted into two sites of origin (meadow and sand dune).

Source of variation	May 1984			August 1984			November 1984			May 1985		
	df	MS	P	df	MS	P	df	MS	P	df	MS	P
Population	2	50.405	0.037	2	68.208	0.003	2	51.386	0.049	2	9.445	0.266
Site	1	493.397	0.001	1	98.535	0.004	1	99.675	0.016	1	19.660	0.099
Population × site	2	184.289	0.001	2	44.016	0.022	2	37.429	0.107	2	6.225	0.414
Total	104	23.759		81	14.964		55	21.308		46	7.536	

vember 1984 site effects were significant (Table 3). The dune site proved to be the site where significantly (P <0.05) the highest tiller production per survivor was found (Fig. 2).

Statistically significant population effects were also found in May, August and November 1984 (Table 3). The sand dune population had significantly (P <0.05) the highest tiller production compared to the other two populations. The meadow and salt marsh populations did not differ significantly.

Population x site interactions was also found in May, August and November 1984 (Table 3, Fig. 2), indicating local adaptation.

Generative production. In the sand dune site significantly more inflorescences were produced compared to the meadow site (Table 4, Fig. 2).

In the summer of 1984 the populations produced significantly more inflorescences than in the summer of 1985 (Table 4, Fig. 2).

The sand dune population did have the highest inflorescence production compared with the other populations (P <0.05). The meadow and salt marsh populations did not differ significantly.

Strongly significant interactions were detected between population and site respectively population and year, indicating local adaptation respectively differential influences of summer periods on population inflorescence production (Table 4, Fig. 2).

Table 4. Analysis of inflorescence number over two years of three populations of *Agrostis stolonifera* transplanted into two sites of origin (meadow and sand dune).

Source of variation	df	MS	P
Population	2	11.637	0.001
Site	1	7.567	0.001
Year	1	8.595	0.001
Population × site	2	7.213	0.001
Population × year	2	6.893	0.001
Site × year	1	0.000	0.999
Total	117	1.058	

Table 5. Repeatabilities (r) of salt tolerance in three populations of *Agrostis stolonifera* L.

Population	r
Meadow	0.68
Salt marsh	0.74
Sand dune	0.70

4.2 Salt tolerance and reciprocal transplant

The populations differed significantly in salt tolerance (P <0.001) when compared with the variation between individuals within populations. The salt marsh population was most tolerant to salt, followed by the dune population, the meadow population was least tolerant.

There were significant differences between individuals within populations in salt tolerance (Table 5).

The frequency distribution of tolerance is shown in Fig. 4A. The differences in salt tolerance can be related to survival of these individuals, when transplanted into the salt marsh site (Fig. 4B). There was some evidence of a positive relationship for the meadow population between salt tolerance and survival in the salt marsh two months after transplanting; after that date insufficient plants survived to allow comparisons. No such relationship was found for either salt marsh or dune population. (Fig. 4B).

5. Discussion

5.1 Reciprocal transplant experiment

The reciprocal transplant method can be used to investigate potential responses of a population to disturbance; survival and fecundity, which are known to be of primarily importance in the analysis of life histories (Stearns, 1976) can be used as criteria. In this transplant study the environmental component determined survival and fecundity more than did the population (genetical) component. This finding was in agreement with previous reports on for example *Ranunculus repens* (Lovett Doust, 1981) and *Plantago lanceolata* (Antonovics & Primack, 1982). However, significant population x site interactions for survival

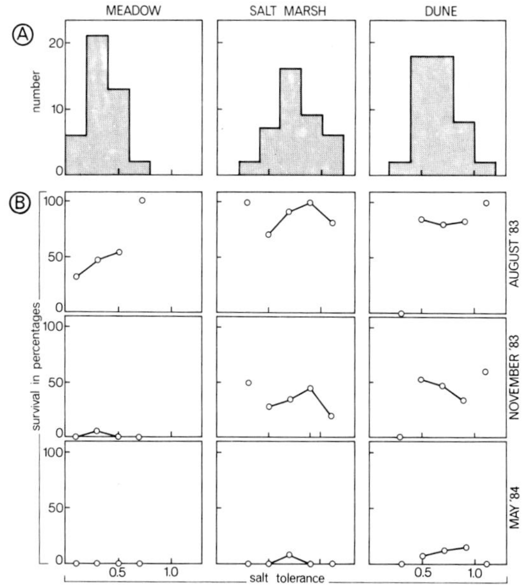

Fig. 4. A) The frequency distribution of salt tolerance among individuals of three populations of *Agrostis stolonifera*. B) The relationship between salt tolerance and survival on three dates, when transplanted into a salt marsh site. Salt tolerance classes with more than three individuals were connected with a solid line.

and fecundity demonstrated the existence of more or less distinct populations of *A. stolonifera* (Tables 2, 3, & 4); the populations are not so differentiated that they only survive and reproduce in their own native environment.

Significant seasonal variations in survival (Fig. 3) were found in both the meadow and dune sites; the summer was the crucial time for survival as in the work on natural populations of *Ranunculus repens* by Sarukhán and Harper (1973). Survival and fecundity in the meadow was not influenced by the same factors as at the dune site. Summer drought was probably the most important factor in the dune site. This was most apparent when the hazard rates of the alien populations were compared with the home population in the summer of 1984; a period with high temperatures. Interference and other biotic factors may be of primarily importance in the meadow site.

The extremely low survival in the salt marsh site may have been a result of high summer temperatures in 1984 and consequent increases in salinity of the topsoil layer.

The use of only the adult phase of the life cycle to investigate the intensity of natural selection probably influenced the estimation of it (Hickey & McNeilly, 1975; McGraw & Antonovics, 1983). However, only in the sand dune population the seed and seedling phases are a normal part of the life cycle, whereas in the meadow and salt marsh population the prevailing mode of reproduction is vegetative propagation. An underestimation of the force of natural selection is therefore only to be expected in the sand dune population.

The analyses of life history characteristics and salt tolerance were carried out on the ramet level. Because *A. stolonifera* is a clonal perennial it might be possible to sample more than once the same genet within a population. The consequences could be that the variability in the different populations is underestimated, depending upon the degree of clonal duplication. Especially in the meadow population this might be of significance.

5.2 Salt tolerance and reciprocal transplant experiment

There was considerable variation in salt tolerance between individuals within the meadow population (Table 5). This might point at a maritime origin for this population of *A. stolonifera* or it may reflect naturally occurring genetic variation present in the species.

The finding of similar variation in both sand dune and salt marsh populations was perhaps less unexpected since both populations have been influenced by the sea resulting in selection for salt tolerance; selection for drought tolerance in the sand dune habitat may affect characteristics governing osmoregulation which may affect salt tolerance.

5.3 Concluding remarks

The meadow population of *A. stolonifera* reproduced almost exclusively by vegetative means (Fig. 5). The responses of such a population to disturbance is therefore primarily restricted to phenotypic plasticity. On the other hand the sand

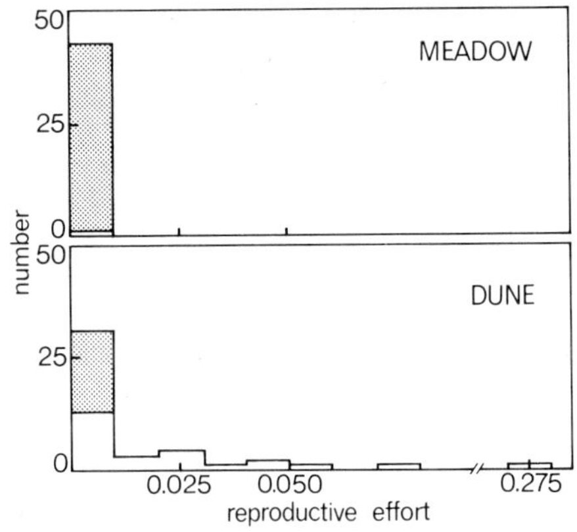

Fig. 5. The frequency distribution of reproductive effort (calculated as the ratio generative biomass and total aboveground biomass) of individuals within two populations of *Agrostis stolonifera*. Shaded area indicates number of individuals which did not flower (Kik, unpublished results).

dune population of *A. stolonifera* has a strong sexual component (Fig. 5) and can respond to disturbance by genetic as well as plastic responses. It could therefore be hypothesized that within the species *A. stolonifera* in a mesic environment (meadow) the production of more identical individuals is at a premium, whilst in a more abiotically harsh environment (sand dune) the generation of genetic variation is at a premium. The production of large numbers of seed could, however, also be a mechanism of spreading the risk of becoming extinct.

Acknowledgements

I would like to thank Prof. Dr. J. van Andel, my colleagues of the Laboratory of Plant Ecology, Dr. W. Joenje, Agricultural University of Wageningen and Dr. R.W. Snaydon, University of Reading (U.K.) for helpfull discussions and comments on earlier versions of this manuscript; Mr. J.J. de Wiljes and Mr. W.A. van Hal for soil chemical analysis; Mrs. Th.E. Linders, Mr. S. Nijdam, Mr. J. Franke, Mr. J.J. Hogendorf for technical assistance; Mr. E. Top for constructing the plotter; Mrs. C.W. Biewenga for typing various stages of the manuscript; Mr. E. Leeuwinga for preparing the figures and Mrs. J. Purdell-Lewis for correcting the english text.

References

Ahmad, I. & Wainwright, S.J. (1977). Tolerance to salt, partial anaerobiosis, and osmotic stress in *Agrostis stolonifera*. New Phytol. 79: 605–612.

Antonovics, J. (1978). The nature of limits to natural selection. Ann. Mis. Bot. Garden 63: 224–247.

Antonovics, J. & Primack, R.B. (1982). Experimental ecological genetics in *Plantago*. VI. The demography of seedling transplants of *P. lanceolata*. J. Ecol. 70: 55–71.

Aston, J.L. & Bradshaw, A.D. (1966). Evolution in closely adjacent plant populations. II. *Agrostis stolonifera* in maritime habitats. Heredity 21: 649–664.

Bakker, J.P., Dekker, M. & Vries, Y. de (1980). The effects of different management practices on a grassland community and the resulting fate of seedlings. Acta Bot. Neerl. 29: 469–482.

Bakker, J.P., Bie, S. de, Dallinga, J.H., Tjaden, P. & Vries, Y. de (1983). Sheep-grazing as a management tool for heathland conservation and regeneration in the Netherlands. J. Appl. Ecol. 20: 541–560.

Bradshaw, A.D. (1965). Evolutionary significance of phenotypic plasticity in plants. Adv. Genet. 13: 115–155.

Cochran, W.G. & Cox, G.M. (1957). Experimental designs. Wiley, New York.

Cullen, J.M., Weiss, P.M. & Werne, G.R. (1978). A plotter for plant population studies. New Phytol. 81: 443–448.

Dingle, H. & Hegmann, J.P. (1982) (Eds.). Evolution and genetics of life histories. Springer Verlag, New York - Heidelberg - Berlin.

Falconer, D.S. (1981). Introduction to Quantitative Genetics. Longman Ltd., New York.

Gehan, E. (1975). Statistical methods for survival time studies. In: Staquet, M.J. (Ed.). Cancer therapy: prognostic factors and criteria of response, pp 7–35. Raven Press, New York.

Grootjans, A.P. (1985). Changes of groundwater regime in wet meadows. Thesis, University of Groningen.

Hannon, N. & Bradshaw, A.D. (1986). Evolution of salt tolerance in two co-existing species of grass. Nature 220: 1342–1343.

Heukels, H. & Ooststroom, S.J. van (1977). Flora van Nederland. Wolters-Noordhoff, Groningen.

Hickey, D.A. & McNeilly, T. (1975). Competition between metal tolerant and normal plant populations: A field experiment on normal soil. Evolution 29: 464–485.

Hofstee, J. & Fien, H.J. (1971). Analyse methoden voor grond, gewas, water en bodemvocht. Rijksdienst voor de IJsselmeerpolders, Lelystad/Kampen.

Lovett Doust, L. (1981). Population dynamics and local specialization in a clonal perennial (*Ranunculus repens*). II. The dynamics of leaves and a reciprocal transplant-replant experiment. J. Ecol. 69: 757–768.

McGraw, J.B. & Antonovics, J. (1983). Experimental ecology of *Dryas octopetala* exotypes. I. Ecotypic differentiation and life-cycle stages of selection. J. Ecol. 71: 879–898.

Pegtel, D.M. (1983). Ecological aspects of a nutrient-deficient wet grassland (*Cirsio-Molinietum*). Verh. der Gesellschaft für Ökologie (Mainz, 1981) Band X: 217–228.

Rozema, J. & Blom, B. (1977). Effects of salinity and inundation on the growth of *Agrostis stolonifera* and *Juncus gerardii*. J. Ecol. 65: 213–222.

Sarukhán, J. & Harper, J.L. (1973). Studies on plant demography: *Ranunculus repens* L., *R. bulbosus* and *R. acris* L. I. Population flux and survivorship. J. Ecol. 61: 675–716.

Silander, J.A. & Antonovics, J. (1979). The genetic basis of the ecological amplitudo of *Spartina patens*. I. Morphometric and physiological traits. Evolution 33: 1114–1127.

Stearns, S.C. (1976). Life-history tactics: a review of the ideas. Q. Rev. Biol. 51: 3–47.

Steiner, A.A. (1961). An universal method for preparing nutrient solutions of a certain desired composition. Plant and Soil XV: 134–154.

Waite, S. & Hutchings, M.J. (1982). Plastic energy allocation patterns in *Plantago coronopus* L. Oikos 38: 333–342.

Wu, L. (1981). The potential for evolution of salinity tolerance in *Agrostis stolonifera* L. and *Agrostis tenuis* Sibth. New Phytol. 89: 471–486.

CHAPTER 16

A physiological analysis of genotypic variation in relative growth rate: Can growth rate confer ecological advantage?

H. Lambers & P. Dijkstra

Detailed contents

1. Introduction

Species from contrasting habitats differ in their relative growth rate (RGR) (Grime & Hunt, 1975). Species from nutrient-poor, acid soils, e.g., *Agrostis setacea* and *Deschampsia flexuosa,* tend to have inherently low growth rates compared with species from nutrient-rich sites, e.g., *Agrostis stolonifera* and *Urtica dioica* (Clarkson, 1967; Rorison, 1968). A frequently quoted paper to support this contention is that by Bradshaw et al. (1964). However, these authors measured *yield,* rather than *growth rate,* of a number of species from rich and poor habitats. Differences in final yield may simply reflect differences in initial size, or differences in the duration of the maximum relative growth rate, as demonstrated by Van de Dijk (1980) in his work on two subspecies of *Hypochaeris radicata,* one from a fairly nutrient-rich roadside, the other from a poor heathland. For this reason, and also since RGR may show ontogenetic drift (Van Andel & Jager, 1981), a careful growth analysis, rather than only one harvest, is required.

Grime & Hunt (1975) compared the growth rate, during the exponential phase, of 132 species from a range of habitats in Britain. Woody species and fine-leaved grasses tended to have a low RGR, whilst annuals generally exhibited high rates. Jarvis & Jarvis (1964) also found lower growth rates in woody species (*Betula verrucosa, Populus tremula, Pinus sylvestris* and Picea abies) compared with *Helianthus annuus* (Table 1a). Fast growing species were usually associated with 'disturbed' and/or 'productive' habitats, whilst slow growing species originated from 'stable and unproductive' sites (Grime & Hunt, 1975). A high potential RGR would thus seem to be characteristic of species from 'disturbed' habitats.

Ingestad and coworkers, followed by other Swedish workers, have compared the RGR of seedlings of various deciduous and evergreen tree species, under both limiting and optimum nutrient conditions. The relative growth rate of the ev-

J. van Andel et al., (eds.), Disturbance in Grasslands. ISBN 90–6193–640–3.

ergreen tree species *Pinus sylvestris*, *Picea abies* and *Pinus contorta* never exceeded 74 mg g^{-1} d^{-1}, even under conditions of optimum nutrient supply (Ingestad & Kahr, 1985). The growth rate of the deciduous species *Betula alba and* Alnus incana was as high as 250 mg g^{-1} d^{-1} at the same photosynthetic photon flux density and under conditions of optimum nutrient supply (Ingestad, 1981). For *Populus simonii* and *Paulownia tomentosa* an RGR of 190 and over 250 mg g^{-1} d^{-1} was found (Jia and Ingestad, 1984). Ericsson (1981) found RGR's over 200 mg g^{-1} d^{-1} for three *Salix* species and Waring et al. (1985) found a rate of 161 mg g^{-1} d^{-1} for the deciduous *Salix aquatica* under optimum nutrient conditions. Considering that RGR tends to decrease with increase in plant size (e.g. Dijkstra & Lambers, 1986), and that the weight of the evergreen plants was lower than that of their deciduous tree seedlings, Jarvis & Jarvis' (1964) data also suggest a lower RGR for evergreens (Table 1a). Taken together, the data demonstrate a large variation in RGR among tree seedlings, with a bias towards low values for evergreens.

Table 1a. The relative growth rate (RGR; mg g^{-1} d^{-1}) of two deciduous tree species (*Betula verrucosa* and *Populus tremula*), two evergreen tree species (*Pinus sylvestris* and *Picea abies*) and one crop species (*Helianthus annuus*) grown in nutrient solution. The net assimilation rate (NAR; g m^{-2} day^{-1}) and leaf area ratio (LAR; m^2 kg^{-1}) for the two evergreens were not available from the source. The final dry weight of the seedlings (mg $plant^{-1}$) is included to show that the RGR of the evergreen and deciduous tree seedlings must not be compared directly. Data from Jarvis and Jarvis (1964).

Species	Final dry weight	RGR	NAR	LAR
B. verrucosa	237	117	7.3	16.2
P. tremula	676	131	6.9	19.4
H. annuus	598	239	10.6	22.6
P. sylvestris	54	101	–	–
P. abies	39	64	–	–

Table 1b. The relative growth rate (RGR; mg g^{-1} d^{-1}), net assimilation rate (NAR; g m^{-2} day^{-1}), leaf area ratio (LAR; m^2 kg^{-1}), specific leaf area (SLA; m^2 kg^{-1}) and leaf weight ratio (LWR; g g^{-1}) of one shade-avoiding species (*Galinsoga parviflora*) and two shade-tolerant species (*Stachys sylvatica* and *Urtica dioica*), grown at full sunlight. Data from Corré (1983).

Species	RGR	NAR	LAR	SLA	LWR
G. parviflora	339	8.1	42	70	0.60
S. sylvatica	245	6.6	37	61	0.61
U. dioica	280	7.0	40	59	0.68

It is the aim of this paper to compile information about genotypic differences in RGR and their morphological, physiological or biochemical basis. A second aim, possibly intricately interwoven with the first, is to analyse the ecological importance of a low relative growth rate. What advantage have slow growing species that they have not been outcompeted by the species with a high potential growth rate? Fast growing species may be more competitive in a rich environment, but growth rate is by no means the only factor determining the competitive ability of a species (Roush & Radosevich, 1985). The selective advantage of an inherently low growth rate is not immediately obvious. For an ecophysiologist there is the additional intriguing question about the basis of inherent differences in RGR. We believe that through a careful analysis of the physiological, biochemical and morphological background of variation in RGR, further insight can be gained into the correlation between a high RGR and the extent of 'disturbance' of the natural habitat.

2. The functional growth analysis

A functional growth analysis, based on experiments in which plants are grown under standard conditions and harvested at regular intervals, can provide the first clues towards an understanding of variation in RGR between genotypes. The relative growth rate (RGR; in mg g^{-1} (total plant) d^{-1}) can be calculated (e.g. Hunt, 1982):

$$RGR = 1/W \times (dW)/(dt) \quad (1)$$

where W is the weight (g) of the investigated plants, and dW is the dry weight increment (mg), during the interval dt (days).

Differences in RGR may be due to a 'physiological' component, the net assimilation rate (NAR;

cal' component, the net assimilation rate (NAR; in g m^{-2} (leaf area) d^{-1}), which is largely the balance of photosynthesis and respiration in the entire plant:

$$NAR = 1/LA \times (dW)/(dt) \quad (2)$$

where LA is the leaf area of the investigated plants (m^2). Differences in RGR may also be due to a 'morphological' component, the leaf area ratio (LAR; m^2 (leaf area) kg^{-1} (total plant)).

$$LAR = LA/W \quad (3)$$

RGR is the product of NAR and LAR:

$$RGR = NAR \times LAR \quad (4)$$

The leaf area ratio can be further analysed. Variation may be due to a difference in 'investment' in leaf biomass, i.e. a difference in leaf weight ratio (LWR; g(leaf weight) g^{-1}(total plant weight), or to a difference in specific leaf area (SLA; m^2 (leaf area) kg^{-1}(leaf weight):

$$LAR = LWR \times SLA \quad (5)$$

A functional growth analysis is the first step in any analysis of morphological, physiological or biochemical factors determining RGR. However, surprisingly few such comparisons have been made between species or genotypes. Some examples are given below of analyses of RGR of different species and genotypes, most of them based on the above functional growth analysis.

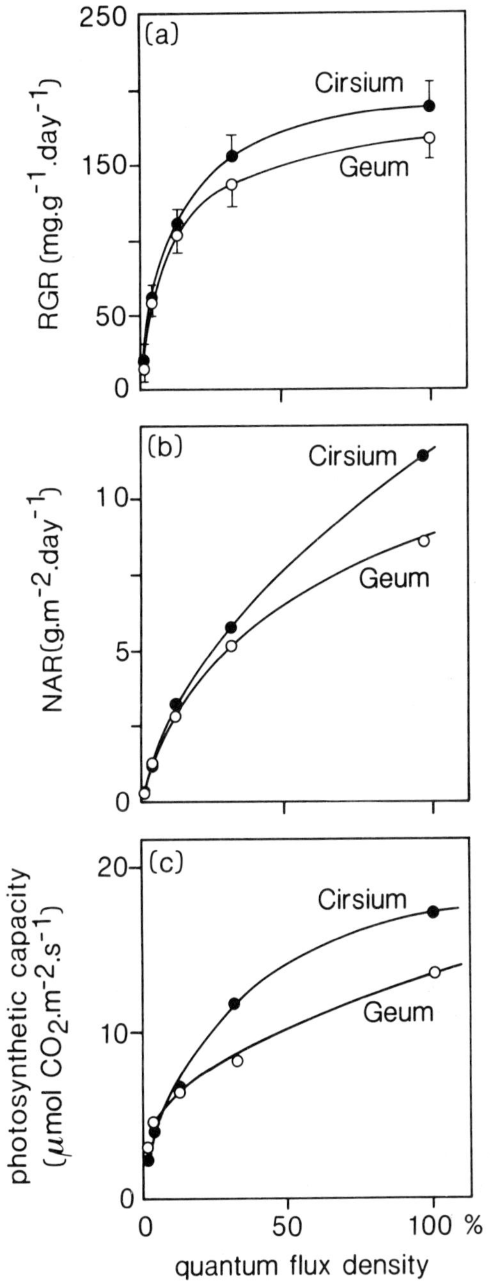

Fig. 1. The relative growth rate (a), the net assimilation rate (b), and the photosynthetic capacity (c) of the shade-avoiding species, *Cirsium palustre*, and the shade-tolerant species, *Geum urbanum*, at different quantum flux densities (data from Pons, 1977a). A quantum flux density of 100% refers to full sunlight in July, 1975 in De Bilt, the Netherlands.

3. Some analyses of inherent differences in relative growth rate

Pons (1977a) compared the growth of the shade-avoiding *Cirsium palustre* and the shade-tolerant *Geum urbanum,* an early and late successional species from the field layer after coppicing of ash trees. At a high quantum flux density the growth rate of the early successional, shade-avoiding species was significantly greater than that of the shade-tolerant species (Fig. 1a). The difference in RGR was apparently caused by a difference in NAR (Fig. 1b), which was reflected in a greater photosynthetic capacity (Fig. 1c). The 50%

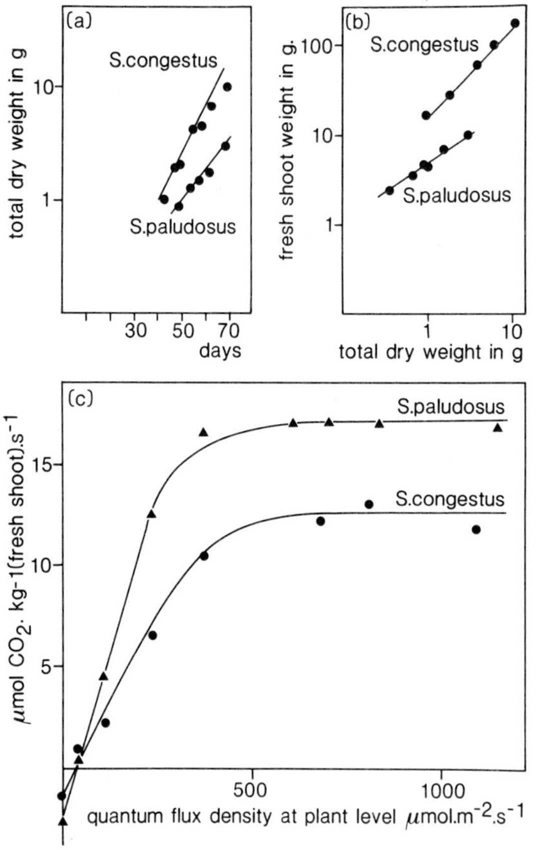

Fig. 2. The total dry weight increment (a), fresh shoot weight as a function of total plant dry weight (b), and net CO_2 assimilation at a range of quantum flux densities of *Senecio congestus*, a pioneer species from nutrient-rich soil, and *S. paludosus*, a species later in the succession (data from Van Dobben, 1967).

greater LWR in *Cirsium* was more than compensated for by a reduction in SLA, so that the LAR was slightly lower than in *Geum*. A similar observation was made on ecotypes of *Dactylis glomerata* from Norway and Portugal (Eagles, 1967). At low temperatures (5–10°C) the Norwegian ecotype had both a higher RGR and a higher NAR, but a lower LAR, in comparison with the one from Portugal. At 30°C, the RGR and NAR of the Portugese ecotype were superior, whilst the LAR was lower than that of the Norwegian one.

Van Dobben (1967) compared the growth of two *Senecio* species: *S. congestus* and *S. paludosus* (Fig. 2a). *S. congestus* is a pioneer winter annual on bare wet soils. It is very productive under conditions of ample nitrogen supply, but declines under poor nutritional conditions, even in the absence of competitors (Bakker, 1960). *S. paludosus* is a perennial from the reed beds along the great rivers in the Netherlands. Since data on leaf area are lacking, the rate of photosynthesis of the two species is expressed per g(fresh weight of shoot) (Fig. 2c). The fast-growing *S. congestus* photosynthesized less, and the difference in growth rate between the two species was apparently explained by the larger shoot fresh weight per total plant dry weight (Fig. 2c), which probably reflected a greater LWR, i.e. a difference in investment of mass in photosynthetic tissue.

Our own analysis of the growth of two subspecies of *Plantago major* has also led to the conclusion that the 'morphological' component offers an explanation for variation in RGR. Subspecies *major* is found in fairly undisturbed habitats, e.g. road verges and lawns, whereas subspecies *pleiosperma* is found in less predictable or more disturbed habitats, e.g. river banks, which are flooded in winter and occasionally also in summer, and arable land (Van Dijk, 1984). The subspecies also differ in the number and weight of seeds produced. Subspecies *pleiosperma* has an inherently higher RGR than subspecies *major* (Kuiper, 1983; Dijkstra & Lambers, 1986), also under nutrient-limited conditions (Poorter & Lambers, 1986).

Two inbred lines were chosen for a detailed growth analysis (Fig. 3). Although subspecies *pleiosperma* had a higher RGR (Fig. 3a), it had a lower NAR (Fig. 3b). However, this lower NAR was more than compensated by a higher LAR, largely due to a higher SLA and to a minor extent a higher LWR (Figs. 3c, 3d). We therefore conclude that the high growth rate of subspecies *pleiosperma* is associated with a large investment in photosynthetically active area (a high LAR) and not with the carbon exchange rate (photosynthesis or respiration).

A comparison of the growth of seedlings of the tree species *Betula verrucosa* and *Populus tremula* with that of the crop species *Helianthus annuus* (Jarvis & Jarvis, 1964) showed that both

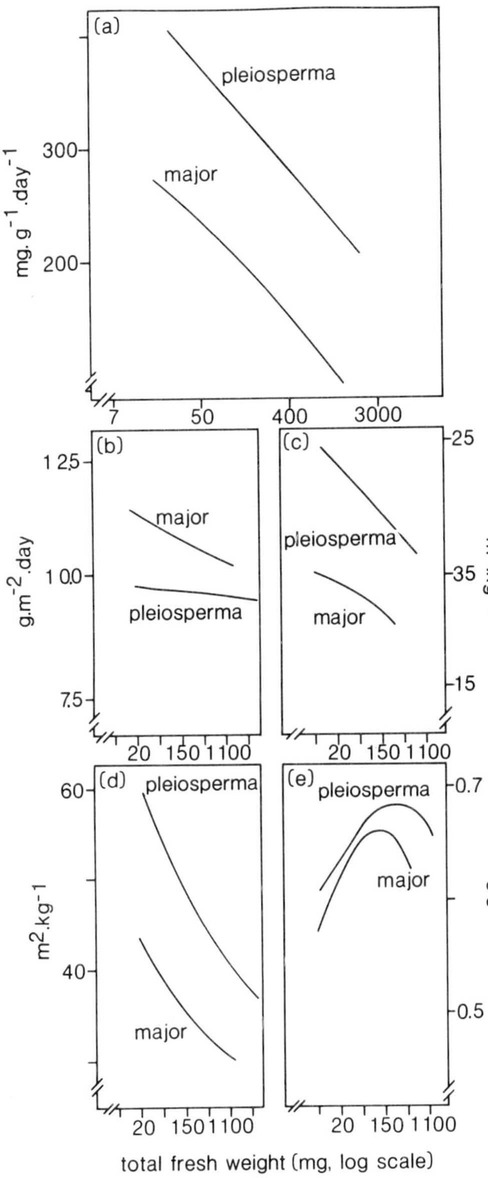

Fig. 3. The relative growth rate (total dry weight) (a), the net assimilation rate (dry weight) (b), the leaf area ratio (c), the specific leaf area (d), and the leaf weight ratio (e) of two subspecies of *Plantago major* (*P. major major* and *P. major pleiosperma* as related to total fresh weight (data from Dijkstra & Lambers (1986) and unpublished data from P. Dijkstra).

NAR and LAR contributed to the higher RGR of the crop species (Table 1a), though the physiological component was quantitatively the most important. A comparison of the shade-avoiding species *Galinsoga parviflora* and the shade-tolerant species *Stachys sylvatica* and *Urtica dioica* (Corré 1983), also showed that the difference in RGR was associated with both the 'physiological' and the 'morphological' component (Table 1b).

Thirdly, differences in growth rate have been found between populations of *Lolium perenne* cv. S23, selected for rate of mature leaf respiration in the dark (Wilson, 1982). A negative correlation was found between the rate of mature leaf respiration and the crop growth rate, but earlier studies (Wilson, 1975) showed no correlation with such characters as photorespiration and photosynthesis or various morphological characteristics. A growth analysis has shown that only NAR, and not LAR, explained the difference in RGR (Wilson, 1982).

4. Variation in net assimilation rate

Some of the above-mentioned case studies have shown that variation in RGR can be explained by the NAR. A further analysis of this 'physiological' component is therefore appropriate.

Genotypic differences in NAR might be due to a difference in: (i) photosynthetic rate, or (ii) respiration rate, or perhaps (iii) losses of material, e.g. volatile compounds from the leaves or exudates from the roots.

4.1 Photosynthesis

Variation in photosynthetic rate could be due to differences in carboxylation efficiency or in leaf conductance to CO_2 diffusion. However, since leaf conductance itself is largely controlled by the leaf's photosynthetic activity (Farquhar & Sharkey, 1982), only aspects of the carboxylation efficiency will be discussed here.

Evergreen tree species tend to have a lower RGR than deciduous ones (cf. Jia & Ingestad, 1984; Ingestad & Kahr, 1985). The difference in growth rate between these groups of species leads to the following question. Is the rate of photosynthesis in evergreen tree species lower than that in trees with a deciduous habit? Numerous reports suggest that the photosynthetic rate is lower in evergreens than in deciduous species (e.g.

Schulze, 1982). If measurements are made on plants in their natural habitat, differences in photosynthetic rate may reflect environmental differences, e.g. the nutrient status of the soil, rather than inherent differences. Karlsson (1985) avoided this problem by measuring rates of photosynthesis of fertilized and irrigated plants, when comparing deciduous (*Vaccinium uliginosum*) and evergreen (*V. vitis-idaea*) dwarf shrubs in their natural environment. Even when fertilized and irrigated, the peak rate of photosynthesis in mid-summer (expressed per g or m^{-2} of leaf) was less in the evergreen than in the deciduous species.

Recent studies of early successional (fast growing) and late successional (slow growing) species (Oberbauer et al., 1985) support the hypothesis that early successional species have a higher photosynthetic capacity. *Ochroma lagopus*, a widespread pioneer of large clearings in the wet forests of Costa Rica, had a photosynthetic rate four times greater than that of *Pentaclethra macroloba,* an abundant climax species. In addition, the LAR of *Ochroma lagopus* was double that of *Pentaclethra macroloba.* Both NAR and LAR are therefore likely to contribute to a high RGR in the pioneer.

The higher NAR in the shade-avoiding *Cirsium palustre* in comparison with the shade-tolerant *Geum urbanum* (Pons, 1977a) was due to a greater photosynthetic activity per unit area (Fig. 1c). This higher photosynthetic activity was correlated with a higher mass per unit area (a lower SLA), especially at a high quantum flux density. There were two layers of palisade parenchyma cells in *Cirsium,* but only one in *Geum.* Thus, *Cirsium* probably has more photosynthetic enzymes and pigments per unit area, giving it a growth advantage at high quantum flux densities. However, in shaded conditions, the potential for higher photosynthesis will not be fully used and so a low SLA cannot give a similar advantage, at least not of a comparable magnitude (Björkman, 1981).

Also in the two subspecies of *Plantago major* (Fig. 3), the lower NAR was a reflection of a lower rate of photosynthesis per m^{-2}, whilst dark respiration was about the same for both subspecies (Table 2). As was the case in the *Cirsium* leaves referred to above (Pons, 1977a), the higher NAR correlated with the lower SLA: more mass and presumably more photosynthetic enzymes and pigments per unit leaf area.

4.2 Respiration

Respiration is a component of NAR; some 50% of the photosynthates produced daily may be lost in the respiration of the various plant parts (Lambers, 1985). The respiratory component is too often forgotten, although there are large variations between species in respiration, especially the respiration of roots (Lambers, 1979). This variation is partly explained by a difference in the contribution of the alternative path (Lambers et al., 1983), a non-phosphorylating mitochondrial electron transport pathway (Laties, 1982; Lambers, 1985).

Root respiration tends to decline sharply upon the formation of a storage organ, such as a tap root (Lambers & Van de Dijk, 1979; Steingröver, 1981). Variation in NAR between sugar-beet and spinach-beet (Thorne & Evans, 1964), which is often carelessly ascribed to a difference in rate of photosynthesis, is more likely due to respiration (cf. Frier, 1977; Steingröver, 1981). Storage roots respire considerably less than fibrous roots via the non-phosphorylating, alternative path (Steingröver, 1981). Differences between some genotypes in NAR may therefore be due to differences in their capacity to form storage organs.

Leaves of shade-tolerant genotypes have an inherently low respiration rate, both when expressed on a leaf dry weight basis (Pons, 1977b)

Table 2. Photosynthesis and dark respiration (μmol CO_2 m^{-2} s^{-1}) of subspecies of *Plantago major*, when grown under identical conditions in a growth room (the same plants as used in Figure 3). Subspecies pleiosperma is fast growing and subspecies major is slow growing. Data from Dijkstra and Lambers (1986).

Subspecies	Photosynthesis	Respiration
pleiosperma	10.2	1.3
major	12.5	1.7

and when expressed per unit leaf area (Pons, 1977a). This is probably related to their lower concentration of protein.

Robson (1982b) concluded that the difference in respiration between populations of *Lolium perenne* (Wilson 1982) could account for differences in their seedling growth, and larger differences at a later stage, after cutting (Wilson, 1982; Robson, 1982a). The higher rate of respiration might point to a large contribution of the non-phosphorylating, alternative path, since this path, unlike the cytochrome path, is not coupled to the phosphorylation of ADP (Lambers, 1985). However, Day et al. (1985) demonstrated that, in both the mature leaves and the roots of the *Lolium* populations, there was no difference in the contribution of the alternative path (Table 3). This suggests that the ATP requirement in mature leaves of the fast respiring population was higher than in those of the other population. So far it is not clear why the energy requirement should be higher, but increased protein turnover (A.J. Barneix, unpublished results), or cycling of amino acids between roots and leaves (H.D. Cooper & A.J. Barneix, unpublished results) have been excluded as factors causing a different energy requirement. The difference in respiration and yield disappears when growth occurs at a low nutrient supply (Pilbeam et al., 1986), which may offer an explanation why the slow growing, fast respiring genotypes still occur in the population (Robson et al., 1983).

Table 3. Respiration (μmol O_2 g^{-1} (fresh weight) h^{-1}) in mature leaves, in meristematic parts of the leaves and in roots of a fast respiring, slow growing and a slow respiring, fast growing population of *Lolium perenne*. KCN is an inhibitor of the cytochrome path; in its presence the capacity of the alternative path is determined. SHAM, salicylhydroxamic acid, is an inhibitor of the alternative path; respiration in the presence of this inhibitor occurs via the cytochrome path, whilst inhibition by SHAM provides the activity of the alternative path. In the presence of both inhibitors the 'residual component' of respiration is determined. Data from Day et al. (1985).

Tissue	Population	
	Fast respiring	Slow respiring
Mature leaves		
Control	7.3	5.2
+KCN	8.0	5.4
+SHAM	7.3	5.4
+KCN and SHAM	1.0	0.8
Leaf meristems		
Control	11.7	11.9
+KCN	2.9	2.9
+SHAM	11.7	11.0
+KCN and SHAM	1.6	1.4
Roots		
Control	4.0	3.8
+KCN	2.3	2.8
+SHAM	2.5	2.5
+KCN and SHAM	0.3	0.4

In seedlings of *Stephanomeria exigua* ssp. *coronaria* and its derivative *S. malheurensis*, the difference in NAR correlates with a difference in respiration (Gottlieb, 1978). When grown at a day/night temperature of 20/12°C or 12/5°C, the NAR was significantly higher in *S. malheurensis*, whilst the two species had a closely similar net rate of CO_2-assimilation. Leaf respiration, expressed on an area basis, was highest in *S. exigua*; this was a reflection of a lower SLA, since there was no difference in leaf respiration between the two species on a dry weight basis.

4.3 Loss of volatiles and exudates

Losses of carbon might occur via exudation from the roots or volatilization from the leaves; such losses are thought to be of importance in specific environments. Exudation from the roots may have allelopathic effects on neighbouring plants. Although allelopathic effects often occur in the litter, there are some classical examples of excretion from intact tissue (e.g. Harborne, 1982).

Jarvis (1964) provided evidence for the loss of substances from the roots of *Deschampsia flexuosa*. These substances, which are also present in the litter, inhibited the growth of *Betula verrucosa* seedlings. No information is available about the quantity of material lost from *Deschampsia* roots, but this might offer an explanation for slow growth of *Deschampsia flexuosa*.

Loss of volatile compounds from leaves might also offer an explanation for slow growth. In the evergreen shrub, *Ledum groenlandicum*, current year leaves may allocate 10% of all recently assimilated carbon to volatiles (Prudhomme, 1983). This peak value was found for current year leaves in mid-summer, though losses at other times were considerably smaller. Volatile losses, often of a terpenoid nature, are known to have allelochemical effects (Baas, 1985).

5. Variation in leaf area ratio

Some of the examples already quoted demonstrate that variation in RGR is explained by differences in carbon allocated to the leaf area. Such variation in LAR may be due to a difference in the amount of weight per unit area of leaves (SLA). Alternatively, it may stem from a difference in investment in the weight of leaves, relative to that in the rest of the plant (LWR). A combination of SLA and LWR might explain variation in LAR. Various examples show that either SLA or LWR are responsible for inherent differences in LAR.

5.1 Specific leaf area

Oberbauer et al. (1985) found a higher SLA in the pioneer species, *Ochroma lagopus*, than in the climax species, *Pentaclethra macroloba*. Both species are trees in the wet forest of the Costa Rican Atlantic lowlands, and the authors suggest that the pioneer has a higher RGR.

Corré (1983) found a 40% greater SLA in the faster growing, shade-avoiding *Galinsoga parviflora* than in the slower growing, shade-tolerant *Stachys sylvatica* (Table 1b).

Karlsson (1985) found a greater SLA in the deciduous *Vaccinium uliginosum,* than in the evergreen *V. vitis-idaea*, when studied in their natural habitat. The difference was smaller in older leaves, and was marginal in the youngest leaves, when plants were fertilized for one year, but reappeared upon the second year of fertilization. Roetman & Sterk (1986) found the highest SLA in microspecies of *Taraxacum* that had a high RGR. They concluded that the difference in SLA and LWR, rather than in NAR, accounted for the differences in growth rate between the microspecies.

If a high SLA in fast-growing species were associated with a low NAR, one might assume that less of the photosynthesizing components are present per unit area. In the experiments with *Cirsium palustre* and *Geum urbanum* (Pons, 1977a; Fig. 1) and *Plantago major* (Fig. 3), a high NAR was indeed correlated with a low SLA. DeJong and Doyle (1985) also found a negative correlation between photosynthesis and SLA. However, in the examples of Oberbauer et al. (1985), Corré (1983; Table 1b) and Karlsson (1985), NAR or photosynthesis was lower in the species with a low SLA. What then is the chemical composition of this extra weight per unit area in the slow growing species? They may be antiherbivore compounds, such as tannins, phenols and lignin. For example, the concentration of carbon-based antiherbivore compounds in the deciduous *Betula glandulosa* is considerably less than in the evergreen, *Ledum groenlandicum* (Prudhomme 1983; Table 4). Similarly, the resin content of the early successional trees, *Populus balsamifera* and *P. tremuloides,* is considerably less than that of the late successional, slow growing black spruce, *Picea mariana,* from the same area (Bryant et al., 1983; Table 5).

More information supporting the hypothesis that early successional species contain less digestibility reducing compounds than late successional species is presented by Cates & Orians

Table 4. Concentration of lignin (mg g^{-1}) and the % allocation of ^{14}C to organic acids, alkaloids and volatile compounds (essential oils) in *Betula glandulosa* (deciduous) and *Ledum groenlandicum* (evergreen). Data from Prudhomme (1983).

Compound(s)	B. glandulosa	L. groenlandicum Current year leaves	1st year	2nd year
Lignin	ca. 10	15–20	15–20	10–25
Total phenolics	ca. 5	5–25	ca. 10	5–20
Organic acids	1–4	1–3	1	1–2
Alkaloids	0.1–0.6	0.1–0.4	0.1–0.3	0.1–0.3
Volatile compounds	–	0–10	–	–

(1975). These authors offered leaves from 100 plant species from three growth forms and from different stages in the succession to two slug species, *Ariolimax columbianus* and *Arion ater*. Slugs of both species preferred early successional annuals rather than early successional perennials, but both these groups were significantly more palatable than later successional species. Grime et al. (1968) investigated the leaf palatability of fresh leaves and leaf extracts, using the snail *Cepaea nemoralis*. The most palatable species were generally associated with disturbed habitats and fertile soils. Unfortunately, in most of these studies, material was collected from plants grown in their natural habitat. Palatability and production of secondary compounds may be determined more by environmental factors than by genetic differences. In particular, the nitrogen supply affects the concentration of secondary compounds, e.g. tannins and lignin in *Salix aquatica* (Waring et al., 1985; Table 6), phenolics in *Betula pubescens* (Haukioja et al., 1985), volatile terpenes in *Heterotheca subaxillaris* (Mihaliak & Lincoln, 1985). There is therefore a need for similar experiments, in which plants are grown under controlled conditions. In the absence of such data the statement that early successional species make a lesser commitment to defense is attractive, but lacks proof.

5.2 Leaf weight ratio

Plants tend to invest more biomass in roots. Relative to above-ground organs, when water and/or nutrients are scarce (Brouwer, 1963, 1983). It is to be expected that some species from dry or nutrient-poor habitats inherently invest more biomass in roots, and therefore have a lower LWR and lower shoot : root ratio than species from more fertile habitats. This could offer an explanation for the slow growth of species which normally grow in dry or nutrient-poor sites.

Differences in RGR between *Senecio* species (Fig. 2), was indeed explained by a different LWR (Van Dobben, 1967). *S. congestus,* from N-rich soils, invested relatively less in roots. Similarly evergreen tree species, which tend to have a low RGR, are reported to have a lower LWR than deciduous trees (Van den Driesche & Wareing, 1966).

The shoot : root ratio is more commonly measured than LWR. Bradshaw et al. (1964) found that *Cynosurus cristatus* (RGR = 154 mg g^{-1} d^{-1}; Grime & Hunt, 1975) had the lowest shoot : root ratio, at all levels of N, while *Nardus stricta* (RGR = 71 mg g^{-1} d^{-1}; Grime & Hunt, 1975) had one of the lowest shoot : root ratio's at most levels of N.

Variation in LWR exists in *Sonchus arvensis* (Pegtel, 1976); ecotypes from arable land invest more dry matter in thickened storage roots in comparison with the coastal ecotypes. The coastal ecotypes, though investing somewhat

Table 5. Resin content (% of the dry weight) in mature tissue of two pioneer trees (*Populus balsamifera, P. tremuloides*), a tree species with a wide distribution (*Betula papyrifera* ssp. *humilis*) and a late successional species from poor soils (*Picea mariana*). The latter species has a slower growth rate than the former three. Data from Bryant et al. (1983).

Species	Resin content
P. tremuloides	ca. 7
P. tremuloides	ca. 7
B. papyrifera	ca. 3
P. mariana	ca. 15

Table 6. The growth analysis and composition of *Salix aquatica* plants during exponential growth. Plants were grown under optimum nutrient supply or with nutrient limitation. Data from Waring et al. (1985).

Item	Nutrient supply	
	High	Low
RGR, mg g^{-1} d^{-1}	161	55
NAR, g m^{-2} d^{-1}	13.8	12.0
LAR, m^2 kg^{-1}	11.7	4.6
LWR, g g^{-1}	0.56	0.37
SLA, m^2 kg^{-1}	20.8	12.5
Net photosynthesis,		
μmol, CO_2 m^{-2} s^{-1}	7.9	8.3
Dark respiration		
μmol, CO_2 m^{-2} s^{-1}	1.7	1.1
N-concentration, g m^{-2}	2.15	1.40
mg g^{-1}	50.4	19.1
Tannins, relative units	0.65	1.00
Lignin, mg g^{-1}	208	245

more in fibrous roots, have a higher LWR; more fibrous roots may be more advantageous with the less favourable nutrient supply of sand dunes compared with arable land. The low LWR of ecotypes from arable land may be associated with the fact that vegetative reproduction, from pieces of storage roots, is favoured under those conditions (Pegtel, 1976). However, the lower LWR of *Sonchus* ecotypes from arable land was more than compensated by a greater SLA, so that their RGR was still higher than that of the coastal ecotypes.

A rather similar ecotypic differentiation for shoot : root ratio, which is likely to be also indicative of LWR in this rosette plant, was found in *Hypochaeris radicata:* Genotypes from relatively fertile roadsides have larger shoot : root ratio's than genotypes from a nutrient-poor heath (Van de Dijk, 1980). Similarly, microspecies of *Taraxacum* from fertile sites had a larger LWR, when grown in climate rooms (Roetman & Sterk, 1986), and this may be the reason for the larger RGR of the microspecies from fertile habitats.

6. The ecological significance of a low relative growth rate

6.1 The significance of RGR

It is easy to see the advantage of an inherently high RGR for a species in productive and competitive environments; fast growth should allow the species to overshadow its slower growing neighbours, provided their architecture is similar. This hypothesis was verified by Pons (1979), in experiments with slow and fast growing weeds competing with rice, *Oryza sativa*; reductions in the yield of rice were linearly correlated with the growth rate of the weeds. However, although *Cyperus difformis* had the highest RGR, it was not the most competitive species; its absolute growth rate was inferior to that of other weeds, because of its smaller seeds. The most competitive species combined a reasonably high RGR with large seeds, so that its absolute growth rate was higher than that of the other species tested.

An inherently high RGR, though potentially important, is only one of the factors determining competitive ability. Roush & Radosevich (1985) compared the growth and competitive ability ('aggressivity') of four summer annual weeds, well adapted to the semi-arid climate of the Central Valley of California. All four had the same RGR, but the more aggressive species (*Echinochloa crus-galli* and *Amaranthus retroflexus*) had the highest NAR and lowest LAR (Table 7). In this semi-arid climate, where the quantum flux density is high, selection for a low LAR and low shoot : root ratio presumably increased the competitive ability, as it increases the plant's potential to extract water from the soil. When a low LAR was combined with a high NAR, the species had a very strong competitive ability. The C_4-species, *E. crus-galli* and *A. retroflexus*, had the highest NAR. The large difference in NAR between the other two species may be associated with differences in SLA; *Solanum nodiflorum* had the larger SLA and the lower NAR. A comparison of *Axonopus compressus* and *Oplismenus compositus* also shows similarity in RGR, despite differences in both NAR and LAR, when grown at a high quantum flux density (Hofstra & Stienstra, 1977). Although *A. compressus*, a C_4-species, had the higher NAR, this was virtually compensated for by a lower LAR.

Table 7. The relative growth rate (RGR), net assimilation rate (NAR), leaf area ratio (LAR) and shoot/root ratio of four aggressive weed species (*Echinochloa crus-galli, Amaranthus retroflexus, Chenopodium album* and *Solanum nodiflorum*) from a semi-arid region in California. The species are arranged from left to right in the order strongest to weakest competitor. The values for RGR are not significantly different. Data from Roush & Radosevich (1985).

Item	Echinochloa	Amaranthus	Chenopodium	Solanum
RGR ($mg\ g^{-1}\ d^{-1}$)	255	262	232	255
NAR ($g\ m^{-2}\ d^{-1}$)	24.7	21.6	19.4	10.9
LAR ($m^2\ kg^{-1}$)	12.8	14.7	15.6	24.1
Shoot/root ratio ($g\ g^{1}$)	0.032	0.055	0.072	0.072

Chapin (1980) defended the hypothesis that an inherently low growth rate is advantageous in saving nutrients in infertile environments. Fast growers would rapidly deplete the resources and then would loose in competition with the slow growers. Rapidly growing ruderals, such as *Urtica dioica*, certainly perform very poorly at low phosphate supply (Rorison, 1968). However, the growth of the slow-growing *Agrostis setacea*, from phosphate-poor sites, is very similar to that of the fast-growing *Agrostis stolonifera,* from richer habitats, when grown at low phosphate supply (Clarkson, 1967). Similarly, the slow growing *Deschampsia flexuosa* (poor, acid soils) does not grow faster than the fast-growing *Rumex acetosa* (relatively rich soils) under relatively nutrient-deficient conditions (Rorison, 1968).

We believe that it is more likely that the RGR is not the target of natural selection under unfavourable conditions, but rather the components (NAR, LWR and SLA). A low NAR, LWR of SLA may all have distinct advantages in a specific environment and may change under the selective pressure. Low values for these parameters subsequently lead to a low RGR.

6.2 The significance of NAR

A low NAR, if the result of a low rate of photosynthesis, may reflect an efficient investment of nitrogen. Photosynthetic capacity and carboxylating efficiency strongly correlate with the N-concentration in the leaf (e.g. Mooney et al., 1981; Evans, 1983; Hunt et al., 1985; Wong, 1985). Thus, except when a large amount of nitrogen is used for N-based toxins, a low capacity of photosynthesis reflects a low amount of N per m^2. Generally, N-based toxins accumulate to high levels in seeds only, whilst their concentration tends to be lower in leaves (Harborne, 1982). A second exception are C_4-plants which have a higher rate of photosynthesis per unit N in the leaf. This is partly due to anatomical and biochemical features which lead to suppression of the oxygenase activity of Rubisco (Osmond et al., 1982), and also to the difference in kinetics (k_{cat}) between Rubisco from C_3- and C_4-plants (Seemann et al., 1984).

When we restrict ourselves to either C_3- or C_4-species, which invest similar amounts of N in N-based toxins, the question about differences in NAR and photosynthesis transforms into one about variation in leaf N concentration, i.e. is there an advantage associated with a low N concentration in the leaf?

Even the photosynthesis of sun leaves is only light saturated for 20% of the entire daytime and many plants invest more N into leaf enzymes than they can actually use under average climatic conditions (Schulze, 1982). If N is not limiting, this may not be so crucial, since more N in the leaves does lead to a higher gain of carbon, even if it is not proportionally higher. However, when N is the dominant factor limiting plant growth, a high N concentration is a trait that is unlikely to be selected for. Species may differ remarkably in their capacity to adjust their photosynthetic capacity to the prevailing quantum flux density; even within one species, *Lolium perenne*, there is large genotypic variation in the extent to which the photosynthetic capacity of the leaves acclimates to the level of radiation during growth (Wilson & Cooper, 1969). Poor acclimation of the photosynthetic capacity will lead to inefficient utilization of N, as illustrated in Table 6 (Waring et al., 1985) for *Salix aquatica.* The higher N concentration in the leaves of *S. aquatica* did not coincide with a larger NAR or rate of photosynthesis at a low quantum flux density (approx. 0.3 mmol $m^{-2}\,s^{-1}$).

It is likely that the lower NAR in *Sonchus arvensis* ecotypes from arable land, compared to that in coastal ecotypes (Pegtel, 1976), is a reflection of the light environment in the natural habitat. In the open dune vegetation, more light is available so that a greater investment of N in the photosynthetic apparatus per unit leaf area leads to greater photosynthetic activity, but in the ecotype from arable land investment of N is in more leaves, rather than highly efficient leaves.

There may be a second disadvantage for a high N concentration, at least for plants in unproductive environments. The efficiency with which a consumer can convert ingested plant material into its own biomass is linearly correlated with the N% in the plant tissues (Mattson, 1980). Sometimes the degree of herbivory increases with N-application and increased N concentration in the leaves, e.g. in *Calluna vulgaris,* attacked by *Lochmaea suturalis* (Brunsting and Heil 1985). Herbivore performance (larval size, growth rates, survivorship) may even strongly correlate with the N concentration of the attacked plant material (Lincoln et al., 1982). But such a correlation is not general, since carbon based antiherbivore compounds also control the extent of herbivore attack (Lincoln et al., 1982; Bryant et al., 1983). A larger number of leaves, with a low N concentration and hence a lower rate of photosynthesis, may therefore be a way of spreading and lowering the risk of destruction by herbivores. This second explanation is the more speculative, and probably less relevant, than the first, for which a substantial amount of evidence is becoming available (e.g. DeJong & Doyle, 1985).

A low NAR may also be a reflection of a high rate of respiration. Crop growth rate in *Lolium perenne* genotypes is inversely correlated with the rate of mature leaf respiration (Wilson, 1982). It is hard to see a specific advantage for a high respiration rate, in the absence of a biochemical explanation for this phenomenon. If it is a reflection of an efficiency in the respiratory metabolism, there is probably no advantage at all. However, we cannot yet exclude the possibility that there is a specific process requiring a major amount of ATP, thus causing a high respiration rate.

6.3 The significance of LWR

A low LWR, if correlated with a low shoot : root ratio, can be significant for absorption of ions from a nutrient-poor soil. In coastal ecotypes of *S. arvensis* such a low LWR has indeed been found, also when plants were grown under the same conditions as the ecotype from arable land (note that for the calculation of LWR, the weight of fibrous roots was taken, rather than that of the entire root system; Pegtel, 1976). Similarly, an ecotype of *Hypochaeris radicata* from a poor heathland had a lower LWR than an ecotype from a more fertile roadside, at a range of nitrate supply (Van de Dijk, 1980). The difference in LWR between *Senecio* species (Van Dobben, 1967; Fig. 2) can also be interpreted in the same vein. Alternatively, a low LWR, such as found in *Sonchus arvensis* ecotypes from arable land (Pegtel, 1976), can be advantageous in an environment where vegetative reproduction from pieces of roots is important.

A low LWR was found in two C_4-species, *Axonopus compressus* and *Setaria plicata,* when compared with the C_3-grass, *Oplismenus compositus* (Hofstra & Stienstra, 1977). Since a low LWR was compensated for by a high NAR, the RGR of the three species was virtually the same. The large investment of biomass in roots is likely to contribute to the occurrence of the C_4-species in open habitats, which are avoided by *Oplismenus*. A large investment of dry weight in roots was also found in competitive weeds from semi-arid locations in California (Roush & Radosevich, 1985). The competitive ability of four weed species was stronger when relatively more material was invested in roots (Table 7). A comparison of two *Solidago* species, differing in their competitive ability, along a moisture gradient, led to the conclusion that LAR was a more important component of growth than photosynthesis or water use efficiency (Potvin and Werner 1983, and their unpublished data cited therein). *S. canadensis* has a higher LAR and occurs at moist locations, in comparison with *S. juncea,* from drier sites. Presumably the latter species has a lower RGR, but data are lacking. Even so, it is likely that the selective force in the natural habitat was on a low LAR, rather than a low RGR, although this may have been its additional effect.

6.4 The significance of SLA

A low SLA is likely to be advantageous in an environment where the photon flux density is high (Pons, 1977b; Corré, 1983). It is to be expected that at a high quantum flux density, N is invested more efficiently in one leaf with a high N concentration, compared to an investment in two leaves, each of the same size, but with half the N concentration. The two leaves with half the N concentration are likely to have similar costs for their basic structures (vascular tissue, epidermal tissue, etc.) as the one leaf with a high N concentration. A low SLA could also reflect a large investment in secondary metabolites, protecting the leaf against herbivory, in thick cell walls, protecting against trampling or grazing, or a thick cuticle, reducing the loss of water. To our knowledge, no data pointing to genotypic differences in SLA related to variation in secondary compounds are available in the literature.

7. Conclusion

Genotypes from infertile habitats may have an inherently low RGR in comparison with genotypes from fertile or 'disturbed' sites, though some studies show otherwise. Also, the competitive ability of a genotype is not necessarily linked to a high RGR.

Genotypes from infertile or dry habitats tend to invest more biomass in roots, especially those involved in uptake of water and nutrients. If this is compensated for by a high NAR, as in some C_4-weeds, RGR may still be high. However, a low LWR, or a low shoot to root ratio, is likely to be selected for in dry and infertile habitats and then cause a low RGR. A low RGR may also be a reflection of a large investment in palatability-reducing compounds.

An advantage of a low NAR, possibly a reflection of a low N concentration in the leaves, is likely in genotypes from a shaded environment, or in those growing in a dense canopy. Insufficient compensation by LAR will cause a low RGR in such genotypes. Herbivory might be an additional force in the selection for a low N concentration in the leaves. Surrounding plants may be a selective pressure leading to the production of volatiles or the exudation of allelopathic substances from the roots. Such losses could also explain a low NAR.

Selective pressure on NAR, LWR or SLA is likely to occur in specific environments and these components may affect RGR of a species. Slow growth is thus considered an indirect, and not even a necessary, effect of the selective forces. Selection is not for a specific growth rate, as often suggested in the literature. The correlation between degree of 'disturbance' and growth rate can therefore only be understood by taking into account the components which lead to a certain growth rate.

Acknowledgements

Various colleagues have given their constructive comments on earlier drafts of this manuscript. We like to thank them all, in alphabetical order: Marion Cambridge, Henk Konings, Thijs Pons, Hendrik Poorter. This research was supported in part by the Foundation for Fundamental Biological Research (BION) which is subsidized by the Netherlands Organization for the Advancement of Pure Research (Z.W.O.). Grassland species Research Group Publ. no. 125.

References

Baas, W.J. (1985). Naturally occurring seco-ring-A-triterpenoids and their possible biological significance. Phytochemistry 24: 1875–1889.

Bakker, D. (1960). Senecio congestus (R. Br.) in the Lake-IJsel polders. Acta Bot. Neerl. 9: 235–259.

Björkman, O. (1981). Responses to different quantum flux densities. In: Encyclopedia of Plant Physiology, N.S., Vol. 12A. O.L. Lange, P.S. Nobel, C.B. Osmond & H. Ziegler (Eds.), pp 57–107. Springer-Verlag, Berlin.

Bradshaw, A.D., Chadwick, M.J., Jowett, D. & Snaydon, R.W.D. (1964). Experimental investigations into the mineral nutrition of several grass species. IV. Nitrogen level. J. Ecol. 52: 665–676.

Brouwer, R. (1963). Some aspects of the equilibrium between overground and underground plant parts. Meded. Inst. Biol. Scheikd. Onderzoek Landbouwgewassen 213: 31–39.

Brouwer, R. (1983). Functional equilibrium: sense or nonsense? Neth. J. Agric. Sci. 31: 335–348.

Brunsting, A.M.H. & Heil, G.W. (1985). The role of nutrients in the interactions between a herbivore beetle and some competing plant species in heathland. Oikos 44: 23–26.

Bryant, J.P., Chapin, F.S. & Klein, D.R. (1983). Carbon/nutrient balance of boreal plants in relation to vertebrate herbivory. Oikos 40: 357–368.

Cates, R.G. & Orians, G.H. (1975). Successional status and

the palatability of plants to generalized herbivores. Ecology 56: 410–418.

Chapin, F.S. (1980). The mineral nutrition of wild plants. Annu. Rev. Ecol. Syst. 11: 233–260.

Clarkson, D.T. (1967). Phosphorus supply and growth rate in species of Agrostis L. J. Ecol. 55: 111–118.

Corré, W.J. (1983). Growth and morphogenesis of sun and shade plants. III. The combined effects of light intensity and nutrient supply. Acta Bot. Neerl. 32: 277–294.

Day, D.A., De Vos, O.C., Wilson, D. & Lambers, H. (1985). Regulation of respiration in the leaves and roots of two Lolium perenne populations with contrasting mature leaf respiration rates and crop yields. Plant Physiol. 78: 678–683.

DeJong, T.M. & Doyle, J.F. (1985). Seasonal relationships between leaf nitrogen content (photosynthetic capacity) and leaf canopy light exposure in peach (Prunus persica). Plant Cell Environ. 8: 701–706.

Dijkstra, P. & Lambers, H. (1986). Photosynthesis and respiration of two inbred lines of Plantago major L. differing in relative growth rate. In: Biological control of photosynthesis. R. Marcelle, H. Clijsters & Poucke, M. van (Eds.), pp 251–255. Martinus Nijhoff Publishers, The Hague.

Eagles, C.F. (1967). The effect of temperature on vegetative growth in climatic races of Dactylis glomerata in controlled environments. Ann. Bot. 31: 31–39.

Ericsson, T. (1981). Effects of varied nitrogen stress on growth and nutrition in three Salix clones. Physiol. 423–430.

Evans, J.R. (1983). Nitrogen and photosynthesis in the flag leaf of wheat (Triticum aestivum L.). Plant Physiol. 72: 297–302.

Farquhar, G.D. & Sharkey, T.D. (1982). Stomata and photosynthesis. Annu. Rev. Plant Physiol. 33: 317–345.

Frier, V. (1977). The relationship between photosynthesis and tuber growth in Solanum tuberosum L. J. Exp. Bot. 28: 999–1007.

Gottlieb, L.D. (1978). Allocation, growth rates and gas exchange in seedlings of Stephanomeria exigua ssp. coronaria and its recent derivative S. malheurensis. Am. J. Bot. 65: 970–977.

Grime, J.P. & Hunt, R. (1975). Relative growth-rate: Its range and adaptive significance in a local flora. J. Ecol. 63: 393–422.

Grime, J.P., MacPherson-Stewart, S.F. & Dearman, R.S. (1968). An investigation of leaf palatability using the snail Cepaea nemoralis L. J. Ecol. 56: 405–420.

Harborne, J.B. (1982). Introduction to ecological biochemistry. Academic Press, London.

Haukioja, E., Niemela, P. & Siren, S. (1985). Foliage phenols and nitrogen in relation to growth, insect damage, and ability to recover after defoliation, in the mountain birch Betula pubescens ssp. tortuosa. Oecologia 65: 214–222.

Hofstra, J.J. & Stienstra, A.W. (1977). Growth and photosynthesis of closely related C_3 and C_4 grasses as influenced by light intensity and water supply. Acta Bot. Neerl. 26: 63–72.

Hunt, R. (1982). Plant growth curves. The functional approach to growth analysis. Edward Arnold, London.

Hunt, E.R., Jr., Weber, J.A. & Gates, D.M. (1985). Effects of nitrate application on Amaranthus powelii Wats. II. Stomatal response to vapor pressure difference is consistent with optimization of stomatal conductance. Plant Physiol. 79: 614–618.

Ingestad, T. (1981). Nutrition and growth of birch and grey alder seedlings in low conductivity solutions and at varied relative rates of nutrient addition. Physiol. Plant. 52: 454–466.

Ingestad, T. & Kahr, M. (1985). Nutrition and growth of coniferous seedlings at varied relative nitrogen addition rate. Physiol. Plant. 65: 109–116.

Jarvis, P.G. (1964). Interference by Deschampsia flexuosa (L.) Trin. Oikos 15: 56–78.

Jarvis, P.G. & Jarvis, M.S. (1964). Growth rates of woody plants. Physiol. Plant. 17: 654–666.

Jia, H. & Ingestad, T. (1984). Nutrient requirements and stress response of Populus simonii and Paulownia tomentosa. Physiol. Plant. 62: 117–124.

Karlsson, P.S. (1985). Effects of water and mineral nutrient supply on a deciduous and evergreen dwarf shrub: Vaccinium uliginosum L. and V. vitis-idaea L. Holarctic Ecol. 8: 1–8.

Kuiper, D. (1983). Genetic differentiation in Plantago major: Growth and root respiration and their role in phenotypic adaptation. Physiol. Plant. 57: 222–230.

Lambers, H. (1979). Efficiency of root respiration in relation to growth, morphology and soil composition. Physiol. Plant. 46: 194–202.

Lambers, H. (1985). Respiration in intact plants and tissues: Its regulation and dependence on environmental factors, metabolism and invaded organisms. In: Encyclopedia of Plant Physiology, N.S. Vol. 18. R. Douce and D.A. Day, (Eds.), pp 418–473. Springer-Verlag, Berlin.

Lambers, H. & Dijk, S.J. Van de (1979). Cyanide-resistant root respiration and taproot formation in two subspecies of Hypochaeris radicata. Physiol. Plant. 45: 235–239.

Lambers, H., Day, D.A. & Azcon-Bieto, J. (1983). Cyanide-resistant respiration in roots and leaves. Measurements with intact tissues and isolated mitochondria. Physiol. Plant. 58: 556–563.

Laties, G.G. (1982). The cyanide-resistant, alternative path in higher plant respiration. Annu. Rev. Plant Physiol. 33: 519–555.

Lincoln, D.E., Newton, F.S., Ehrlich, P.R. & Williams, K.S. (1982). Coevolution of the checkerspot butterfly Euphydras chalcedona and it larval food plant Diplacus aurantiacus: Larval response to protein and leaf resin. Oecologia 52: 216–223.

Mattson, W.J., Jr. (1980). Herbivory in relation to plant nitrogen content. Annu. Rev. Ecol. Syst. 11: 119–161.

Mihaliak, C.A. & Lincoln, D.E. (1985). Growth pattern and carbon allocation to volatile leaf terpenes under nitrogen-limiting conditions in Heterotheca subaxillaris (Asteraceae). Oecologia 66: 423–426.

Mooney, H.A., Field, C., Gulmon, S.L. & Bazzaz, F.A. (1981). Photosynthetic capacity in relation to leaf position in desert versus old-field annuals. Oecologia 50: 109–112.

Oberbauer, S.F., Strain, B.R. & Fetcher, N. (1985). Effect of CO_2-enrichment on seedling physiology and growth of two tropical tree species. Physiol. Plant. 65: 352–356.

Osmond, C.B., Winter, K. & Ziegler, H. (1982). Functional significance of different pathways of CO_2 fixation in photosynthesis. In: Encyclopedia of Plant Physiology, N.S. Vol. 12B. O.L. Lange, P.S. Nobel, C.B. Osmond & H. Ziegler (Eds.), pp 479–547. Springer-Verlag, Berlin.

Pegtel, D.M. (1976). On the ecology of two varieties of Sonchus arvensis L. PhD Thesis, University of Groningen.

Pilbeam, C.J., Robson, M.J. & Lambers, H. (1986). Respiration in mature leaves of Lolium perenne as affected by nutrient supply, cutting and competition. Physiol. Plant. 66: 53–57.

Pons, T.L. (1977a). An ecophysiological study in the field layer of ash coppice. II. Experiments with Geum urbanum and Cirsium palustre in different light intensities. Acta Bot. Neerl. 26: 29–42.

Pons, T.L. (1977b). An ecophysiological study in the field layer of ash coppice. III. Influence of diminishing light intensity during growth on Geum urbanum and Cirsium palustre. Acta Bot. Neerl. 26: 251–263.

Pons, T.L. (1979). Growth rates and competitiveness to rice of some annual weed species. Proc. 7th Asian-Pacific Weed Sci. Conf. Suppl. Vol.: 65–70.

Poorter, H. & Lambers, H. (1986). Growth and competitive ability of a highly plastic and a marginally plastic genotype of Plantago major in a fluctuating environment. Physiol. Plant. 67: 217–222.

Potvin, M.A. & Werner, P.A. (1983). Water use physiologies of co-occurring goldenrods (Solidago juncea and S. canadensis): Implications for natural distribution. Oecologia 56: 148–152.

Prudhomme, T.I. (1983). Carbon allocation to antiherbivore compounds in a deciduous and an evergreen species. Oikos 40: 344–356.

Robson, M.J. (1982a). The growth and carbon economy of selection lines of Lolium perenne cv. S23 with differcing rates of dark respiration. 1. Grown as simulated swards during a regrowth period. Ann. Bot. 49: 321–329.

Robson, M.J. (1982b). The growth and carbon economy of selection lines of Lolium perenne cv. S23 with differcing rates of dark respiration. 2. Grown as young plants from seed. Ann. Bot. 49: 331–339.

Robson, M.J., Stern, W.R. & Davidson, I.A. (1983). Yielding ability in pure swards and mixtures of lines of perennial ryegrass with contrasting rates of 'mature tissue' respiration. In: Efficient Grassland Farming. A.J. Corral (Ed.), pp 291–292. British Grassland Society, Hurley.

Roetman, E. & Sterk, A.A. (1986). Growth of microspecies of different sections of Taraxacum in climatic chambers. Acta Bot. Neerl. 35: 5–22.

Rorison, I.H. (1968). The response to phosphorus of some ecologically distinct plant species. New Phytol. 67: 913–923.

Roush, M.L. & Radosevich, S.R. (1985). Relationships between growth and competitiveness of four annual weeds. J. Appl. Ecol. 22: 895–905.

Schulze, E.-D. (1982). Plant life forms and their carbon, water and nutrient relations. In: Encyclopedia of plant physiology, N.S., Vol. 12B., O.L. Lange, P.S. Nobel, C.B. Osmond & H. Ziegler (Eds.), pp 615–676. Springer-Verlag, Berlin.

Seeman, J.R., Badger, M.R. & Berry, J.L. (1984). Variations in the specific activity of ribulose-1,5-bisphosphate carboxylase between species utilizing different photosynthetic pathways. Plant Physiol. 74: 791–794.

Steingröver, E. (1981). The relationship between cyanide-resistant root respiration and the storage of sugars in the taproot in Daucus carota L. J. Exp. Bot. 32: 911–919.

Thorne, G.N. & Evans, A.F. (1964). Influence of tops and roots on net assimilation of sugar beet and spinach beet and grafts between them. Ann. Bot. 28: 499–508.

Van Andel, J. & Jager, J.C. (1981). Analysis of growth and nutrition of six plant species of woodlands clearings. J. Ecol. 69: 871–882.

Van de Dijk, S.J. (1980). Two ecologically distinct subspecies of Hypochaeris radicata L. II. Growth response to nitrate and ammonium, growth strategy and formative aspects. Plant and Soil 57: 111–122.

Van den Driesche, R. & Wareing, P.F. (1966). Nutrient supply, dry matter production and nutrient uptake of forest tree seedlings. Ann. Bot. 30: 657–672.

Van Dijk, H. (1984). Genetic variability in Plantago species in relation to their ecology. II. Quantitative characters and allozyme loci in P. major. Theor. Appl. Genet. 68: 43–52.

Van Dobben, W.H. (1967). Physiology of growth in two Senecio species in relation to their ecological position. Mededeling Inst. Biol. Scheikd. Onderzoek Landbouwgewassen 346: 75–83.

Waring, R.H., McDonald, A.J.S., Larsson, S., Ericsson, T., Wiren, A., Arwidsson, E., Ericsson, A. & Lohammar, T. (1985). Differences in chemical composition of plants grown at constant relative growth rates with stable mineral nutrition. Oecologia 66: 157–160.

Wilson, D. (1975). Variation in leaf respiration in relation to growth and photosynthesis of Lolium. Ann. Appl. Biol. 80: 323–338.

Wilson, D. (1982). Response to selection for dark respiration rate of mature leaves in Lolium perenne and its effects on growth of young plants and simulated swards. Ann. Bot. 49: 303–312.

Wilson, D. & Cooper, J.P. (1969). Effect of light intensity during growth on leaf anatomy and subsequent light-saturated photosynthesis among contrasting Lolium genotypes. New Phytol. 68: 1125–1135.

Wong, S.C. (1985). Leaf conductance in relation to rate of CO_2 assimilation. I. Influence of nitrogen nutrition, phosphorus nutrition, photon flux density, and ambient partial pressure of CO_2 during ontogeny. Plant Physiol. 78: 821–825.

CHAPTER 17

Ecological significance of size variation within populations

A. Biere

Detailed contents

1. Introduction

Within most plant populations, individuals span a wide range of sizes. Factors which may determine size differences within populations include (1) maternal effects, (2) age differences, (3) genetic variation, (4) heterogeneity of resources, (5) differential effects of herbivores, parasites or pathogens, and (6) competition (Weiner & Thomas, 1986).

Frequency distributions of sizes are often characterized by the occurrence of many small individuals but only a few large ones, which contain most of the population's biomass (e.g. Obeid et al., 1967; Ogden, 1970). Such size distributions have been referred to as 'size hierarchies' (Harper, 1977; Weiner & Solbrig, 1984).

Interest in size hierarchies stems from two different fields of biology. On the one hand evolutionary biologists have considered the implications of the unequal distribution of sizes, since individual size is often positively correlated with both fecundity (e.g. Werner, 1975; Sohn & Policansky, 1977) and survivorship (e.g. Watkinson et al., 1983), so that only a small fraction of the population will contribute substantially to future generations (Leverich & Levin, 1979). The evolutionary significance of this phenomenon remains unclear. Although several studies in controlled environments have shown significant among-genet variation for plant size (e.g. Antlfinger et al., 1985; Waller, 1985), no differences in allelic frequencies between large and small sized individuals have been demonstrated in natural populations (e.g. Gottlieb, 1977); this has been interpreted as indicating the importance of small scale environmental heterogeneity. In addition, loose or negative correlations among fitness-components may occur (viz. Antonovics, 1976) or fitness at one stage may be poorly correlated with that at another stage. Indeed, juvenile size appears to be negatively correlated with adult fitness in *Geranium carolinianum* (Roach, 1985), suggesting that selection for size characters may be weak.

On the other hand, most agronomists have been interested in size hierarchies from the viewpoint of crop productivity. Many studies initiated

J. van Andel et al., (eds.), Disturbance in Grasslands. ISBN 90–6193–640–3.

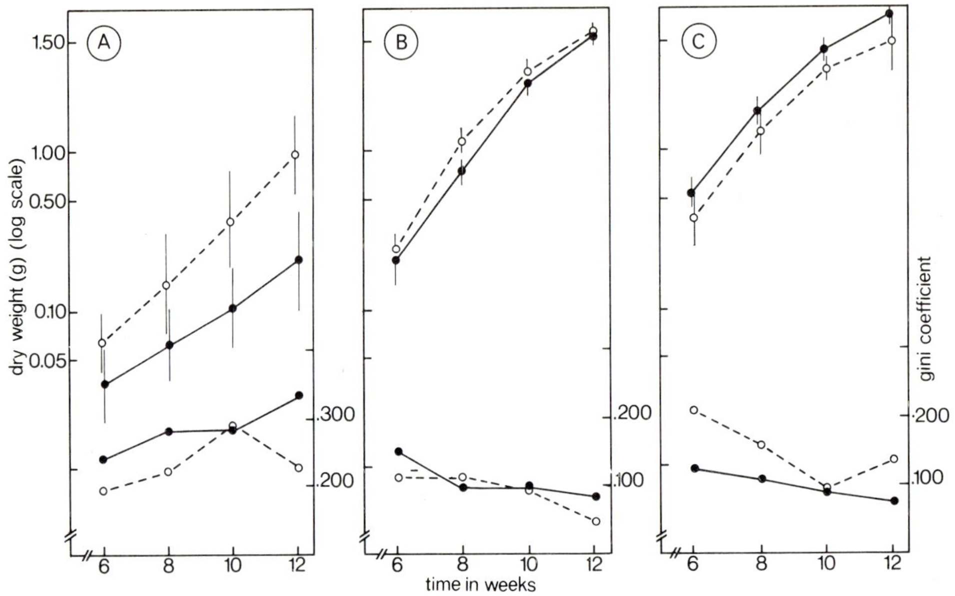

Fig. 2. The dry weight (upper lines) and size inequality (lower lines) of populations of individually grown plants of A: *Caltha palustris*, B: *Anthoxanthum odoratum* and C: *Festuca pratensis* on fertile (●) and infertile (○) soils. Bars indicate s.e. of the mean.

conditions, i.e. having a large $W_{fertile}$:$W_{infertile}$ ratio, tend to have low inequality under these conditions, i.e. a low $G_{fertile}$:$G_{infertile}$ ratio, and vice versa. Generalizing, in environments that are unfabourable for biomass production, high inequality is displayed; this contrasts markedly with the predictions.

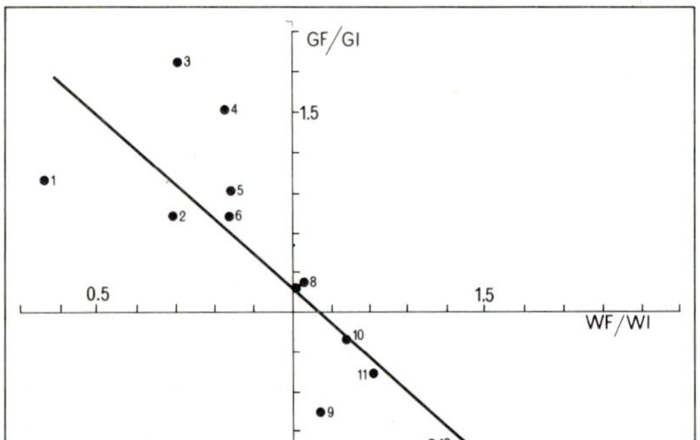

Fig. 3. Relationship between the ratio of dry weight produced on fertile (W_f) and infertile (W_i) soils and the ratio of Gini coefficients. Average values of 4 harvest data. 1. *Caltha palustris* 2. *Crepis paludosa* 3. *Carex acutiformis* 4. *Poa trivialis* 5. *Lychnis flos-cuculi* 6. *Anthoxanthum odoratum* 7. *Myosotis palustris ssp. palustris* 8. *Plantago lanceolata* 9. *Rumex acetosa* 10. *Cardamine pratensis* 11. *Holcus lanatus* 12. *Festuca pratensis*.

In a further study, sixteen plants of each species were grown individually on seven different levels of soil fertility, ranging from 5 to 320 ppm nitrogen, for seven weeks. *Rumex acetosa* (Fig. 4b) represents the average response of all species; size inequality was lowest at the nutrient concentration giving the highest yield (80 ppm N) and increased as conditions became less favourable for biomass production, either because of sub- or supra optimal conditions. The response of *Myosotis palustris ssp. palustris* (Fig. 4c) will be considered further in section 3.3.

3.2 Competing plants

Nutrient supply may influence size inequality among competing plants by modifying the relative importance of two different processes operating during intraspecific competition.

1. Dominance and suppression. Several authors have reported a positive correlation between individual size and RGR among competing plants (e.g. Ford, 1975). Thus, in contrast to isolated plants, the increase of size differences with time tends to continue, until growth ceases (Fig. 5a), though it may ultimately decrease if density dependent mortality occurs (Weiner & Thomas,

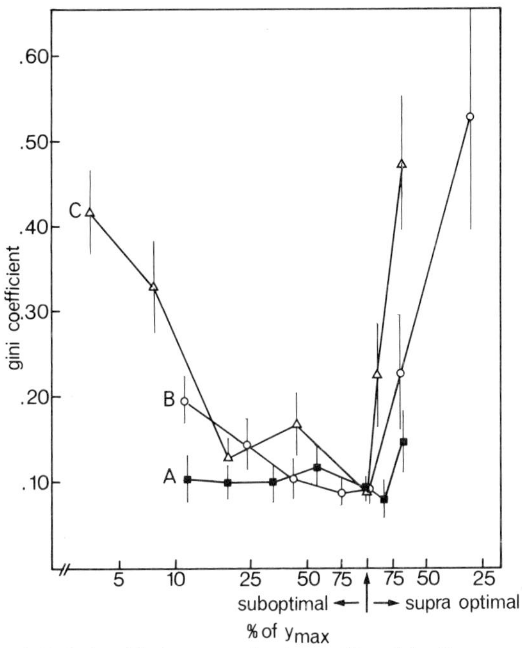

Fig. 4. Relationship between size inequality of A. *Festuca pratensis*, B. *Rumex acetosa* and C. *Myosotis palustris ssp. palustris* and the yield, expressed as a percentage of that produced at the highest yielding nutrient level for each species. The yield was varied by applying nitrogen concentrations ranging between 5 and 320 ppm. Bars represent s.e. of the mean.

1986). Data such as those of Ford (1975) have been interpreted as supporting the idea of 'dominance and suppression' (Harper, 1977), i.e. the larger individuals are able to exploit a disproportional large part of the resources and suppress the growth of smaller neighbours; competition is asymmetric in nature.

2. Resource depletion. In the case of *Festuca paradoxa* (Turner & Rabinowitz, 1983; Fig. 5b), size inequality among competing plants did not increase to a larger extent than that among isolated plants. This has been interpreted as showing that competition is more symmetric, i.e. neighbours reduce resources proportionally ('resource depletion'). Following once more the model of Koyama & Kira (1956), this will cause a delay in the generation of size inequality in the same way as low nutrient supply might do (see above), resulting in similar or lower size inequality among competing plants after a given period of growth (Weiner & Thomas, 1986). However, the observations of Turner & Rabinowitz (1983) were re-

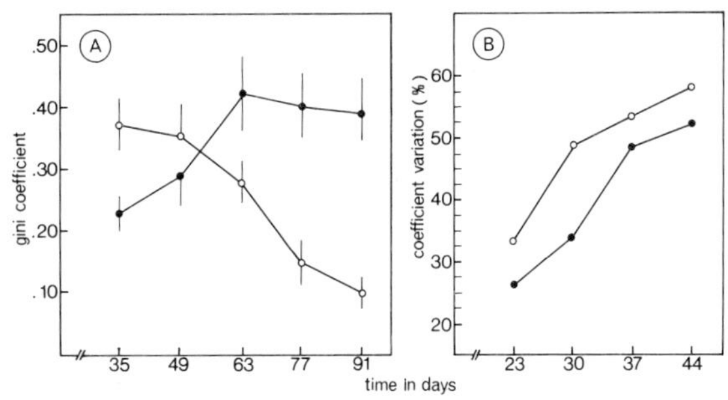

Fig. 5. Size inequality among competing (●) and isolated plants (○) of A. *Urtica dioica* (van Andel et al., 1984) and B *Festuca paradoxa* (Turner and Rabinowitz, 1983). Bars indicate s.e. of the mean.

stricted to a short period of growth, which limits their interpretation (cf. Fig. 5a and b). Other data supporting the idea of resource depletion were produced by Newbery & Newman (1978). They demonstrated that small individuals in monocultures of *Plantago lanceolata, Holcus lanatus* and *Lolium perenne* were at a competitive advantage compared with their larger neighbours, resulting in more equal sizes. Competition in their experiment was mainly for nutrients; the results in these conditions contrast markedly with those predicted by dominance and suppression.

These models represent two extemes; intermediate situations might have aspects of both. For example, Waller (1985) observed that size inequality in populations of *Impatiens capensis* was lowest at intermediate densities; dominance and suppression models predict lowest inequality at low densities, resource depletion models at high densities.

It has been argued (Newbery & Newman, 1978; Turner & Rabinowitz, 1983; Weiner, 1985) that the extent to which dominance and suppression or resource depletion operates, depends on which resources are limiting, e.g. light or nutrients. A small difference in size may imply a large difference in light interception, whereas in competition for nutrients the uptake is more or less in proportion to size (Andrews & Newman, 1970). As a result, lowering the amount of nutrient supply may not only cause a delay of the process of dominance and suppression, due to an overall reduction of average plant size, but may also cause a shift towards more symmetric competition, due

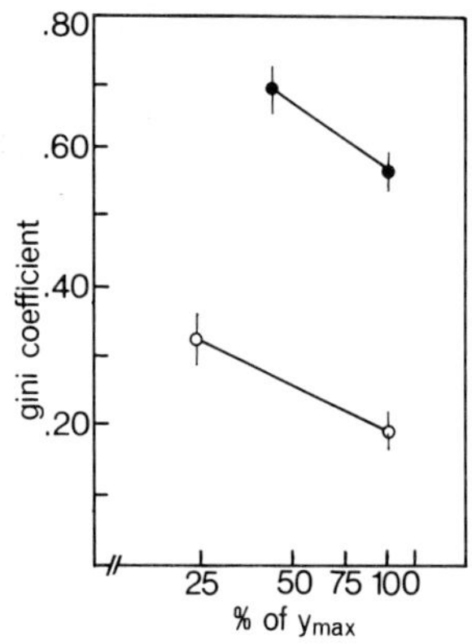

Fig. 6. Relationship between size inequality among isolated (○) and competing plants (●) of *Myosotis palustris ssp. palustris* and percentual yield (see Fig. 4), when grown at different nutrient levels.

to the relative increase of root compared with shoot competition. Both would result in less size inequality.

In the present study, size inequality among competing plants of *Myosotis palustris ssp. palustris* was larger than that among isolated plants (Fig. 6), indicating the importance of dominance and suppression. However, size inequality was not reduced at low levels of nutrient supply, either for individually grown plants (Fig. 4c) or for competing plants (Fig. 6).

3.3 Causes of differences in size inequality

Apparently, plant sizes at moderately low levels of nutrient supply are not more uniformly distributed than those at higher levels, where yield is at a maximum; this is in contrast to predictions. At extremely low (and high) nutrient levels, size inequality is even higher than in conditions of maximum yield. This corresponds with other reports of high inequality in populations exposed to severe 'stress', whether moisture, light, temperature or nutrient supply (Duncan et al., 1973; Pederson, 1968).

Two factors may increase size inequality at low levels of nutrient supply. Firstly, genotypes may differ in their tolerance of extremes, though having the same optimum (Fig. 7a) or a few genotypes may have different optima (Fig. 7b). In both cases this would result in an increase in inequality at the two extremes and in significant genotype x environment interactions. At sib-family level, no indication of such interactions was apparent in populations of *Myosotis palustris ssp. palustris* (Table 1). Though shoot dryweight production differed significantly among sib-families, the increase in inequality at low nutrient levels (Table 2) was due to an increase within sib-families, rather than to differences in performance among families. It must be noted however that *Myosotis palustris ssp. palustris* is an outbreeding species and that results might have been different if genotypes and not sib-families had been used.

Secondly, differences in requirements among seedlings may be more pronounced under unfavourable conditions. For example, Austin (1966) found that seeds of watercress (*Rorippa nasturtium aquaticum* L. Hayek), raised under high phosphorus conditions all germinated and grew equally well, regardless of the nutrition of their parents; under conditions of P-deficiency however, seedlings form parents raised under high levels of phosphorous supply performed better than seedlings from P-deficient parents resulting in differences in phenotypic response (cf. differences at sub-optimal and optimal nutrient levels in Fig. 7a).

Similar differences in performance at low and high levels of nutrient supply could result from variation in seed size within populations, but no indication of such differences was apparent in *Myosotis palustris ssp. palustris*, when grown as isolated plants. The ratios of average plant weight between individuals from small (<0.5

Table 1. Analysis of variance of shoot dry weights (square root transformed) of individually grown plants of *Myosotis palustris ssp. palustris*.

Source of variation	d.f.	Mean squares	P
Co-factor			
Seed size	1	74.215	.104 ns.
Main effects			
A. Sibfamily	1	113.152	.001 ***
B. Nutrient level	1	8035.731	.001 ***
Interactions			
A×B	6	25.417	.482 ns.
Total	115	99.756	

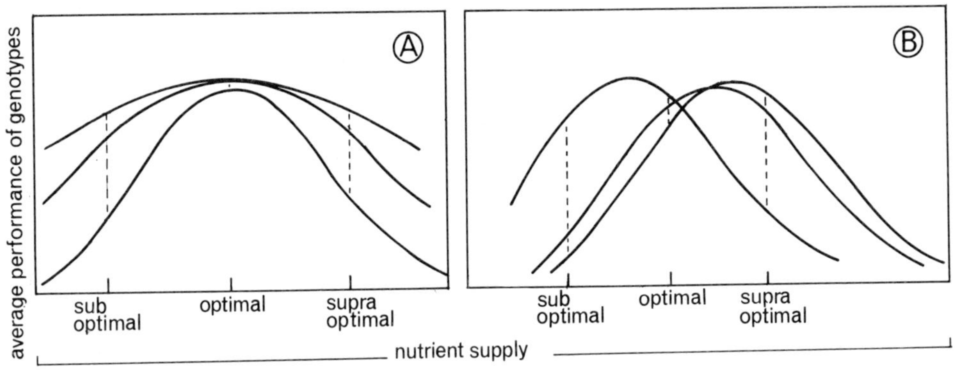

Fig. 7. Schematic representation of the effect of genotypic differences on size inequality at different levels of nutrient supply. For explanation see text.

mm) and large (>0.5 mm) seeds were 0.92 and 0.89, respectively at low and high levels of nutrient supply. However, when plants were grown in intraspecific competition, the weight ratio of seedlings from small compared with large seeds was 0.18 at the low nutrient level, which was significantly lower than the ratio of 0.36 at the higher level. Although this might offer an explanation for the enhanced size inequality at the low nutrient level, it appeared that size differences within the group of individuals that emerged from large seeds were yet so pronounced (G=.645) that the observed differences in performance between size groups hardly contributed to the increased inequality of the entire population (G=.682, Fig. 6).

4. Ecological significance of size distributions

4.1 Size variability and population yield

One of the major difficulties in assessing the ecological significance of size variability per se is that size inequality is greatest in conditions where low yield occurs and vice versa. If we wish to study the direct effect of size inequality in terms of population yield, we must establish populations differing in size inequality in each of the environments studied.

Table 2. Gini coefficients for shoot weight of different sib-families of individually grown plants of *Myosotis palustris ssp. palustris*.

Sibfamily	Nutrient level Opt.	Subopt.	Sibfamily	Nutrient level Opt.	Subopt.
1	.109	.456	5	.126	.333
2	.057	.237	6	.105	.205
3	.201	.322	7	.443	.248
4	.007	.127	A11	.196	.324

In a preliminary study, variable (mixed aged) and uniform (even aged) populations of *Lychnis flos-cuculi*, which is a rosette forming perennial meadow species, were grown in 1:1 mixtures with *Anthoxantum odoratum*, both under relatively favourable and suboptimal levels of nutrient supply in a glasshouse. In addition, variable and uniform populations of *Lychnis* were grown in pure stands under favourable nutritional conditions. In order to create different age structures, seeds were germinated at three different data and populations were composed of either equal numbers of pregrown seedlings from the resulting three size classes, or only of seedlings from the medium size class. Start of germination was minutely timed to reach similar average initial fresh weight for both types of populations. All *Anthoxantum* seeds were germinated simultaneously with seeds of the medium *Lychnis* size class. Each individual was preweighed and located in a marked position in a regular pattern in 43×37×8 cm flats, each containing 168 individuals; two edge rows (88 plants) were discarded from analysis.

Initial values of Gini coefficients for uniform populations of *Lychnis* ranged from 0.11 to 0.13, for variable populations from 0.25 to 0.29. Size inequality in uniform populations remained lower

Table 3. Shoot biomass of uniform and variable populations of *Lychnis flos-cuculi*, grown in single species stands or with *Anthoxanthum odoratum*. The proportion of *Lychnis* in the total shoot biomass is also shown.

Nutrient-supply	Stand type	Size structure *Lychnis*	Shoot biomass (mg)		Proportion of *Lychnis* Init.	Final	Change in proportion
Favourable	Single	Uniform	308 (18)	380 (27)	33.5	44.8	
		Variable	308 (9)	n.d.	n.d.		
Favourable	Mixed	Uniform	252 (24)	463 (23)	38.4	35.8	0.93 (.04)
		Variable	285 (62)	444 (34)	37.5	38.9	1.04 (.09)
Suboptimal	Mixed	Uniform	220 (27)	473 (36)	39.1	31.8	0.81 (.10)
		Variable	251 (21)	464 (24)	37.8	35.1	0.93 (.05)

Standard error of the mean in parentheses. N.d.: not determined. Initial shoot dry weight values calculated from linear regressions on initial total fresh weight ($n=20$, $r>.96$ $p<.001$)

throughout the experiment. In monospecific stands, shoot yield of uniform and variable *Lychnis* populations was identical, i.e. final population yield in a monospecific stand under favourable conditions was independent of the initial distribution of sizes (Table 3). The lower yield of *Lychnis* compared to *Anthoxanthum* in monocultures merely reflects a difference in starting capital; in fact the yield ratio (*Lychnis/Anthoxanthum*) increased in the course of the experiment (Table 3).

In mixed stands, the situation was markedly different. The contribution of *Lychnis* to the total shoot biomass tended to decrease with time (Table 3), but was significantly greater for variable than for uniform populations (Tables 3, 4). Apparently, populations with high size inequality are able to capture a larger part of the resources that are available to both species, than populations with low inequality.

In section 3 we saw that plant size was more variable in nutrient conditions that were unfavourable for biomass production. We now observe a competitive advantage for large size variation under these conditions, but also under favourable conditions since there was no nutrient x size structure interaction (Table 4). Apparently, large size variation is not particularly suitable to the unfavourable conditions in which it is displayed, nor is low size variation particularly suitable to favourable conditions. We must note however, that the difference between favourable and suboptimal levels of nutrient supply in this study was small; although the yield ratio (*Lychnis/Anthoxanthum*) differed significantly among nutrient levels, the effect of nutrient supply on total shoot biomass was negligible (Table 3).

4.2 The individual in the population

Considering the growth of individuals in the *Lychnis* populations of the previous section more in detail, may provide additional insight in the mechanisms underlying the generation of size differences, discussed in section 3.

The negative correlation between plant size and RGR in *Lychnis*, when grown individually (Fig. 8a) would counteract a divergence of sizes with time. However, in monospecific stands and in favourable nutrient conditions, this was overruled by competition between neighbours, which

Table 4. Analysis of variance of the final:initial proportion of *Lychnis flos-cuculi* in the total shoot biomass, when grown in mixtures with *Anthoxanthum odoratum*.

Source of variation	d.f.	Mean squares	P
Main effects			
A. Nutrient level	1	.039	.037 *
B. Size structure	1	.037	.042 *
Interactions			
A×B	1	.000	.923 n.s.
Total	11	.011	

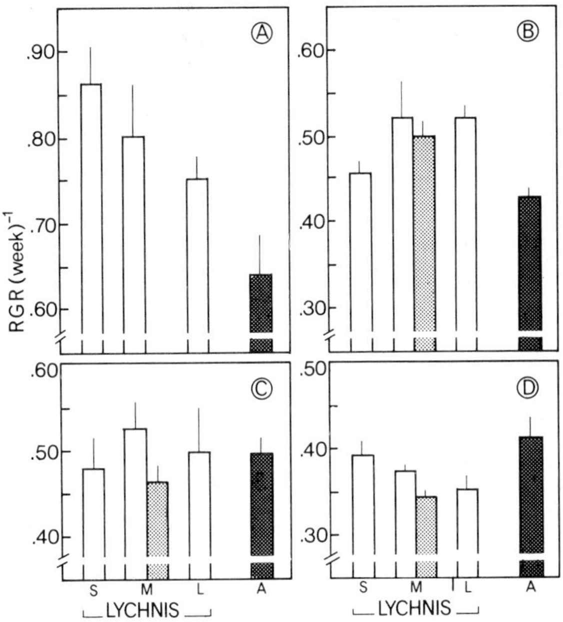

Fig. 8. The RGR of small (s), medium (m) and large (l) size groups in mixed size stands of *Lychnis flos-cuculi,* and of medium sized *Anthoxanthum adoratum* plants (A, dark shades bars) when grow as A. Single plants on fertile soils, B. Single species stands on fertile soils, C. Species mixtures on fertile soils, and D. Species mixtures on infertile soils. In the case of medium size, the RGR of uniform stands are also shown (shaded bars).

reduced the growth of smaller individuals (Fig. 8b; table 5), as predicted by the dominance and suppression model; Ford (1975) obtained similar results.

In species mixtures, a marked difference in the performance of small individuals was observed at different levels of soil fertility. In favourable conditions, the contribution of small *Lychnis* individuals to the total shoot yield of this species decreased with time (Table 5; Fig. 8c), as was observed in monospecific stands. However, at low levels of nutrient supply, small individuals increased their contribution to the total biomass at the expense of larger individuals (Table 5; Fig. 8d), so that sizes would converge with time. Similar results were obtained by Newbery & Newman (1978). Competition was evident even in these low nutrient conditions; overall yield was reduced by a factor four compared with isolated plants grown in these conditions.

Assuming that the lower level of nutrient supply implies greater importance of root competitions, these results are in accordance with the concept that belowground competition is more symmetric in nature (see section 3.2), i.e. that in conditions of low nutrient supply resource depletion is more important and the process of dominance and suppression less important than in favourable nutrient conditions. The results seem to contrast with the greater divergence of sizes observed in populations of *Myosotis palustris ssp. palustris* at lower levels of nutrient supply (section 3); however, in *Myosotis*, this was only observed at levels that resulted in severe yield reduction, which did not occur in the *Lychnis* experiment.

5. Concluding remarks

5.1. Size variation and competitive ability

The consequences of size variability within plant populations have mostly been considered for generative characters in annual plant species. Important generative aspects are the evolutionary consequences of unequal seed production among individuals, resulting from differences in plant size, and the agricultural consequences of non-uniformity in crops for grain yield. Size variation might be responsible for a reduction of grain yield. For example, Glenn & Daynard (1974) found that grain yield in plots of overplanted corn that were uniformly thinned, was generally higher than that in non-uniformly thinned plots.

Although the effect of size distribution on the

Table 5. Ratio of final:initial contribution of three differently sized groups of *Lychnis flos-cuculi* in the total shoot biomass of *Lychnis* populations, when grown in single species stands or together with *Anthoxanthum odoratum*.

Stand type	Nutrient supply	Initial size group Small	Medium	Large
Single	Favourable	0.77 (.04)	1.09 (.15)	1.02 (.09)
Mixed	Favourable	0.85 (.07)	1.15 (.02)	0.96 (.01)
	Suboptimal	1.16 (.11)	1.06 (.03)	0.92 (.03)

Standard error of the mean in parenthesis.

individual's chance of survival has been well documented, data concerning the consequences of size variation for vegetative characters like population biomass or crop yield are scarce. However, the available data indicate that size variation either has little effect or results in enhanced biomass yield. For example, in self thinning stands of radish (*Raphanus sativus* L.) starting from similar biomass-density combinations, mixed age stands had higher biomass yield at a given density of survivors than uniform stands (Westoby and Howell, 1986).

In the experiments with the perennial species *Lychnis flos-cuculi* (section 4), size variation did not influence biomass yield in single species stands on fertile soils. The enhanced growth of large and intermediate sized individuals, compared with uniform stands of medium-sized individuals, were exactly counteracted by reduced growth of small individuals (Fig. 8b; table 5). However, large size variation did increase the contribution of *Lychnis* in mixture with *Anthoxanthum*, indicating that size variation may influence the competitive ability of a species, enabling it to gain a greater proportion of the resources when grown in species mixtures. In the case of *Lychnis* and *Anthoxanthum*, the latter species was dominant; in these conditions large size variation in *Lychnis* would promote coexistence.

5.2 Size variation in unfertilized grasslands

Snaydon & Howe (1986) have stressed the importance of belowground competition in permanent grasslands; they showed that the performance of seedlings in root competition with established grass species was severely reduced, whereas the effect of shoot competition on seedlings was small. In unfertilized grasslands, we can expect root competition to play an even greater role; also, after cessation of fertilizer application as part of management practices (Bakker, Chapter 12), belowground competition may become more important. Data presented in part 4.2. partly support the view (Newbery & Newman 1978; Turner & Rabinowitz, 1983; Weiner, 1985) that under conditions of intense competition for soil factors, competition is more symmetric, i.e. the exploitation of resources is in proportion to size rather than one-sided as assumed in the dominance and suppression model (but see section 3.3 in case of nutrient stress). Small individuals may even have a relative growth advantage over larger plants (Fig. 8d).

In experiments of Newman & Newbery (1978) this led to a convergence of sizes, both between size groups in single-species stands, and between pairs of species in mixtures. As a result, no single species became completely dominant and stable mixtures were reached. This implies that under conditions of intense belowground competition coexistence of established species is promoted, i.e. species richness is increased.

Acknowledgements

The author gratefully thanks Mr R. Lindeboom and Mr L. Ronde who performed part of the experiments, Mr J. Franke and Mr S. Nijdam for technical assistence and Mr E. Leeuwinga for drawing the graphs.

The investigations were supported by the Foundation for Fundamental Biological Research (BION), which was subsidized by the Netherlands Organization for Fundamental Research (ZWO).

References

Van Andel, J. & Jager, J.C. (1981). Analysis of growth and nutrition of six plant species of woodland clearings. J. Ecol. 69: 871–82.

Van Andel, J., Nelissen, H.M.J., Wattel, E., van Valen, T.A. & Wassenaar, A.T. (1984). Teil's inequality index applied to quantify population variation of plants with regard to dry matter allocation. Acta Bot. Neerl. 33: 161–75.

Andrews, R.E. & Newman, E.I. (1970). Root density and competition for nutrients. Oecol. Plant. 5: 319–34.

Antlfinger, A.E., Curtis, W.F. & Solbrig, O.T. (1985). Environmental and genetic determinants of plant size in *Viola sororia*. Evolution 39: 1053–64.

Antonovics, J. (1976). The nature and limits to natural selection. Ann. Missouri Bot. Gard. 63: 224–47.

Austin, R.B. (1966). The growth of watercress (*Rorippa nasturtium aquaticum* L. Hayek) from seed as affected by the phosphorus nutrition of the parent plant. Plant and Soil XXIV: 113–20.

Burdon, J.J. & Harper, J.L. (1980). Relative growth rates of individual members of a plant population. J. Ecol. 68:953–57.

Duncan, W.G., Hammond, L.C., Robertson, W.K. & Lundy, H.W. (1973). Effects of environmental stress and planting accuracy on plant-to-plant variability in corn (*Zea mays* L.). Soil and Crop Sci. Soc. Florida Proc. 32: 1–4.

Evans, G.C. (1972). The Quantitative Analysis of Plant Growth. Blackwell, Oxford.

Ford, E.D. (1975). Competition and stand structure in some even-aged plant monocultures. J. Ecol. 63: 311–33.

Gini, C. (1912). Variabilita e Mutabilita. Bologna.

Glenn, F.B. & Daynard, T.B. (1974). Effects of genotype, planting pattern, and plant density on plant-to-plant variability and grain yield of corn. Can. J. Plant Sci. 54: 323–30.

Gottlieb, L.D. (1977). Genotypic similarity of large and small individuals in a natural population of the annual plant *Stephanomeria exigua ssp. coronaria* (Compositae). J. Ecol. 65: 127–34.

Harper, J.L. (1977). The Population Biology of Plants. Academic Press, London.

Hunt, R. (1982). Plant Growth Curves. University Park, Baltimore.

Hunt, R. (1984). Relative growth rates of cohorts of ramets cloned from a single genet. J. Ecol. 72: 299–305.

Koyama, H. & Kira, T. (1956). Intraspecific competition among higher plants VIII. Frequency distributions of individual plant weight as affected by the interaction between plants. J. Inst. Polytech. Osaka City Univ. Series D 7: 73–94.

Leverich, W.J. & Levin, D.A. (1979). Age-specific survivorship and reproduction in *Phlox drummondii*. Am. Nat. 113: 881–903.

Morris, E.C. & Myerscough, P.J. (1984). The interaction of density and resource levels in monospecific stands of plants: a review of hypotheses and evidence. Aust. J. Ecol. 9: 51–62.

Newbery, D.Mc.C. & Newman, E.I. (1978). Competition between grassland plants of different initial sizes. Oecologia (Berl.) 33: 361–80.

Obeid, M., Machin, D. & Harper, J.L. (1967). Influence of density on plant-to-plant variation in fiber flax, *Linum usitatissimum*. Crop Sci. 7: 471–73.

Ogden, J. (1970). Plant population structure and productivity. Proc. N.Z. Ecol. Soc. 17: 1–9.

Pederson, D.G. (1968). Environmental stress, heterozygote advantage and genotype-environment interaction in *Arabidopsis*. Heredity 23: 127–38.

Roach, D.A. (1985). Genetic and environmental constraints to selection repsonse for juvenile characters in *Geranium carolinianum*. In: Jacquard, P., Heim, G. & Antonovics, J. (Eds.). Genetic Differentiation and Dispersal in Plants, pp 191–201. Springer-Verlag, Berlin.

Sen, A. (1973). On Economic Inequality. Clarendon, Oxford.

Snaydon, R.W. & Howe, C.D. (1986). Root and shoot competition between established ryegrass and invading grass seedlings. J. Ecol. 23: 667–74.

Sohn, J.J. & Policanski, D. (1977). The cost of reproduction in the mayapple *Podophyllum peltatum* (Berberidaceae). Ecology 58: 1366–74.

Sokal, R.R. & Rohlf, F.J. (1981). Biometry, 2nd. Edn. Freeman, New York.

Soria, L.L. (1985). On suppression in plant populations: an approach based on the Shannon index. Acta Oecologia Oecol. Plant. 6: 223–31.

Theil, H. (1967). Economics and Information Theory. North. Holland Publ. Comp., Amsterdam.

Turner, M.D. & Rabinowitz, D. (1983). Factors affecting frequency distributions of plant mass: the absence of dominance and suppression in competing monocultures of *Festuca paradoxa*. Ecology 64: 469–75.

Waller, D.M. (1985). The genesis of size hierarchies in seedling populations of *Impatiens capensis* Meerb. New Phytol. 100: 243–60.

Watkinson, A.R., Lonsdale, W.M. & Firbank, L.G. (1983). A neighbourhood approach to self-thinning. Oecologia (Berl.) 56: 381–84.

Weiner, J. (1985). Size hierarchies in experimental populations of annual plants. Ecology 66: 743–52.

Weiner, J. & Solbrig, O.T. (1984). The meaning and measurement of size hierarchies in plant populations. Oecologia (Berl.) 61: 334–36.

Weiner, J. & Thomas, S.C. (1986). Size variability and competition in plant monocultures. Oikos 47: 211–22.

Werner, P.A. (1975). Predictions of fate from rosette size in teasel (*Dipsacus fullonum* L.). Oecologia (Berl.) 20: 197–201.

Westoby, M. & Howell, J. (1986). Influence of population structure on self-thinning of plant populations. J. Ecol. 74: 343–59.

White, J. & Harper, J.L. (1970). Correlated changes in plant size and number in plant populations. J. Ecol. 58: 467–85.

CHAPTER 18

Population responses of *Rhinanthus angustifolius* to disturbance of grassland communities

P.C. de Hullu & A.P. Grootjans

Detailed contents

1. Introduction

*Rhinanthus angustifolius** is an annual hemi-parasitic plant that thrives in grasslands under a wide variety of soil conditions, for example in mown and grazed grasslands on wet or moist clay or peat soils with a low K and P-status. Kruijne et al. (1976) and Ellenberg (1974) characterized its habitat as not shaded, moist, not acid and with low N-availability.

The species is locally common in the Netherlands, especially in some nature reserves but its frequency has declined since 1950 (Fig. 1). This decline has probably been caused by increasing fertilizer use and drainage. In principle, the species has a wide ecological amplitude (Fresco, 1980). If fertilizer use is not too heavy, only extreme wet conditions appear to limit the species distribution in valley meadows (Fresco, 1980). It

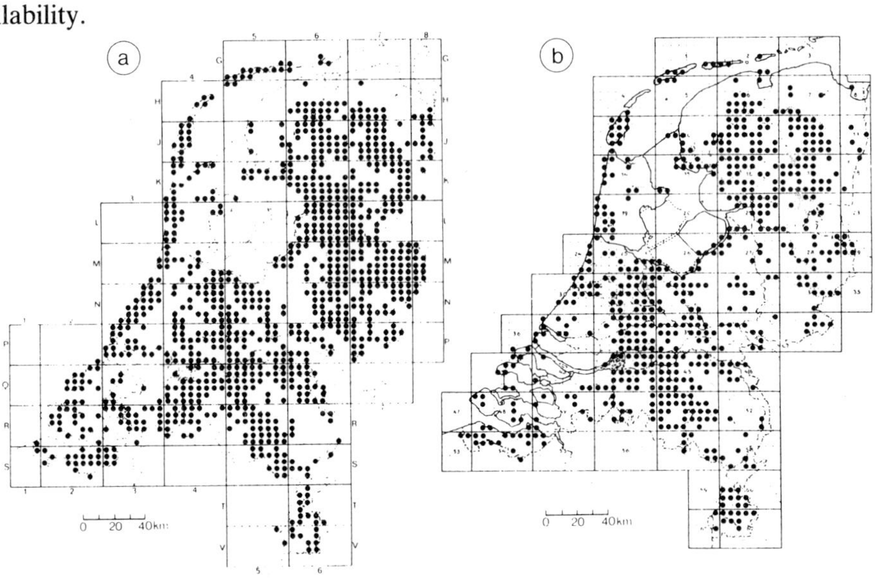

Fig. 1. The distribution of *Rhinanthus angustifolius* in the Netherlands. Presence-absence data are plotted in a 5 × 5 km grid before (a) and after (b) 1950 (after Van den Ham - in prep.).

J. van Andel et al., (eds.), Disturbance in Grasslands. ISBN 90–6193–640–3.

* (Nomenclature follows Van der Meijden et al., 1983. Flora van Nederland. Wolters-Noordhoff, Groningen).

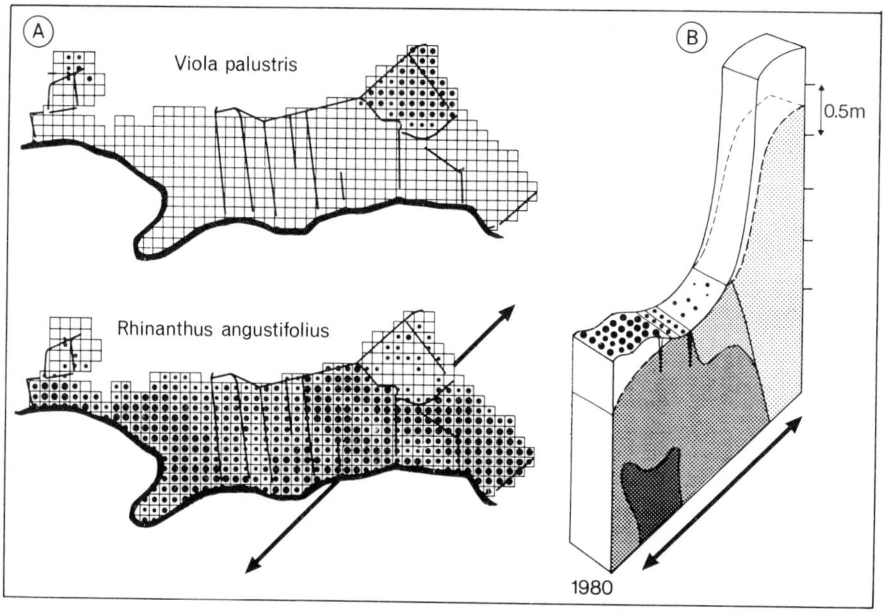

Fig. 2. a) The distribution of *Rhinanthus angustifolius* and *Viola palustris* in a nature reserve ('Estate Mensinge') Roden. b) Water quality and level shown in a cross section; electrical conductivity in μS/cm. <150, 150–225, 225–375 375–525; ----=water level. *Rhinanthus angustifolius* is rare, when base-poor groundwater occurs on the valley flank, but is abundant, when base-rich groundwater arises from the deep aquifer.

avoids the wet small sedge vegetation at the valley slope (Fig. 2a), characterised by *Viola palustris*. However, these areas are also nourished by base poor ground water (Fig. 2b), so differences in soil nutrient availability may also influence the distribution pattern of *Rhinanthus angustifolius*. This paper aims to analyse the environmental conditions of establishment, growth and maintenance of *Rhinanthus angustifolius* populations with particular emphasis on changes in management and drainage as an example of the effects of disturbance on grassland communities.

2. Response to management

In the hayfields of the river valleys of the nature reserve 'Drentsche Aa', which was formerly intensively cropped and fertilized, a secondary succession was induced following the cessation of fertilizer use and a single cutting per year.

The vegetation changed from a Poo-Lolietum to a Lolio-Cynosuretum/Juncetum acutiflori association (Van Duuren et al., 1981), with a gradual change in vegetation structure (De Hullu et al., 1985; De Hullu, 1985).

Under such conditions populations of *Rhinanthus angustifolius* can increase rapidly. This phenomenon has been recorded in a series of permanent plots in the 'Drentsche Aa' nature reserve; high numbers are maintained for some years, but than decline and level off (Fig. 3).

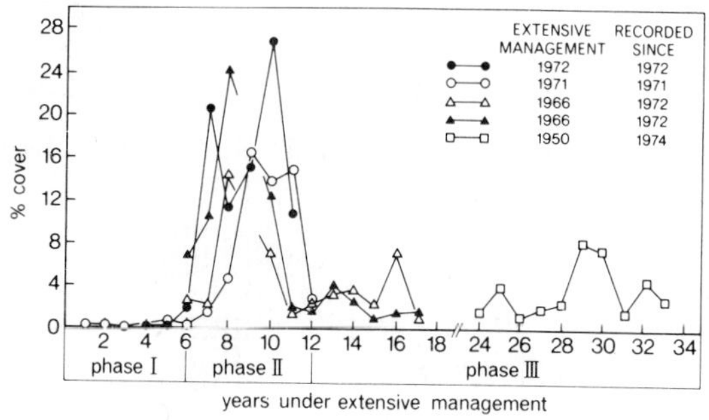

Fig. 3. The percentage cover of *Rhinanthus angustifolius* when colonizing a grassland in a successional series. The data are from permanent plots in grasslands acquired by the State at various dates since 1950, and subjected to low intensity management thereafter (after ter Borg, 1985).

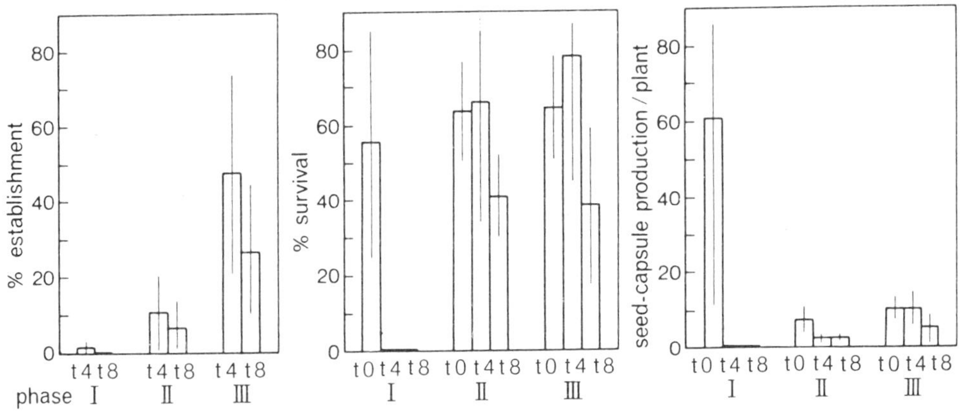

Fig. 4. a) Percentage establishement, b) percentage survival, and c) number of seed capsules of *Rhinanthus angustifolius* in three stages of succession (I-III). t0=naturally occurring seedlings, t4=seed sown 4 weeks after the emergence of naturally occuring seedlings, t8=seed sown 8 weeks after. (After de Hullu, 1987).

3. Light reduction

Since fields were acquired (and management changed) at different times, different vegetation types were available for a comparative study (Ter Borg, 1985; De Hullu et al., 1985).

The following succession stages with different numbers of *Rhinanthus angustifolius* plants, were used: stage I- Poo-Lolietum/Lolio-Cyniosuretum (4 years after agricultural use) with a dense vegetation structure and 4 plants.m^{-2} of *Rhinanthus angustifolius*; stage II- Lolio-Cynosuretum (8 years after agricultural use) with intermediate vegetation density and 300 plants.m^{-2} of *Rhinanthus angustifolius*; stage III- Lolio-Cynosuretum/Juncetum acutiflori (34 years after agricultural use) with a low vegetation density early in the season and 60 plants.m^{-2} of *Rhinanthus angustifolius*. The seedling growth and survival was studied in the successional stages by planting ten pre-germinated seeds of *Rhinanthus angustifolius* (originating from a stage II vegetation) into ten random plots, each 20×20 cm. The seeds were planted on 14th April and 8th May, four weeks and eight weeks respectively after emergence of most of the natural seedlings. Four weeks delay of emergence resulted in a very low percentage (5%) of established seedlings in stage I (Fig. 4a), and no seedlings survived to maturity (Fig. 4b). About 10% of the germinated seeds became established in stage II and about 50% in stage III (Fig. 4a), if delayed by 4 weeks, and about the same proportion of seedlings survived to maturity as for natural seedlings (Fig. 4b); they also produced a similar number of seed capsules per plant (Fig. 4c). An emergence delay of 8 weeks resulted in a reduction in establishment, survival and seed production (Fig. 4a-c), compared with 4 weeks delay. The differences were not significant, due to a high variability among the plants. We conclude that early germination of *Rhinanthus angustifolius* is important for survival and seed production, especially in early successional stages (stage I), probably because of the dense vegetation late in the season.

The effect of shade on the development of seedlings and young plants was investigated in a controlled environment chamber (14 hours of light at 320 W.m^{-2}, temp. 21° C during the day or at night, air humidity 70% during the day, and 90% during the night). Four pre-germinated seeds of *Rhinanthus angustifolius* (orginating from stage II) were planted three weeks after planting a single host seedling of *Lolium perenne* or *Deschampsia flexuosa; Deschampsia flexuosa* is shade tolerant and *Lolium perenne* is shade intolerant. After the *Rhinanthus angustifolius* seedlings had emerged, the pots were divided in five experimental groups, which were subjected to severe shading (80 W.m^{-2}) for various periods, in an attempt to simulate the light intensity under dense vegetation. The five experimental groups were: (i) grown eight weeks in full light (control); (ii) grown four weeks after emergence in full

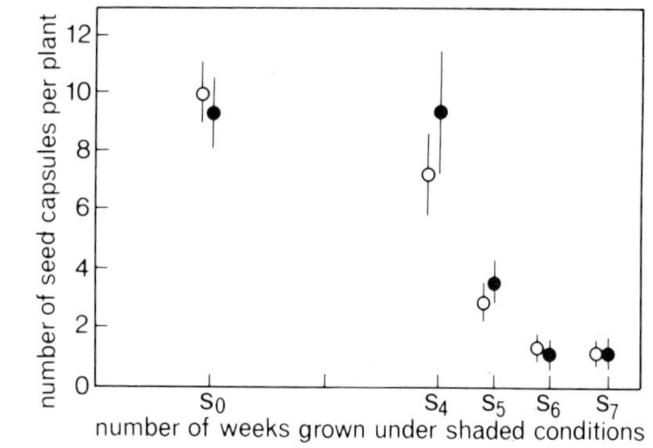

Fig. 5. The effect of shading for 4–7 weeks (S4–S7), after 4–1 weeks of full light on the number of seed capsules per plant of *Rhinanthus angustifolius* compared to plants grown for all eight weeks in full light (So). *Rhinanthus angustifolius* was grown on either *Lolium perenne* (●) or *Deschampsia flexuosa* (○) as the host species. Bars indicate S.E. of mean (After de Hullu, 1987).

light, then four weeks under shade; (iii) three weeks light then five weeks shade; (iv) two weeks light then six weeks shade; (v) one week light then seven weeks shade. There was no significant effect of host species (*Lolium perenne* or *Deschampsia flexuosa*) on seed production of *Rhinanthus angustifolius* (Fig. 5). Shading for more than 4 weeks significantly reduced the number of seed capsules, but shading for the first four weeks after emergence had no significant effect compared with plants grown in light during the complete experimental period (Fig. 5). There was no significant hostspecies x shading interaction suggesting that light had a direct effect on the growth and seed capsule production of *Rhinanthus angustifolius*.

4. The response to drainage

Several studies of the effect of drainage on *Rhinanthus angustifolius* have been carried out. First the effects of drainage in a species-rich fen meadow in the Drentsche Aa river valley system was investigated. The meadow was drained in autumn 1975 and in the same year sold to the nature conservancy. The previous mowing regime (annual cutting) was maintained, but mowing was about

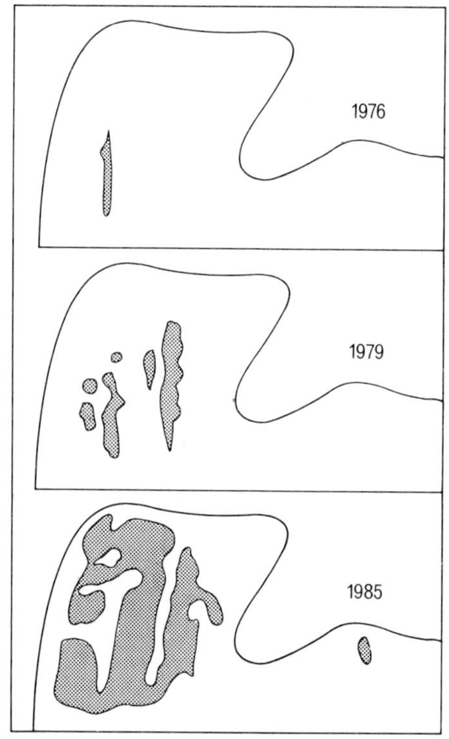

Fig. 6. The distribution of *Rhinanthus angustifolius* in a field after drainage in 1975.

one month later and grazing and application of fertilizers (previously modest) ceased. Within two years after drainage had begun. *Holcus lanatus* dominated the vegetation while *Caltha palustris* and *Carex nigra* declined. *Rhinanthus angustifolius* and *Rhinanthus minor* which were rare in 1976 spread from the ditches, and by 1985 *Rhinanthus angustifolius* was abundant over much of the area (Fig. 6). However, *Rhinanthus angustifolius* failed to establish in tall herb vegetation, containing *Urtica doica, Anthriscus sylvestris* and *Filipendula ulmaria*, and in very wet places, containing *Carex aquatilis* or *Carex rostrata*. The changes in management and drainage can both be regarded as disturbance of the existing community, to which *Rhinanthus angustifolius* responded positively.

The effects of water level and water quality on the distribution of *Rhinanthus angustifolius* were studied in more detail in the catchment area of the Drentsche Aa nature reserve in a series of hay-meadows with a variety of vegetation types

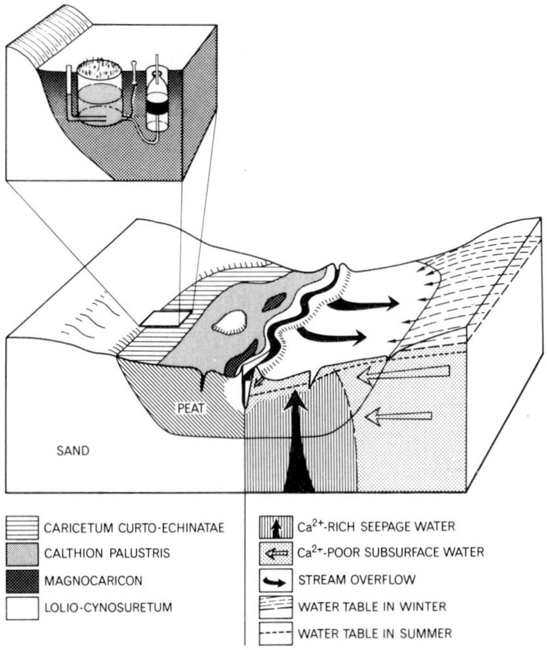

Fig. 7. The distribution of vegetation types and watertypes in a river valley system in Roodzand (Drentsche A-area). Four lysimeters were installed in the Caricetum curto echinatae. The place of the lysimeters used in the experiments is also indicated.

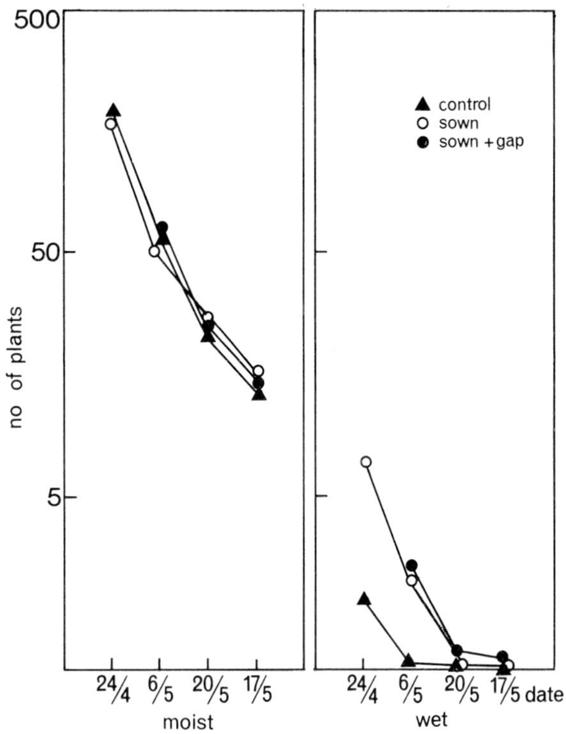

Fig. 8. The number of *Rhinanthus angustifolius* plants during April-May in wet and moist plots which were untreated, or sown with 780 seeds m², or sown into plots with changed vegetation structure.

(Fig. 7). Three treatments were applied: (i) seeds of *Rhinanthus angustifolius* were sown (500 seeds per 80×80 cm² quadrat) into the existing vegetation; (ii) the same number of seeds were sown in 80×80 cm² quadrat in which gaps (17 cm diameter) were created by removing the vegetation and turning the soil surface; (iii) control (no gaps created and no seeds sown).

The study was made in the moist part of the vegetation gradient (Calthion community) and in the wet part (small sedge community). The number of seedlings and surviving plants were considerably greater in the moist than in the wet part of the gradient (Fig. 8). Neither sown seeds nor creating gaps affected the number of *Rhinanthus angustifolius* plants in the moist (Calthion vegetation), but sown seed increased the number in the wet (small sedge) vegetation. This indicates that in the moist vegetation seed availability was not the limiting factor though it was important in the wet vegetation type. In the experiment the water in the wet vegetation was base-poor; in the moist type it was base-rich. This may be another factor influencing plant germination and survival. We therefore investigated the effect of water tables per se on the establishment and survival in the small sedge community, with a base-poor water type (Fig. 7). Four lysimeters (60 cm diameter) were installed in 1977; two had a water level at 0 cm, and two had water levels between −30 and −40 cm. In each treatment, the lysimeters were sown with 500 *Rhinanthus angustifolius* seeds in November 1983. More seedlings and surviving plants occurred in the lysimeters with a water level at −30 cm than in those with a water level at the surface (Fig. 9), indicating that high water tables restrict the distribution, and that groundwater quality may be relatively unimportant.

5. Discussion

Rhinanthus angustifolius is hemi-parasitic; al-

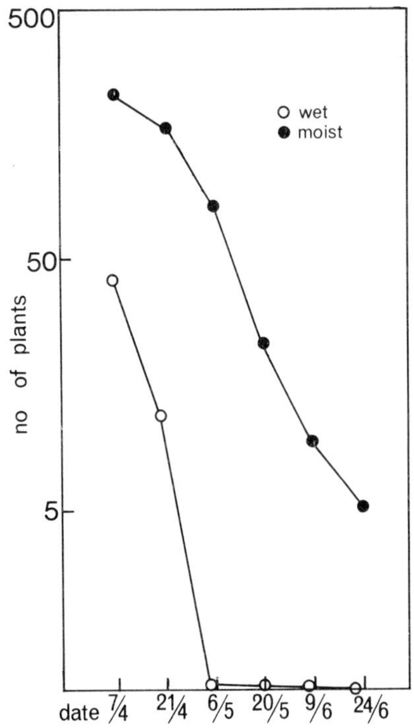

Fig. 9. The number of *Rhinanthus angustifolius* plants surviving in lysimeters with the water level at the surface (wet) or at −30 cm (moist).

though it is not host-specific, performance does depend on the host species (Ter Borg, 1972). In general more productive, faster growing grasses are better hosts (De Hullu, 1985), increasing the seed production per plant. Seed production is important because population growth is largely dependent on seed production in recent years, since seeds are short-lived and no seed bank is formed. The abundance of *Rhinanthus angustifolius* therefore depends on recent abiotic and biotic conditions, making it a good indicator for changes in biotic or abiotic conditions. In the Netherlands *Rhinanthus angustifolius* does not establish in hay-fields with a high water table, though it is often found in hay-fields which are flooded regularly; apparently the time of flooding can influence the distribution considerably. When flooding occurs in the winter, emergence is not affected (Ter Borg, 1972), but when the fields are flooded later in the season (late spring) hardly any *Rhinanthus angustifolius* plant emerges. The species is then restricted to the higher, usually more sandy, areas or to the higher valley flanks. From these higher places it colonizes the lower areas in subsequent years. Establishment and growth is not limited when the water tables are lower (Fresco, 1980), though this does not necessarily imply that abundance increase following drainage, since the vegetation density usually increased temporally, following drainage, because of increased mineralization (Grootjans, 1985). This can limit the growth and establishment of *Rhinanthus angustifolius*, despite the availability of good hosts. Drainage is often a component of intensified agriculture and is usually accompanied by intensive use of fertilizers. Under such conditions *Rhinanthus angustifolius* is less abundant because the vegetation is too dense. However, when no fertilizers are used following drainage, the abundance of *Rhinanthus angustifolius* can increase considerably, once the effects of increased mineralization have decreased and the vegetation becomes less dense. In the Nature Reserve of 'Drentsche Aa', the number of species starts to increase about 3–8 years after drainage, when the vegetation becomes more open and variable. In this phase, populations of *Rhinanthus angustifolius* can grow explosively if mowing is not too early to prevent plants from setting seed. A similar population explosion of *Rhinanthus angustifolius* occurs 1–5 years after fertilizer use ceases.

When no further disturbance occurs, population numbers decline and level off after about 15 years. Population explosions of *Rhinanthus angustifolius* in the Netherlands are nearly always the result of some disturbance of grassland, such as changes in the hydrology or fertility.

Acknowledgements

We wish to thank Jan Brinkman and Flos Engels for their contribution in the field experiments, Staatsbosbeheer Drenthe for their permission to study at the nature reserve 'Drentsche Aa'. The investigations were supported by the Foundation for Fundamental Biological Research (BION), which was subsidized by the Netherlands Organization for Fundamental Research (ZWO).

References

Borg, S.J. ter (1972). Variability of *Rhinanthus* (Schönh.) Oborny in relation to the environment. Ph.D. Thesis, Groningen.

Borg, S.J. (1985). Population biology and habitat relations of some hemi-parasitic *Scrophulariaceae*. In: J. White (Ed.). The Population Structure of Vegetation. Vol. III. Handbook of Vegetation Science. Junk, The Hague.

Duuren, L. van, Bakker, J.P. & Fresco, L.F.M. (1981). From intensively agricultural practices to hay-making without fertilization. Effects on moist grassland communities. Vegetatio 47: 241–258.

Ellenberg, H. (1974). Zeigerwerte der Gefässpflanzen Mitteleuropas. Scripta Geobotanica IX, Göttingen. 97 pp.

Fliervoet, L. (1984). Canopy structure of Dutch grasslands. Ph.D. Thesis, Utrecht.

Fresco, L.F.M. (1980). Ecological response curves of *Rhinanthus serotinus*; a synecological study. Acta Bot. Neerl. 29: 533–539.

Grootjans, A.P. (1985). Changes in groundwater regimes in wet meadows. Ph.D. Thesis, Groningen.

Grootjans, A.P., Schipper, P.C. & Windt, H.J. van der (1985). Influence of drainage on N-mineralization and vegetation response in wet meadows. I. *Calthion palustris* stands. Oecologia Plantarum 6: 403–417.

Ham, R. van der (in prep.). In: Atlas van de Nederlandse Flora 3. J. Mennema, A.J. Quené-Boterenboord & C.L. Plate (Eds.).

Hullu, E. de (1985). Population dynamics of *Rhinanthus angustifolius* in a succession series. Ph.D. Thesis, Groningen (The Netherlands).

Hullu, E. de, Brouwer, T. & Borg, S.J. ter (1985). Analysis of the demography of *Rhinanthus angustifolius* population. Acta Bot. Neerl. 34: 5–22.

Hullu, E. de (1987). The effect of emergence time in relation to shoot and root development of the host vegetation on the population dynamics of *Rhinanthus angustifolius*. Vegetatio (accepted).

Hullu, E. de & Braak, C.J.F. ter (1987). The influence of different micro-habitats on the population dynamics of *Rhinanthus angustifolius*. Vegetatio (accepted).

Kruijne, A.A., Vries, D.M. de & Mooi, H. (1967). Bijdrage tot de oecologie van de Nederlandse graslandplanten. Pudoc, Wageningen.

Meijden, R. van der, Weeda, E.J., Adema, F.A.C.B. & Jonckheere, G.J. (1983). Plantengemeenschappen in Nederland, Thieme, Zutphen.

CHAPTER 19

Pollination and pollen flow disturbed by honeybees in bumblebee-pollinated *Rhinanthus* populations?

M.M. Kwak

Detailed contents

1. Introduction

One of the most immediate responses to disturbance is a change in population density. Change in plant population density may have an effect on visiting and pollinating insects which may influence both fecundity and gene flow. The management regime of the grasslands has a major influence on the density and distribution of plant species. Changes in the management in the river valley grasslands of the Drentse Aa from intensive grassland production to mowing once a year without the applications of fertilizers has caused large changes in the density and distribution of *Rhinanthus angustifolius*. Initially the density rose than fall becoming very abundant locally (de Hullu et al., 1985). This study considers the effect of population density on visitation pollination and gene exchange.

For many plants, pollinator visits may be a limiting resource for reproduction (Bierzychudek, 1981). If a plant is visited, the number of flowers visited may also affect its overall success in setting outcrossed seeds. Revisitation may increase seed set depending on the species (Kwak & Jennersten, 1986).

Optimal foraging theory predicts that a pollinator's foraging behaviour, i.e. whether to forage in a patch and when to leave a patch or an inflorescence is the result of natural selection maximizing the net rate of energy intake whilst foraging (e.g. Pyke et al., 1977). It has often been suggested that plant density affects the energy intake rate, because the time spent moving between flowers increases with plant spacing (Levin & Kersten, 1969; Heinrich & Raven, 1972; Waddington, 1980; Zimmerman, 1981). In patches of higher flower density, pollinator visitation should be greater (Silander, 1978; Thomson, 1981) and within a patch pollinators may be attracted to plants with a greater number of flowers (Wilson & Ratcke, 1974; Wilson & Price, 1979; Schaffer & Schaffer, 1979; Davis 1981). Schmitt (1983), however, found that in *Senecio*, which requires

J. van Andel et al., (eds.), Disturbance in Grasslands. ISBN 90–6193–640–3.

insect pollination for full seed set, flowering density had no effect on seed production. Flowering phenology patterns and differences among pollinators in foraging behaviour might have more important consequences for seed set and gene flow patterns than plant density or plant size. Wilson et al. (1979) found that even single, isolated, flowering stems of *Phlox* were on average pollinated as effectively as other stems and that the risk of total failure also did not change consistently. It seems likely that most flowers on isolated plants are fertilized by pollen from the same plant (Levin, 1974; Wilson et al., 1979).

The annual grassland species *Rhinanthus angustifolius* C.C. Gmeling (Rinanthoideae, Scrophulariaceae) (= *R. serotinus* (Schönh.) Oborny) offers a good opportunity to study the influence of the density of plants and of flowers on pollinator visitation.

R. angustifolius is visited mainly by bumblebees to which it is well adapted (Kwak, 1977, 1979a) although in some years honeybees have been observed as visitors (Kwak, 1980). Worker bumblebees of 5 species are common visitors, and queens of some late-emerging species also visit *Rhinanthus*. Depending on their tongue length, bumblebees are either legitimate visitors (pollen and nectar foraging, nototribically pollinating, i.e. the essential parts of the flower contact the dorsal side of the insect's body; Fig. 1A), primary or secondary nectar robbers (no pollinators) or pollen foragers in an inverted position (sternotribically pollinating, i.e. the essential parts of the flower contact the ventral side of the insect's body (Fig. 1B). Nectar robbery implies that a hole is made and used to obtain nectar, by passing the opening used by pollinators. In this paper bumblebees are the hole makers (primary robbers) and honeybees (and also bumblebees) are the hole users (secondary robbers) (terminology follows Inouye, 1980). Specializations occur amongst the individuals of short-tongued bumblebees and honeybees namely in nectar foraging (as robbers) or pollen foraging (as sternotric pollinators).

The aim of this study was to investigate the effect of flower density on the chances of a flower being visited and pollinated by various insect species in populations of *R. angustifolius*. In addition, an attempt was made to analyse the effects of various visitors on pollen flow. Observations were made on the number of flowers per axis visited and the distance flown between flowers and between plants of various densities. Fluctuations in the occurrence of honeybees from year to year will be considered as a disturbance during pollination.

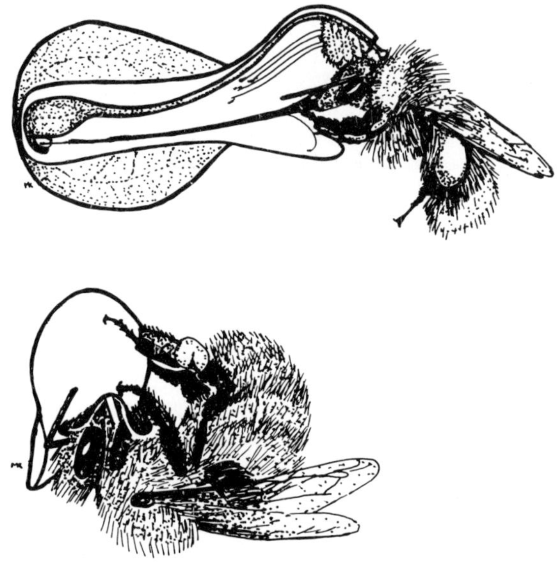

Fig. 1. A A *Bombus hortorum* worker, visiting *Rhinanthus angustifolius* legitimately, nototribically pollinating. Note the position of the stigma, anthers and nectary.
B A honeybee, pollen collecting on a *Rhinanthus angustifolius* flower, sternotribically pollinating.
(These drawings are based on the author's photographs).

2. Material and methods

2.1 Visitation and pollination

Three adjacent meadows in the State Nature Reserve 'Stroomdallandschap Drentsche Aa', 30 km south of Groningen (The Netherlands), 53°01′ N, 6°39′ E, were selected for study (Fig. 2). Populations of *R. angustifolius* in these meadows differ in flower density (m^{-2}) (Table 1). Observations of bee visits were made on 3rd July, 1985, during peak flowering of the populations. The maximum day temperature was 23.5° C, with

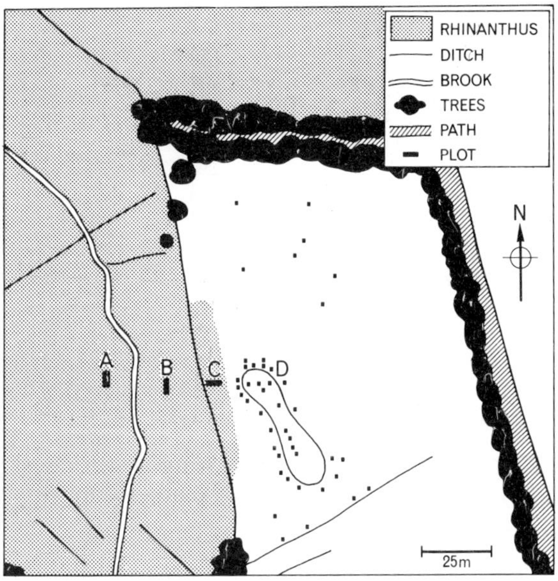

Fig. 2. Location of study plots in the field. See Table 1 for population characters.

some cloud. Observations were made in four plots (Table 1), two in very high densities (A and B), one in a medium density (C) and one in a very low density (D). The size of observation quadrats (2 m × 1 m) was such that all visits within the quadrat could be recorded by a single observer, even when several visits occurred simultaneously. The quadrats were observed for 15 minutes, 5–7 times between 10.00 and 18.00 hours, when both bumblebees and honeybees are active (Kwak, 1980). In the very low density plot too few plants were present to use quadrats, so observations were made by walking along 70

Table 1. Plant density and flowering of *Rhinanthus angustifolius* on the 3rd July 1985 at each of four plots indicated in Fig. 3. Plot A has been managed at low intensity (no fertilizer and annual cuts) since 1971, plot B since 1973, and plot C and D since 1982.

Plot	A	B	C	D
Anthesis stage	Peak	After peak	Peak	Peak
Plant (m^{-2})	158	125	32	0.2
Flowering axes (m^{-2})	252	200	168	1
Flowers per axis	3.28	3.64	3.76	3.28
Flowers (m^{-2})	826	728	632	3.6
% perforated flowers	29.3	31.9	43.6	0

plants with 950 flowers on 264 m^2 (D); 4 walks took 10 minutes. The species of bee, the number of visits, the position adopted, and the number of flowers per axis visited were recorded for each insect visitor entering a quadrat. The number of flowering axes and flowers in an observation plot were counted.

2.2 Nectar and robbery

The amount of nectar present in flowers in the same phase of anthesis was measured 1 to 3 times per day by inserting a 1 μl capillary tube (Drummond Microcaps) into the nectary. The number of flowers containing a hole in calyx and corolla (made by robber bumblebees) were counted.

2.3 Flight distances

On the 1st June 1982, observations were made on the distances flown by *B. hortorum* and *A. mellifera* within two *Rhinanthus* populations one of a high density (472 flowers m^{-2}) and one of a low density (42 flowers m^{-2}). Each visitor was followed for as long as possible and the number of flowers visited per axis, the flight distance between two successively visited flowers and plants were noted. Flight distances were estimated by eye and occasionally checked with the measured distances. In addition, the direct distance between the first and last flower visited was measured. Total foraging time per visitor was measured with a stopwatch, and the number of *Rhinanthus* flowers visited per hour calculated.

3. Observations

3.1 Visitation and pollination

Although seven bumblebee species visit *R. angustifolius* in the area studied (Kwak, 1979a) only a few individuals of four species were observed in 1985 (Table 2); honeybees were frequent visitors (Fig. 3). However, bumblebees accounted for a disproportionate number of flower visits because of their more rapid foraging speed; bumblebees

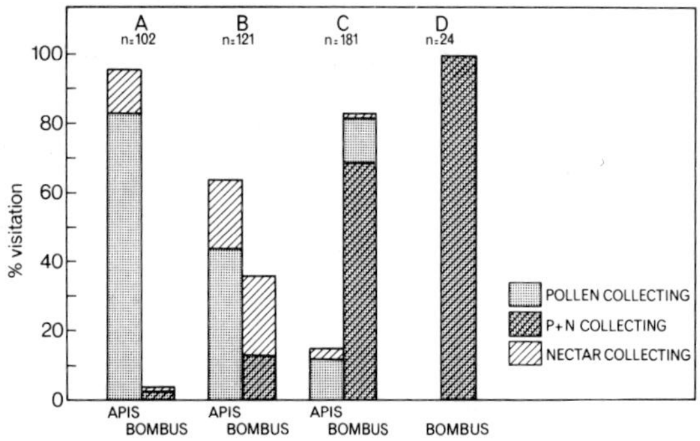

Fig. 3. The percentage of visits of honeybees and bumblebees in four observations plots (A-D) of *Rhinanthus angustifolius* on the 3rd July 1985.

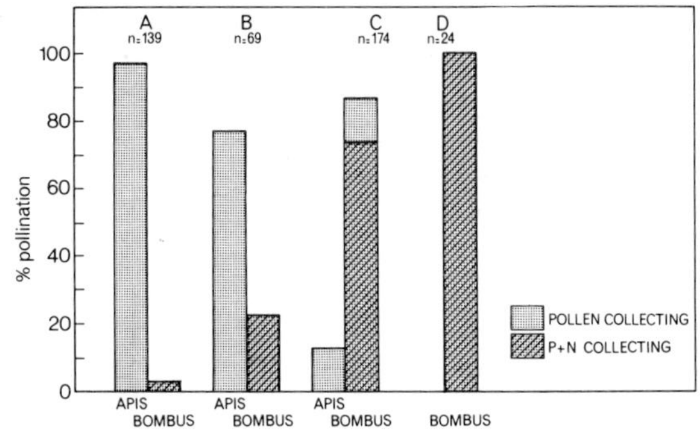

Fig. 4. The percentages of pollination of visited flowers by honeybees and bumblebees in four observation plots (A-D) of *Rhinanthus angustifolius* on the 3rd July 1985.

visited aproximately 3 times as many flowers per unit time as honeybees (see also Table 5). In the rest of this paper only the number of visits will be considered, since the crucial factor is the frequency of pollinating events, rather than the number of visitors. In fact in *Rhinanthus* only pollen collecting visits are important in relation to pollination (Fig. 4). Flowers were most likely to be visited on plot C (9.6%), and last likely (3.8%) on plot D (Table 3). However, the proportion of flowers visited by honeybees and bumblebees visited for nectar or pollen differed between plots (Fig. 3), so that actual pollination (nototribic and sternotribic visits) was greatest on plot C and lowest on plot B and D. No distinction is made between the possible difference in pollination efficiency of pollen collecting honeybees and various bumblebee species.

Assuming on eight hour activity day, the proportion visited flowers would be 30, 77, 54 and 62% per day in order of increasing flower density but the proportions pollinated would be 30, 74, 30 and 54% respectively, due to nectar robbery.

The plot with the medium flower density (632 flowers m^{-2}) therefore had both the highest chance of being visited and pollinated; the plot with the lowest density (3.6 flowers m^{-2}) had the lowest chance.

Honeybees were the main pollinators on high flower density, both honeybees and bumblebees were present on the medium density plots and bumblebees only were pollinators on the low density plot.

3.2 Nectar collection

The amount of nectar present at a given time depends on the balance between the rate of removal and the rate of supply, and is therefore, to some

Table 2. Visitor species, observed number of individuals and total observation time in different populations of *R. angustifolius* (Q=queen, W=worker; * not in observation plot).

	A (75 min)	B (75 min)	C (90 min)	D (40 min)
Apis mellifera	10	11	3	
Bombus terrestris W	1	6	8	
B. pratorum W		*		
B. hortorum Q	2	3	1	3
B. pascuorum Q			3	
B. muscorum Q				1
Psithyrus spec.			1	

Table 3. The proportion (%) of flowers of *Rhinanthus angustifolius* visited and pollinated by bees; observations standarized for one hour observation/day; * due to a worker *Bombus terrestris* collecting pollen sternobribically and subsequently nectar robbing on the same flower.

	A	B	C	D
% of visited flowers pollinated by Apis	83.3	43.8	12.2	
Bombus	2.5	13.2	84.0	100
% of visited flowers robbed	14.2	42.9	54.2*	0
% flowers visited	7.8	6.7	9.6	3.8
% flowers actually pollinated	6.7	3.8	9.2	3.8

extent an indication of visitation frequency. In the morning, plants on plot A have significantly more nectar than the other plots (Table 4), but in the afternoon the differences between the four plots were not significant. There was significantly less nectar in robbed flowers than in unrobbed flowers ($P < 0.05$), but no significant location × robbing interaction.

Honeybees are completely dependant on the robber activities of short-tongued bumblebees, and must therefore find robbed flowers which containing nectar. Plot A, with the highest nectar content, and a moderate perforation percentage (29), should have been favourable for nectar robbers. Indeed in plot A most of the nectar robber visits were made by *A. mellifera*. But totally (*Apis* and *Bombus* robber visits togheter) most of the robber visits were made on plot C. Small amounts of nectar on plot D may be the result of foraging activities of nototribically visiting bumblebees.

Assuming an eight hour activity day, nearly all perforated flowers were visited once a day on plot C (a value of 0.95), 0.72 on plot B, but not on plot A (0.30). Plot D had no perforated flowers. Honeybee robbers were not most numerous in the highest density areas, as was found the case with pollen collecting honeybees but were most abundant on plot B, which was past the peak of flowering.

3.3 Flight distances

Despite the tenfold difference in flower density

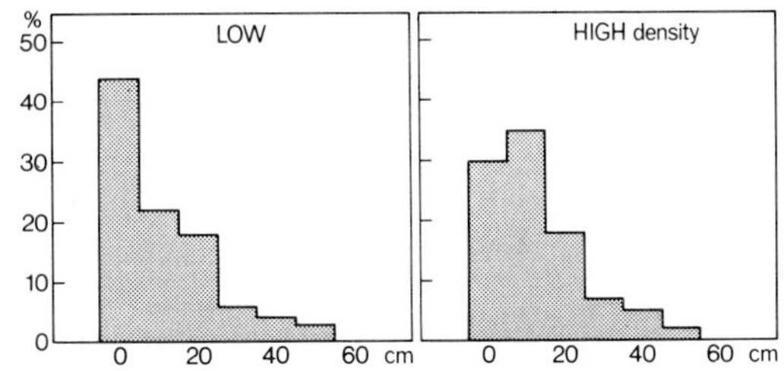

Fig. 5. Frequency distributions of distances flown by *Bombus hortorum* between two flowers of *Rhinanthus angustifolius*. (low density: 42 flowers m^{-2}, high density: 472 flowers m^{-2}).

and twenty fold difference in plant density between populations, there was no significant difference in the distance flown by either *B. hortorum* and *A. mellifera* between flowers or plants (Table 5). However, at the low plant density, there was a greater tendency ($P < 0.05$) for to fly less then 10 cm (Fig. 5), i.e. within the same plant. The direct distance between the 1st and 100th flower visited by A. mellifera was significantly greater at low density than at high density (Table 5). Foraging speed, i.e. number of flowers visited per hour, was also influenced by density but foragers were influenced in opposite ways; the number of visits was greater at low density for *A. mellifera*, but at high density for *B. hortorum* (Table 5). The foraging speed of *A. mellifera* was always lower than that of *B. hortorum*.

3.4 Visits per axis

Both pollen collecting honeybees and bumblebees visited more flowers per axis on low density

Table 4. Amount of nectar (μl±SE) in flowers of *Rhinanthus angustifolius* during the morning and afternoon of the 3rd July 1985. Values for robbed and unrobbed flowers are also given together with the percentages flowers visited by nectar robbers (*Apis* and *Bombus*) during an eight hour activity day. (NA=no value available).

	A	B	C	D
morning	1.43±0.58 (n=10)	0.16±0.04 (n=10)	0.28±0.06 (n=10)	NA
afternoon	0.41±0.15 (n=10)	0.15±0.02 (n=20)	0.35±0.10 (n=10)	0.17±0.05 (n=10)
unrobbed	NA	0.16±0.04 (n= 9)	0.49±0.14 (n=13)	NA
robbed	NA	0.13±0.02 (n=11)	0.10±0.02 (n= 7)	NA
% robbed flowers	29.3	31.9	43.6	0
% visited by *Apis* and *Bombus* robbers	8.9	22.9	41.4	0
% visited by *Apis* robbers	8.1	10.6	2.1	0

Table 5. The distance (cm) between sequent flowers and plants, and the direct distance between the 1st and 100th flower. The number of flowers visited per hour per axis is also shown.

	n	Distance (cm) between Flowers	Visit 1 and visit 100	Plants	Flowers per hour	Visits per axis
Low density+						
Apis mellifera	6	15.6	1230*	19.2	440*	1.32
Bombus hortorum	6	16.8	920	28.8	712*	1.62
High density++						
Apis mellifera	1	14.0	490*	17.4	369*	1.24
Bombus hortorum	11	15.8	730	22.3	927*	1.51

+: 42 flowers m^{-2}, 4.72 flowers per axis, 8.9 plants m^{-2}, mean distance between two plants 33.6 cm, between adjacent flowers 15.4 cm.

++: 472 flowers m^{-2}, 2.81 flowers per axis, 168.6 plants m^{-2}, mean distance between plants 7.7 cm, between adjacent flowers 4.6 cm.

*: Difference between densities significant at P=0.05 (last column (visits per axis not tested, see Table 6)).

than on high density populations (Table 5), and pollen collecting honeybees visited less flowers per axis (1.11) than pollen collecting bumblebees (1.21) ($p < 0.01$ lumping data of plots A, B and C for both groups, Table 6). Robber bumblebees and robber honeybees never visited more than one flower per axis despite flower density differences (Table 6).

Table 6. The number of flowers visited by honeybees and bumblebees per axis foraging on *Rhinanthus angustifolius* plots (A-D). n=number of visited flowers; bold values based on more than 20 visits.

	A (n=162)	B (n=121)	C (n=181)	D (n=24)
A. mellifera pollen	1.15	1.15	1.00	
nectar	1.00	1.00	**1.00**	
B. terrestris pollen			1.21	
nectar	1.00	1.00	1.00	
p+n			1.23	
B. hortorum p+n	1.00	1.14	1.00	2.20
B. muscorum p+n				2.17
B. pascuorum p+n			1.25	

4. Discussion

4.1 Visitors to Rhinanthus

Flowers of *R. angustifolius* are more easily visited by bumblebees than by honeybees, since honeybees can only collect nectar after nectar robbing bumblebees have visited. Fossel (1974, 1977) showed that honeybees are common visitors of *Rhinanthus* in Alpine meadows, but were absent at higher altitudes (Kwak, 1980). The 1985 season came after an extremely hard winter (deep frost) and a late, cool spring. This would influence the hibernation of the queens (Heinrich, 1979b) and the development of the colony, result in fewer bumblebees. Visscher & Seeley (1982) found that honeybees routinely forage several kilometers from the hives but the presence of honeybee hives in the close vicinity of dense *Rhinanthus* stands does not necessarily mean that they actually visit the species (Kwak, 1979a).

There are few studies on the interactions of bee species under natural conditions (Donovan, 1980). Eickwort and Ginsberg (1980), concluded that honeybees influence the foraging patterns of native bees, while Schaffer et al. (1979, 1983) concluded that honeybees predominate in the most productive habitats, *Xylocopa* in the least productive habitats and *Bombus* in patches of intermediate productivity. There are also differences in foraging behaviour, with *Apis* preferring to forage at those times of the day when resources were most abundant. Holmes (1964) believed that the absence of bumblebees is due to honeybees crowding on the flowers. The number of honeybee individuals on *Rhinanthus* was greater than might be expected from the number of visits, since the foraging speed of honeybees is rather low (Table 5). Heinrich (1979b) has estimated that a single honeybee hive can reduce the population of *B. vosnesenskii* reproductives by 38,400 but this is probably an underestimate, because no account is taken of the food resources required for rearing the workers, that are used to make those reproductives.

Resource utilization by bumblebees species is affected by the presence of other bumblebee species. Inouye (1978) found that removal of other bumblebee species increased the number of flowers visited per flower stalk and the number of flower stalks visited before the bumblebees left the patch. Both aspects have implications for pollen flow.

4.2 Flower density and visitation

Honeybees made more visits in high densities of *Rhinanthus* than in low densities. This agrees with Ginsberg (1983) who concluded that honeybees had the greatest effects on wild bees in dense flower patches. He considered *A. mellifera* to be a specialist in foraging in high density patches. A relation existed between the density of flowers and the percentage of *A. mellifera* visitation (Fig. 3) and pollination (Table 3); the highest flower density had the highest honeybee-visitation and pollination and the very low density had no honeybees at all. This means that the relationship was reversed in the case of bumblebee visitation (Fig. 3) and pollination (Fig. 4). Pollen collecting honeybees seem to be attracted to high density plots.

4.3 Flower density and robbery

The relationship between flower density and robbery is more complex than that between flower density and pollen collecting. Nectar-collecting honeybees are dependant on the holes made by short-tongued bumblebees, so they cannot forage in their absence. The small amount of nectar available in plot B was not only the result of nectar foraging activities of legitimate and robber visitors, since the morning value, i.e. before most visits had occurred, was already low. The small amount of nectar in plot B may have been due to the late stage of flowering. The greater proportion of robbery in plot B cannot be explained by different numbers of perforated flowers/m^2 (A 242 and B 232 perforated flowers/m^2). High robbery by honeybees in plot B may have been caused by the fact that honeybees remained in plot B, as it was the first to flower. If only nectar robbers were present, a decreased seed set would have been expected. However, Heinrich and Raven (1972) suggest that robbers may increase pollination when other pollinators are present, because they cause functional pollinators to pay more visits. In *Rhinanthus* the functional pollinators (especially medium-tongued and long-tongued nectar-collecting and also pollen collecting bumblebees) may perhaps developed a discriminating behaviour, resulting in the avoidance or perforated flowers since these flowers contain less nectar. It is surprising that the effect of robbers on the behaviour of functional pollinators and seed set in unknown (see review Inouye, 1983). Kendall & Smith (1975, 1976) found that if no holes were present in both *Vicia faba* and *Phaseolus coccineus*, honeybees might enter the flowers in the proper way. Flowers robbed by short-tongued bumblebees, and subsequently by honeybees, produced fewer pods than those visited in the proper way. Robbery not only alters the standing crop or production rate of nectar but also the composition of nectar. In particular, Willmer (1980) found that nectar robbery caused leakage of nitrogeneous solutes form damaged plant cells into the remaining nectar in *Phaseolus* and *Lamium*. A legitimate visit may effect the amino acid content for subsequent foragers, by the deposition of pollen in the remaining nectar. The effect of change in nectar composition on foragers is unknown.

4.4 Amount of nectar and bee visits

Nectar amounts influence several aspects of foraging: for example more flowers may be visited per inflorescence and more time may be spent in flowers on nectar-rich plants (Zimmerman, 1983). Waddington (1981) found that, when bumblebees visited inflorescences with copious nectar, they picked up more pollen and flew shorter distances than when they visited inflorescences with little nectar. Thompson & Plowright (1980) found that deposited significantly more grains on flowers which contained large amounts of nectar than on drained flowers. Waddington (1983) stated that flight distance is inversely related to the volume of nectar obtained during recent floral visits. Pyke (1978) has shown that bumblebee queens fly greater distances after encounting low-yielding plants and that they increase their rate of turning after encountering high-yielding plants. Heinrich (1979a) found that bumblebees tended to move through depleted areas and to concentrate in nectar-rich areas of *Trifolium repens*. Although nectar in *Rhinanthus* differed between the plots, the mean number of flowers/axis visited by honeybees was almost the same (Table 6). The difference between the number of flowers/axis visited by pollen or nectar collectors agree with Zimmerman (1982). He also found that pollen collecting bumblebees are significantly more likely to visit more flowers per inflorescence than are nectar gathering bumblebees to revisit flowers.

4.5 Flower density and pollination

In general the chance of flowers being visited and pollinated varies with the flowering season as well as with density. Stephenson (1982) found that flowers of *Catalpa speciosa* that opened during the final phase of flowering had the highest probability of being outcrossed. He suggested that the peak period of flowering plays an important role in attracting and rewarding pollinators, but that a period of rapid decline in flower number increases the number of pollinators per open flower and leads to a high percentage of outcrossed flowers. Zimmerman (1980) found that when the bee: flower ratio in *Polemonium* was relatively high, large number of seeds per ovary were set. No correlation existed between the number of visits by pollinators and seed set. The bee: flower ratio and seed set were lower later in the flowering season and a very strong correlation existed between pollinator service and seed set.

Assuming an activity period of eight hours per day, pollination would be 54, 30, 74 and 30% for plots A-D respectively. Flower longivity was about four days, so during the life span of a flower it may have been visited 1.2–3 times. Pollination should be greater in plots with a high percentage of bumblebee pollination (plot C and D, respectively 74 and 30% pollination) because bumblebees start foraging before 10.00 hours and continue after 18.00 hours. Still the observed pollination values are rather low compared to earlier studies (Kwak, 1979b); this is likely to have a large effect since *R. angustifolius* needs repeated pollinations, each pollination resulting in a still higher seed set, if the visits occur within a certain time interval (Kwak & Jennersten, 1986).

4.6 Flower density and pollen flow

There have been few investigations into the effects of changes in floral abundance throughout the flowering period on pollination movements and pollen flow (Stephenson, 1982). In this investigation the movements of pollinators was tracked in only one phase of the flowering period. It may not be possible to extrapolate these findings to the entire flowering period.

The habit of the plants, especially the number of axes bearing flowers, is of great consequence to the pollen flow. Widely branched specimens of *R. angustifolius* are present at very low plant densities in the beginning of the colonization. At high plant densities, both the number of axes and flowers per plant are lower than in low densities. According to de Hullu et al. (1985) the habit of *R. angustifolius* is not influenced by the density of the species but by how long the meadows have been subjected to low intensity management. Plants with a large number of flowers (up to

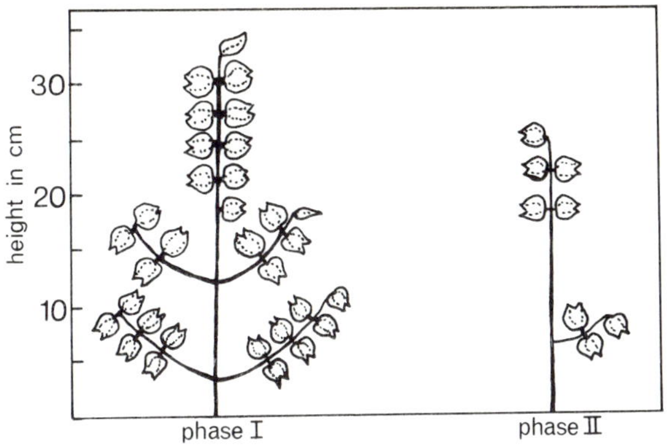

Fig. 6. Habit; (branching and capsule production) of *Rhinanthus angustifolius* at the beginning of colonization (phase 1) and after about 9 years of low management when density was greatest (phase II); Phase I: 97.5 plants m^{-2}, 30.8 flowers per plant, 9.4 flowers per main axis, 4.2 branches per plant, plant height 32.4 cm. Phase II: 513.3 plants m^{-2}, 7.2 flowers per plant, 5.2 flowers per main axis, 0.6 branches per plant, plant height 25.0 cm. All characters except plant height differed significantly.

61.4±48.4, data of de Hullu et al., 1985) were present shortly after the change in management (plant density 4.4±3.7) (Fig. 6). The number of flowers per plant decreased in each subsequent year (to 6–8 flowers per plant) and plant density increased (to 271±93). At low plant densities, insects can and do visit more flowers per plant than at high densities. As *R. angustifolius* is self fertile, seed set in low plant densities will be at least partly the result of self pollination though bumblebees are still the vectors.

Pollen flow in *Rhinanthus* is restricted to a small area within a population. Bumblebee flights between populations are more likely at low plant density. Only the very first flowers will be pollinated with pollen from the other population, due to dilution and the grooming activity of the foragers.

The seed dispersal component of gene flow in *Rhinanthus* may be important for the colonization of meadows when the management is changed. About 90% of the seeds of undisturbed plants were deposited within 25 cm of the parent plant. Mowing during seed ripening stimulates seed shedding, and hay-making may have resulted in a greater seed dispersal. The greatest distances observed were 12–15 meters, although seeds may have covered greater distances with hay or by machinery (ter Borg, 1985). It is clear that mowing late in the season, which is often the practice, will lower the seed dispersal and in those populations the gene flow is determined to a larger extent by the pollen flow.

4.7 Disturbance

The presence of honeybees in *Rhinanthus* populations which are usually pollinated by bumblebees can be considered to be a new disturbance, introduced by man. The presence of honeybees causes changes in the chances of pollination and in the direction and distance of pollen flow. The degree to which this disturbance occurs varies from year to year. In *R. angustifolius* it also varies with flower (and plant) density, with higher flower densities having more honeybees visits then lower densities.

Honeybees may compete with bumblebees. *Rhinanthus* is an important source of pollen and nectar for bumblebee colonies during its flowering period. Because the possible competition between honeybees and native bees Brantjes (1976) has argued that honeybees should not be kept in nature conservation areas. Considered from the point of view of *Rhinanthus*, pollen-collecting honeybees, increase outbreeding, by increasing pollen transfer between flowers of different plants. However, honeybees are more area restricted foragers than bumblebees at high plant (flower) density and may therefore increase inbreeding. The significance of inbreeding and outbreeding on the seed set of *R. angustifolius* is not known, yet. After change in management of hay-meadows, density of *R. angustifolius* changes, resulting in changes in attraction of insect pollinators. Although the course of the density curve is more or less predictable, the presence and number of honeybees may fluctuate largely from year to year. As honeybees are present, they are attracted to high density populations.

CHAPTER 20

Disturbance, hybridization and hybrid speciation

K.M. Urbanska

Detailed contents

1. Introduction

The evolutionary forces which operate within a given plant population both produce and sort out genetic variation. In higher plants, evolution is mainly concerned with the restructuring of polygenic variation and the fixation of new adaptive gene combinations. In this way, evolution proceeds from the molecular level of mutation to the adaptation of individuals in populations.

Local plant populations are not totally isolated groups of individuals, but are units within a more extensive population system, belonging to a race and to a species (Grant, 1985). Speciation leads to the formation of different, reproductively isolated, sets of adaptive gene combinations and can thus be broadly defined as evolutionary changes occurring at the level of population systems. Disturbance of physical or biological processes results in temporal and/or spatial changes in population structure and dynamics, because selection pressures are altered. For this reason, disturbance has great evolutionary significance; differential mortality and reproduction are not only essential demographic processes, but may also lead to changes in adaptive traits of given populations.

The present paper deals with the rôle of disturbance in the speciation process. Because of the limits of space allotted, not all aspects of this complex problem are considered, and the main objectives are (1) to outline briefly the principal patterns and mechanisms involved in speciation of higher plants, and (2) to focus attention on the evolutionary effects of hybridization in relation to disturbance; case studies are usually presented.

2. Plant speciation

Attempts to classify the various mechanisms of plant speciation have been based either on the actual speed of the process or on the genetic pool(s) involved. For instance Valentine (1962) distinguished between gradual and abrupt speciation; the former process being long and influenced by

J. van Andel et al., (eds.), Disturbance in Grasslands. ISBN 90–6193–640–3.

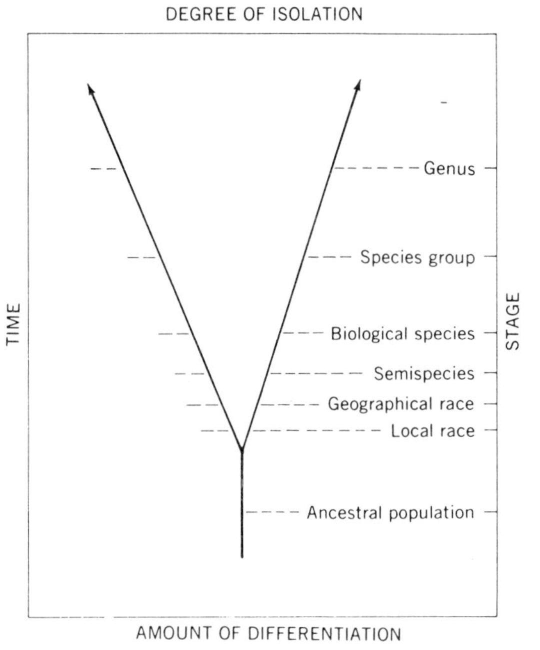

Fig. 1. Stages of evolutionary divergence. Reproduced from Grant (1985). Copyright (C) 1985 Columbia University Press. By permission.

slowly changing environmental conditions, the latter one is rapid and frequently associated with sudden changes in selective pressures. Grant (1963, 1977, 1985) essentially distinguished between primary speciation and hybrid speciation. Primary speciation is considered as evolutionary divergence proceeding from a polymorphic population to the species level without participation of other gene pools (Fig. 1). On the other hand, hybrid speciation is preceded by natural hybridization of the parent populations which evolved different adaptive norms; in this sense, hybridization is considered as a reversal in the process of evolutionary divergence.

Three phases characteristic of primary speciation are: genetic polymorphism, local racial variation and geographical racial variation. The best-known is geographical speciation proceeding through all three steps mentioned; this process has been particularly well-documented in *Gilia* sect. *Arachnion* (Grant & Grant, 1956). A pathway which bypasses geographical races formation is called by Grant 'quantum speciation'; in this case, a new species may develop from a local race under influence of strong selective pressures and/or genetic drift. Quantum speciation has been reported, for example, in the genus *Clarkia* (Lewis, 1962, 1966; Lewis & Lewis, 1955; Lewis & Roberts, 1956; Vasek, 1968). The third pathway is represented by sympatric speciation (Grant, 1963, 1971, 1981), which bypasses both local races and geographical races, a new species developing directly out of polymorphic variation.

Hybrid speciation stabilizes hybrids, protecting them from break-up due to segregation. The best-known mechanism involved are amphiploidy and various forms of asexual reproduction (see e.g. Babcock & Stebbins, 1938; Clausen, Keck & Hiesey, 1940, 1945; Gould & Kapadia, 1964; Rousi, 1965; Grant, 1971, 1977, 1981, 1985; Urbanska, 1974, 1985; Dobzhansky et al., 1977). Permanent translocation heterozygosity (e.g. Cleland, 1962; Grant, 1964; Lewis & Schweykowski, 1964) and permanent odd polyploidy (Täckholm, 1922; Hurst, 1931; Darlington, 1937; Fagerlind, 1940; Smith-White, 1948, 1955, 1959; Urbanska, 1977a, 1978) are less frequent mechanisms. Little is known about recombinational speciation (Gerassimova, 1939; Grant, 1954a, 1954b, 1958, 1963, 1966).

Grant (1981) considered hybrid speciation as the origin of a new species directly form a natural hybrid and argued that formation of hybrid races is not necessarily phase of the speciation process. In our opinion, hybrid race formation should be considered part of the process of hybrid speciation. Grant (1985) maintained that the process of evolutionaty divergence may be arrested at any stage and 'divergence from races to biological species continues only so long as the evolutionary forces producing that divergence continue to act'; these considerations concerning the primary speciation seem also to apply to the hybrid speciation.

The slow and conservative process of geographical speciation corresponds to the concept of gradual speciation proposed by Valentine (1962). On the other hand, the accelerated and radical process of quantum and sympatric speciation, like that of hybrid speciation, is abrupt.

Factors influencing speciation in plants, such as

variation-producing forces (mutation, gene flow and recombination) and variation-fixing processes (selection and genetic drift), operate in various combinations and are interlocked in different pathways. Variation also exists in spatial conditions accompanying speciation, so that speciation may be allopatric, quasi-sympatric or wholly sympatric. The modes of speciation were recently characterized by Grant (1985) according to pathways and field features, but this schema also clearly takes into account differences between primary speciation and hybridization.

3. Genetic and ecological control of natural hybridization

Natural hybridization occurs at various taxonomical levels; the general distinction between intraspecific, interspecific and supraspecific hybrids is accepted by most authors (Stace, 1975). The primary genetic control of natural hybridization is based upon the compatibility of two potential hybrid parents and incompatibility may be expressed by both pre-mating and post-mating syndroms. Incompatibility barriers are not necessarily related to karyological differentiation; for example, numerous cases of inter-racial sterility occur at a single ploidy level, whereas interspecific of intergeneric hybrids may be formed between taxa with widely different chromosome numbers and/or chromosome morphology. Isolating mechanisms, which preclude natural hybridization between at least partly inter-compatible taxa, are often affected by external factors (Baker, 1951, Levin, 1971). In the wild, most plants are isolated by several kinds of incomplete barriers operating in cooperation; the stronger the isolation barriers, the stronger the intervention needed to result in break-down. Disturbance is most likely to affect geographical and ecological isolation barriers and so cause hybridization.

Hybridization between native and alien introduced taxa can occur when man introduces new plants, sometimes not intentionally. An especially interesting case was reported in *Spartina* (Huskins, 1930; Marchant, 1967, 1968). Human disturbance can also result in the formation of intermediate or open habitats that are free to be newly colonized. In extreme ecosystems (e.g. high-alpine vegetation belt), disturbance by man may result in an entirely new ecological situation; it seems that succession in such sites may follow a course different from that occuring in neighbouring undisturbed areas (e.g. Meisterhans, 1982; Urbanska, 1986; Urbanska, 1986a; Urbanska & Schütz, 1986).

Ecological factors controlling natural hybridization may be partly related to habitat preferences and dispersal range of potential parent taxa and their offspring. According to the generally accepted view of Kerner (1894–95), hybrid plants can become established only if they are adaptively superior to their parents in ancestral habitats, or if they find a convenient ecological niche not inhabited by the parent taxa. In a closed and stable community, no niches are available to hybrids and the stabilizing selection is operating in favour of established parent genotypes. If, however, the stability has been disturbed, selective pressures become less effective and new niches may appear; in this case, hybrid establishment may be possible. Most authors agree that the environmental conditions in a stable closed community usually operate against natural hybridization but disturbed habitats can favour hybrids (e.g. Stebbins, 1950, 1971; Grant, 1963, 1971, 1981, 1985).

The establishment and survival of the first hybrid generation is the necessary prerequisite for further evolutionary development. However, subsequent hybrid generations may require further site differentiation for their successful establishment. Anderson (1948) termed the ecological disturbance which is often indispensable to establishment of hybrids as 'hybridization of the habitat'. In a 'hybridized' habitat, numerous new microniches may become available and enable hybrids to survive and undergo further evolution.

The vast majority of the present open and 'hybridized' habitats are the result of human intervention; many are dependent upon a regular management if they are not to dissapear. This disturbance or management usually involves manipulation of the moisture content or nutrient content of the soil resulting in changes in the vegetation cover (cf. Chapter 5).

4. Evolutionary effects of hybridization – case studies

Natural hybridization can be considered as an evolutionary 'catalyst' with three main effects:
a) formation of new taxa, which may be homogamic, agamic or heterogamic;
b) blurring of boundaries between taxa as a result of hybrid swarm formation;
c) enrichment of the gene pool of a given species through introgression.

In homogamic taxa of hybrid origin, normal sexual reproduction is the rule; on the other hand, agamic taxa reproduce asexually and heterogamic taxa are characterized by production of gametes that are genetically distinct and function only in precise combination (Grant, 1971, 1981). The subsexual reproduction occuring in heterogamic plants is usually influenced by balanced lethals systems and most frequently associated with high sterility (e.g. Cleland, 1962; Fagerlind, 1940; Urbanska, 1977a).

The development of hybrid swarms and of introgression is dependent upon at least partial fertility of the first generation of hybrids. Introgression has been considered as the results of selective interbreeding between the F_1 hybrids and one or both parents (Anderson & Hubricht, 1938; Anderson, 1949, 1953; Heiser, 1973; Stace, 1975). In most cases, repeated backcrossing only occurs to one of the parental species. Formation of hybrid swarms results from sexual reproduction of hybrids and gene exchange between individuals representing various hybrid generations.

Formation of hybrid swarms and introgression are closely related phenomena. Stace (1975) argued that hybrid swarms and introgression differed only in degree, so that introgressed populations could be considered to be an extreme form of a hybrid swarm. While this opinion is essentially correct, it should be pointed out that the term 'introgression' has been rather arbitrarily used to cover any case of hybridization; it seems that only well-documented cases should be classified as introgression, and the more general term of hybrid swarm formation should be used when a given situation has not been unequivocally assessed but when fertile hybrids occur. Selective factors operating in the process of introgression apparently act in favour of hybrids resembling one of the parents, whereas selection in sites inhabited by hybrid swarms seems to be rather relaxed.

I shall now consider in more detail selected examples of natural hybridization and its consequences. In each of the cases presented, disturbance caused by various human activities has been involved, though its evolutionary effects are rather varied.

4.1 Hybrid speciation by amphiploidy in Tragopogon *L.*

Tragopogon dubius, T. porrifolius and *T. pratensis* are quite widely distributed in the Old World. They are diploid (2n=12), and their chromosome sets are to some extent differentiated (Fig. 2). Natural hybridization between *T. dubius* and *T. pratensis* has been reported from Europe (Hess, Landolt & Hirzel, 1972) and hybrids between *T. pratensis* and *T. porrifolius* have been observed in disturbed sites (Britton, 1917; Britton & Todd, 1910; Ellis, 1931, 1932; Stace, 1975). These various hybrids are highly sterile and apparently do not persist in Europe.

The three species of *Tragopogon* have been introduced to North America in the post-Columbian time, and have become naturalized in a large part of the continent. They each contain racial variation in chromosome morphology at the diploid level (Ownbey & McCollum, 1954).

Extensive natural hybridization between species of *Tragopogon* has been found in Washington and Idaho. This part of Northwestern Pacific America originally carried natural grasslands, being too arid to support forest, but the rolling hills of Washington and Idaho (Fig. 3) are now largely used for crop production (Fig. 4).

Each of the three species of *Tragopogon* hybridizes readily with the others in North America;

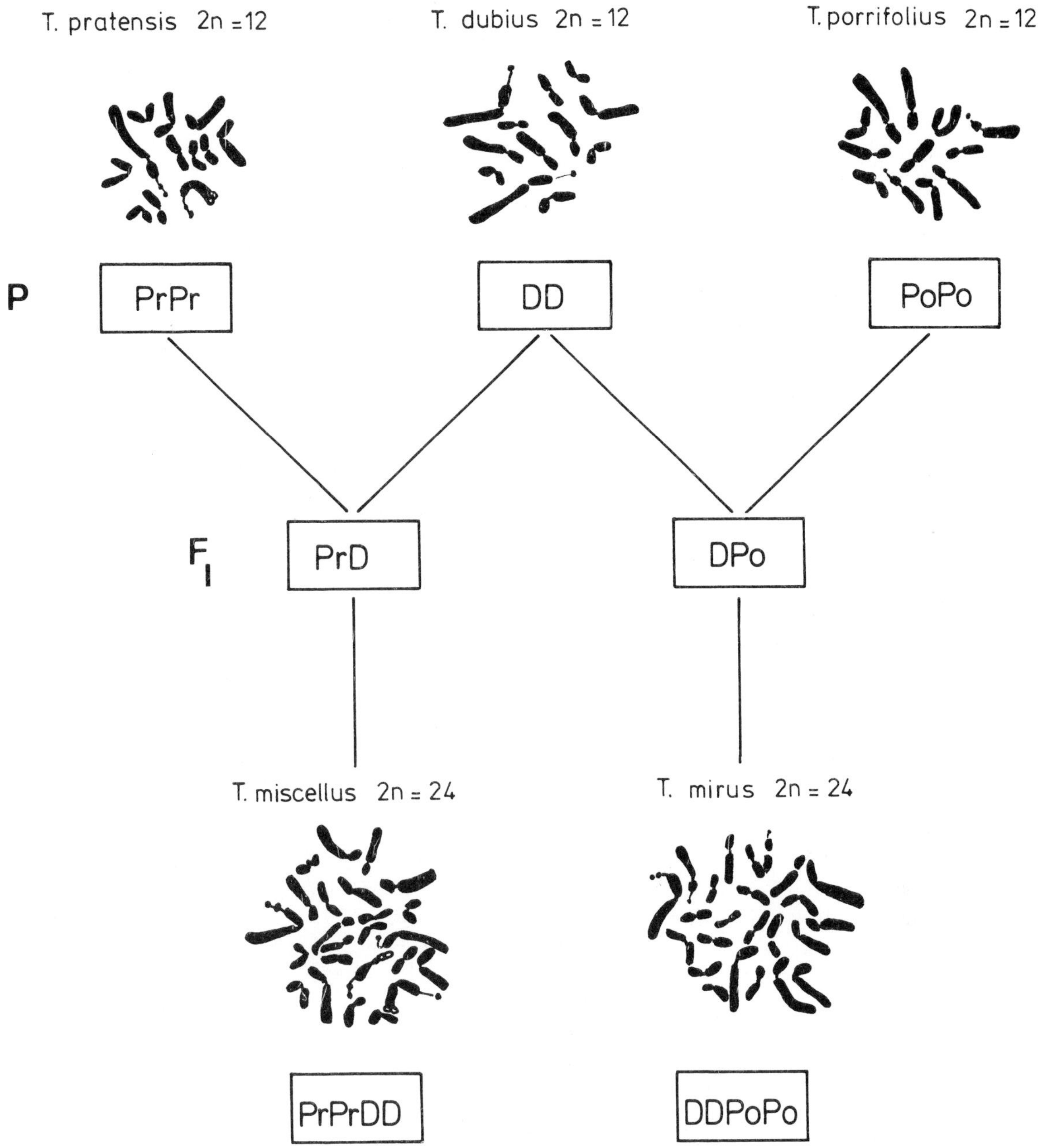

Fig. 2. Amphiploidy in *Tragopogon* L. P = parents; F_1 = first hybrid generation. Chromosomes partly redrawn from Ownbey and McCollum (1954).

wherever two or three taxa occur, hybrids of the first generation are frequent. These hybrids, like those reported from Europe, have the same chromosome number (2n=12) as their parents, are highly sterile, and are intermediate only in that they combine certain dominant characters of the parental species involved (Ownbey, 1950; Ownbey & McCollum, 1953). In 1949, Marion Ownbey found four small amphiploid populations (2n=24); further colonies have been discovered since (Mastrogiuseppe, personal communication). The amphiploid species *T. mirus* and *T. miscellus* occur mainly in man-influenced sites, e.g. roadsides, waste places, pastures, farmyards

Figs. 3–4. The area of extensive hybridization in *Tragopogon*. 3. Palouse Hills, Washington. 4. Crop fields in the surroundings of Pullman, Washington, with only small patches of the original grassland remaining. Photos courtesy of Joy D. Mastrogiuseppe, Washington State University, Pullman.

or town parks. They obviously represent independent instances of speciation in two different combinations: *T. mirus* is an amphiploid of *T. dubius* x *T. porrifolius*, whereas *T. miscellus* is the result of hybridization between *T. dubius* and *T. pratensis*, followed by doubling of the chromosome number (Fig. 2). The origin of the two tetraploid species has been confirmed by chromosomal evidence (Ownbey & McCollum 1953); morphologically, the amphiploids correspond to the highly sterile diploid hybrids, except for their larger size, more luxurious growth and high fertility (Ownbey, 1950).

Fig. 5. Man-made hay meadows at Urnerboden (1310 m a. s. 1., the Canton of Uri, Central Switzerland). Freshly manured plots in the background and in the middle. Photo courtesy of Stephan Zimmerli, Geobotanical Department, SFIT Zurich.

Of the five *Tragopogon* species that occur in Washington and Idaho, *T. pratensis* is the most restricted in ecological preferences; it grows in rather moist sites. *T. dubius* seems to have the widest ecological amplitude, whereas *T. porrifolius* is rather infrequent. *T. miscellus* is the rarer of the two amphiploids but *T. mirus* can form quite dense stands and seems to out-compete both parental species at some sites (Mastrogiuseppe, personal communication).

4.2 Agamy, heterogamy and amphiploidy in Cardamine *L.*

Cardamine rivularis is a mountain taxon of the *C. pratensis* group *Cardamine rivularis* is a mountain taxon of the *C. pratensis* group. It is karyologically differentiated into diploids (2n=16) and tetraploids (2n=32) which have rather different distributions (Landolt & Urbanska, 1971; Urbanska & Landolt, 1974; Urbanska & Landolt, unpubl.). *C. amara* is rather widely distributed both in lowland as well as at higher altitudes; it is also karyologically differentiated with diploid (2n=16) and tetraploid (2n=32) populations (Urbanska & Landolt, 1972; Urbanska & Landolt, unpubl.). *C. rivularis* and *C. amara* have distinctly different chromosomes (Urbanska & Landolt, 1972; Urbanska, 1977a, 1977b).

C. rivularis and *C. amara* are highly inter-incompatible; Lövkvist (1956, 1957) obtained no experimental hybrids, and the five plants that I have obtained did not flower (Urbanska, 1977a). natural hybridization between the two taxa is therefore likely to be rare.

In Switzerland, a single triploid clone formed as a direct product of hybridization between diploid *C. rivularis* and tetraploid *C. amara*, was found in the surroundings of St. Moritz; the sterile hybrid was weak and *C. amara* clearly dominated the site (Landolt & Urbanska, 1972). On the other hand, another hybrid discovered at Urnerboden (Central Switzerland) was exceedingly successful. Forest in this site (1310 m a. s. l.) was cleared several hundred years ago. During 1901–1905 part of the area was subdivided and the plots were fenced and turned into manured, drained hay meadows (Zimmerli, 1983, 1984,

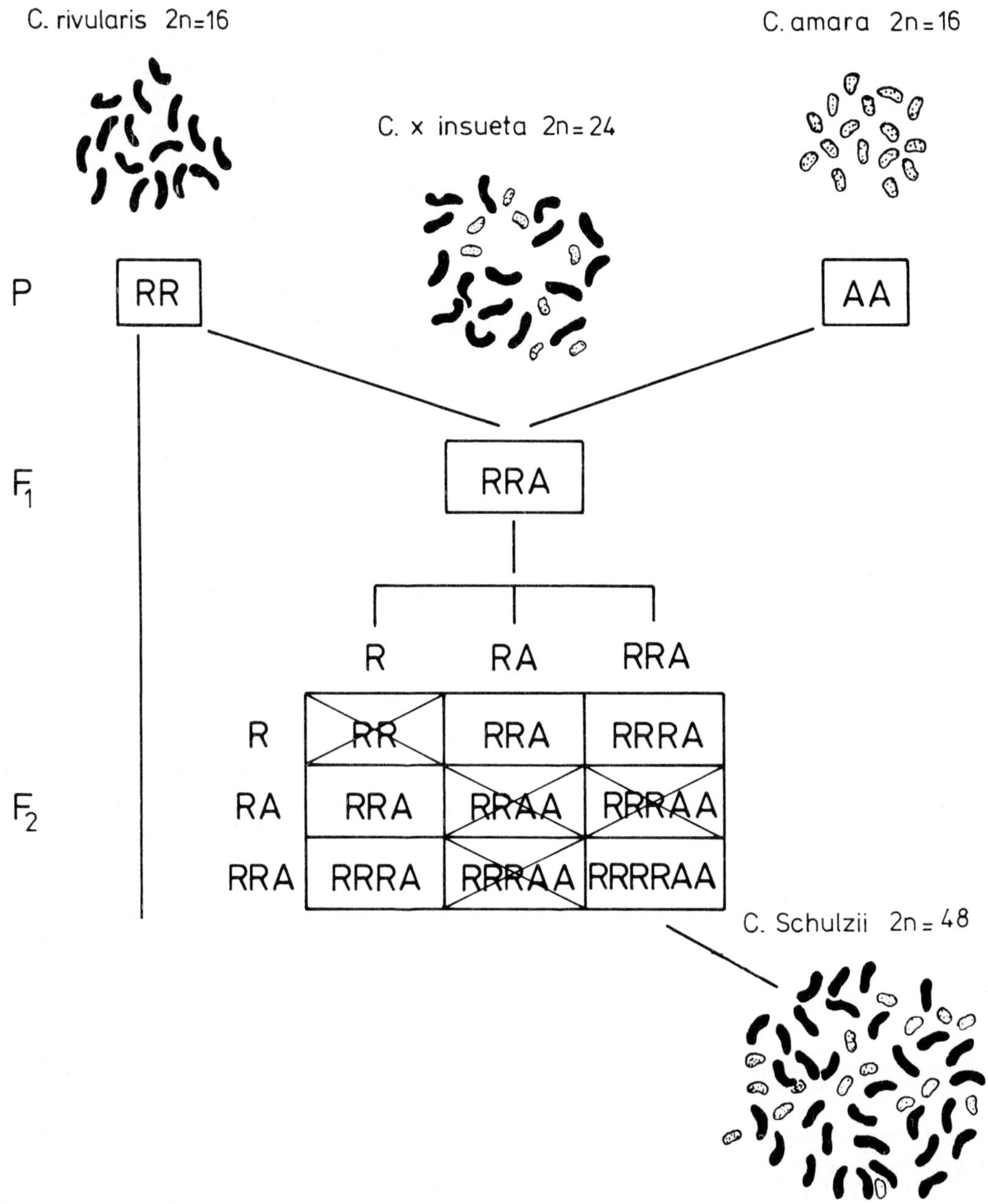

Fig. 6. Hybridization in *Cardamine* L. at Urnerboden. P = parents; F_1 = first hybrid generation; F_2 = second and further hybrid generations. Partly from Urbanska (1978).

1985). These plots have received rather different management resulting in different vegetation (Fig. 5).

Both *Cardamine rivularis* and *C. amara* at Urnerboden are represented by diploid populations. An exceedingly vigorous triploid hybrid, *C. insueta* (2n=24) is also present; it carries two genomes of *C. rivularis* and a single genome of *C. amara* (Urbanska & Landolt, 1972; Urbanska, 1977a, 1977b, 1978, 1980, 1981, 1984) and has obviously been formed with a part of an unreduced gamete of *C. rivularis* (Fig. 6).

C. insueta reproduces mostly by asexual means, spreading by short runners with supplementary rosettes. However, it also produces abundantly plantlets upon the surface of rosette

Figs. 7–9. Cardamine insueta (2n=24) and *C. Schulzii* (2n=48) from Urnerboden. 7. Flower of *C. insueta* with sterile, non-dehiscent anthers. 8. Triploid hybrid abundantly flowering in hay meadows. 9. Flower of the amphiploid *C. Schulzii* with fertile anthers bearing large amounts of pollen. From Urbanska (1981).

leaves and in the axils of the tall flower stalks. The resulting ramets are often shed when the hay meadows are being cut, and the propagules easily become established (Urbanska, 1977a, 1980, 1984).

Although *C. insueta* flowers profusely (Fig. 8), it is mostly male-sterile with non-dehiscent anthers (Fig. 7). However, some plants produce a small amount of viable pollen. The limited reproduction by seed in *C. insueta* usually follows a most unusual pattern of heterogamy. Two kinds of gametes are mostly produced as a result of polarized segregation; one carries a single genome of *C. rivularis* (R) and the other carries one genome each of *C. rivularis* and *C. amara* (RA). Of the four possible combinations of R and RA (Fig. 6) only the combination RRA is viable and it seems that true breeding in *C. insueta* is controlled by a balanced lethals system strikingly

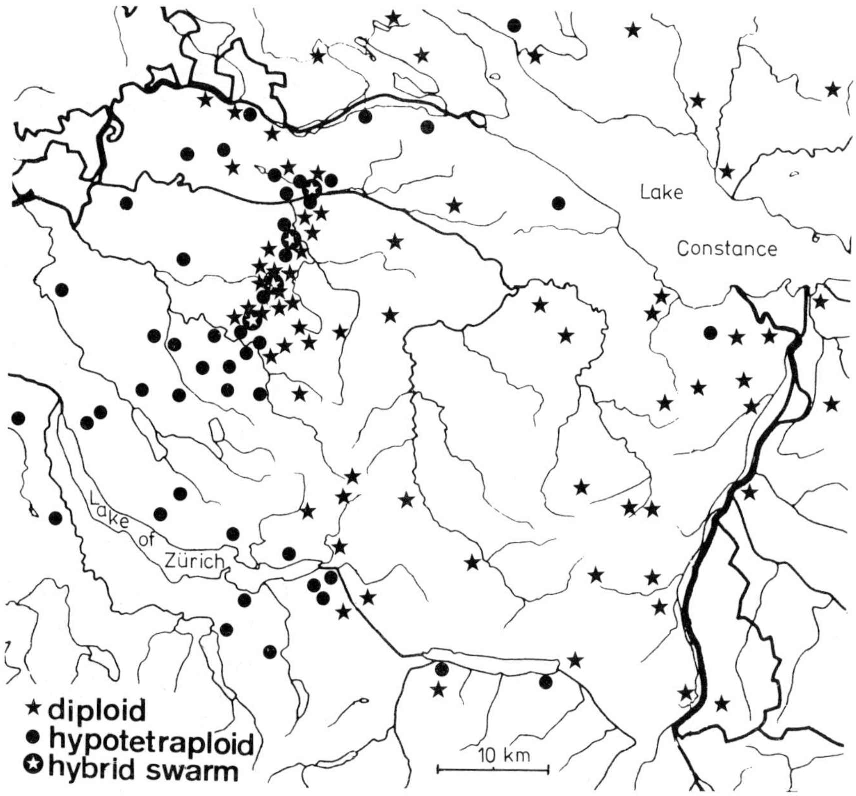

Fig. 10. The distribution of parental populations and inter-racial hybrids of *Cardamine pratensis* s. str. in NE Switzerland showing the contact zone between diploid (2n=16) and hypotetraploid (2n=30) race. Each marking may represent several local populations. Heteroploid populations are not marked. After unpublished data of K.M. Urbanska and E. Landolt.

similar to that occurring in *Oenothera*.

Apart form the polarized gametes R and RA, *C. insueta* occasionally produces unreduced gametes with the genomic constitution RRA (Fig. 6). The amphiploid *C. Schulzii* discovered at Urnerboden in 1974 (Urbanska, 1977b, 1978, 1979, 1980, 1981, 1984; Urbanska & Landolt, 1974a), apparently originated by doubling of the triploid chromosome set of *C. insueta*; it is hexaploid (2n=48) with the genomic constitution RRRRAA (Fig. 6), morphologically rather similar to *C. insueta* but much larger than the hybrid and highly fertile (Fig. 9). The amphiploid produces viable seed and grows vigoriously (Urbanska, 1977b, 1980, 1984).

C. insueta is exceedingly successful in man-made hay meadows (Fig. 8); its reproductive strategy seems to be directly adapted to the conditions occurring in its representative biotope (Urbanska, 1977a; Urbanska & Landolt, 1978). However, it is obviously dependent upon the regular management of hay meadows and flower production is reduced in unfavourable conditions (Zimmerli, 1983, 1984, 1986). *C. Schulzii* does not occur within hay meadows; it apparently originated in a drainage ditch a subsequently spread into neighbouring pastures (Urbanska, 1977b; Urbanska & Landolt, 1978). It seems that a recent ditching at Urnerboden has further promoted the expansion of the amphiploid (Urbanska, unpubl.). Neither of the original parents of *C. insueta* and *C. Schulzii* has succeeded in the man-made sites at Urnerboden. *C. rivularis* is mostly confined to poor swamp and some parts of the open pastures, but does not occur in hay meadows. *C. amara* occurs quite abundantly upon stream banks; small clones, occasionally growing in hay meadows are weak and probably represent the remains of past plots as the present-day water table in meadows seems unfavourable

Fig. 11. The old orchards with fertilized meadows in the Canton of Thurgau, NE Switzerland, where hybrid swarms of *Cardamine pratensis* s. str. are abundant. Photo taken by the author.

to the taxon (Urbanska & Landolt, 1978; Urbanska, 1980, 1981, 1984).

4.3 Hybrid swarm formation in Cardamine pratensis *s. str.*

The *Cardamine pratensis* group consists of numerous taxa; karyological and geographic/ecological differentiation is not always associated with distinct morphological characters (e.g. Lövkvist, 1956; Landolt, 1984; Landolt & Urbanska, 1971; Urbanska & Landolt, 1974; Urbanska & Landolt, unpubl.).

Fertilized meadows in Switzerland are inhabited by various chromosomal races of *C. pratensis* s. str. which seem to follow some regional distribution trends (Urbanska and Landolt, unpubl.). Populations of *C. pratensis* s. str. are usually homoploid i.e. consist of individuals representing one ploidy level.

In NE Switzerland, hypotetraploid populations (2n=30) occur quite frequently in the area of Zurich-Winterthur but are replaced by diploid colonies (2n=16) further to the north-east. The narrow zone of contact between the diploid and the hypotetraploid race (Fig. 10) corresponds to the area where fertilized meadows (cut and/or grazed) frequently accompany old orchards (Fig. 11). Homoploid, heteroploid and mixed populations form there a small-scale mosaic pattern. Homoploid colonies are represented either by diploids or hypotetraploids, no appreciable differences in frequency of a given type being noted. Heteroploid populations consist of 16- and 30-chromosomic plants, whereas the mixed colonies are characterized by occurrence of various cytotypes representing one or both parental races as well as various hybrids (2n=16, 18, 23, 23–24, 30, 38, 46, 54), Some of the plants belonging to the hybrid swarm apparently have been formed with a part of unreduced gametes, while some aneuploid differentiation was also noted (Urbanska & Landolt, unpubl.).

4.4 Introgressive hybridization in Carduus *L.*

Carduus nutans (2n=16) and *C. acanthoides* (2n=22) are of European origin and, in spite of

Figs. 12–15. Introgressive hybridization in *Cardus* L. in Southern Ontario. 12. *C. nutans* (2n=16). 13. *C. acanthoides* (2n=22). 14. Grazing-land in Grey County where hybridogenous populations are frequent. 15. *C. acanthoides*-like hybrid issued from introgression. Photos courtesy of Dr. Suzanne Warwick, Biosystematics Research Institute, Ottawa.

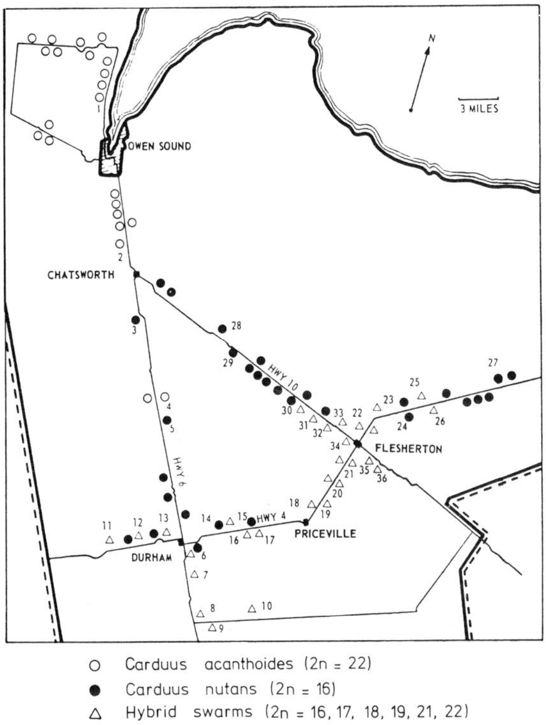

Fig. 16. The distribution of parental species and interspecific hybrids of *Carduus* L. in Grey County, Ontario. After data by Mulligan and Moore (1961).

the difference in chromosome number, natural hybridization between them occurs in Europe (Focke, 1881). *C. nutans* (Fig. 12) and *C. acanthoides* (Fig. 13) were introduced to Canada long ago; both species are abundant in rough pastures of Southern Ontario (Fig. 14) and extensive hybridization occurs in this area (Moore & Mulligan, 1956, 1964; Mulligan & Moore, 1961; Warwick, unpubl.).

All chromosome numbers between 2n=16 and 2n=22 have been found in the natural hybrids, though Moore & Mulligan (1956) suggest that both ecological and gametic selection is acting in favour of plants with the chromosome number and morphology close to *C. acanthoides* (Fig. 15). It seems that both introgresssion and segregation in late hybrid generations strongly influence the development of hybridogenous populations. The introgressed populations are similar to *C. acanthoides* but show an unusual degree of variability. They are very successful and the survival of the orginal parent species may be endangered (Moore & Mulligan, 1956; Mulligan & Moore, 1961).

C. nutans usually occupies rather low-lying grassy pastures, that have a higher soil moisture content, while the rolling stony hills are the territory of *C. acanthoides*. Detailed studies have shown that some areas in Ontario were first inhabited by *C. nutans*, and *C. acanthoides* invaded later (Mulligan & Moore, 1961). The spread of *C. acanthoides* may have also been promoted by the expansion of gravel pits; the distribution pattern of hybrid population strongly supports this (Fig. 16).

The area of Grey County in Ontario was surveyed in 1952, 1957 and 1962, though some plots could not have been included in the last survey because they were strongly altered or destroyed by human intervention. The overall impression was that the abundance of *Carduus* has greatly declined, probably because of competition from other species, both in pastures and along roadsides; wild carrot (*Daucus carota*) is especially abundant in the latter sites, perhaps because of the increasing use of herbicides (Moore & Mulligan, 1964).

Recent studies on *Carduus* in Southern Ontario (Warwick, personal communication) have shown that still more sites have disappeared because grazing-land has either been cultivated or destroyed by gravel extraction. Some hybrid colonies are still present, and further data on their performance are awaited with interest.

5. Concluding remarks

Contributors to this workshop have agreed to define disturbance as 'change in conditions interfering with a normal functioning of a given biological system'; this interference will obviously be reflected in changes of selective pressures. Changes in abiotic conditions can have radical effects on mortality percentages and/or fecundity rates. On the other hand, biotic disturbance such as that caused by invasion of a new taxon, may initially be quite modest. However, in the long run it may lead to pronounced changes resulting from either competition or hybridization, as shown by the examples presented in this paper.

The disturbance which gives rise to hybridization and its subsequent evolutionary development, may sometimes represent a truly isolated event. More frequently, however, various kinds of disturbance seem to accompany each other or follow each other in a short sequence of events. This situation is quite well illustrated by *Tragopogon* and *Carduus* in North America where the introduction of Old World species was closely associated with other kinds of human interference which promoted the establishment of hybrids. The spectacular hybrid speciation in *Cardamine* at Urnerboden is also inseparable from the combined influence of various kinds of disturbance viz. draining, fencing, manuring and cutting.

Not only the type of disturbance and its complexity but also its temporal pattern is of a great evolutionary importance. A single initial disturbance may result in the temporary formation of intermediate sites or, if strong enough, may result in open sites. On the other hand, recurrent disturbance seems essential for the maintenance of intermediate and hybridized habitats. The term 'recurrent' does not seem to be sufficiently precise, however; the regularity with which disturbance, e.g. management practices, are carried out may prove decisive for the course and speed of evolution. For example, the management of fertilized meadows inhabited by hybrid swarms of *Cardamine pratensis* s. str. follows a long-established annual pattern of management. As a result, conditions in the old orchards have become fairly stabilized, because the pattern of management is quite predictable and becomes *the norm* to which plants are adapted. In such cases, it is disputable whether the pattern can any longer be considered disturbance (Chapter 1 of this volume). On the other hand, the management of hay meadows at Urnerboden although recurrent is erratic in some plots. The hybrid *C. insueta* only succeeds in the plots that are regularly manured and cut, because the triploid plants are directly adapted to these conditions. On the contrary, the plots that are left without management show pronounced changes in the performance of the hybrids: curiously enough, in this context, non-management in hay meadows at Urnerboden represents the disturbance.

Still another aspect of recurrent disturbance relates to a situation where initial changes in life conditions stimulate evolutionary development but, after some time, become more and more extensive and finally become destructive. The fate of some hybrid colonies of *Carduus* in Canada clearly shows that the exploitation of the gravel pits first promoted the expansion of *C. acanthoides* and its hybrids, but eventually led to elimination of numerous populations. Spraying with herbicides is obviously a separate and unexpected disturbance with lethal effects as far as the *Carduus* plants are concerned.

The complex speciation processes occurring in higher plants are frequently, but not invariably, influenced by sudden changes in environmental conditions and by resulting abrupt changes in selective pressures. Geographical speciation does not generally seem to be disturbance-dependent, although local races may sometimes be formed under the influence of disturbance. On the other hand, accelerated speciation in general, and the hybrid speciation in particular, seems to be rather closely tied to an *initial* disturbance.

If recurrent changes in conditions become representative of a given ecosystem, so that disturbance becomes the norm, it does not interfere with a normal functioning of populations; on the contrary, it may be necessary for the successful functioning. In these cases, evolutionary processes such as speciation are not caused by disturbance, but the absence of such event(s) may alter the population response and accordingly the course of evolution.

Acknowledgements

Ms. Joy D. Mastrogiuseppe, Curator of the Marion Ownbey Herbarium, Washington State University, Pullman, provided valuable data on *Tragopogon*, Dr. Susanne I. Warwick, Biosystematics Research Institute, Ottawa permitted me to use her unpublished results on *Carduus*. S. Zimmerli, Geobotanical Department SFIT Zurich provided the photo of hay meadows at Urnerboden, Prof. Dr. Elias Landolt critically read an early draft of the manuscript. Drawings were carefully made by Ms. E. Wohlmann, and Mr. R. Graf prepared the photographs. All this help is gratefully acknowledged. Linguistic revision of the manuscript by Dr. R.W. Snaydon is greatly appreciated.

References

Anderson, E. (1948). Hybridization of the habitat. Evolution 2: 1–9.

Anderson, E. (1949). Introgressive Hybridization. Wiley, N. York.

Anderson, E. (1953). Introgressive hybridization. Biol. Rev. 28: 280–307.

Anderson, E. & Hubricht, L. (1938). Hybridization in *Tradescantia*. III. The evidence of introgressive hybridization. Amer. J. Bot. 25: 396–402.

Babcock, E.B. & Stebbins, G.L. (1938). The American species of *Crepsis*. Their interrelationships and distribution as affected by polyploidy and apomixis. Carnegie Inst. Washington Publ. 504.

Baker, H.G. (1951). Hybridization and natural gene-flow between higher plants. Biol. Rev. 26: 302–337.

Briggs, D. & Walters, S.M. (1969). Plant Variation and Evolution. McGraw-Hill Comp., N. York.

Britton, C.E. (1917). *Tragopogon porrifolius* x *pratensis*. Rep. B. E. C. 4: 576–577.

Britton, C.E. & Todd, W.A. (1910). *Tragopogon porrifolius* x *pratense*. J. Bot. Lond. 48: 203–204.

Clausen, J., Keck, D.D. & Hiesey, W.M. (1940). Experimental studies on the nature of species. I. Effects of varied environments on western American plants. Carnegie Inst. Wash. Publ. 520.

Clausen, J., Keck, D.D. & Hiesey, W.M. (1940). Experimental studies on the nature of species. II. Plant evolution through amphiploidy and autoploidy, with examples from the Madiinae. Carnegie Inst. Wash. Publ. 564.

Cleland, R.E. (1962). The cytogenetics of *Oenothera*. Adv. Gen. 11: 147–237.

Darlington, C.D. (1937). Recent Advances in Cytology. IInd Ed. London, Churchill.

Dobzhansky, Th., Ayala, F.J., Stebbins, G.L. & Valentine, J.W. (1977). Evolution. Freeman Comp., S. Fransisco.

Ellis, A.E. (1931). *Tragopogon minor* x *porrifolius*. Rep. B. E. C. 9: 272–274.

Ellis, A.E. (1932). *Tragopogon porrifolius* x *pratensis* (*minus* Mill.) Rep. B. E. C. 9: 566–567.

Fagerlind, F. (1940). Sind die *canina*-Rosen agamospermische Bastarde? Svensk Bot. Tidskr. 34: 334–354.

Gerassimova, H. (1939). Chromosome alterations as a factor of divergence of forms. I. New experimentally produced strains of *C. tectorum* which are physiologically isolated from original forms owing to reciprocal translocation. Compt. Rend. Acad. Sci URSS 25: 148–154.

Gould, F.W. & Kapadia, Z.J. (1964). Biosystematic studies in the *Bouteloua curtipendula* complex. II. Taxonomy. Brittonia 16: 182–207.

Grant, V. (1954a). Genetic and taxonomic studies in *Gilia*. IV. *Gilia achillaefolia*. Aliso 3: 1–18.

Grant, V. (1954b). Genetic and taxonomic studies in *Gilia*. VI. Interspecific relationships in the Leafy-stemmed Gilias. Aliso 3: 35–49.

Grant, V. (1958). The regulation of recombination in plants. Cold Spring Harbor Symp. Quant. Biol. 23: 337–363.

Grant, V. (1963). The Origin of Adaptations. Columbia Univ. Press, N. York.

Grant, V. (1964). The architecture of the Germplasm. Wiley, N. York.

Grant, V. (1966). The origin of a new species in *Gilia* in a hybridization experiment. Genetics 54: 1189–1199.

Grant, V. (1971). Plant Speciation. Irst Ed. Columbia Univ. Press, N. York.

Grant, V. (1977). Organismic Evolution. Freeman Comp., S. Francisco.

Grant, V. (1981). Plant Speciation. IInd. rev. Ed. Columbia Univ. Press, N. York.

Grant, V. (1985). The Evolutionary Processes. A Critic Review of Evolutionary Theory. Columbia Univ. Press, N. York.

Grant, A. & Grant, V. (1956). Genetic and taxonomic studies in *Gilia*. VIII. The Cobwebby Gilias. Aliso 3: 203–287.

Gustafson, A. (1946–47). Apomixis in higher plants. Lunds Univ. Arsskr. 42–43: 1–370.

Hess, H., Landolt, E. & Hirzel, R. (1972). Flora der Schweiz. Vol. III. Birkhäuser, Basle.

Heiser, C.B. (1973). Introgression re-examined. Bot. Rev. 39: 347–366.

Hurst, C.C. (1931). Embryo-sac formation in diploid and polyploid species of *Rosaceae*. Proc. Roy. Soc. B. 109: 126–148.

Huskins, C.L. (1930). The origin of *Spartina townsendii*. Genetica 12: 531–538.

Kerner, A. (1894–95). Natural History of Plants. Translation. Blackie and Son, London.

Landolt, E. (1984). Ueber die Artengruppe der *Cardamine pratensis* L. in der Schweiz. Diss. Bot. 72: 481–497.

Landolt, E. & Urbanska, K.M. (1971). Zytotaxonomische Untersuchungen an *Cardamine pratensis* L. s. l. im Bereich der Schweizer Alpen und des Jura. Ber. Dtsch Bot. Ges. 11: 683–690.

Levin, D.A. (1971). The origin of reproductive isolating mechanisms in flowering plants. Taxon 19: 167–171.

Lewis, H. (1962). Catastrophic selection as a factor in speciation. Evolution 16: 257–271.

Lewis, H. (1966). Speciation in flowering plants. Science 152: 167–172.

Lewis, H. & Lewis, M. (1955). The genus *Clarkia*. Univ. Calif. Publ. Bot. 20: 241–392.

Lewis, H. & Roberts, M. (1956). The origin of *Clarkia lingulata*. Evolution 10: 126–138.

Lewis, H. & Schweykowski, J. (1964). The genus *Gayophytum (Onagraceae)*. Brittonia 16: 343–391.

Lövkvist, B. (1956). The *Cardamine pratensis* complex – outlines of its cytogenetics and taxonomy. Symb. Bot. Upps. 14: 231 pp.

Lövkvist, B. (1957). Experimental studies in *Cardamine amara*. Bot. Not. 110: 423–441.

Marchant, C.J. (1967). Evolution in *Spartina (Gramineae)*. 1. The history and morphology of the genus in Britain. J. Linn. Soc. Bot. 60: 1–24.

Marchant, C.J. (1968). Evolution in *Spartina (Gramineae)*. 2. Chromosomes, basic relationships and the problem of *S.* x *Townsendii* agg. J. Linn. Bot. 60: 381–409.

Meisterhans, E. (1982). Entwicklungsmöglichkeiten für Vegetation und Boden auf Skipistenplanierungen. Fachbeitr. Schweiz. MAB-Inform. 10: 13–26.

Moore, R.J. & Mulligan, G.A. (1956). Natural hybridization between *Carduus acanthoides* and *C. nutans* in Ontario. Can. J. Bot. 34: 71–85.

Moore, R.J. & Mulligan, G.A. (1964). Further studies on natural selection among hybrids of *Carduus acanthoides* and *C. nutans*. Can. J. Bot. 42: 1605–1613.

Mulligan, G.A. & Moore, R.J. (1961). Natural selection among hybrids between *Carduus acanthoides* and *C. nutans* in Ontario. Can. J. Bot. 39: 269–279.

Ownbey, M. (1950). Natural hybridization and amphiploidy in the genus *Tragopogon*. Amer. J. Bot. 37: 487–499.

Ownbey, M. & McCollum, G.D. (1953). Cytoplasmic inheritance and reciprocal amphiploidy in *Tragopogon*. Amer. J. Bot. 40: 788–796.

Ownbey, M. & McCollum, G.D. (1954). The chromosomes of *Tragopogon*. Rhodora 56: 7–21.

Rousi, A. (1965). Biosystematic studies on the species aggregate *Potentila anserina* L. Ann. Bot. Fennici 2: 47–112.

Smith-White, S. (1948). Polarized segregation in the pollen mother cells of a stable triploid. Heredity 2: 119–129.

Smith-White, S. (1955). The life history and genetic system of *Leucopogon juniperinus*. Heredity 9: 79–91.

Smith-White, S. (1959). Cytological evolution in the Australian flora. Cold Spring Harbor Symp. Quant. Biol. 24: 273–289.

Stace, C.A. (1975). Hybridization and the Flora of the British Isles. Academic Press, London.

Stebbins, G.L. (1950). Variation and Evolution in Plants. Columbia Univ. Press, N. York.

Stebbins, G.L. (1971). Chromosomal Evolution in Higher Plants. E. Arnold, London.

Stebbins, G.L. (1980). Polyploidy in plants: unsolved problems and prospects. In: Lewis, W.H. (Ed.), Polyploidy: Biological Relevance. Plenum Press, N. York.

Täckholm, G. (1922). Zytologische Studien über die Gattung *Rosa*. Acta Horti Berg. 7: 97–381.

Urbanska, K.M. (1974). L'agamospermie, système de reproduction important dans la spéciation des Angiospermes. Bull. Soc. bot. Fr. 121: 329–346.

Urbanska, K.M. (1977a). Reproduction in natural triploid hybrids (2n=24) between *Cardamine rivularis* Schur and *C. amara* L. Ber. Geobot. Inst. ETH Stiftung Rübel 44: 42–85.

Urbanska, K.M. (1977b). An autoallohexaploid in *Cardamine* L., new to the Swiss flora. Ber. Geobot. Inst. ETH Stiftung Rübel 44: 86–103.

Urbanska, K.M. (1978). Ségrégation polarisée chez les hybrides naturels triploides (2n=24) entre *Cardamine rivularis* Schur (2n=16) and *C. amara* L. (2n=16). Soc. bot. Fr., Actualités Bot. 1–2: 91–93.

Urbanska, K.M. (1979). Reproductive strategies in a hybridogenous population of *Cardamine* L. Proc. Inst. Symp. Repr. Flow. Pl. Christchurch p. 46.

Urbanska, K.M. (1980). Reproductive strategies in a hybridogenous population of *Cardamine* L. Oecol. Plant. 1: 137–150.

Urbanska, K.M. (1981). Reproductive strategies in some perennial Angiosperms. Vierteljahrschr. Naturforsch. Ges. Zürich 126: 269–284.

Urbanska, K.M. (1984). Plant reproductive strategies. In: Grant, W.F. (Ed.), Plant Biosystematics. Academic Press Canada.

Urbanska, K.M. (1985). Some life history strategies and population structure in asexually reproducing plants. Bot. Helv. 95: 81–97.

Urbanska, K.M. (1986). High altitude revegetation research in Switzerland – problems and perspectives. Veröff. Geobot. Inst. ETH Stiftung Rübel 87: 155–167.

Urbanska, K.M. (1986a). Behaviour of Alpine plants and high altitude revegetation research. Proc. 7th HAR Workshop Fort Collins, 214–226.

Urbanska, K.M. & Landolt, E. (1972). Natürliche Bastarde zwischen *Cardamine amara* L. and *C. rivularis* Schur aus den Schweizer Alpen. Ber. Geobot. Inst. ETH Stiftung Rübel 41: 88–101.

Urbanska, K.M. & Landolt, E. (1974). Biosystematic investigations in *Cardamine pratensis* L. s. 1. I. Diploid taxa from Central Europe and their fertility relationships. Ber. Geobot. Inst. ETH Stiftung Rübel 42: 42–139.

Urbanska, K.M. & Landolt, E. (1974a). Hybridation naturelle entre *Cardamine rivularis* Schur et *C. amara* L., ses aspects cytologiques et génétiques. Acta Soc. helv. Sci nat. 1974: 89–90.

Urbanska, K.M. & Landolt, E. (1978). Recherches démographiques et écologiques sur une population hybridogène de *Cardamine* L. Ber. Geobot. Inst. ETH Stiftung Rübel 45: 30–53.

Urbanska, K.M. & Schütz, M. (1986). Reproduction by seed

in alpine plants and revegetation research above timberline. Bot. Helv. 96: 43–60.

Valentine, D.H. (1962). La taxonomie expérimentale: tendances actuelles. Rev. Cytol. Biol. Vég. 25: 255–266.

Vasek, F.C. (1968). The relationship between two ecologically marginal, sympatric *Clarkia* populations. Amer. Nat. 102: 25–40.

Zimmerli, S. (1983). Einfluss der Bewirtschaftung auf die Entwicklung und Struktur der *Cardamine*-Population auf dem Urnerboden. Diploma Thesis SFIT Zürich.

Zimmerli, S. (1984). The influence of management upon the development and the structure of the *Cardamine* populations at Urnerboden. Ber. Geobot. Inst. ETH Stiftung Rübel 51: p. 38.

Zimmerli, S. (1986). Einfluss der Bewirtschaftung auf die Entstehung und Struktur der *Cardamine*-Population auf dem Urnerboden. Veröff. Geobot. Inst. ETH Stiftung Rübel 87: 141–154.

EPILOGUE

We have defined disturbance as 'a change in conditions which interferes with the normal functioning of a biological system'. This implies that 'disturbance' is an umbrella-term encompassing a large set of phenomena. What is the ecological value of such a term?

Ecological value

The usefulness of terms which indicate a general concept depends on their operational value. This holds for disturbance as well as for many other ecological concepts, for example niche, strategy or syndrome, fitness, optimization. Similarly, the 'species' concept is useful due to its operational value, independent of whether a species can be regarded as a 'natural kind' (sensu Rosenberg, 1985). We have, therefore, attempted to define 'disturbance' in terms of responses of biological systems, rather than regarding particular causes (e.g. fire) or mechanisms (e.g. removal of biomass) as disturbances. It is the response of a biological system that indicates whether or not a change in conditions should be considered to cause disturbance of that system. The concept of disturbance thus challenges us to define normal functioning, or optimality and the limits of resistance of biological systems. The responsive characteristics are the parameters to detect both optimality and disturbance. Inasmuch as a characteristic of a plant, a population or a community is appropriate to define an optimal response, it is also suitable to detect disturbance. In this way we have been able to operationalize the concept of disturbance for several biological systems.

Individuals and populations

The components of the relative growth rate of plants, rather than the resulting RGR, appeared to be of adaptive value with regard to nutrient supply (Chapter 16). We can, therefore, utilize these parameters to detect both optimality and disturbance. This does not mean, however, that the components of RGR are suitable to detect all types of disturbances. We have, for example, seen that a meadow population of *Agrostis stolonifera* appeared capable of persisting due to phenotypic plasticity of the clonal growth of individual plants, whereas sexual reproduction is the important adaptive mechanism in a dune population (Chapter 15). Population differentiation as a result of e.g. disruptive selection, can be considered normal adaptive functioning of a population, provided that the resulting populations become as well adapted to the new environmental conditions as the original population was to the former conditions. But this is only part of the ecological reality. If selection causes an irreversible decrease of genetic variation in a population, such a population may become adapted to the new environmental conditions, but at the same time less capable of coping with still another disturbing agent (Chapter 14). We have seen that it is not necessarily the application of fertilizers, herbicides or other agro-chemicals per se that cause disturbance of a population, but rather the dosage, the exposure time and the frequency of the applications. Interactive effects (e.g. synergism) demand particular attention.

Populations and species

Conservation of 'species', an important aim of nature management, should be operationalized by indicating how the local populations, being the representatives of a species, can be maintained. Providing we have knowledge about optimum variability of a population (Chapters 14 and 17), we can also detect disturbance. Recovery of the numbers of a reduced population does not necessarily imply recovery of genetic variability. Therefore, conservation of a 'species' demands knowledge of the quality of the component populations, e.g. their capability to cope with disturbing agents. One should keep in mind, however, that conservation of a population may at the same time result in disturbance of the plant community concerned (see Chapter 18). These two objectives, maintenance of a community and maintenance of populations of particular species, can be combined only if one aims at conserving populations of species characteristic for a community.

Communities and ecosystems

Which characteristics of a plant community are suitable to detect normal or disturbed functioning? Disturbance may be indicated by a deviation of a known progressive succession, considered normal. If one, however, aims at maintaining a certain steady state, the maintenance of characteristic species can be used as a criterion. If the aim is a high species diversity, normal functioning is defined by this parameter. Herbivory cannot, in general, be regarded as a cause of disturbance of a grassland community, because herbivores, as a rule, do not fully exhaust their own food supply (Chapter 9). Overgrazing may cause disturbance, detectable by, for example, retrogressive succession, decreasing species diversity, extinction of characteristic species, decreasing net primary production. It is the criterion for 'normal functioning' that determines the optimum grazing intensity, and can also be used to define overgrazing (causing disturbance of the grassland community).

Plant communities are part of an ecosystem. Local environmental factors are influenced by larger-scale processes, including transport, e.g. tidal fluctuations, subsurface waterflows, acid deposition. Deviations of normal tidal fluctuations may result in disturbance of a local community (Chapter 7). Diminuation of groundwater discharge form deep aquifers may disturb a local community or a gradient of communities (Chapter 5), despite adequate local management (Chapter 12). Therefore, management of communities necessitates ecosystem management. This type of community-oriented management may conflict with population-oriented management. Debates about optimality and disturbance are useful only if these two aspects are focussed on the same ecological entity.

Perspective

Our study of disturbance in grasslands, presented in this volume, can help in evaluating our state of knowledge about normal functioning of populations, communities and ecosystems. Operationalization of the general term 'disturbance' for a particular biological system necessitates defining criteria for optimality and the ranges of resistance. It is not a change in conditions as such that causes disturbance, because adaptation and (micro)evolution should be considered normal biological phenomena. We recognise the challenge implied in quantifying the role of disturbance in evolution, and indeed the wider question of whether change in conditions can usefully be contrasted as 'normal' or 'abnormal' on an evolutionary time scale. We have attempted to contribute to the development of the general concept of disturbance by making the term 'disturbance' operational for particular systems. Other ecological concepts may be developed in a similar way.

Reference

Rosenberg, A. (1985). The Structure of Biological Science. Cambridge University Press, New York.

Subject index

Species index

Communities index